Chiral Sulfur Ligands
Asymmetric Catalysis

RSC Catalysis Series

Series Editor:
Professor James J Spivey, *Louisiana State University, Baton Rouge, USA*

Advisory Board:
Krijn P de Jong, *University of Utrecht, The Netherlands*, James A Dumesic, *University of Wisconsin-Madison, USA*, Chris Hardacre, *Queen's University Belfast, Northern Ireland*, Enrique Iglesia, *University of California at Berkeley, USA*, Zinfer Ismagilov, *Boreskov Institute of Catalysis, Novosibirsk, Russia*, Johannes Lercher, *TU München, Germany*, Umit Ozkan, *Ohio State University, USA*, Chunshan Song, *Penn State University, USA*

Titles in the Series:
1: Carbons and Carbon Supported Catalysts in Hydroprocessing
2: Chiral Sulfur Ligands: Asymmetric Catalysis

How to obtain future titles on publication:
A standing order plan is available for this series. A standing order will bring delivery of each new volume immediately on publication.

For further information please contact:
Sales and Customer Care, Royal Society of Chemistry, Thomas Graham House, Science Park, Milton Road, Cambridge, CB4 0WF, UK
Telephone: + 44 (0)1223 432360, Fax: + 44 (0)1223 420247, Email: sales@rsc.org
Visit our website at http://www.rsc.org/Shop/Books/

Chiral Sulfur Ligands
Asymmetric Catalysis

Hélène Pellissier

CNRS and Paul Cézanne University Aix-Marseille III, Institute of Molecular Sciences, Marseille, France

RSCPublishing

RSC Catalysis Series No. 2

ISBN: 978-1-84755-924-1
ISSN: 1757-6725

A catalogue record for this book is available from the British Library

© Hélène Pellissier 2009

Published by The Royal Society of Chemistry,
Thomas Graham House, Science Park, Milton Road,
Cambridge CB4 0WF, UK

Registered Charity Number 207890

For further information see our web site at www.rsc.org

In loving memory of Cyril and Valérie Pellissier
Dedicated to their so little children, Tess and Clément

Preface

The preparation of chiral compounds is an important and challenging area of contemporary synthetic organic chemistry, mainly in connection with the fact that most natural products are chiral and their physiological or pharmacological properties depend upon their recognition by chiral receptors, which will interact only with molecules of the proper absolute configuration. The use of chiral drugs in enantiopure form is now a standard requirement for virtually every new chemical entity, and the development of new synthetic methods to obtain enantiopure compounds has become a key goal for pharmaceutical companies. Indeed, the growing economic importance of chiral compounds has spurred major research efforts towards the selective preparation of chiral compounds. The synthesis of optically active chiral compounds, which play an important role in medicine and materials, is one of the most fascinating aspects of modern organic synthesis. Over the last three decades an explosive growth of research in the field of asymmetric synthesis has occurred. Asymmetric synthesis constitutes one of the main strategies to gain access to enantioenriched compounds, involving the use of either chiral auxiliaries or catalysts derived preferentially from cheap chiral pool sources. In particular, asymmetric catalysis of organic reactions to provide enantiomerically enriched products is of central importance to modern synthetic and pharmaceutical chemistry.

In this context, the development of chiral ligands for asymmetric catalytic reactions is a subject of considerable interest in the field of asymmetric synthesis. Extensive efforts have been made for the development of new advantageous chiral ligands applicable in a broad variety of reaction types and their preparation continues to be an important area of synthetic organic research. Indeed, the ligand design is becoming an increasingly important part of the synthetic activity in chemistry. This is, of course, because of the subtle control that ligands exert on the metal centre to which they are coordinated. The impressive number of reports dealing with the use of chiral

RSC Catalysis Series No. 2
Chiral Sulfur Ligands: Asymmetric Catalysis
By Hélène Pellissier
© Hélène Pellissier 2009
Published by the Royal Society of Chemistry, www.rsc.org

sulfur-containing ligands for asymmetric catalysis is well representative of the success of such ligands to promote numerous catalytic transformations. Nowadays, these ligands have become fairly renowned competitors to more usual phosphorus- or nitrogen-containing ligands. Indeed, chiral sulfur-containing ligands represent a highly efficient class of ligands applicable in a broad variety of reaction types, also in view of some peculiar properties, which differentiate them from more popular ligands such as phosphorus- or nitrogen-containing ligands. In particular, the stereoelectronic assistance of organo-sulfur functionalities and the possibility of stereocontrol through stereogenic sulfur atoms can provide interesting results in many applications. The key advantages of these new types of ligands are their easy synthesis, mostly starting from readily available commercial compounds, and their high stability, which allows easy storage and handling, especially compared to phosphine derivatives. Whereas the synthesis of chiral phosphorus ligands is often complex and difficult, the preparation of chiral sulfur compounds is much more convenient. Consequently, a wide diversity of chiral sulfur-containing ligands, more than forty different classes, is easily available either directly from the chiral pool or by facile modifications of other heteroatomic ligands. Many of the sulfur-containing ligands have only recently appeared in the literature, and it is clear that their application in asymmetric catalysis will undoubtedly increase in the near future for a number of additional asymmetric reactions, due in part to their easy synthesis. Moreover, due to their great stability, asymmetric heterogeneous catalysis seems to be a good way to develop the potential of these ligands in an economic and environmentally friendly manner.

The goal of this book is to provide to researchers and professionals of academic and industrial laboratories a broad overview of the works concerning the use of chiral sulfur-containing ligands in one of the most important focal areas in organic synthesis, asymmetric catalysis. The book is divided into ten chapters corresponding to the different types of reactions based on the use of complexes containing chiral sulfur ligands, such as allylic substitution, conjugate addition, addition of organometallic reagents to aldehydes, addition of organozinc reagents to ketones, Diels–Alder reaction, cyclopropanation, Heck-type reactions, hydrogenation, hydrogen transfer and miscellaneous reactions.

Contents

RSC Catalysis Series No. 2
Chiral Sulfur Ligands: Asymmetric Catalysis
By Hélène Pellissier
© Hélène Pellissier 2009
Published by the Royal Society of Chemistry, www.rsc.org

Abbreviations

Ac: acetyl
Acac: acetylacetone
Ad: adamantyl
AIBN: 2,2′-azobisisobutyronitrile
Ar: aryl
BDMPB: 2,6-bis(dimethylphenoxy)borane
BDPP: 2,4-bis(diphenylphosphino)pentane
BINAM: 1,1′-binaphthalenyl-2,2′-diamine
BINAP: 2,2′-bis(diphenylphosphino)-1,1′-binaphthyl
Binas: 1,1′-binapthalene-2,2′-dithiol
BINOL: 1,1′-bi-2-naphthol
BITIANP: 2,2′-bis(diphenylphosphino)-3,3′-bi(benzo[b]thiophene)
BITIOP: 4,4′-bis(diphenylphosphino)-3,3′-bithiophene
Bn: benzyl
Boc: tert-butoxycarbonyl
Box: bisoxazoline
BSA: bis-(trimethylsilyl)acetamide
Bu: butyl
Bz: benzoyl
Cbz: benzyloxycarbonyl
Chiraphos: 2,3-bis(diphenylphosphine)butane
Cod: cyclooctadiene
Cp: cyclopentadienyl
CPG: controlled-pore glass
Cy: cyclohexyl
DABCO: 1,4-diazabicyclo[2.2.2]octane
DAIB: dimethylamino isoborneol
Dba: (E,E)-dibenzylideneacetone
de: diastereomeric excess
Dec: decyl
DIBAL: diisobutylaluminium hydride
DIOP: 2,3-(isopropylidenedioxy)-2,3-dihydroxy-l,4-
               bis(diphenylphosphanyl)butane)
DIOS: 2,3-O-isopropylidene-2,3-dihydroxy-1,4-
               bis(methylsulfinyl)butane

DMF:	dimethylformamide
DMSO:	dimethylsulfoxide
DOSP:	N-p-dodecylbenzenesulfonylprolinate
DPEN:	1,2-diphenylethylenediamine
Dppe:	1,2-bis(diphenylphosphine)ethane
Dppp:	1,3-bis(diphenylphosphine)propane
EDA:	ethyl diazoacetate
ee:	enantiomeric excess
Et:	ethyl
Fm:	fluorenylmethyl
Fu:	furyl
Hept:	heptyl
Hex:	hexyl
HMPA:	hexamethylphosphoramide
HOCSAC:	bis(camphorsulfonyl)-substituted trans-cyclohexane-1,2-diamine ligand
L:	ligand
Me:	methyl
Menth:	menthyl
Mes:	mesyl
MTBE:	methyl tert-butyl ether
NADH:	nicotinamide adenine dinucleotide
NADPH:	nicotinamide adenine dinucleotide phosphate
Naph:	naphthyl
Nbd:	norbornadiene
NMO:	N-methylmorpholine N-oxide
NOBIN:	2-amino-2-hydroxy-1,1'-binaphthalene
Non:	nonyl
Nu:	nucleophile
Oct:	octyl
Pent:	pentyl
Ph:	phenyl
PHOX:	phosphinooxazoline
Piv:	pivalate
PMHS:	polymethylhydrosiloxane
Pr:	propyl
Py:	pyridine
SES:	2-(trimethylsilyl)ethanesulfonyl
Siam:	bis(sulfinyl)imidoamidine
Suc:	succinimide
TADDOL:	a,a,a',a'-tetraphenyl-2,2-dimethyl-1,3-dioxolane-4,5-dimethanol
TBAB:	tetra-n-butylammonium bromide
TBDPS:	tert-butyldiphenylsilyl
TBS:	tert-butyldimethylsilyl
TEA:	triethylamine
Tf:	trifluoromethanesulfonyl

TFA:	trifluoroacetic acid
Tfbb:	tetrafluorobenzobarrelene
THF:	tetrahydrofuran
Thio:	thiophene
TIPS:	triisopropylsilyl
TMBTP:	4,4′-bis(diphenylphosphino)-2,2′,5,5′-tetramethyl-3,3′-bithiophene
TMS:	trimethylsilyl
Tol:	tolyl
Tr:	triphenylmethyl (trityl)
Ts:	4-toluenesulfonyl (tosyl)

General Introduction

The preparation of chiral compounds is an important and challenging area of contemporary synthetic organic chemistry.[1] Asymmetry is ubiquitous in every part of Nature and has a great impact on many fields, not only chemistry but also even the arts. In pharmaceuticals, as well as in related industries,[2] asymmetry plays an important role, since both enantiomers of a determinate drug do not necessarily have the same activity. The disastrous incident of thalidomide, in which each enantiomer has a totally different biological effect in humans, has had an impact on the society. The public demands for avoiding similar tragedies have been transferred first to the pharmaceutical and related companies, mainly through the questions of regulatory agencies, and second to the scientific community, which in turn has to provide highly efficient and reliable methods of asymmetric synthesis. In fact, these demands have already had an important response, since the worldwide sales of single-enantiomer drugs are continuously growing. For example, approximately 50 per cent of the drugs launched from 1992 to 2003 contain at least one stereogenic element and indeed, there appeared to be an increasing trend in the production of such drugs.[3] In addition, the policy of regulatory agencies (FAD, EMEA, *etc.*) on stereoisomers has triggered a move away from the development of racemates to the development of single enantiomer drugs. Moreover, the additional cost of producing a single enantiomer is almost always lower than the development work that is necessary to elucidate the toxicological and pharmaceutical profile of the undesired enantiomer. Among the different strategies of asymmetric synthesis, enantioselective approaches have many advantages compared to other strategies. For example, the annoying attachment and detachment of chiral auxiliaries is not necessary. In addition, the atom efficiency of preparing an enantiomer by enantioselective catalysis is much higher than either the diastereoselective or the racemate resolution approach. The development of efficient methods for enantioselective synthesis remains at the centre of modern-day organic

RSC Catalysis Series No. 2
Chiral Sulfur Ligands: Asymmetric Catalysis
By Hélène Pellissier
© Hélène Pellissier 2009
Published by the Royal Society of Chemistry, www.rsc.org

chemistry; as such methods have many important applications, from the total synthesis of natural products[4] to the preparation of analogues of lead compounds in the pharmaceutical industry. In particular, catalytic asymmetric synthesis is a valuable and general method for preparing optically active substances.[5] In contrast to stoichiometric methods, the chiral information of a ligand molecule is transferred to several product molecules, through a catalytic cycle; in addition, the reactivity of the active species bearing the chiral ligand is, in general, enhanced, compared to that of the nonchiral ligand, due to the so-called "ligand acceleration effect".[6] Among the enantioselective catalytic transformations, those involving carbon–carbon bond formation are probably the most attractive for synthesis, compared to functional group conversions on a given carbon skeleton. Asymmetric catalysis is a topic of increasing interest and is one of the most important focal areas in organic synthesis. In the last three decades, the number and the quality of reports dealing with asymmetric synthesis have undergone a true revolution with the coronation being the awarding of the 2001 Nobel Prize in Chemistry to Professors Sharpless, Knowles, and Noyori for their studies on enantioselective synthesis. In particular, the development of chiral ligands for asymmetric catalytic reactions is a subject of considerable interest in the field of asymmetric synthesis.[7] Much attention has been devoted in the pharmaceutical field to catalytic asymmetric synthesis for the preparation of biologically active chiral compounds with complete optical purity and high efficiency.[8] Extensive efforts have been made for the development of new advantageous chiral ligands applicable in a broad variety of reaction types,[9] and their preparation continues to be an important area of synthetic organic research.[10,11]

Practical asymmetric catalysis using transition-metal complexes was initially inspired by the studies of Kagan[12] and Knowles.[13] Their important results, based on the use of chiral phosphines as ligands for asymmetric hydrogenation, have induced a tremendous amount of work, dealing with the synthesis and use of new chiral phosphine-containing complexes as catalysts. Numerous catalytic asymmetric reactions have been discovered over the last 30 years, often with spectacular results in terms of efficiency and selectivity, allowing the access to numerous biologically important molecules. Nevertheless, the contribution of asymmetric catalysis in the overall production of chiral chemicals is much lower than originally expected, which is surprising given the huge amount of work devoted to this subject. Factors such as the price of the catalyst precursor and the difficulties encountered in the separation and recycling of the catalyst are responsible for this lack of practical application. Only a few processes have, however, permitted high turnovers. Apart from these economic considerations, it is almost impossible to recycle, for example, phosphine-containing catalysts, due to their low stability towards oxidation. Indeed, the chemical and economic characteristics of these catalysts were partly responsible for problems encountered in the development of catalytic asymmetric processes in general. The field of asymmetric catalysis is witnessing an ever-growing interest, and several highly efficient catalytic methods are nowadays known in the literature. Despite the positive results, knowledge in this field is still limited, and much work will be needed to make this methodology a comprehensive and well-established technique. The nature of the ancillary ligands used for a given metal-catalysed process is central, with chiral phosphorus- and nitrogen-based ligands occupying an incontestable leading position. In particular, chiral C_1- and C_2-symmetric phosphorus ligands possessing the axially chiral 1,1′-binaphthyl framework, are among the most widely used chiral ligands.[14] Moreover, oxazolines have also played a key role as efficient ligands for various types of catalytic asymmetric reactions.[15]

New classes of ligands that might offer new opportunities for applications or provide insight into fundamental chemical processes are always of interest. One relatively rare class of ligands is that in which stereogenicity resides not at carbon atoms, but at heteroatomic sites such as sulfur atoms. Compared to ligands having phosphorus, nitrogen or oxygen as donor atoms, sulfur-containing chiral mixed ligands have received much less attention. A possible reason is that sulfur has a tendency to poison transition-metal catalysts. On the other hand, the sulfur moiety creates additional possibilities compared to nitrogen- and oxygen-containing ligands since sulfur can become chiral when coordinated to a metal. Compared to phosphorus, sulfur has less donor and acceptor character. In addition to these electronic considerations, the sulfur atom, in thioether ligands for example, has only two substituents, which can create a less hindered environment than trivalent phosphorus. The formation of mixtures of diastereomeric complexes and the difficulty to control their interconversion in solution have been regarded as a problem for asymmetric induction in catalytic reactions. Nevertheless, in recent years, chiral bidentate S-donor ligands, in particular, have proved to be as useful as other classical asymmetric ligands, especially when combined with other donor atoms.[16] Whereas the synthesis of chiral phosphorus ligands is often complex and difficult, the preparation of chiral sulfur compounds is more convenient. In addition, sulfur possesses higher oxidation states available and can form some compounds that have different functional groups, such as thiol, sulfide, thioamide, sulfoxide, sulfinyl, thiocarbonyl and thiocarbamide. Furthermore, it's empty relatively low-energy d orbitals can accept back-donation of π-electron density from the metal, leading to a stabilisation of the metal–S bond. On complexation to a metal, chirality can be induced at sulfur.[17] These key structural features exemplify very rich coordination chemistry towards transition metals and they serve as powerful stereodirecting ligands in asymmetric synthesis. In addition to the vast knowledge on sulfur-metal interactions in coordination chemistry,[18] the study of sulfur-containing ligands has increased considerably over the last three decades.[19] Indeed, more than 40 different classes of chiral sulfur compounds have been described in the literature, and a large number of useful procedures for the synthesis of enantiomerically pure sulfur compounds have been developed.[20] The coordination chemistry of sulfur ligands has shown a unique variety of structures with most of the transition metals in different oxidation states.[18,21]

Transition-metal complexes of chiral sulfur ligands are nowadays considered as powerful catalysts in a considerable number of reactions, although, they have been generally less investigated than complexes with other donor atoms. Indeed, sulfur donor ligands have been used much less than phosphorus donor ligands in asymmetric homogeneous catalysis,[22] although, in recent decades, the number of studies with chiral sulfur-containing catalytic systems has increased considerably.[19b–e,23]

The goal of this book is to cover the works concerning the use of chiral sulfur-containing ligands in asymmetric catalysis. The book is divided into ten chapters corresponding to the different types of reactions based on the use of complexes containing chiral sulfur ligands, such as allylic substitution, conjugate addition, addition of organometallic reagents to aldehydes, addition of organozinc reagents to ketones, Diels–Alder reaction, cyclopropanation, Heck-type reactions, hydrogenation, hydrogen transfer and miscellaneous reactions.

References

1. M. Nogradi, in *Stereoselective Synthesis*, VCH, Weinheim, 1995.
2. (a) *Chirality in Industry*, A. N. Collins, G. N. Sheldrake, J. Crosby, ed. John Wiley and Sons, Chichester, 1992; (b) *Process Chemistry in the Pharmaceutical Industry*, K. G. Gadamasetti, ed. Marcel Dekker, New York, 1999; (c) H. U. Blaser, F. Spindler and M. Studer, *Appl. Catal. A*, 2001, **221**, 119–143; (d) *Asymmetric Catalysis on Industrial Scale, Challenges*, Approaches and Solutions, H. U. Blaser, E. Schmidt, E., ed. Wiley-VCH, Weinheim, 2004; (e) G. Wu and M. Huang, *Chem. Rev.*, 2006, **106**, 2596–2616.
3. V. Farina, J. T. Reeves, C. H. Senanayake and J. J. Song, *Chem. Rev.*, 2006, **106**, 2734–2793.
4. K. C. Nicolaou and E. J. Sorensen, in *Classics in Total Synthesis, Targets, Strategies, and Methods*, VCH, Weinheim, 1996.
5. (a) *Catalytic Asymmetric Synthesis*, I. Ojima, ed. VCH Publishers, Cambridge, 1993; (b) R. Noyori, in *Asymmetric Catalysis in Organic Synthesis*, John Wiley and Sons, New York, 1994; (c) R. E. Gawley and J. Aubé, in *Principles of Asymmetric Synthesis*, ed. J. E. Baldwin, P. D. Magnus, Pergamon, Oxford, 1996.
6. (a) E. N. Jacobsen, I. Marko, M. B. France, J. S. Svendsen and K. B. Sharpless, *J. Am. Chem. Soc.*, 1989, **111**, 737–739; (b) D. J. Berrisford, C. Bolm and K. B. Sharpless, *Angew. Chem., Int. Ed. Engl.*, 1995, **34**, 1059–1070.
7. *Comprehensive Asymmetric Catalysis*, E. N. Jacobsen, A. Pfaltz, H. Yamamoto, ed. Springer, New York, 1999.
8. R. Noyori, in *Asymmetric Catalysis in Organic Synthesis*, John Wiley and Sons Inc., New York, 1994.
9. J. Seyden-Penne, in *Chiral Auxiliaries and Ligands in Asymmetric Synthesis*, John Wiley and Sons, New York, 1995.

10. (a) M. McCarthy and P. J. Guiry, *Tetrahedron*, 2001, **57**, 3809–3844; (b) *Chiral Auxiliaries and Ligands in Asymmetric Synthesis*, ed. J. Seyden-Penne, Wiley Interscience, New York, 1995, Comprehensive Asymmetric Catalysis, ed. E. N. Jacobsen, A. Pfaltz, H. Yamamoto, Springer, Berlin, Vol. I–III, 1999.

11. (a) I. Ojima, in *Catalytic Asymmetric Synthesis II*, Wiley-VCH, New York, 2000; (b) E. N. Jacobsen, A. Pfaltz and H. Yamamoto, in *Comprehensive Asymmetric Catalysis*, Springer, Berlin, 1999, **Vols 1–3**.

12. H. B. Kagan and T. -P. Dang, *J. Am. Chem. Soc.*, 1972, **94**, 6429–6433.

13. W. S. Knowles and M. J. Sabacky, *Chem. Commun.*, 1968, 1445–1446.

14. (a) M. Mc Carthy and P. J. Guiry, *Tetrahedron*, 2001, **57**, 3809–3844; (b) H. Shimizu, I. Nagasaki and T. Saito, *Tetrahedron*, 2005, **61**, 5405–5432; (c) J. M. Brunel, *Chem. Rev.*, 2005, **105**, 857–897.

15. (a) A. K. Ghosh, P. Mathivanan and J. Capellio, *Tetrahedron: Asymmetry*, 1998, **9**, 1–45; (b) P. Braunstein and F. Naud, *Angew. Chem., Int. Ed. Engl.*, 2001, **40**, 680–689; (c) G. S. K. Wong and W. Wu, in *Chemistry of Heterocyclic Compounds*, ed. D. C. Palmer, John Wiley and Sons, New York, 2004, pp 331–528, Part B, Chapter 5; (d) A. Ghosh, G. Bilcer and S. Fidanze, in *Chemistry of Heterocyclic Compounds*, ed. D. C.Palmer, John Wiley and Sons, New York, 2004, pp 529–594, Part B, Chapter 5.

16. (a) K. Boog-Wick, P. S. Pregosin and G. Trabesinger, *Organometallics*, 1998, **17**, 3254–3264; (b) S.-L. You, X.-L. Hou, L.-X. Dai, Y.-H. Yu and W. Xia, *J. Org. Chem.*, 2002, **67**, 4684–4695; (c) D. A. Evans, K. R. Campos, J. R. Tedrow, F. E. Michael and M. R. Gagné, *J. Org. Chem.*, 1999, **64**, 2994–2995; (d) G. J. Dawson, C. G. Frost, C. J. Martin, J. M. J. Williams and S. J. Coote, *Tetrahedron Lett.*, 1993, **34**, 7793–7796; (e) C. G. Forst and J. M. J. Williams, *Tetrahedron: Asymmetry*, 1993, **4**, 1785–1788; (f) C. G. Frost, G. Christopher and J. M. J. Williams, *Tetrahedron Lett.*, 1993, **34**, 2015–2018.

17. P. J. Walsh, A. E. Lurain and J. Balsells, *Chem. Rev.*, 2003, **103**, 3297–3244.

18. S. G. Murray and F. R. Hartley, *Chem. Rev.*, 1981, **81**, 365–414.

19. (a) P. F. Kelly and J. D. Woollins, *Polyhedron*, 1986, **5**, 607–632; (b) A. M. Masdeu-Bulto, M. Diéguez, E. Martin and M. Gomez, *Coord. Chem. Rev.*, 2003, **242**, 159–201; (c) I. Fernandez and N. Khiar, *Chem. Rev.*, 2003, **103**, 3651–3705; (d) H. Pellissier, *Tetrahedron*, 2007, **63**, 1297–1330; (e) M. Mellah, A. Voituriez and E. Schulz, *Chem. Rev.*, 2007, **107**, 5133–5209.

20. M. Mikolajczk, J. Drabowicz and P. Kielbasinski, in *Chiral Sulfur Reagents*, CRC Press, USA, 1997.

21. (a) S. E. Livingstone, *Quart. Rev.*, 1965, **19**, 386–425; (b) E. W. Abel, S. K. Barghava and K. G. Orrell, in *Progress in Inorganic Chemistry*, ed. S. L. Lippard, Wiley, New York, 1984, p. 1; (c) G. Dance, *Polyhedron*, 1986, **5**, 1037–1104; (d) K. G. Orrell, *Coord. Chem. Rev.*, 1989, **96**, 1–48; (e) P. J. Blower and J. R. Dilworth, *Coord. Chem. Rev.*, 1987, **76**, 121–185; (f) J. R. Dilworth and J. Hu, *Adv. Inorg. Chem.*, 1994, **40**, 411–459; (g) K. Dance and K. Fisher, *Prog. Inorg. Chem.*, 1994, **41**, 637–803.

22. (a) L. H. Pignolet, (Ed.), in *Homogeneous Catalysis with Metal Phosphine Complexes*, Plenum Press, New York, 1983; (b) M. C. Simpson and D. J. Cole-Hamilton, *Coord. Chem. Rev.*, 1996, **155**, 163–207; (c) B. Cornils, W. A. Herrmann, (Ed), in *Applied Homogeneous Catalysis with Organometallic Compounds*, VCH, Weinheim, 1996; (d) M. Rakowski DuBois, *Chem. Rev.*, 1989, **89**, 1–9.
23. J. C. Bayon, C. Claver and A. M. Masdeu-Bulto, *Coord. Chem. Rev.*, 1999, **193–195**, 73–145.

CHAPTER 1
Allylic Substitution

1.1 Introduction

Carbon–carbon bond formation is one of the most important reactions in synthetic organic chemistry. One useful and popular method is the palladium-catalysed allylation,[1] *e.g.* the Tsuji–Trost reaction,[2] of which asymmetric versions have been extensively studied over the last decade.[3] Several classes of chiral ligands, such as N/N ligands,[4] biphosphines,[1d,2a] monodentate phosphines,[5] and P/N mixed donor ligands[6] have been extensively studied and proven to be effective ligands for Pd-catalysed asymmetric allylic substitution reactions. The Tsuji–Trost reaction is by far the most intensively studied reaction performed in the presence of sulfur-containing ligands over the past few years. Since sulfur is a soft complexation site and palladium a soft metal, the resulting complexes are expected to be strong complexes. In addition, retrodonation of π-electron density from the metal atom to the empty relatively low-energy d orbitals of the sulfur can also increase the strength of the Pd–S bond.[7] Strategies for controlling enantioselectivity in palladium-catalysed asymmetric reactions have depended on the design and application of chiral ligands. Many of the efficient homo- and heterodonor chiral ligands, such as N/N- (*e.g.* bis(oxazolines)[8]), P/P- (*e.g.* Trost's P/P ligands[9]), and P/N- (phosphinooxazolines[10]) types have been exploited. It must be noted that mixed P/N ligands have played a dominant role among the heterodonor ligands. A particular efficient method of C–C bond formation was opened up by the reaction of carbon nucleophiles with allylpalladium complexes, the generation of which is *in situ* accomplished and requires only a catalytic amount of the transition metal. Considerable efforts have been devoted to study the reaction between allylic substrates and nucleophiles catalysed by chiral palladium complexes. The palladium-catalysed allylic substitution is one of the catalytic

RSC Catalysis Series No. 2
Chiral Sulfur Ligands: Asymmetric Catalysis
By Hélène Pellissier
© Hélène Pellissier 2009
Published by the Royal Society of Chemistry, www.rsc.org

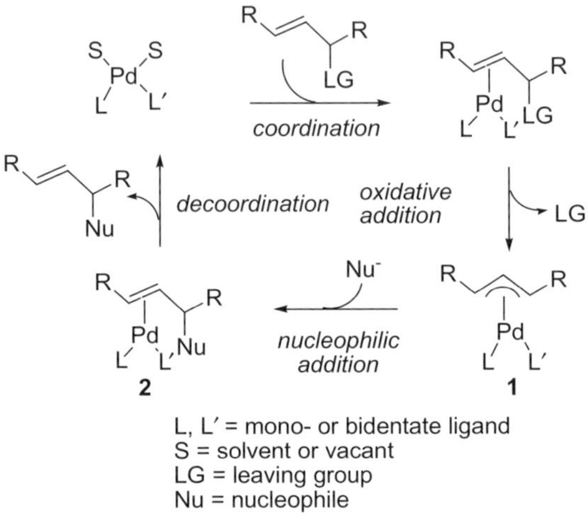

Scheme 1.1   Mechanism for Pd-catalysed allylic substitution with soft nucleophiles.

homogeneous processes that has attracted most attention in recent decades and for which the catalytic cycle is well established (Scheme 1.1).[11]

This is due in part to the relative ease of isolating catalytic intermediates, especially the palladium allylic species **1** (Scheme 1.1), although some related Pd(0) species (**2**) have also been characterised in solution.[12] The enantioselectivity of the process with soft nucleophiles (derived from conjugated acids with pKa < 25) is controlled by the external nucleophilic attack on the more electrophilic terminal allylic carbon of **1**. The chemo-, regio-, diastereo-, and enantioselectivities of this process have been widely analysed and the results applied to the synthesis of target molecules.[13] Since the first enantioselective catalytic process, described by Trost *et al.* in 1977,[14] which implicated bidentate ligands with phosphorus as the donor atoms, the enantioselective allylic alkylation reaction catalysed by Pd has been of great interest in recent years, involving many chiral ligands and allowing excellent enantioselectivities.[15] The catalysts often consist of a palladium complex containing a chiral chelate ligand but they can also be generated *in situ*. The mechanism of this palladium-mediated allylic reaction is reasonably well understood.[16] A chiral Pd(0) olefin complex oxidatively adds the prochiral allylic acetate to afford an isolable η³-allylic cationic compound, which is then attacked by the nucleophile. One of the most widely investigated sulfur ligands have been the S/N-donor type,[17] often derived from a chiral oxazoline moiety.[18] Oxazolines are known to have several advantages as sources of chirality, the main one being that they are readily accessible from homochiral amino alcohols and have proved to be effective catalysts in a variety of reactions.[8,18de] Furthermore, these ligands are easily modifiable and can incorporate different donor atoms in the side chains of the heterocyclic ring.[19] Other combinations of donor atoms with sulphur[20] have also been explored, such as chiral S/P-donor ligands,[21] chiral S/O-donor

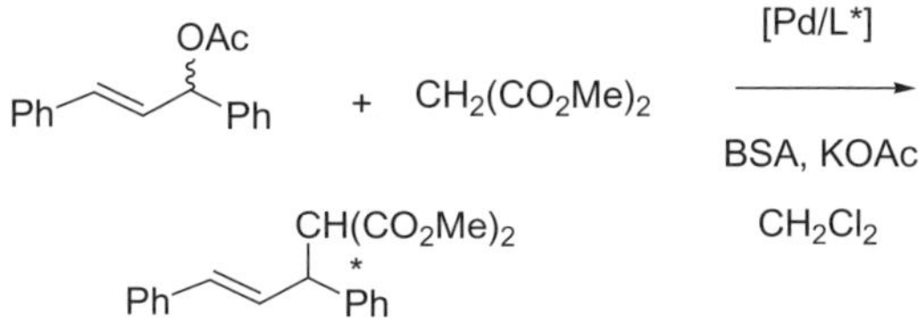

Scheme 1.2 Test reaction: Pd-catalysed allylic alkylation of 1,3-diphenylpropenyl acetate with dimethyl malonate.

ligands,[22] and chiral bis(sulfoxides) ligands.[23] These processes are commonly catalysed by palladium systems, but other metals such as rhodium, platinum, molybdenum, tungsten, nickel or iridium are also efficient.[1d,11,24] The fact that a mixture of diastereomers can be obtained upon coordination of the ligand (thioether for example) to a metal can cause a decrease of stereoselectivity if the relative rates of the intermediates are similar. In spite of this feature, however, excellent enantioselectivities have been achieved. To evaluate the selectivity of a new chiral ligand for allylic substitutions, the reaction usually performed, called the test reaction in the text, is the transformation of *rac*-1,3-diphenylprop-2-enyl acetate with dimethyl malonate in the presence of *N,O*-bis(trimethylsilyl) acetamide (BSA) and a base (Scheme 1.2).

Two general strategies based on the type of ligands can be applied to catalyse the asymmetric allylic alkylation. The first strategy is based on the use of chiral C_2-symmetric ligands, which results in catalytic systems with restricted numbers of diastereomeric transition states and consequently enantioselection. Indeed, the use of this type of ligands leads to the formation of only one (π-allyl)-palladium complex when a symmetric allylic substrate is used. The nucleophilic attack will take place preferentially at the site at which the allyl functionality has the strongest steric interaction with the ligand. The second strategy is based on the use of mixed heterodonating ligands containing strong and weak donor heteroatom pairs, thus giving rise to different electronic properties associated with each metal-heteroatom. In this case, two different (π-allyl)palladium complexes may be obtained, with each one having the possibility of being attacked by the nucleophile at two different sites, leading to four possible transition states. Two effects are responsible for the stereochemical outcome. The first one is the electronic effect: if the two donor atoms exhibit sufficiently different donor–acceptor strengths, the number of possible transition states is reduced. Due to the *trans*-effect, the nucleophile would thus preferentially attack the allylic system on the carbon possessing a greater positive charge character, *e.g.* on the carbon situated *trans* to the best π-acceptor. The second effect is the steric effect: the final stereoselectivity is determined by the chiral moiety of the ligand favouring one of the two π-allyl conformations. The following sections of this chapter successively deal with homodonor S/S ligands, heterodonor S/P, S/N, S/C, S/O ligands, sulfur-containing P/N, P/O, N/N, N/O and P/P ligands, and then sulfur-containing ferrocenyl ligands.

1.2 S/S Ligands

Relatively few chiral dithiother ligands have been used for Pd-catalysed allylic substitutions, even though the coordinating capability of thioether donors in transition-metal complexes is known. These chiral homodonor ligands generally provided only modest asymmetric inductions for these reactions even if some of them were really performing. An inherent characteristic of thioether ligands is that, upon coordination to the metal, the sulfur atom becomes stereogenic. While the close proximity of the chiral sulfur centre to the coordination sphere of the transition metal may be beneficial,[25] the low inversion barrier of the sulfur–metal bond may account for the scarce use of dithioethers in asymmetric catalysis.[26] Hence, any attempts to incorporate a thioether into a chiral ligand must firstly address stereocontrol at the sulfur atom. Such a control may be accomplished by steric bias such as the involvement of efficient catalysts based on chiral mixed S/P ligands.

Unlike other homo- and heterodonor chiral ligands, the S/S-type ligand has hardly been involved in spite of having advantages such as lower cost, toxicity and oxidation potential. In 2001, Gomez *et al.* reported the first example employing C_2-symmetric S/S-type ligands **3–10** (Scheme 1.3) for the test reaction.[27] For unexplained reasons, the major focus in academia has been on this particular allylic alkylation, although this system does not seem to have any industrial importance. These workers showed that the enantioselectivity could

L* = **3**: 0% ee = 0%
L* = **4**: 74% ee = 27% (*R*)
L* = **5**: 100% ee = 13% (*S*)
L* = **6**: 100% ee = 42% (*S*)
L* = **7**: 100% ee = 81% (*S*)
L* = **8**: 100% ee = 30% (*S*)
L* = **9**: 80% ee = 76% (*S*)
L* = **10**: 56% ee = 76% (*S*)

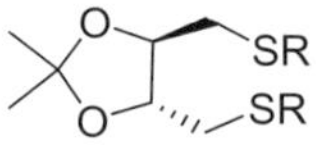

L* = **3**: R = Ph
L* = **4**: R = *i*-Pr

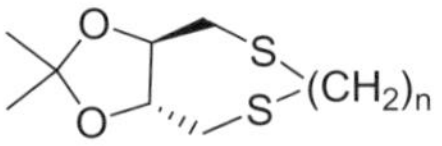

L* = **5**: n = 2
L* = **6**: n = 3

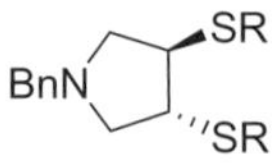

L* = **7**: R = Ph
L* = **8**: R = *i*-Pr
L* = **9**: R = 2-Naph
L* = **10**: R = o-(*t*-Bu)C$_6$H$_4$

Pd = [Pd(C$_3$H$_5$)Cl]$_2$

Scheme 1.3   Test reaction with dithioether ligands.

be increased to high values with the appropriate combination of the chiral backbone rigidity and the substituent at the sulfur atom (Scheme 1.3). The modest asymmetric induction observed ($\leq 81\%$ ee) was due to the donor sites being insufficiently different for discrimination between both terminal allylic carbons in the intermediate. In all cases, solid structures of complexes and structural studies in solution provided these authors with proof of S/S-coordination.

In order to rationalise these results, these workers have studied more examples of palladium systems in which systematic changes of the chelate ring size and the electronic and steric effects of the sulfur substituents were performed.[28] In the course of this systematic study, novel chiral dithioether ligands were shown to afford high activities and excellent selectivities in all palladium-catalysed allylic reactions (Scheme 1.4). The study of the allylic intermediates, which were fully characterised both in solution and in the solid state, has demonstrated that the selectivity in the palladium-catalysed allylic alkylation containing homodonor dithioether ligands could be controlled by the thermodynamics of the palladium diastereomer formation (high-energy barrier between Pd isomers, as is the case for these new ligands) or by the kinetics of the nucleophilic attack (low-energy barrier among the palladium species) depending on the nature of the metallacycle.

In 2003, Nakano *et al.* planned to synthesise novel chiral S/S-type ligands having a borneol backbone and without C_2-symmetry.[29] These ligands were readily prepared from the reactions of mercaptoisoborneol or mercaptoborneol with phenylthiobenzaldehydes with good yields. The chiral sulfideoxathiane ligands **11–14** were shown to give excellent enantioselectivities (up to 99% ee) in

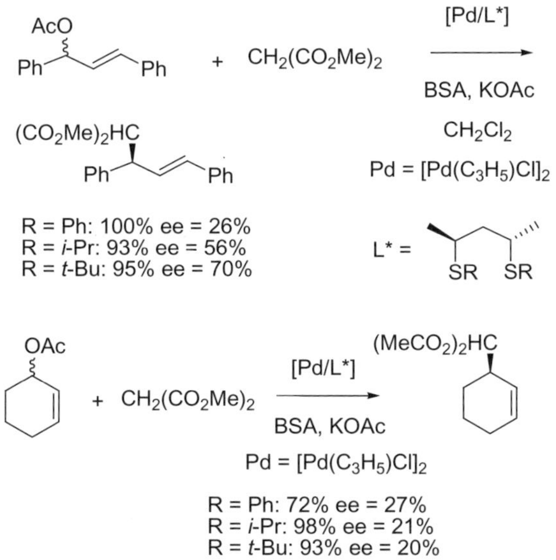

Scheme 1.4 Pd-catalysed allylic alkylations with dithioether ligands.

the palladium-catalysed allylic alkylation of 1,3-diphenyl-2-propenyl acetate with a range of alkyl malonate nucleophiles (Scheme 1.5). The presence of the bulky linked 2,6-dimethylphenylthio moiety proved necessary for achieving a high level of enantioselectivity. Semiempirical molecular orbital calculations were performed in order to propose an explanation for the high performance of ligand **12** compared to **11**. Geometry optimisation and energy calculations were in accordance with a control of the catalyst conformation by the steric

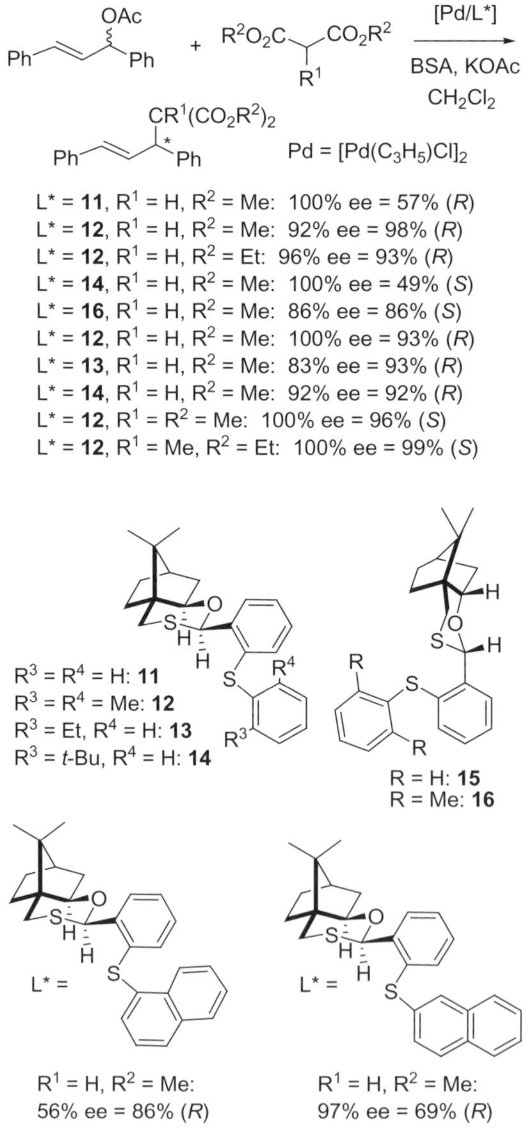
OAc
Ph
Ph
+
R2O2C
CO2R2
R1
[Pd/L*]
BSA, KOAc
CH2Cl2
CR1(CO2R2)2
Ph
Ph
Pd = [Pd(C3H5)Cl]2
L* = 11, R1 = H, R2 = Me: 100% ee = 57% (R)
L* = 12, R1 = H, R2 = Me: 92% ee = 98% (R)
L* = 12, R1 = H, R2 = Et: 96% ee = 93% (R)
L* = 14, R1 = H, R2 = Me: 100% ee = 49% (S)
L* = 16, R1 = H, R2 = Me: 86% ee = 86% (S)
L* = 12, R1 = H, R2 = Me: 100% ee = 93% (R)
L* = 13, R1 = H, R2 = Me: 83% ee = 93% (R)
L* = 14, R1 = H, R2 = Me: 92% ee = 92% (R)
L* = 12, R1 = R2 = Me: 100% ee = 96% (S)
L* = 12, R1 = Me, R2 = Et: 100% ee = 99% (S)
R3 = R4 = H: 11
R3 = R4 = Me: 12
R3 = Et, R4 = H: 13
R3 = t-Bu, R4 = H: 14
R = H: 15
R = Me: 16
L* =
L* =
R1 = H, R2 = Me:
56% ee = 86% (R)
R1 = H, R2 = Me:
97% ee = 69% (R)

Scheme 1.5 Pd-catalysed allylic alkylations with sulfideoxathiane ligands.

hindrance generated by the two methyl substituents of the phenyl ring. Bulkier ligands, containing linking 1-naphthylthio or 2-naphthylthio moieties, were also investigated for the test reaction and gave contrasted results, since the first ligand gave a moderate reactivity and a good enantioselectivity, whereas the second ligand gave, inversely, an excellent yield and a moderate enantioselectivity (Scheme 1.5).[30]

As previously mentioned, an inherent characteristic of the thioether ligands is that, upon coordination to the metal, the sulfur atom becomes stereogenic. While the close proximity of the chirality to the coordination sphere of the transition metal may be beneficial, the low inversion barrier of the sulfur–metal bond may be responsible for the poor results observed. In this context, Khiar *et al.* have reported the synthesis of C_2-symmetric bis(thioglycosides) as new ligands for the test reaction (Scheme 1.6).[31] The sugar residue was intended to provide a well-defined chiral environment, while the control of the sulfur configuration was expected, due to stereoelectronic factors acting at the anomeric centre. The best enantioselectivity was obtained with the ligand bearing a pivalate protecting group. Both enantiomers of the allylated product were prepared with an enantioselectivity of 90% ee by using inexpensive natural sulfur-modified D-sugars. Exploiting the fact that α-D-arabinose is almost the mirror image of β-D-galactose, these authors demonstrated that their sulfur derivatives behaved as pseudoenantiomers in the Pd-catalysed allylic substitution.[32] They further succeeded in synthesising the corresponding Pd(II) complex and observed by in-depth NMR studies the formation of a unique *anti* diastereomer with C_2-symmetry in solution, thus indicating a real control of the sulfur configuration by the sugar backbone. X-Ray analyses further indicated that the sugar residues were placed in a pseudoaxial orientation (in the *exo*-anomeric conformation).[33] As a result of n–σ* hyperconjugative delocalisation, this effect is strong enough for an efficient stereochemical control over the sulfur configuration, as both atoms possess (*S*) absolute configurations.

In 2008, Skarzewski and Wojaczynska studied the test reaction in the presence of chiral C_2-symmetric S/S-donor five- and six-membered cyclic ligands depicted in Scheme 1.7, providing moderate activity and enantioselectivity.[34] The best enantioselectivity (42% ee) was observed when (1*R*,2*S*)-bis(phenylsulfenyl)cyclopentane was involved as the ligand, whereas the corresponding

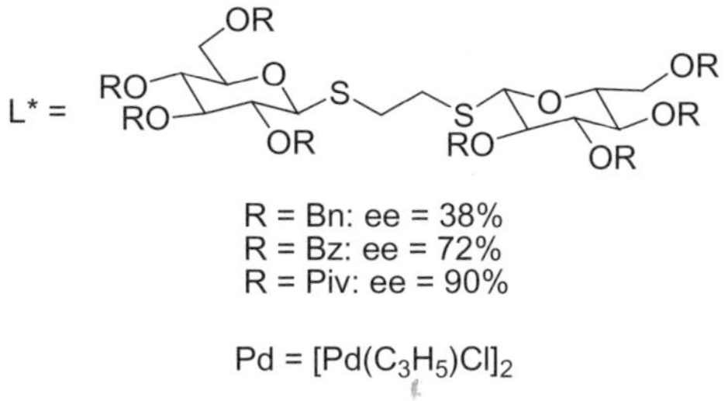

$$L^* =$$

R = Bn: ee = 38%
R = Bz: ee = 72%
R = Piv: ee = 90%

Pd = [Pd(C₃H₅)Cl]₂

Scheme 1.6 Test reaction with C_2-symmetric bis(thioglycosides) ligands.

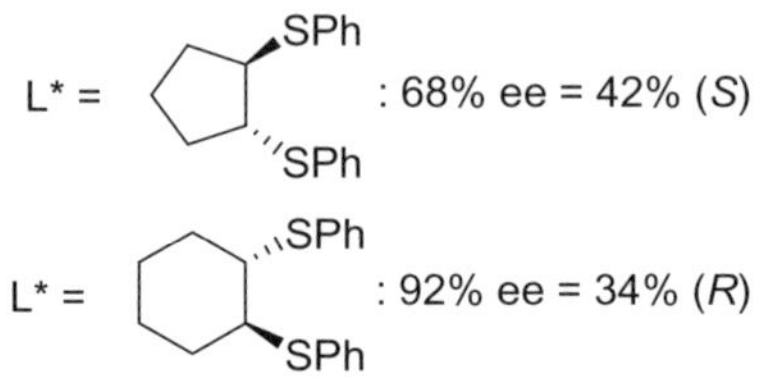

L* = (five-membered cyclopentane ring with SPh and SPh substituents) : 68% ee = 42% (*S*)

L* = (six-membered cyclohexane ring with SPh and SPh substituents) : 92% ee = 34% (*R*)

Scheme 1.7   Test reaction with C_2-symmetric S/S-donor five- and six-membered cyclic ligands.

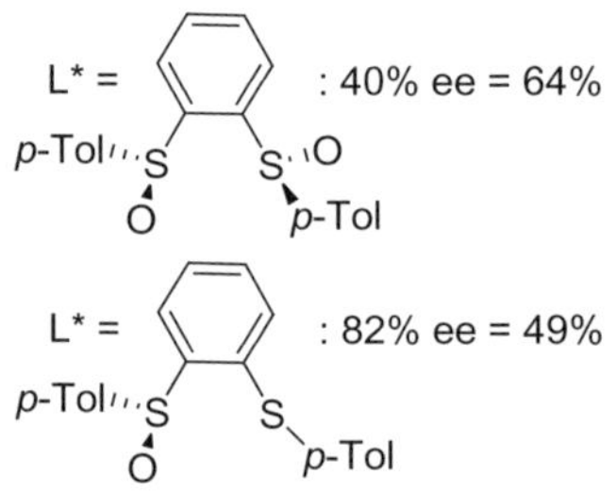

L* = (benzene ring with *p*-Tol–S(O)– and –S(O)–*p*-Tol substituents) : 40% ee = 64%

L* = (benzene ring with *p*-Tol–S(O)– and –S–*p*-Tol substituents) : 82% ee = 49%

Scheme 1.8   Test reaction with C_2-symmetric bis(sulfoxides) and monosulfoxide ligands.

six-membered ligand gave the best activity combined with a lower enantioselectivity, as shown in Scheme 1.7.

Although the use of chiral sulfoxides as chiral controllers in asymmetric synthesis is well documented, their utilisation as ligands in asymmetric catalysis has met with little success. In 2005, Khiar *et al.* reported the synthesis of C_2-symmetric bis(sulfoxides) but, surprisingly, when used as chiral ligands in the palladium-catalysed asymmetric alkylation of 1,3-diphenylpropenyl acetate with dimethyl malonate, they were completely inactive,[35] whereas the corresponding C_2-symmetric bis(thioethers) afforded the (*R*)-isomer with 42% ee. Another example of the application of this type of ligands for the asymmetric Pd-catalysed allylic substitution was reported by Shibazaki *et al.*, providing moderate enantioselectivities of up to 64% ee (Scheme 1.8).[23] These authors have studied the chelating ability of the chiral bis(sulfoxides), depicted in Scheme 1.8, for palladium, rhodium and ruthenium. X-ray crystal-structure analyses of the palladium complex indicated that this ligand coordinated Pd(II) through the sulfur atom and that the complex had a C_2-symmetry. In addition, the corresponding unsymmetric monosulfoxide ligand, depicted in Scheme 1.8, was also investigated for the same reaction, giving a higher yield but a lower enantioselectivity.

In 2004, Shi *et al.* reported Pd-catalysed asymmetric allylic substitutions using axially chiral S/S- and S/O-heterodonor ligands based on the binaphthalene backbone.[36] The test reaction was performed in the presence of

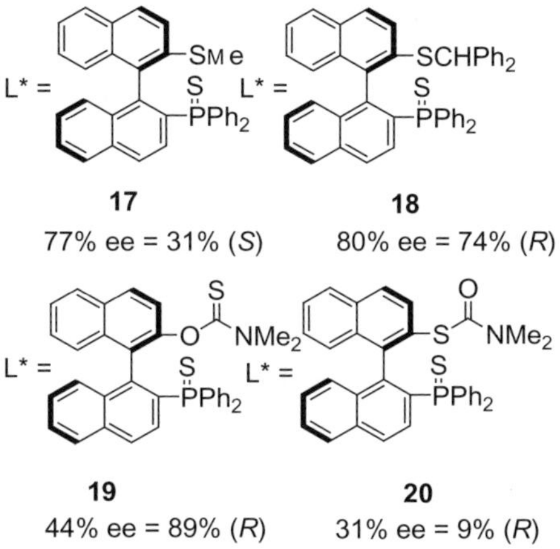
L* =
SMe
S
PPh2
L* =
SCHPh2
S
PPh2
17
77% ee = 31% (S)
18
80% ee = 74% (R)
L* =
S
O
NMe2
S
PPh2
L* =
O
S
NMe2
S
PPh2
19
44% ee = 89% (R)
20
31% ee = 9% (R)

Scheme 1.9 Test reaction with S/S- and S/O-heterodonor ligands with binaphthalene backbone.

these ligands combined with $[Pd(\eta^3\text{-}C_3H_5)Cl]_2$ as the palladium precursor, BSA, and different additives (KOAc or LiOAc) in various solvents. As shown in Scheme 1.9, moderate to high enantioselectivities could be reached with the S/S-coordinating ligands **17–19**, depending on the nature of the solvent and the additive used. The coordination of ligand **17** to Pd was studied by NMR analyses, suggesting the formation of a S,S-heterodonor complex. No explanation was proposed concerning the differences observed in the configuration of the product. As shown in Scheme 1.9, the corresponding S/O-chelate **20** proved to be less efficient in terms of both activity and selectivity for the test reaction.

In conclusion, S/S-coordinating ligands have proved to be efficient ligands for the Pd-catalysed asymmetric allylic substitution process, since high levels of enantioselectivity have been reached in some cases, thus becoming competitors of the generally most efficient nitrogen- or phosphorus-containing ligands.

1.3 S/P Ligands

The incorporation of the C_2-symmetry into the chiral ligand design is a well-recognised strategy for restricting the number of diastereomeric transition states in metal-catalysed enantioselective processes.[37] Equally powerful stereochemical restrictions may also be achieved with chiral ligands lacking C_2-symmetry through the use of electronic effects such as the *trans* influence. Such effects are a natural consequence of the use of chiral bidentate ligands equipped with strong and weak donor heteroatom pairs (*e.g.* PR_3/NR_3, PR_3/SR_2). Such electronic effects have the potential to influence both the stability and reactivity of the intervening diastereomeric reaction intermediates in the catalytic cycle. While mixed P/N-bidentate ligands have been applied in enantioselective palladium-catalysed nucleophilic alkylation, chiral thioether-containing donor ligands and, more generally, S/P ligands have been less well developed. In 1999, Evans *et al.* reported a new class of mixed S/P ligands incorporating a metal-bound thioether as a chiral control element and a diarylphosphinite moiety as a strong donor heteroatom in asymmetric catalysis. The utility of these thioether-phosphinite ligands **21–23** was illustrated in the palladium-catalysed allylic alkylation involving enol-malonate and amine nucleophiles (Scheme 1.10).

After a systematic variation of the ligand substituents at sulfur, phosphorus, and the ligand backbone, the S/P ligand **22a** was found to be optimal in the palladium-catalysed allylic substitution of 1,3-diphenylpropenyl acetate with dimethyl malonate or benzylamine.[38] A similar optimisation of the mixed S/P ligand for the palladium-catalysed allylic substitution of cycloalkenyl acetates showed that the ligand **21b** afforded the highest enantioselectivities (91–97% ee). The application of this methodology to heterocyclic substrates was developed as an efficient approach to the enantioselective synthesis of 3-substituted piperidines and dihydrothiopyrans (Scheme 1.11). These authors could, furthermore, prove the contribution of the sulfur in the coordination of the palladium atom by X-ray analysis of crystals of these chiral organometallic complexes. These structures also showed the relative electronic impact of the

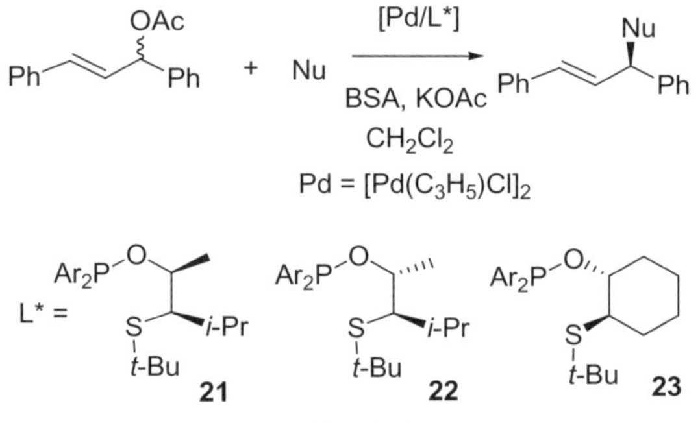

a) Ar = Ph; b) Ar = 1-Naph

L* = **21a**, Nu = $CH_2(CO_2Me)_2$: 91% ee = 93%
L* = **22a**, Nu = $CH_2(CO_2Me)_2$: 98% ee = 97%
L* = **23a**, Nu = $CH_2(CO_2Me)_2$: 94% ee = 95%
L* = **21b**, Nu = $CH_2(CO_2Me)_2$: 28% ee = 91%
L* = **22b**, Nu = $CH_2(CO_2Me)_2$: 30% ee = 94%
L* = **23b**, Nu = $CH_2(CO_2Me)_2$: 69% ee = 92%
L* = **21a**, Nu = $BnNH_2$: 99% ee = 96%
L* = **22a**, Nu = $BnNH_2$: 95% ee = 97%
L* = **23a**, Nu = $BnNH_2$: 95% ee = 95%
L* = **21b**, Nu = $BnNH_2$: 78% ee = 90%
L* = **22b**, Nu = $BnNH_2$: 66% ee = 95%
L* = **23b**, Nu = $BnNH_2$: 89% ee = 93%

Scheme 1.10   Mixed S / P ligands for Pd-catalysed allylic alkylations and aminations.

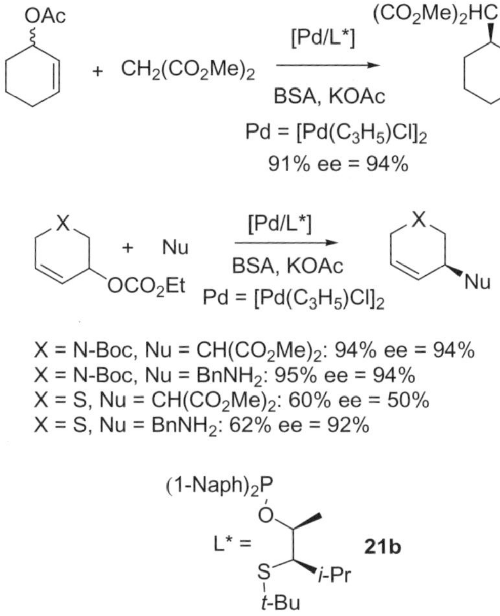

Scheme 1.11 Mixed S / P ligands for Pd-catalysed allylic substitutions of cycloalkenyl acetates.

heteroatom P- and S-donors, since the Pd–C bond *trans* to the phosphinite functionality was longer than the Pd–C bond *trans* to the thioether group.

In order to involve organosulfur functionality as an alternative enantio-controllable coordinating element in chiral phosphine ligands, Hiroi *et al.* have reported the synthesis of (*S*)-proline-derived phosphines bearing organosulfur groups, **24–28**, and their successful use as chiral ligands in the test reaction (Scheme 1.12).[39] The best enantioselectivity was obtained by using ligand **24g** that presented a sterically hindered naphthyl group around the sulfur atom. Unexpected results were obtained with ligand **24e** bearing a phenyl group, which provided the corresponding product with similar activity, enantioselectivity and configuration than those obtained by using ligands **26c** and **29**. In these three cases of ligands, the authors proposed an N/P-coordination, referring to the low coordination ability of the aromatic sulfenyl groups compared with other alkyl sulfenyl functions. In the other cases of ligands, a nine-membered S/P-chelate was proposed to be formed by coordination of the organosulfur functionality and the phosphine group to the palladium catalyst. The nucleophilic attack occurred *trans* to the sulfenyl group, which is the better π-acceptor.

The test reaction was also performed by RajanBabu *et al.* in the presence of monophospholanes bearing a pendant *t*-BuS group, demonstrating that the chirality of the C3 and C5 oxygens played a crucial role in the asymmetric induction (Scheme 1.13).[40]

L* = **24a**: 65% ee = 62% (*S*)
L* = **24b**: 72% ee = 72% (*S*)
L* = **24c**: 68% ee = 77% (*S*)
L* = **24d**: 54% ee = 31% (*S*)
L* = **24e**: 42% ee = 41% (*R*)
L* = **24f**: 76% ee = 84% (*S*)
L* = **24g**: 59% ee = 88% (*S*)
L* = **25**: 72% ee = 60% (*S*)
L* = **26a**: 76% ee = 82% (*R*)
L* = **26b**: 74% ee = 87% (*R*)
L* = **26c**: 69% ee = 42% (*R*)
L* = **27**: 69% ee = 59% (*R*)
L* = **28**: 76% ee = 79% (*R*)
L* = **29**: 44% ee = 30% (*R*)

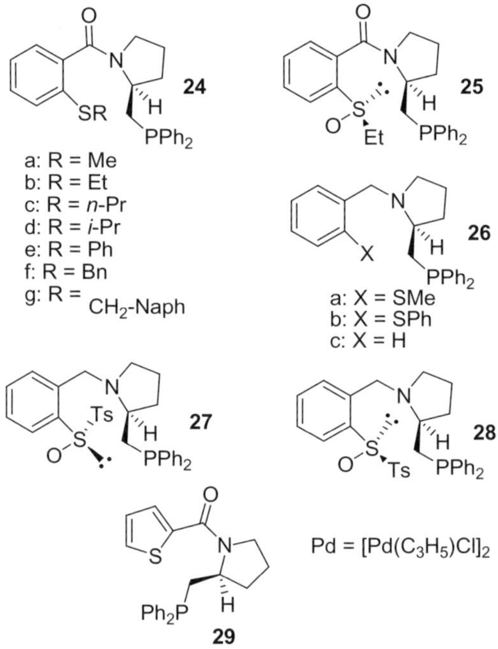

Scheme 1.12   Test reaction with (*S*)-proline-derived sulfur-containing phosphine ligands.

In 2001, Imamoto *et al.* reported the preparation of novel chiral S/P-bidentate ligands containing a chirogenic centre at the phosphorus atom and their stereoinduction capability in palladium-catalysed asymmetric allylic substitution reactions (Scheme 1.14).[41]

The preparation of BINAP reported in 1980 has marked a landmark in asymmetric catalysis and has illustrated the peculiar stereorecognitive properties inherent with the axially chiral 1,1′-binaphthalene framework. Since then, a great deal of work has been devoted to the preparation of binaphthalene-templated ligands of related design. These efforts have resulted in the

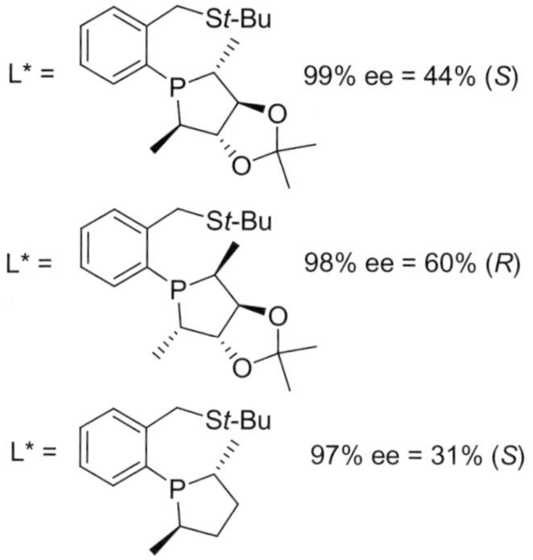

L* = 99% ee = 44% (*S*)

L* = 98% ee = 60% (*R*)

L* = 97% ee = 31% (*S*)

Scheme 1.13 Test reaction with monophospholanes with pendant *t*-BuS group.

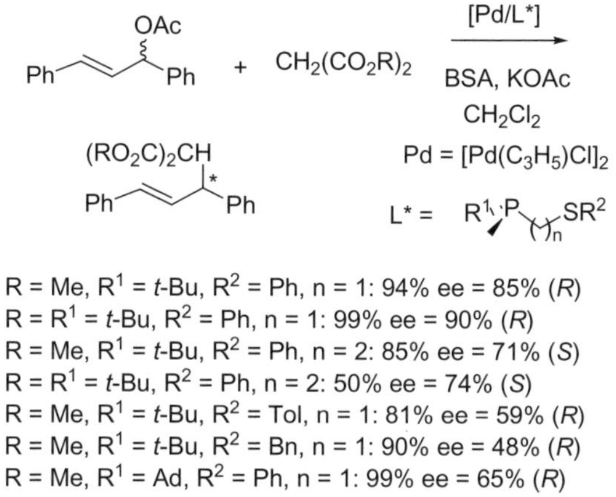

$$Ph \overset{OAc}{\underset{Ph}{\diagdown}} + CH_2(CO_2R)_2 \xrightarrow[\substack{BSA,\ KOAc \\ CH_2Cl_2}]{[Pd/L^*]}$$

$$(RO_2C)_2CH$$
$$Ph \overset{*}{\diagdown} Ph$$

Pd = [Pd(C$_3$H$_5$)Cl]$_2$

L* = R^1-P$\diagdown$SR^2 (n)

R = Me, R^1 = *t*-Bu, R^2 = Ph, n = 1: 94% ee = 85% (*R*)
R = R^1 = *t*-Bu, R^2 = Ph, n = 1: 99% ee = 90% (*R*)
R = Me, R^1 = *t*-Bu, R^2 = Ph, n = 2: 85% ee = 71% (*S*)
R = R^1 = *t*-Bu, R^2 = Ph, n = 2: 50% ee = 74% (*S*)
R = Me, R^1 = *t*-Bu, R^2 = Tol, n = 1: 81% ee = 59% (*R*)
R = Me, R^1 = *t*-Bu, R^2 = Bn, n = 1: 90% ee = 48% (*R*)
R = Me, R^1 = Ad, R^2 = Ph, n = 1: 99% ee = 65% (*R*)

Scheme 1.14 Test reaction with P-chirogenic S / P ligands.

generation of a library of bidentate binaphthyl ligands, featuring equal (or diverse) substituents with the same (or different) donors on the 2,2'-positions of the binaphthalene backbone. In this context, the groups of Kang[42] and Gladiali[43] have developed axially chiral S/P-heterodonor ligands based on a binaphthalene backbone, which were further submitted to the test palladium-catalysed allylic alkylation, providing moderate to good enantioselectivities (60–91% ee) (Scheme 1.15). In 2004, Shi *et al.* extended the scope of this reaction to various S/P ligands (BINAPs) with different alkyl groups on the sulfur atom.[44] It was demonstrated that an alkyl group on the sulfur atom acted

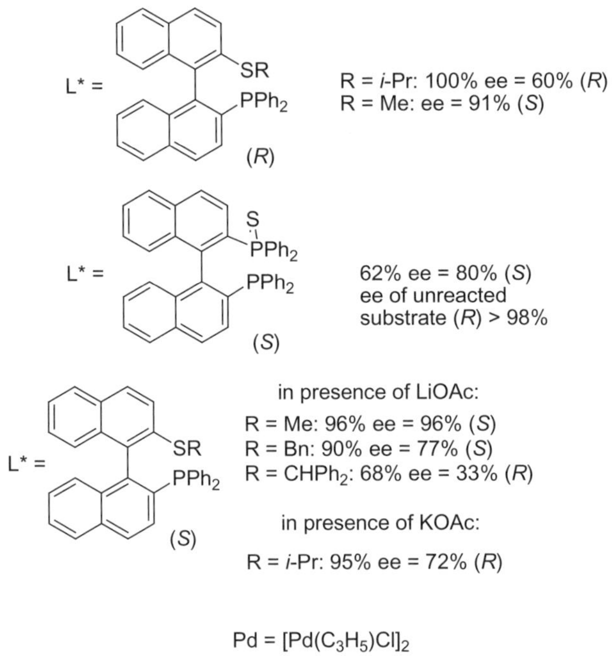

L* =

R = *i*-Pr: 100% ee = 60% (R)
R = Me: ee = 91% (S)

SR
PPh$_2$

(R)

L* =

S
PPh$_2$
PPh$_2$

62% ee = 80% (S)
ee of unreacted
substrate (R) > 98%

(S)

L* =

SR
PPh$_2$

(S)

in presence of LiOAc:
R = Me: 96% ee = 96% (S)
R = Bn: 90% ee = 77% (S)
R = CHPh$_2$: 68% ee = 33% (R)

in presence of KOAc:
R = *i*-Pr: 95% ee = 72% (R)

Pd = [Pd(C$_3$H$_5$)Cl]$_2$

Scheme 1.15   Test reaction with S / P-heterodonor ligands (BINAPs).

as a key factor in forming the reversal of the enantioselectivity. The steric bulkiness of an (S)-alkyl group in the BINAPs was sufficient to control the orientation of the nucleophilic attacks to give the product with a different absolute configuration (Scheme 1.15). According to X-ray analyses and NMR spectroscopic studies, these authors assumed in all cases an S/P-coordination as a metallacycle in a pseudo-boat-seven-membered arrangement. The steric bulkiness of alkyl groups borne by the sulfur atom seemed to be responsible for the observed reversal of the enantioselectivity by favouring one or the other diastereomeric π-allyl complex. In addition, another BINAP-derived S/P ligand was developed by Faller *et al.*, demonstrating that this palladium-catalysed allylic alkylation could also be used to achieve effective kinetic resolution of acyclic allylic acetates, since the unreacted allylic acetate was recovered in >98% ee (Scheme 1.15).[45]

Since carbohydrates constitute an inexpensive and highly modular chiral source for preparing chiral ligands,[46] Claver *et al.* have reported the use of a series of thioether-phosphite[47] and thioether-phosphinite furanoside ligands[48] in the test palladium-catalysed allylic substitution reaction. In the first type of ligand, a systematic variation of the donor group attached to the carbon atom C5 indicated that the presence of a bulky phosphite functionality had a positive effect on the enantioselectivity. Indeed, the enantioselectivity was controlled mainly by the phosphite moiety. This was confirmed by the use of a ligand

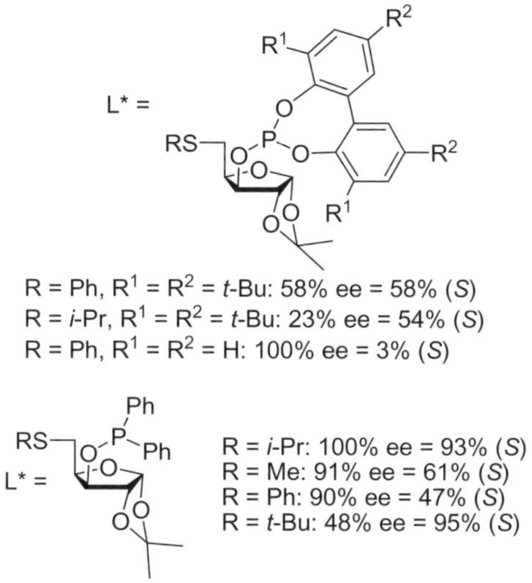

L* =

R = Ph, R¹ = R² = *t*-Bu: 58% ee = 58% (*S*)
R = *i*-Pr, R¹ = R² = *t*-Bu: 23% ee = 54% (*S*)
R = Ph, R¹ = R² = H: 100% ee = 3% (*S*)

L* =

R = *i*-Pr: 100% ee = 93% (*S*)
R = Me: 91% ee = 61% (*S*)
R = Ph: 90% ee = 47% (*S*)
R = *t*-Bu: 48% ee = 95% (*S*)

Scheme 1.16 Test reaction with furanoside thioether-phosphite or -phosphinite ligands.

containing the smaller unsubstituted biphenol moiety, which resulted in a drop of the ee value from 58 to 3% (Scheme 1.16). Interestingly, the thioether-phosphinite ligands showed a much higher degree of enantioselectivity and higher reaction rates than their thioether-phosphite analogues (Scheme 1.16). In this case, the enantioselectivities were strongly dependent on the steric properties of the substituent in the thioether moiety of the carbohydrate backbone.

In 2006, these workers successfully expanded the previous study to several acyclic and cyclic allylic substrates (Scheme 1.17).[49] In all cases, the best enantioselectivity (up to 91% ee) was obtained by using the ligand that contained the more bulky sulfur substituent (*t*-Bu). The methodology was also applied to monosubstituted acyclic substrates but, however, this ligand proved to be inadequate in terms of regioselectivities, whereas a good enantioselectivity of up to 89% ee was obtained.

On the other hand, a novel chiral xylofuranose-based phosphinooxathiane ligand was found by Nakano *et al.* to provide high levels of enantioselectivity (up to 94% ee) in palladium-catalysed asymmetric allylic alkylations and aminations (Scheme 1.18).[50] This ligand was easily prepared from commercially available 1,2-*O*-isopropylidene-D-xylofuranose in five steps. In order to explain the absolute configuration of the product, these authors have proposed a nucleophilic attack occurring at the allyl terminus *trans* to the phosphorus atom, which was the better π-acceptor. These authors have previously reported the synthesis of other phosphinooxathiane ligands, such as norbornane- and pulegone-based phosphinooxathianes, and their successful application to the test reaction.[51] In 2005, these results were further extended to the syntheses of

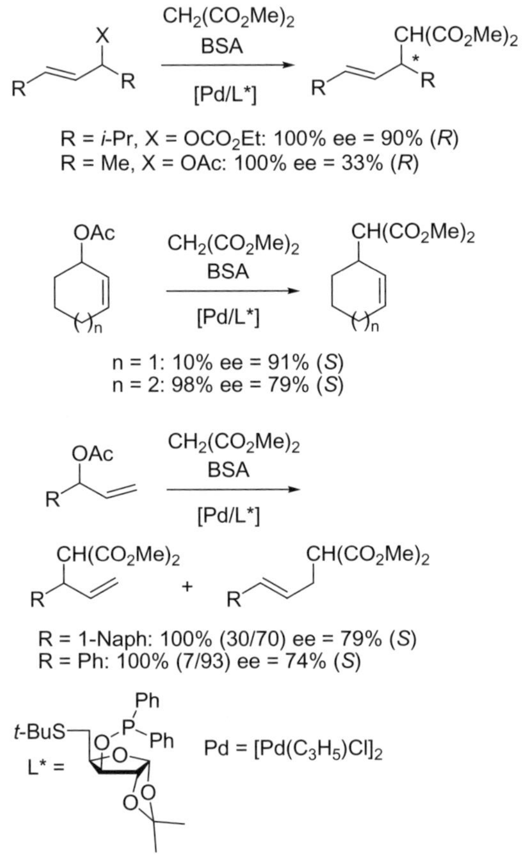
CH2(CO2Me)2
BSA
[Pd/L*]
X
R
CH(CO2Me)2
R = i-Pr, X = OCO2Et: 100% ee = 90% (R)
R = Me, X = OAc: 100% ee = 33% (R)
OAc
n = 1: 10% ee = 91% (S)
n = 2: 98% ee = 79% (S)
R = 1-Naph: 100% (30/70) ee = 79% (S)
R = Ph: 100% (7/93) ee = 74% (S)
t-BuS
L* =
Ph
P
Pd = [Pd(C3H5)Cl]2

Scheme 1.17 Furanoside thioether-phosphinite ligand for Pd-catalysed allylic substitutions of acyclic and cyclic allylic substrates.

related polymer-supported chiral phosphinooxathiane ligands, which were found to provide high levels of enantioselectivity (up to 99% ee) in palladium-catalysed asymmetric allylic alkylations and aminations.[52] In addition, the same authors have examined the abilities of new chiral phosphinooxathiolane ligands derived from (1*R*,2*S*,3*R*)-3-mercaptocamphan-2-ol as catalysts of the test reaction.[53] The best enantioselectivity of 85% ee was obtained by using the phosphinooxathiolane ligand depicted in Scheme 1.18.

In 2005, Khiar *et al.* reported the preparation and use of other novel phosphinite thioglycosides derived from D-galactose, showing that bulky alkyl thioglycosides, such as the *t*-butyl thioglycoside depicted in Scheme 1.19, allowed the syntheses of the expected alkylated and aminated products in, respectively, 96 and 90% ees in the palladium-catalysed allylic substitutions.[54] The importance of the stereochemistry at the anomeric centre was shown, since the analogous α-thioglycoside led to the racemic product. Dynamic and NOE NMR studies revealed that the complex between the ligand and the palladium

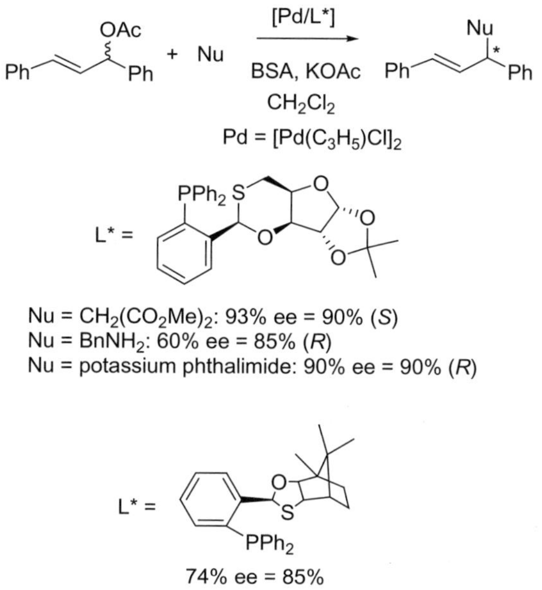
OAc
Ph
Ph
+ Nu
[Pd/L*]
BSA, KOAc
CH2Cl2
Pd = [Pd(C3H5)Cl]2
Nu
Ph
Ph
L* =
PPh2
S
O
O
O
O
Nu = CH2(CO2Me)2: 93% ee = 90% (S)
Nu = BnNH2: 60% ee = 85% (R)
Nu = potassium phthalimide: 90% ee = 90% (R)
L* =
O
S
PPh2
74% ee = 85%

Scheme 1.18 Test reaction with xylofuranose-based phosphinooxathiane ligands.

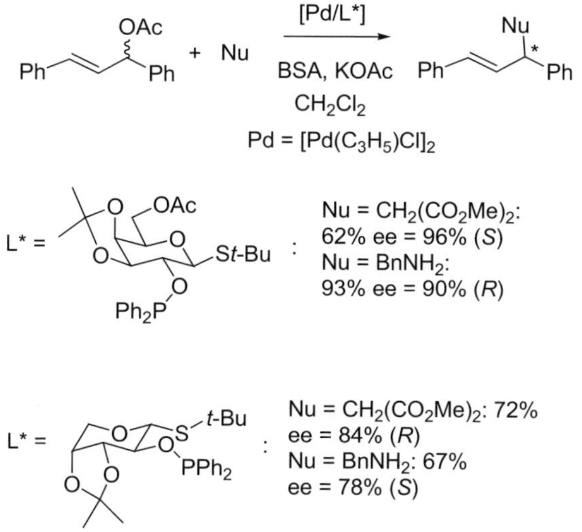
OAc
Ph
Ph
+ Nu
[Pd/L*]
BSA, KOAc
CH2Cl2
Pd = [Pd(C3H5)Cl]2
Nu
Ph
Ph
L* =
O
OAc
O
O
St-Bu
O
Ph2P
Nu = CH2(CO2Me)2:
62% ee = 96% (S)
Nu = BnNH2:
93% ee = 90% (R)
L* =
O
O
S
t-Bu
PPh2
O
O
Nu = CH2(CO2Me)2: 72%
ee = 84% (R)
Nu = BnNH2: 67%
ee = 78% (S)

Scheme 1.19 Test reaction with phosphinite thioglycoside ligands.

was obtained as a single diastereomer, showing an efficient control of the sulfur configuration. In addition, these authors demonstrated that it was possible to generate the corresponding opposite enantiomers in good enantioselectivity (up to 84% ee) by using a phosphinite thioglycoside derived from D-arabinose as the ligand of the test reaction.[55] Thus, even though derived from natural D-sugars, the two ligands behaved as enantiomers in Pd-catalysed asymmetric substitutions including those using benzylamine as the nucleophile (Scheme 1.19).

Although a large variety of structurally diverse ligands have already been tested in asymmetric catalysis, the cyclopropane skeletons have received relatively little attention to date. In this context, Molander *et al.* have developed the synthesis of a series of new chiral cyclopropane-based S/P ligands and have evaluated them in the test palladium-catalysed allylic alkylation reaction.[56] Variations of the ligand substituents at phosphorus, sulfur, and the carbon backbone revealed the ligands depicted in Scheme 1.20 to have the optimal configuration for the test reaction. On the basis of X-ray crystallographic data and NMR studies, the absolute configuration of the products was explained by a nucleophilic attack occurring at the carbon *trans* to the π-accepting phosphorus atom.

In 1995, Pregosin *et al.* reported the synthesis of a new heterobidentate S/P ligand, exo-8-((o-(diphenylphosphino)-benzyl)thio)borneol, in which the chirality was not associated with the phosphine backbone.[21b] The presence of an additional OH group did not allow a good enantioselectivity to be obtained since only 22% ee was obtained for the test reaction combined with 93% yield. These results were further optimised by the same authors who prepared other new ligands in which the chiral centres were placed nearer to the palladium coordination sphere in order to induce a better transfer of the chiral information.[21c] Therefore, the two ligands depicted in Scheme 1.21 and based on β-D-thioglucose provided the expected product of the test reaction with a higher but still moderate enantioselectivity (up to 64% ee).

Recent studies have reported good results by using metals other than palladium. As an example, Pregosin *et al.* have performed rhodium- and iridium-catalysed asymmetric allylic alkylations in the presence of chiral phosphito-thioether ligands.[57] More recently, Takemoto *et al.* have developed an elegant synthesis of β-substituted α-amino acids on the basis of the first iridium-catalysed asymmetric allylic substitutions of diphenylimino glycinates with allylic phosphates by using a chiral bidentate phosphite bearing a 2-ethylthioethyl group as the chiral ligand (Scheme 1.22).[58] Surprisingly, it was shown that the choice of the base and especially the effect of counteractions of the resulting enolate dramatically affected the diastereoselectivity of the reaction.

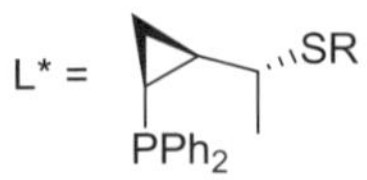

$$L^* = \quad \text{(cyclopropane structure with ...SR and PPh}_2\text{)}$$

R = 2,6-(Me)$_2$Ph: > 95% ee = 74% (*R*)
R = Et: > 95% ee = 91% (*R*)
R = Me: > 95% ee = 93% (*R*)

Scheme 1.20   Test reaction with cyclopropane-based S / P ligands.

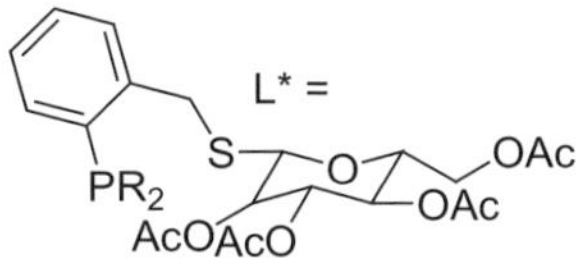

R = Ph: ee = 64%
R = Cy: ee = 53%

Scheme 1.21 Test reaction with β-D-thioglucose-based ligands.

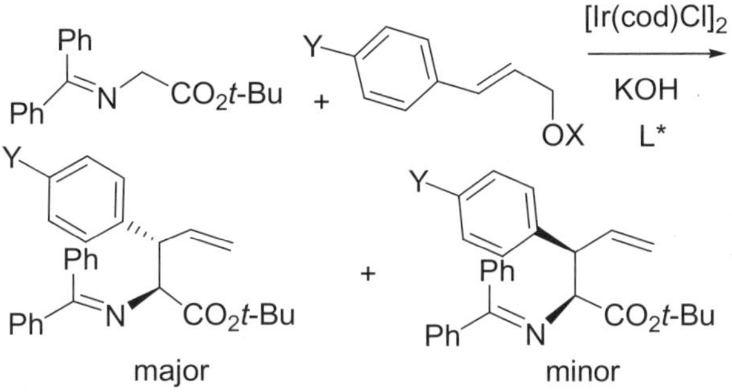

X = Bz, Y = H: 11% (73:27) ee (major) = 93% ee (minor) = 25%
X = P(O)(OEt)$_2$, Y = H: 82% (82:18) ee (major) = 97%
ee (minor) = 66%
X = P(O)(OEt)$_2$, Y = F: 77% (78:22) ee (major) = 97%
ee (minor) = 63%
X = P(O)(OEt)$_2$, Y = CF$_3$: 77% (68:22) ee (major) = 97%
ee (minor) = 68%
X = CO$_2$Me, Y = Me: 63% (83:17) ee (major) = 91%
ee (minor) = 74%

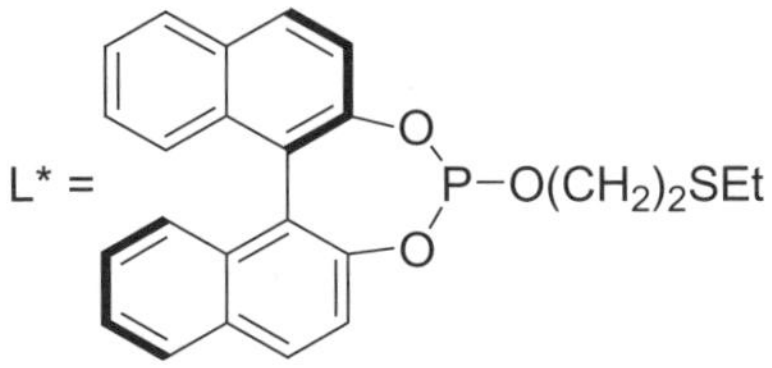

Scheme 1.22 Syntheses of β-substituted α-amino acids through Ir-catalysed asymmetric allylic substitutions.

Thus, a reversal of the diastereoselectivity of the reaction was observed if the enolate was prepared in the presence of a lithiated base. The different behaviour of the base could be attributable to the geometry of the enolate. It was assumed that the use of KOH as a base would give predominantly the *E* enolate, whereas the *Z* enolate would be formed with a lithiated base such as LiN(TMS)$_2$. This methodology was applied to the asymmetric synthesis of quaternary α-amino acids starting from an imino alaninate compound.

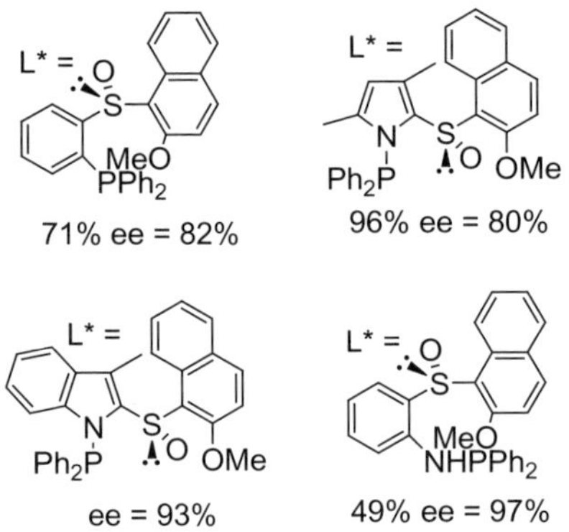

L* =
71% ee = 82%

L* =
96% ee = 80%

L* =
ee = 93%

L* =
49% ee = 97%

Scheme 1.23 Test reaction with β-phosphino sulfoxide and *N*-diphenylphosphano nitrogen-containing five-membered aromatic ligands.

In 1999, Hiroi *et al.* reported the synthesis of chiral β-phosphino sulfoxides for which the chirality was solely introduced by the chiral sulfoxide moiety.[59] These ligands were successfully applied to the test reaction providing up to 82% ee, as shown in Scheme 1.23. The formation of a five-membered chelated palladium complex with coordination between the phosphino function and the sulfinyl group was proved by X-ray analyses. It was assumed that the nucleophile attacked preferentially the allyl substrate *trans* to the better π-acceptor, which was the phosphine group, despite the steric effect of the bulky substituent (2-methoxy-1-naphthyl). In the same context, these workers reported, in 2004, the synthesis of *N*-diphenylphosphano nitrogen-containing five-membered aromatic ligands, which allowed enantioselectivities of up to 93% ee to be obtained (Scheme 1.23).[60] Once again, the coordination of the sulfur atom and the phosphano group was demonstrated by X-ray crystallographic analyses. In addition, a better enantioselectivity (97% ee) was obtained for the test reaction by these authors by using a chiral *o*-(phosphinoamino)phenyl sulfoxide ligand (Scheme 1.23).[61] It must be noted that this last result was, in 1998, the most highly selective example among the asymmetric synthetic methods reported previously by means of a chiral organosulfur ligand as the sole chiral source.

1.4 S/N Ligands

Among chiral sulfur-coordinating ligands, S/N ligands have been the most widely studied to promote the Pd-catalysed allylic substitutions. Thus, a number of cyclic and acyclic aminothioethers as well as iminothioethers have been synthesised and successfully used in these reactions. Among these, the sulfur-oxazoline ligands, which were applied to the test reaction for the first time by Williams *et al.*,[19c] have been the most studied. The ligands developed by Williams *et al.* bearing an alkyl sulfide,[19b] an aryl sulfide,[19b] or a sulfide within a

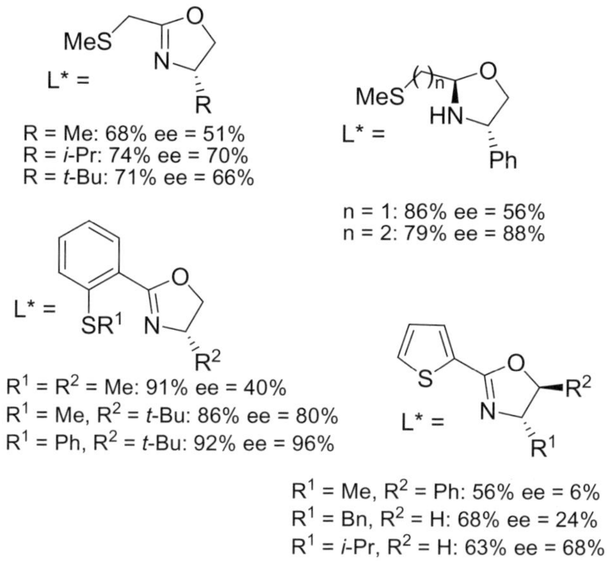

Scheme 1.24 Test reaction with sulfanylmethyl-, aryl-, and thiophene-oxazoline ligands.

thiophene ring, were prepared with good yields by reacting the appropriate nitriles with homochiral aminoalcohols in order to study the control of the stereochemical course of the reaction via the modified binding properties, electronic behaviour and steric environment of both the oxazoline and the sulfur group. As shown in Scheme 1.24, all the sulfanylmethyl-oxazolines were proven to be efficient ligands for the test reaction, providing moderate to high enantioselectivities (up to 88% ee). The highest enantioselectivity was obtained in the case of the ligand with a longer tether that should allow the formation of a six-membered chelate between the bidentate ligand and the metal. In the same context, these authors have shown that the use of a diarylsulfide, such as that depicted in Scheme 1.24, allowed an even higher enantioselectivity (96% ee) to be obtained.[19a] This ligand was also studied by Helmchen *et al.* who compared its efficiency in the test reaction with the corresponding P/N- and Se/N-chelates, finding that the steric course of the reaction was not dependent on the nature of the soft atom, but the reactivity varied strongly as follows: P > Se > S. On the other hand, the thiophene-oxazoline ligands prepared by Williams *et al.* were proven to be less efficient catalysts for the test reaction for both activity ($\leq$68%) and enantioselectivity ($\leq$68% ee), as shown in Scheme 1.24.

In 1994, the same authors reported the synthesis of corresponding oxazolines tethered to sulfoxides by ortholithiation of the parent 2-phenyl oxazoline with butyllithium followed by addition of either (*S*)- or (*R*)-*p*-tolylmenthyl-sulfinate.[19a] Applied as ligands in the test reaction, these diastereomeric ligands gave very different results, as shown in Scheme 1.25, thus demonstrating the

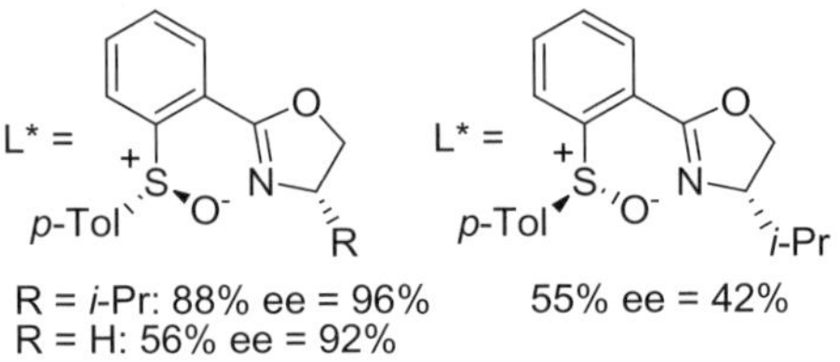

R = *i*-Pr: 88% ee = 96%   55% ee = 42%
R = H: 56% ee = 92%

Scheme 1.25   Test reaction with sulfoxide-oxazoline ligands.

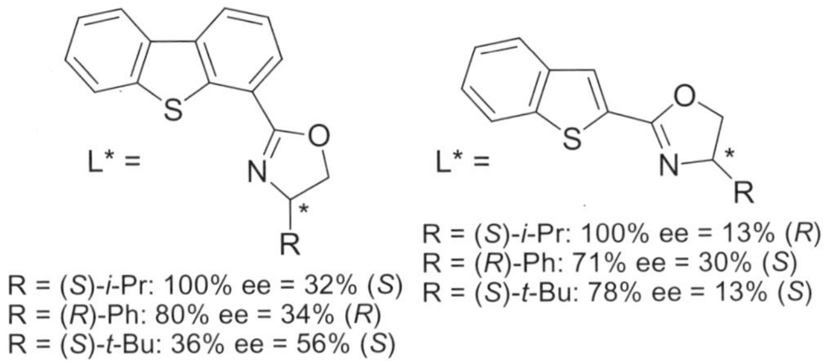

R = (S)-*i*-Pr: 100% ee = 13% (R)
R = (R)-Ph: 71% ee = 30% (S)
R = (S)-*t*-Bu: 78% ee = 13% (S)

R = (S)-*i*-Pr: 100% ee = 32% (S)
R = (R)-Ph: 80% ee = 34% (R)
R = (S)-*t*-Bu: 36% ee = 56% (S)

Scheme 1.26   Test reaction with dibenzothiophene or benzothiophene-oxazoline ligands.

importance of the stereochemistry at the sulfur centre in the determination of the stereochemical outcome of the process. Surprisingly, it was shown that the ligand bearing an achiral oxazoline moiety also provided a chiral product with a high enantioselectivity of 92% ee. In order to complete this study, these authors prepared the corresponding sulfone ligand but, however, its corresponding palladium complex was unreactive for the test reaction. Consequently, the authors concluded from these results that the sulfoxide ligand was more likely to bind through the sulfur atom than through the oxygen one.

Novel asymmetric sulfur-containing oxazoline ligands have been described by Schulz *et al.*[62] Their structure included a dibenzothiophene or a benzothiophene ring as the backbone in which the sulfur atom was enclosed in a strong π-donor structure. These ligands have been prepared in good yields starting from the corresponding dibenzothiophene- or benzothiophene-carboxylic acids and commercially available amino acids. When applied to the test reaction, these ligands led to the expected product with low to good activities combined with low to moderate enantioselectivities, as shown in Scheme 1.26.

In 1999, Ikeda *et al.* reported a new type of sulfur-oxazoline ligands with an axis-fixed or -unfixed biphenyl backbone prepared in good yields by coupling reactions of methoxybenzene derivatives substituted with a chiral oxazoline and a sulfur-containing Grignard reagent.[63] These ligands were subsequently evaluated for the test palladium-catalysed asymmetric allylic alkylation reaction (Scheme 1.27). It was found that the axial chirality of the biphenyl backbone in the ligands exerted a considerable influence on the catalytic activity, and the alkylthio group similarly influenced the enantioselectivity. The ligands with an axis-fixed biphenyl backbone led to moderate enantioselectivities (20–74% ee), whereas the ligands with an axis-unfixed biphenyl backbone showed the highest catalytic activity (93%) and enantioselectivity (82% ee). These ligands, having a free-rotation biphenyl axis, were proven to adopt only one of the two possible diastereoisomers, as shown by ^{1}H NMR, on coordination with bis(acetonitrile)–dichloropalladium(II).

In the same context, novel chiral binaphthalene-core ligands, possessing an axial chirality as the unique chiral source and in which an achiral oxazoline pendant was flanked by a sulfur group, have been prepared by Gladiali *et al.* and successfully involved in the test reaction.[64] The most potent ligand, depicted in Scheme 1.28, was obtained in 85% ee and a high chemical yield through a Ni-catalysed asymmetric coupling between disubstituted naphthalene derivatives.

In order to study the role of the [2.2]paracyclophane-type planar chirality in asymmetric induction, Hou *et al.* have developed the synthesis of novel S/N-[2.2]paracyclophane ligands with planar and central chirality based on the [2.2]paracyclophane backbone.[65] The ligands, bearing two coordinating atoms at both benzylic and benzene ring positions, have given the highest enantioselectivities (up to 94% ee) combined with almost quantitative yields in the test reaction (Scheme 1.29). These authors assumed that the longer tether between

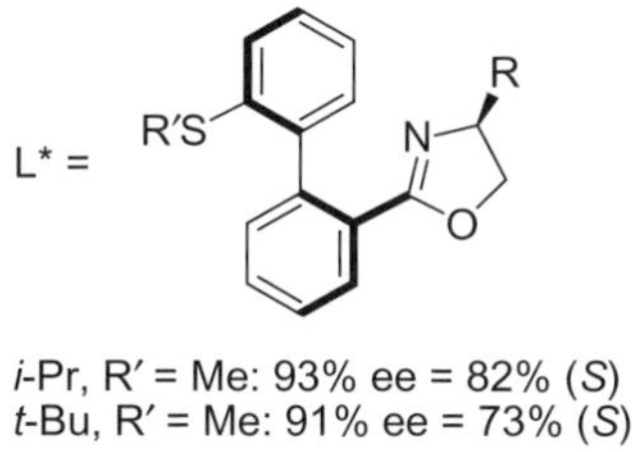

R = *i*-Pr, R' = Me: 93% ee = 82% (*S*)
R = *t*-Bu, R' = Me: 91% ee = 73% (*S*)

Scheme 1.27 Test reaction with axial sulfur-oxazoline ligands with biphenyl backbone.

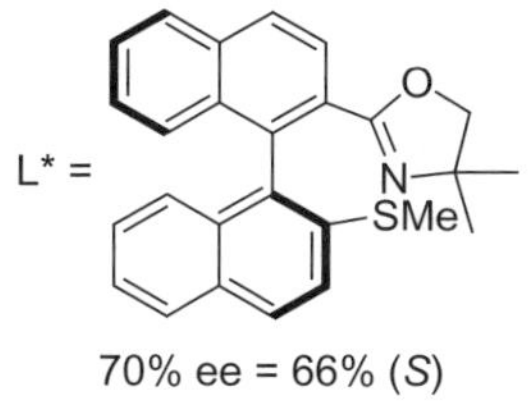

70% ee = 66% (*S*)

Scheme 1.28 Test reaction with binaphthalene-templated S/N ligand with achiral oxazoline pendant.

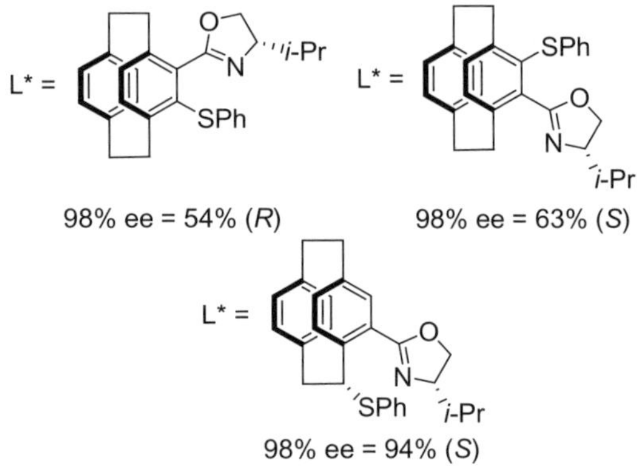

L* = [structure] 98% ee = 54% (R)

L* = [structure] 98% ee = 63% (S)

L* = [structure] 98% ee = 94% (S)

Scheme 1.29   Test reaction with [2.2]paracyclophane-type ligands with oxazoline pendant.

the donor atoms that coordinate the palladium brought the asymmetric environment closer to the allyl species during the reaction.

On the other hand, good enantioselectivities (up to 74% ee) and almost quantitative yields in all cases were obtained for the test reaction using a new class of S/N ligands developed by Ricci *et al.* in 2004 (Scheme 1.30).[66] These ligands included rigid cyclopenta[*b*]thiophene and oxazoline moieties as the sources of chirality, the sulfur atom being part of a strong π-donor structure. In this study, it was shown that whatever the configuration of the cyclopentanyl chiral carbon of the ligand, the product was obtained with the same stereochemistry. This result was explained by the classical *trans* effect for the attack of the nucleophile, leading to the same stable Pd-allyl intermediate in each case. Moreover, the use of a mixture of the two diastereomeric ligands led to the same product in comparable yield and enantioselectivity. Interestingly, it was deduced that the resolution of the diastereomeric mixture was not necessary to obtain asymmetric induction. The chiral centre located in the oxazoline core seemed of major importance for inducing enantioselectivity. In 2005, the same group reported new chiral oxazoline-1,3-dithianes (Scheme 1.30) as new efficient S/N-donating ligands for the test reaction, providing almost quantitative yields and enantioselectivities of up to 90% ee.[67] According to the results, it seemed that the presence of a methyl substituent on the dithianyl ring was not crucial to induce enantioselectivity. However, the presence of a bulky substituent such as a *tert*-butyl group on the oxazoline ring was necessary to obtain a good enantioselectivity (90% ee).

In 1998, Pregosin *et al.* reported the synthesis of various S/N ligands with the nitrogen function arising from an oxazoline group and the S-coordinating atom from a thiosugar moiety. Excellent enantioselectivities were obtained for the test reaction product for all the variously substituted ligands, which all contained stereogenic centres both on the oxazoline core and on the sugar moiety, as shown in Scheme 1.31. The best result was obtained when the thiosugar moiety was substituted by a bulky substituent such as a pivaloyl group (97% ee).

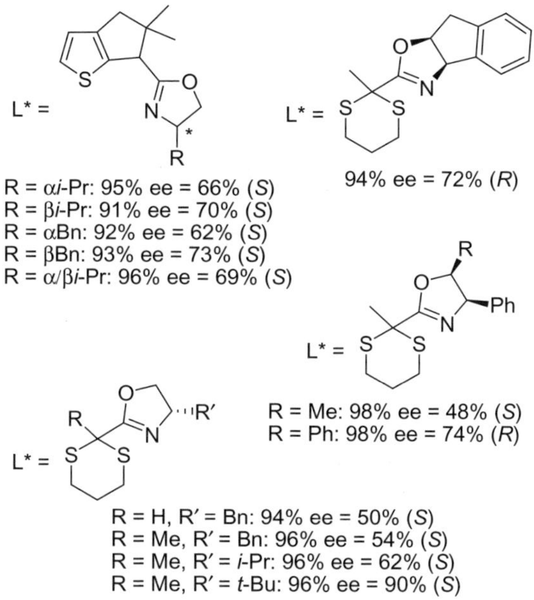

L* =

R = αi-Pr: 95% ee = 66% (S)
R = βi-Pr: 91% ee = 70% (S)
R = αBn: 92% ee = 62% (S)
R = βBn: 93% ee = 73% (S)
R = α/βi-Pr: 96% ee = 69% (S)

L* =

94% ee = 72% (R)

L* =

R = Me: 98% ee = 48% (S)
R = Ph: 98% ee = 74% (R)

L* =

R = H, R' = Bn: 94% ee = 50% (S)
R = Me, R' = Bn: 96% ee = 54% (S)
R = Me, R' = i-Pr: 96% ee = 62% (S)
R = Me, R' = t-Bu: 96% ee = 90% (S)

Scheme 1.30   Test reaction with cyclopenta[*b*]thiophene-alkyloxazoline ligands.

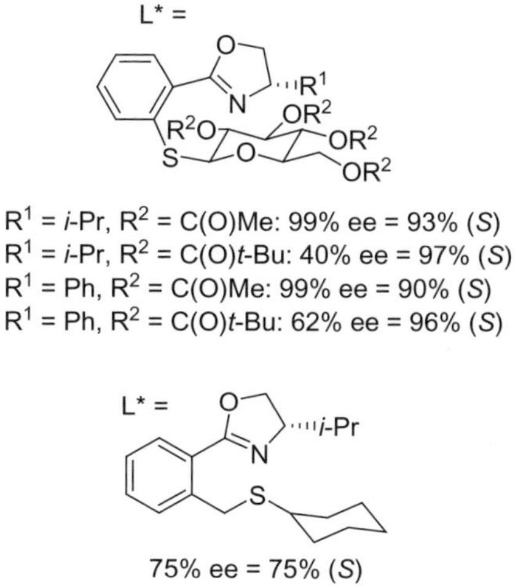

L* =

R^1 = i-Pr, R^2 = C(O)Me: 99% ee = 93% (S)
R^1 = i-Pr, R^2 = C(O)t-Bu: 40% ee = 97% (S)
R^1 = Ph, R^2 = C(O)Me: 99% ee = 90% (S)
R^1 = Ph, R^2 = C(O)t-Bu: 62% ee = 96% (S)

L* =

75% ee = 75% (S)

Scheme 1.31   Test reaction with thioglucose-oxazoline ligands.

In order to gain insight into the origin of the enantioselectivity of the process, these authors prepared a ligand in which the thioglucose function was replaced by a cyclohexyl ring. The application of this ligand to the test reaction led to the obtention of a significantly lower enantioselectivity.

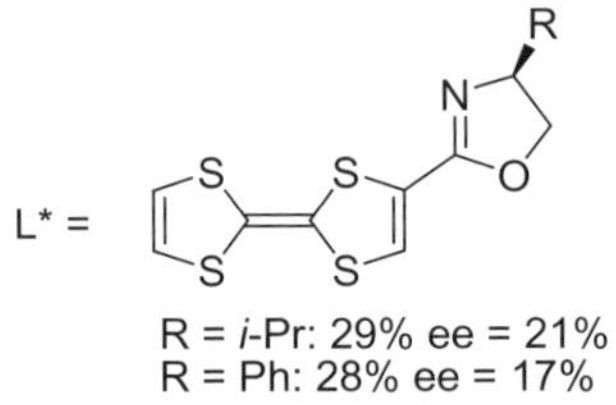

$$L^* =$$

R = *i*-Pr: 29% ee = 21%
R = Ph: 28% ee = 17%

Scheme 1.32   Test reaction with tetrathiafulvalene-oxazoline ligands.

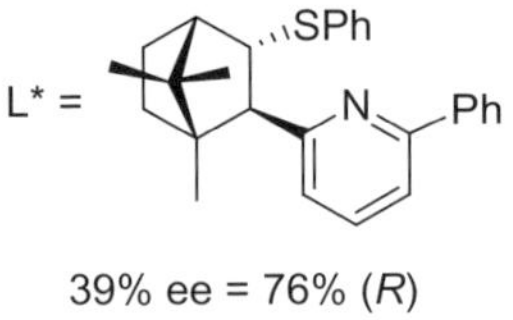

$$L^* =$$

39% ee = 76% (*R*)

Scheme 1.33   Test reaction with (+)-camphor-derived pyridine thioether ligand.

In addition, Bryce and Chesney have developed chiral oxazolines linked to tetrathiafulvalene in order to use these ligands as redox-active ligands.[68] When applied to the test reaction, these ligands gave only low enantioselectivities (≤21% ee), as shown in Scheme 1.32.

In contrast to the large number of chiral pyridine derivatives used as ligands of metal complexes in asymmetric catalysis, only a few examples of chiral sulfur-containing pyridine ligands have so far been reported, such as pyridine thioethers derived from (+)-camphor depicted in Scheme 1.33, which were assessed in the test reaction providing enantioselectivities of up to 76% ee.[69] The related 2,2′-bipyridine thioethers were also prepared but showed a lower stereodifferentiating capability in the test reaction.

The same authors have developed two other sulfur-containing pyridine ligands prepared in a two-step procedure starting from (+)-pinocarvone.[17a] By using two molar equivalents of these ligands compared to the palladium precursor, the test reaction was performed with good yields and high enantioselectivities of up to 83% ee, as shown in Scheme 1.34. Both these two ligands gave a similar enantioselectivity, but the product was obtained with the opposite configuration indicating that the stereodifferentiation was highly sensitive to the stereogenic centre bound to the sulfur atom. More recently, an analogous class of S/N-coordinating ligands, namely chiral 2-(2-phenylthiophenyl)-5,6,7,8-tetrahydroquinolines, was prepared by these authors on the basis of the Kröhnke methodology involving the reaction of α,β-unsaturated ketones with a pyridinium salt.[70] When investigated for the test reaction, the two ligands depicted in Scheme 1.34 gave poorer results in terms of both activity and enantioselectivity compared to the analogous pyridine derivatives.

In the same context, the synthesis of chiral 1-(2-methylthio-1-naphthyl)iso-quinoline depicted in Scheme 1.35 was reported by the same authors on the

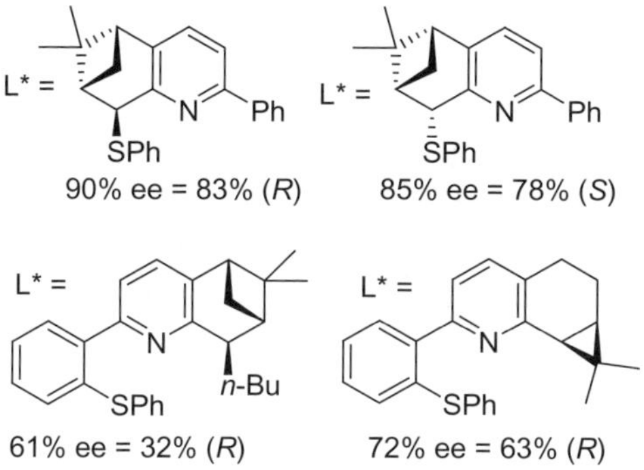

Scheme 1.34   Test reaction with (+)-pinocarvone-derived pyridine thioether ligands and 2-(2-phenylthiophenyl)-5,6,7,8-tetrahydroquinoline ligands.

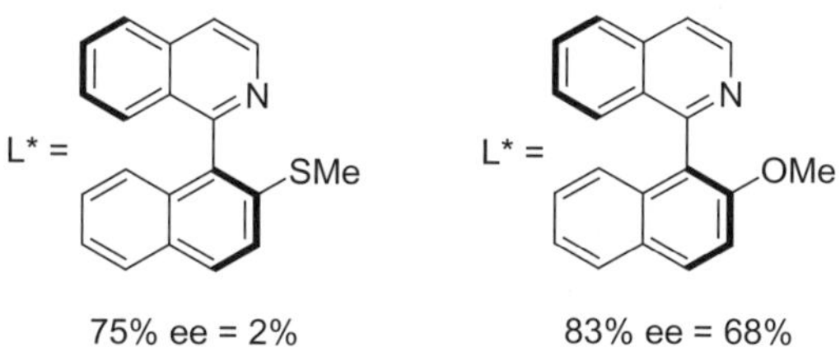

75% ee = 2%   83% ee = 68%

Scheme 1.35   Test reaction with 1-(2-methylthio-1-naphthyl)isoquinoline ligand and its N/O-analogue.

basis of a Suzuki-type coupling and the formation of Pd-diastereomeric complexes was used for the resolution step.[71] Unfortunately, this ligand gave almost no enantioselectivity when applied to the test reaction. The corresponding N,O-chelate depicted in Scheme 1.35 was not found to be a very reactive complex either, since after two weeks, the expected product was isolated in a good yield and a moderate enantioselectivity (68% ee).

In 1998, Kellogg *et al.* developed the synthesis of other pyridine sulfide ligands on the basis of the base-induced addition of 2,6-lutidine to thiofenchone.[72] Pyridine thiols and dithiols thus prepared were further investigated as ligands for the test reaction. As shown in Scheme 1.36, the pyridine thiol ligand was not very active, probably due to a deprotonation of the ligand by the base present in the reaction mixture. Consequently, these authors prepared a series of corresponding thioether derivatives by alkylation of the mono- and dithiol adducts and further investigated their efficiency in the test reaction. In most cases of thioether ligands, higher enantioselectivities (up to 92% ee) and activities (up to 96% yield) were observed, in particular when BSA/KOAc was used as a base compared to NaH. As shown in Scheme 1.36, the best result was obtained with the benzyl mono-thioether ligand that gave even better results with a 96% yield combined with an enantioselectivity of 98% ee when the reaction was carried out in the presence of

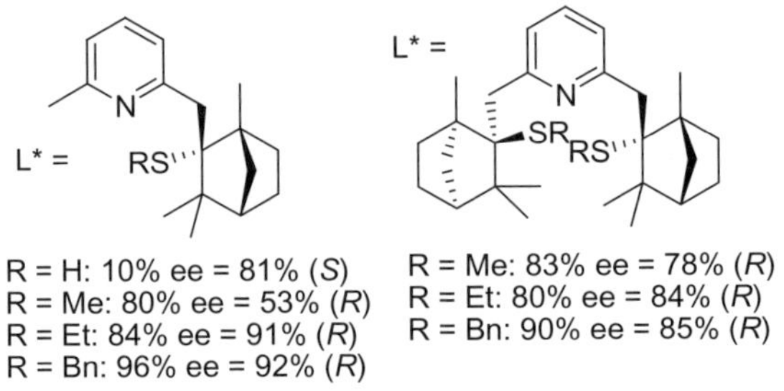
L* =
RS
N
L* =
N
SR
RS
R = H: 10% ee = 81% (S)
R = Me: 80% ee = 53% (R)
R = Et: 84% ee = 91% (R)
R = Bn: 96% ee = 92% (R)
R = Me: 83% ee = 78% (R)
R = Et: 80% ee = 84% (R)
R = Bn: 90% ee = 85% (R)

Scheme 1.36 Test reaction with pyridine thioether and dithioether ligands.

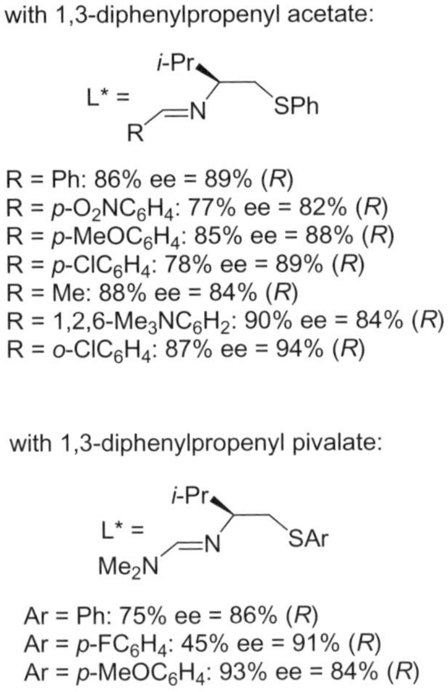
with 1,3-diphenylpropenyl acetate:
L* =
i-Pr
N
SPh
R
R = Ph: 86% ee = 89% (R)
R = p-O2NC6H4: 77% ee = 82% (R)
R = p-MeOC6H4: 85% ee = 88% (R)
R = p-ClC6H4: 78% ee = 89% (R)
R = Me: 88% ee = 84% (R)
R = 1,2,6-Me3NC6H2: 90% ee = 84% (R)
R = o-ClC6H4: 87% ee = 94% (R)
with 1,3-diphenylpropenyl pivalate:
L* =
i-Pr
N
SAr
Me2N
Ar = Ph: 75% ee = 86% (R)
Ar = p-FC6H4: 45% ee = 91% (R)
Ar = p-MeOC6H4: 93% ee = 84% (R)

Scheme 1.37 Imine-sulfide ligands for Pd-catalysed allylic alkylations of 1,3-diphenylpropenyl alkanoates with dimethyl malonate.

acetonitrile instead of dichloromethane. An X-ray structure of a palladium complex was proposed in order to demonstrate that the sulfur atom was coordinated in a rigid six-membered ring in a single absolute configuration due to the steric hindrance generated by the fenchyl methyl groups. High chemical yields and enantioselectivities were also provided by the use of the corresponding C_2-symmetric dithioether ligands (Scheme 1.36).

The test reaction was also investigated by Anderson *et al.* employing chiral imine-sulfide ligands derived from amino acids.[17c,73] The ligand of choice, depicted in Scheme 1.37, (*S*)-*N*-2′-chlorobenzylidene-2-amino-3-methyl-1-thiophenylbutane readily prepared from (*S*)-valinol, led to an enantioselectivity of 94% ee. These authors were able to isolate and characterise a Pd-allyl intermediate by X-ray diffraction. They proposed a possible mechanism for the chirality transfer, controlled by the steric environment of the chiral imine-sulfur chelate ligand with the reaction occurring *trans* to the imine donor. The corresponding amidine ligands were studied by Morimoto *et al.*, giving excellent results for the palladium-catalysed allylic alkylation of 1,3-diphenylpropenyl pivalate with dimethyl malonate (Scheme 1.37).[74]

In 2005, Kunieda *et al.* developed sterically congested 2-iminothioethers as novel and highly efficient chiral ligands for the test reaction, providing the product in excellent yield and enantioselectivity, as shown in Scheme 1.38.[75]

The first successful use of simple 1,2-aminothioethers as hybrid ligands in the test reaction was reported by Bulman Page *et al.* in 2000.[76] In this work, the involvement of 2-[2-thioethyl]tetrahydroisoquinoline derivatives, readily available from the norephedrine-derived iminium salt, as ligands resulted in quantitative yields and enantioselectivities of up to 72% ee when the sulfur atom was substituted with a *t*-butyl group (Scheme 1.39), whereas poor enantioselectivities were obtained when the sulfur atom was substituted with an aromatic group. These authors explained these results by steric and not electronic considerations. Other heterobidentate sulfide-tertiary amine ligands, incorporating 1,2-aminothioethers derived from ephedrine and pseudoephedrine, have been prepared and used successfully in the test reaction, giving enantioselectivities of up to 89% ee (Scheme 1.39).[77] The enantioselectivities observed did not seem to be greatly dependent upon the nature of the group at the nitrogen atom.

Simpler chiral pyrrolidine thioethers, reported in 2004 by Skarzewski *et al.*, proved to be effective ligands in the test reaction.[78] The sense of the stereoinduction was in agreement with the nucleophilic attack directed at the allylic carbon located *trans* to the sulfur atom in the intermediate complex (Scheme 1.40).

On the other hand, Morimoto *et al.* have investigated the two chiral thioimidazoline ligands, depicted in Scheme 1.41, which were prepared starting from 2-bromobenzonitrile and the corresponding chiral diamines in four steps with 13 and 8% overall yield, respectively.[79] Although the enantioselectivity and the activity for the ligand bearing a cyclohexyl group were modest, an

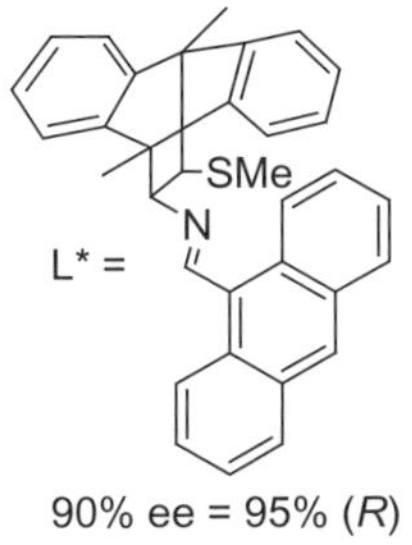

90% ee = 95% (*R*)

Scheme 1.38 Test reaction with congested 2-iminothioether ligand.

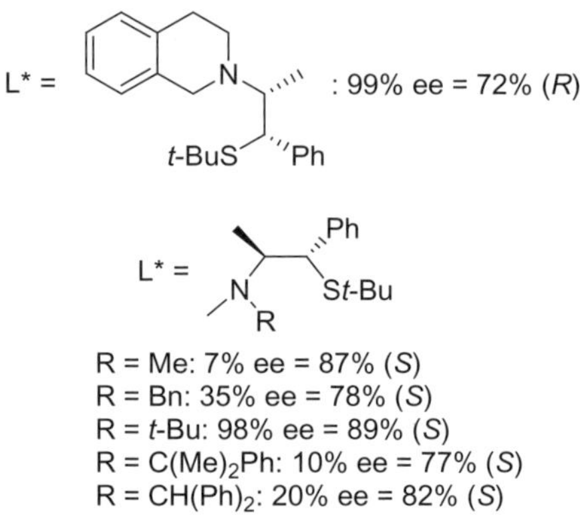

L* =   : 99% ee = 72% (*R*)

t-BuS   ''Ph

L* =   Ph

N   S*t*-Bu
|
R

R = Me: 7% ee = 87% (*S*)
R = Bn: 35% ee = 78% (*S*)
R = *t*-Bu: 98% ee = 89% (*S*)
R = C(Me)₂Ph: 10% ee = 77% (*S*)
R = CH(Ph)₂: 20% ee = 82% (*S*)

Scheme 1.39   Test reaction with sulfide-tertiary amine ligands incorporating 1,2-aminothioethers.

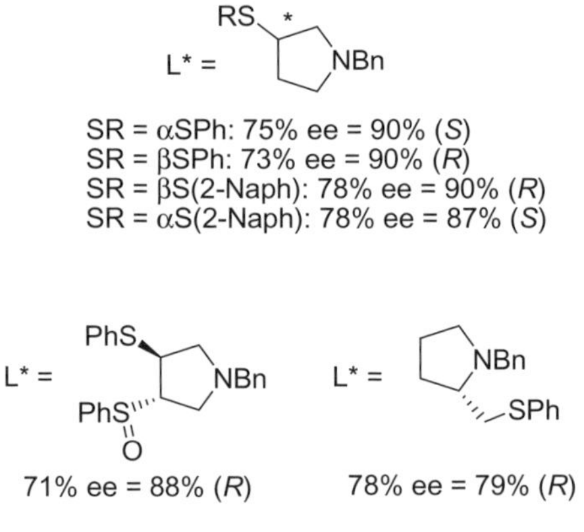

RS   *

L* =   NBn

SR = αSPh: 75% ee = 90% (*S*)
SR = βSPh: 73% ee = 90% (*R*)
SR = βS(2-Naph): 78% ee = 90% (*R*)
SR = αS(2-Naph): 78% ee = 87% (*S*)

PhS   NBn
L* =   NBn   L* =
PhS   SPh
  ‖
  O
71% ee = 88% (*R*)   78% ee = 79% (*R*)

Scheme 1.40   Test reaction with pyrrolidine thioether ligands.

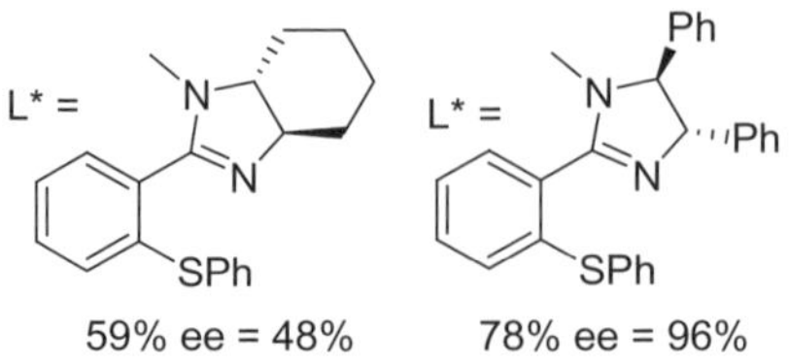

L* =   L* =   Ph

                                                   ''Ph

SPh   SPh
59% ee = 48%   78% ee = 96%

Scheme 1.41   Test reaction with thioimidazoline ligands.

excellent enantioselectivity combined with a good yield were observed with the other ligand, as shown in Scheme 1.41.

More recently, Kim *et al.* have reported the synthesis of analogous chiral imidazolidine ligands, such as thioimidazolidines starting from (*R*,*R*)-1,2-diaminocyclohexane. In order to compare their efficiency with that of the corresponding P/N-analogues, the corresponding phosphinoimidazolidines were also prepared.[80] Submitted as ligands to the test reaction, these two types of ligands gave contrasted results, since the phosphinoimidazolidine ligands were both efficient and enantioselective for this reaction, whereas the corresponding sulfur-containing ligands were much less active and enantioselective (≤45% ee), as shown in Scheme 1.42. In all cases, the reactions were performed in the presence of a ligand/catalyst ratio of 4/1. In addition, these authors examined, in 2004, the efficiency of chiral bidentate thiophene diamine ligands that gave for the same reaction a moderate asymmetric induction and a modest reactivity, as shown in Scheme 1.42.[81]

Some new chiral oxazolidine-thioether and thiazolidine-methanol ligands have been efficiently synthesised by Braga *et al.* from readily available amino acids, such as L-cysteine, *S*-methyl-L-cysteine and L-methionine.[82] These new types of ligands have been further explored in the test reaction, providing in the case of using oxazolidine-thioether ligands excellent yields and good to high enantioselectivities (Scheme 1.43). Subtle changes in the ligand's structure made it possible to clarify the mode of action, determine a ligand structure–catalyst efficiency relationship and undertake a rational improvement of the ligand's structure. This showed the importance of a flexible synthetic strategy for systematic optimisation of the catalyst's structure and an enantiomeric excess of 94% was obtained by increasing the steric bulk at the 2-position of the oxazolidine. Concerning the thiazoline-oxygen ligands they were supposed to act as bidentate ligands and form oxygen/nitrogen complexes with palladium. The large difference between the donor–acceptor strengths of oxygen and sulfur

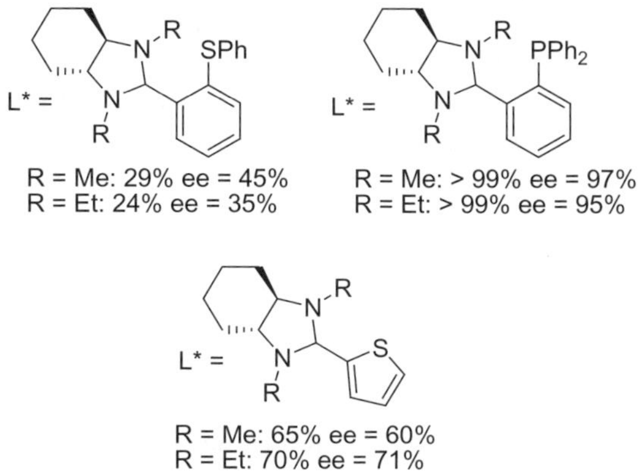

R = Me: 29% ee = 45%
R = Et: 24% ee = 35%

R = Me: > 99% ee = 97%
R = Et: > 99% ee = 95%

R = Me: 65% ee = 60%
R = Et: 70% ee = 71%

Scheme 1.42 Thioimidazolidine ligands compared to phosphinoimidazolidine ligands for the test reaction.

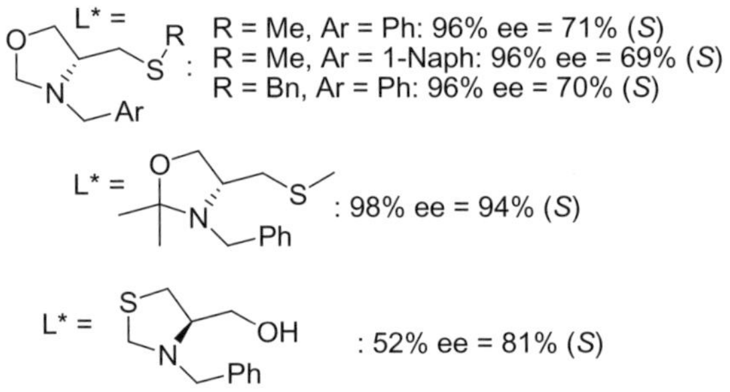
L* =
R
S
N
O
Ar
R = Me, Ar = Ph: 96% ee = 71% (S)
R = Me, Ar = 1-Naph: 96% ee = 69% (S)
R = Bn, Ar = Ph: 96% ee = 70% (S)
L* =
O
N
S
Ph
: 98% ee = 94% (S)
L* =
S
N
OH
Ph
: 52% ee = 81% (S)

Scheme 1.43 Test reaction with oxazolidine-thioether and thiazolidine-methanol ligands.

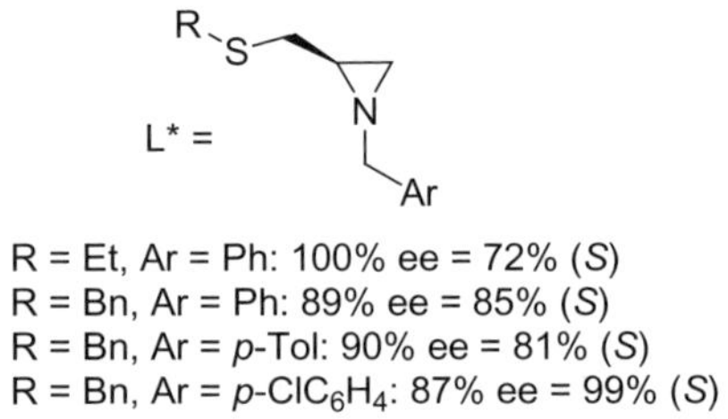
R
S
L* =
N
Ar
R = Et, Ar = Ph: 100% ee = 72% (S)
R = Bn, Ar = Ph: 89% ee = 85% (S)
R = Bn, Ar = p-Tol: 90% ee = 81% (S)
R = Bn, Ar = p-ClC6H4: 87% ee = 99% (S)

Scheme 1.44 Test reaction with aziridine sulfide ligands.

explained the poor yield and long reaction time obtained by using these ligands compared with the almost complete conversion that occurred with oxazolidine-thioether ligands.

These workers have developed another new type of chiral S/N ligands, namely aziridine sulfides, which were easily synthesised in a straightforward synthetic route from inexpensive and readily available (*R*)-cysteine.[83] The efficiency of this sterically and electronically varied set of ligands was then examined as chiral catalysts in the palladium-catalysed test reaction. The alkylated product was obtained in excellent yields and stereoselectivities of up to 99% ee, as shown in Scheme 1.44.

In 2008, these authors described the synthesis of chiral β-thioamides through a short and high yielding synthetic route and with a modular approach by ring-opening reaction of 2-oxazolines.[84] These ligands were further evaluated for the test reaction, providing good to excellent yields and enantioselectivities of up to 91% ee (Scheme 1.45). The corresponding selenium and tellurium analogues were also prepared in this study and evaluated for the same reaction. In all cases, the β-selenium amides have proven to be more efficient than the respective sulfur- and tellurium-containing analogues. This observation of a difference in the behaviour among the chalconide donors provided evidence for the higher ability of selenium to complex with palladium, leading to a superior level of enantioselection (up to 98% ee).

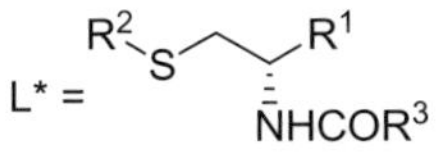
L* =
R2
S
R1
NHCOR3

R¹ = *i*-Pr, $R^2 = R^3$ = Ph: 94% ee = 86% (*R*)
R¹ = *i*-Bu, $R^2 = R^3$ = Ph: 79% ee = 83% (*R*)
R¹ = Bn, $R^2 = R^3$ = Ph: 77% ee = 72% (*R*)
R¹ = *i*-Pr, R² = *o*-ClC_6H_4, R³ = Ph: 79% ee = 74% (*R*)
R¹ = *i*-Pr, R² = *p*-$MeOC_6H_4$, R³ = Ph: 84% ee = 80% (*R*)
R¹ = *i*-Pr, R² = Ph, R³ = *p*-(*t*-Bu)C_6H_4: 94% ee = 91% (*R*)

Scheme 1.45 Test reaction with β-thioamide ligands.

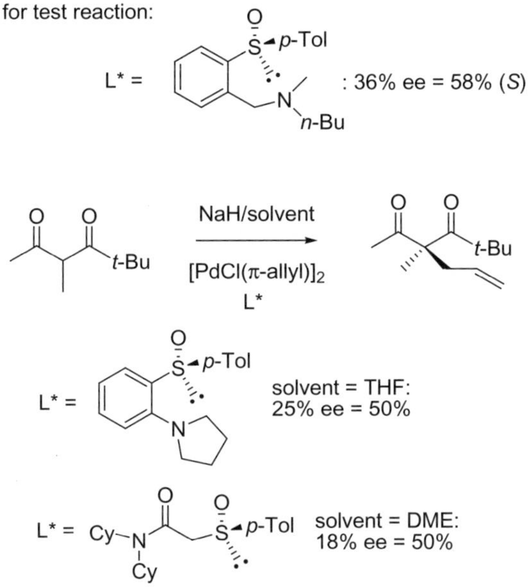
for test reaction:
L* =
p-Tol
n-Bu
: 36% ee = 58% (S)
t-Bu
NaH/solvent
[PdCl(π-allyl)]2
L*
t-Bu
L* =
p-Tol
solvent = THF:
25% ee = 50%
L* = Cy
Cy
p-Tol
solvent = DME:
18% ee = 50%

Scheme 1.46 Test reaction with amino sulfoxide ligands.

The test reaction has also been studied by Hiroi *et al.* by using various chiral amino sulfoxide ligands bearing a chiral sulfinyl functionality as the sole chiral source.[85] Among chiral α-sulfinylacetamides, β- or γ-amino sulfoxides, and β-sulfinyl sulfonamides, the highest enantioselectivity of 58% ee was obtained by using the 2-(aminomethyl)phenyl sulfoxide ligand depicted in Scheme 1.46. In addition, the asymmetric palladium-catalysed allylation of *t*-butyl 2-methylacetoacetate with allyl acetate was examined in the same study, giving in this case the highest enantioselectivities (50% ee) by using either a chiral α-sulfinylacetamide ligand or a β-amino sulfoxide ligand, depicted in Scheme 1.46.

In a different context, Bäckvall *et al.* have reported another type of substitution, the arenethiolatocopper(I)-catalysed substitution reaction of

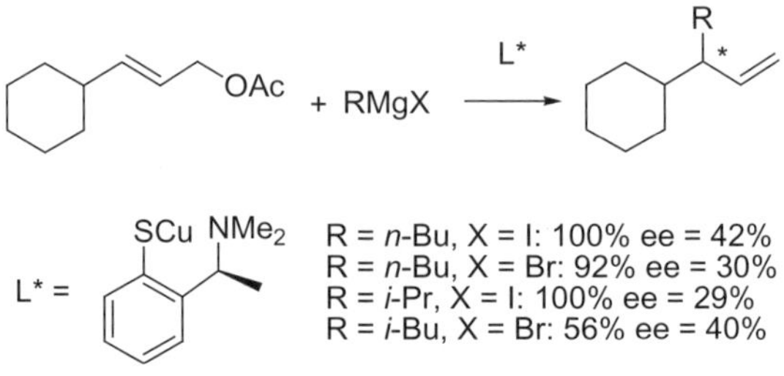

Scheme 1.47 Arenethiolatocopper(I)-catalysed substitution reactions of RMgX with allylic substrates.

Grignard reagents with allylic substrates, providing the corresponding γ-products in enantioselectivities of up to 50% ee (Scheme 1.47).[86]

In conclusion, many various S/N ligands have been developed and successfully investigated in the presence of palladium to catalyse nucleophilic allylic substitutions with both efficiency and enantioselectivity. In spite of advantages such as their stability and rapid synthesis, these ligands are, apart from few exceptions, generally less active and enantioselective than their analogous N/P-counterparts, which have played a dominant role among the heterodonor ligands.

1.5   S/C and S/O Ligands

Relatively few examples of chiral S/C and S/O ligands involved in palladium-catalysed allylic substitution reactions have been reported so far. In 1994, Williams *et al.* reported the preparation of sulfur-containing enantiomerically pure acetals derived from C_2-symmetric diols, such as (*S,S*)-1,2-diphenylethanediol and (*R,R*)-2,3-butanediol and their subsequent use as ligands in the test reaction.[22] In the case of using diphenyl-substituted acetals, the product was obtained in a good yield and an enantioselectivity of 82% ee, whereas the use of a dimethyl-substituted acetal ligand led to a moderate yield and a very low enantioselectivity (< 5% ee), as shown in Scheme 1.48. Moreover, in the case of the most efficient ligand, these authors have prepared and evaluated the corresponding phosphorus analogue for the same reaction (Scheme 1.48), which revealed a lower efficiency but a higher enantioselectivity (88% ee instead of 82% ee) than those obtained with the corresponding sulfur-containing ligand.

In 2007, Fernandez *et al.* demonstrated that transition-metal complexes with heterobidentate S/C ligands based on imidazopyridin-3-ylidene and thioether functionalities could be readily prepared from the corresponding azolium salts by reaction with Ag_2O and transmetalation of the resulting silver carbenes with appropriate metal sources.[87] The cationic Pd(allyl)(carbene-S) complexes have proven to be active catalysts in the test reaction, reaching enantioselectivities of

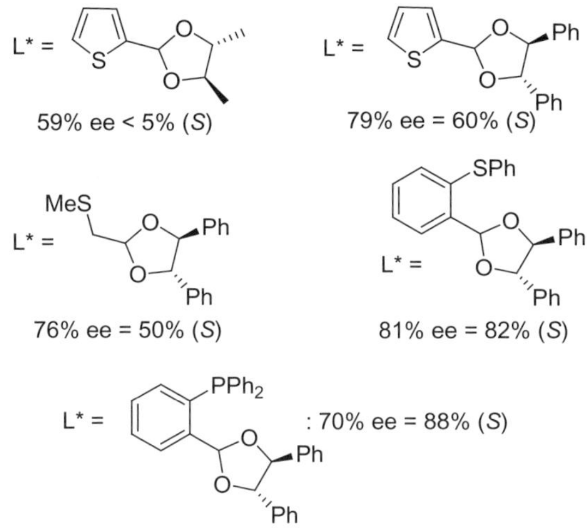

Scheme 1.48 Test reaction with S/O ligands and P/O analogue.

up to 91% ee in the case of the more rigid complex depicted in Scheme 1.49 and in the presence of Bu$_4$NBr as an additive. Structural studies have demonstrated that the formation of the S/C complex proceeded in a stereoselective way involving a preferential coordination of one of the lone pairs of the thioether sulfur atom to the metal, affording a boat-like chelate with *S*-configuration at sulfur and a *trans* relative disposition of the bulky isopropyl and cyclohexyl groups. In addition, thioether-containing imidazolium salts depicted in Scheme 1.49 could be transformed into the corresponding palladium complexes with bidentate S/C ligands by transmetalation of the corresponding silver carbenes with [Pd(allyl)(COD)]$^+$SbF$_6^-$.[88] These catalysts were further evaluated for the test reaction, leading to the alkylated product in a good yield and an enantioselectivity of 82% ee (Scheme 1.49).

1.6 Sulfur-Containing P/N, P/O, N/N, N/O and P/P Ligands

In 2003, a novel class of P/N-sulfinyl-imine ligands, prepared by condensation of commercially available *tert*-butane-sulfinamide with aldehydes or ketones and incorporating chirality at sulfur, was evaluated for the test reaction.[89] The first crystal structure of a Pd-bound sulfinyl-imine provided insight into the binding mode and origins of the stereoselectivity. This structure confirmed that the ligand bound to palladium through phosphorus and nitrogen to form a six-membered ring chelate (Scheme 1.50). The positive effect of the steric hindrance arising from the *tert*-butyl group was demonstrated since no enantioselectivity was observed in the case of using the corresponding ligand that did not bear this substituent.

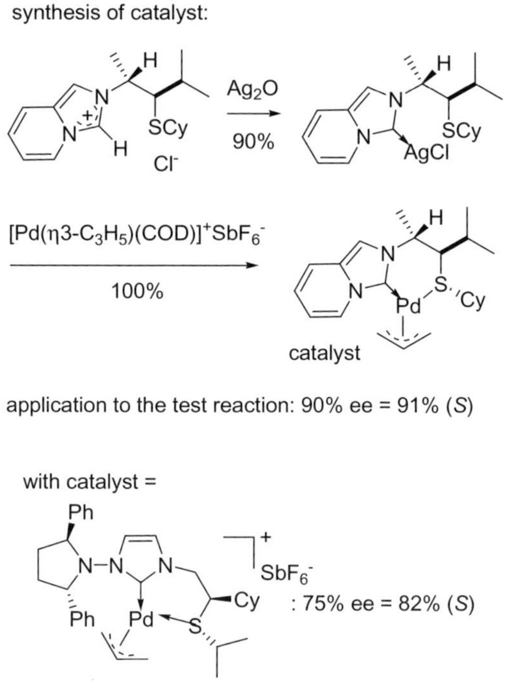

synthesis of catalyst:

application to the test reaction: 90% ee = 91% (*S*)

with catalyst =

: 75% ee = 82% (*S*)

Scheme 1.49   Test reaction with S/C ligands.

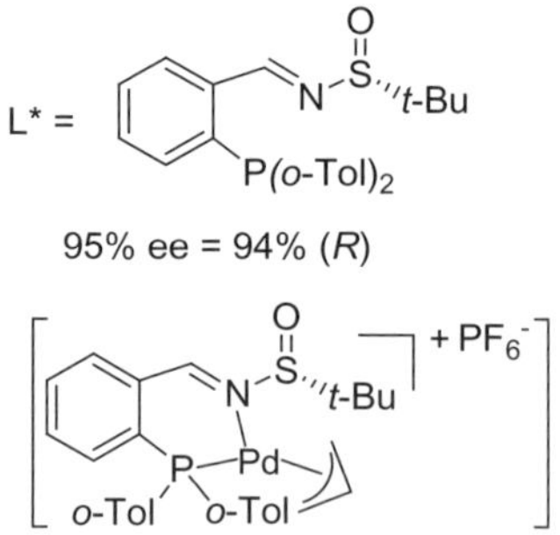

L* =

95% ee = 94% (*R*)

Scheme 1.50   Test reaction with P/N-sulfinyl-imine ligand.

Sulfoximines bearing a chiral sulfur atom have recently emerged as valuable ligands for metal-catalysed asymmetric synthesis.[90] In particular, C_2-symmetric bis(sulfoximines), such as those depicted in Scheme 1.51, were applied to the test reaction, achieving enantioselectivities of up to 93% ee.[91] The most selective ligand ($R^1 = c$-Pent, $R^2 = $ Ph) of the series was also applied to the nucleophilic substitution reaction of 1,3-diphenyl-2-propenyl acetate with substituted malonates, such as acetamido-derived diethylmalonate, which provided the corresponding product in 89% yield and 98% ee.

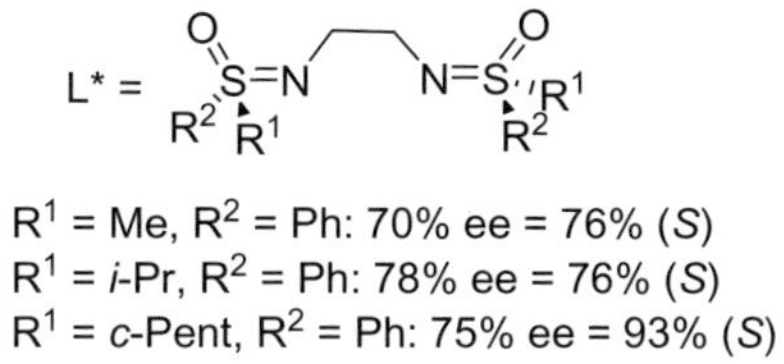

R^1 = Me, R^2 = Ph: 70% ee = 76% (*S*)
R^1 = *i*-Pr, R^2 = Ph: 78% ee = 76% (*S*)
R^1 = *c*-Pent, R^2 = Ph: 75% ee = 93% (*S*)

Scheme 1.51   Test reaction with *C*$_2$-symmetric bis(sulfoximines) ligands.

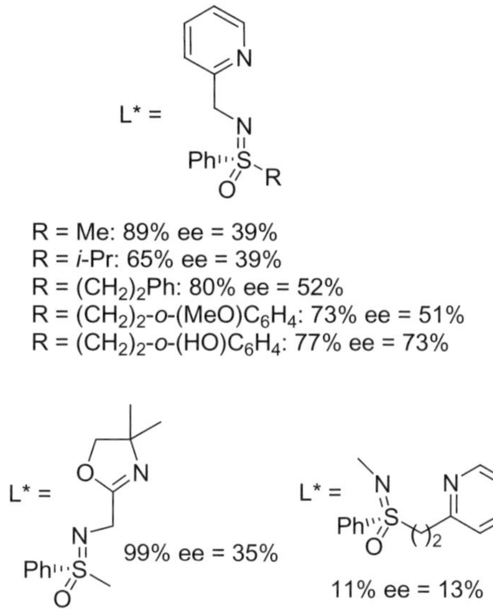

R = Me: 89% ee = 39%
R = *i*-Pr: 65% ee = 39%
R = (CH$_2$)$_2$Ph: 80% ee = 52%
R = (CH$_2$)$_2$-*o*-(MeO)C$_6$H$_4$: 73% ee = 51%
R = (CH$_2$)$_2$-*o*-(HO)C$_6$H$_4$: 77% ee = 73%

99% ee = 35%

11% ee = 13%

Scheme 1.52   Test reaction with sulfoximine ligands.

The same authors have previously reported the preparation of a series of various sulfoximine ligands, which were further investigated in the test reaction.[17d] In almost all the ligands, moderate enantioselectivities were observed for the product of the test reaction, except for a ligand bearing a hydroxyl group on the aryl substituent of the sulfur atom, which gave a 73% ee (Scheme 1.52).

In 2005, Reetz *et al.* prepared, for the first time, BINOL-derived *N*-phosphino sulfoximines through a high yield, one-step phosphorylation of the corresponding sulfoximines.[92] These new ligands were evaluated in the presence of palladium to perform the test reaction, resulting in moderate enantioselectivities (≤63% ee, Scheme 1.53). It was not clear, in this case, whether the ligand was monodentate at phosphorus or whether it underwent a P/O-chelation. The authors have observed that the ligand/Pd ratio greatly influenced the enantioselectivity of the reaction, providing the best enantioselectivity in the case of using a 1/1 ratio.

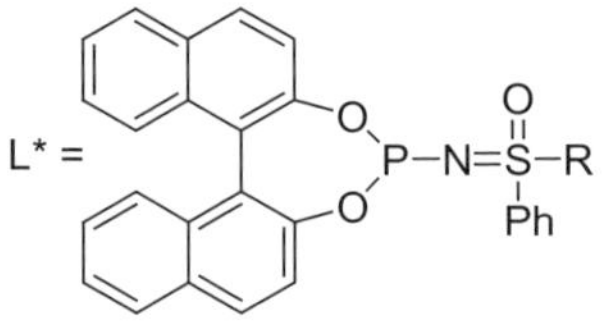

L* =

with R_{BINOL}/S_{sulfur}:
R = Me: 93% ee = 57% (*R*)
R = *t*-Bu: 90% ee = 63% (*R*)
with S_{BINOL}/S_{sulfur}:
R = Me: 95% ee = 35% (*S*)
R = *t*-Bu: 89% ee = 58% (*S*)

Scheme 1.53   Test reaction with BINOL-derived *N*-phosphino sulfoximines.

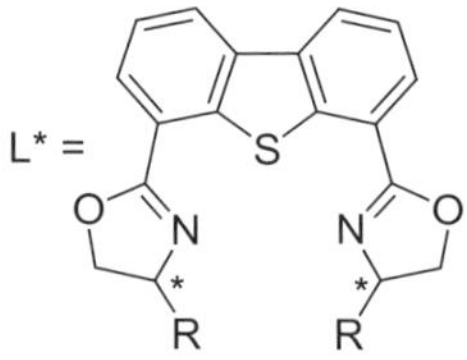

L* =

R = (*R*)-*i*-Pr: 100% ee = 77% (*R*)
R = (*S*)-Ph: 39% ee = 28% (*S*)
R = (*R*)-*t*-Bu: 50% ee = 51% (*R*)

Scheme 1.54   Test reaction with dibenzothiophene-bis(oxazolines) ligands.

New chiral sulfur-containing bis(oxazolines) ligands have been described by Schulz *et al.*[62] Their structure included a dibenzothiophene ring as the backbone in which the sulfur atom was enclosed in a strong π-donor structure. These ligands have been successfully tested in the asymmetric palladium-catalysed test reaction, leading to the expected product with enantioselectivities of up to 77% ee (Scheme 1.54). These authors have observed that the ligands containing chiral fragments with (*S*)-configuration led to the substitution product with (*R*)-configuration as the major isomer. This was in contrast with the results obtained using other C_2-symmetric bis(oxazolines),[93] where the (*S*)-configuration in the chiral ligand led to the (*S*)-product as the major compound. Compared with the total inefficiency of the dibenzofuran analogue to perform the test reaction (no reaction occurred),[94] it was reasonable to assume a probable participation of the sulfur atom in stabilising the complex involved in the selective transformation.

In 2002, Tietze and Lohmann reported the synthesis of various thiophene-, benzothiophene-, and benzofuran-oxazoline/phosphine-containing ligands depicted in Scheme 1.55 based on a simple introduction of the substituents on the heteroaromatic backbone by metalation and subsequent introduction of an aldehyde function, precursor of the oxazoline moiety, or reaction with PPh$_2$Cl.[95] When applied as ligands in the test reaction, these ligands gave good to high enantioselectivities of up to 99% ee. According to the results collected

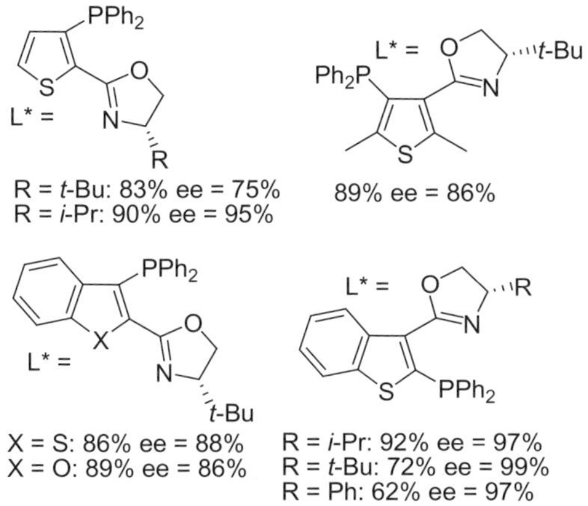

R = *t*-Bu: 83% ee = 75% 89% ee = 86%
R = *i*-Pr: 90% ee = 95%

X = S: 86% ee = 88% R = *i*-Pr: 92% ee = 97%
X = O: 89% ee = 86% R = *t*-Bu: 72% ee = 99%
                     R = Ph: 62% ee = 97%

Scheme 1.55 Test reaction with thiophene- and benzothiophene-oxazoline/phosphine-containing ligands.

in Scheme 1.55, it seemed that the position of the substituents at the heterocyclic skeleton of the ligand was important for the enantioselectivity. Although the sulfur atom was not a chelating atom in these ligands, its replacement by an oxygen atom did not have a strong effect on the enantioselectivity of the test reaction, as shown in Scheme 1.55. This study has been more recently extended by Cozzi *et al.* to other substituted thiophene- and benzothiophene-oxazoline/phosphine-containing ligands, providing similar results.[96]

In the same context, Masson *et al.* have studied new chiral mono- and bis(thiazolines)-containing ligands, analogues of known oxazolines, in order to study the influence of the sulfur atom replacing the oxygen atom on the enantioselectivity and activity in the test reaction.[97] The bis(thiazolines) ligands were easily prepared in an overall yield of 75% through a five-step sequence by using dithioesters as sulfur sources and commercially available aminoalcohols. Evaluated as ligands in the test reaction, these ligands allowed the desired product to be obtained in high yields and moderate to high enantioselectivities, as shown in Scheme 1.56. In 1992, Pfaltz *et al.* reported a more detailed study of the use of analogous bis(oxazolines) ligands, providing similar or lower enantioselectivities than the corresponding bis(thiazolines) ligands (Scheme 1.56).[98] More recently, the same authors reported the synthesis of another series of this type of bis(thiazolines) ligands bearing various substituents on the thiazoline rings in order to make a direct comparison with the known bis(oxazolines) analogues.[99] When applied as ligands in the test reaction, these ligands have proven to be as efficient as the corresponding oxazolines, and even in some cases more active and enantioselective. Moreover, an inversion in the product configuration by using *t*-Bu-substituted ligand was observed, which made the authors consider a possible competition between sulfur and nitrogen for the palladium chelation, leading to the existence of three π-allylic palladium complexes as intermediates of the reaction. However, an X-ray analysis

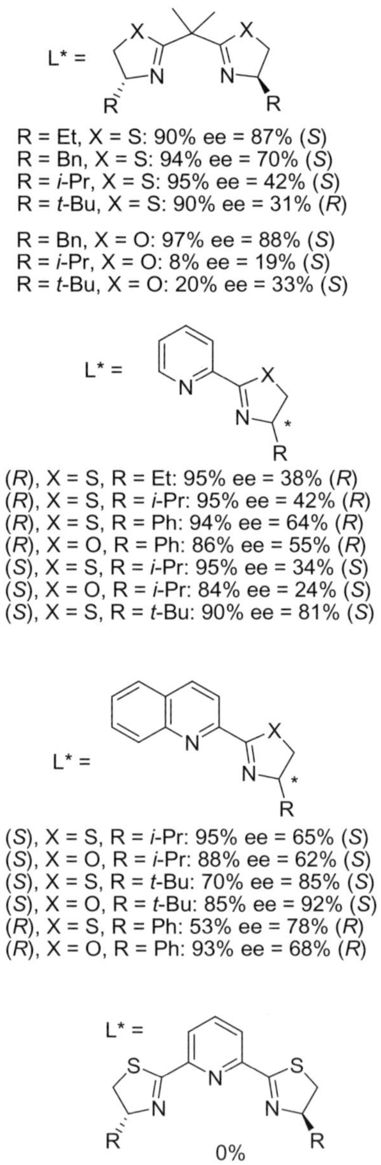

L* =

R = Et, X = S: 90% ee = 87% (*S*)
R = Bn, X = S: 94% ee = 70% (*S*)
R = *i*-Pr, X = S: 95% ee = 42% (*S*)
R = *t*-Bu, X = S: 90% ee = 31% (*R*)

R = Bn, X = O: 97% ee = 88% (*S*)
R = *i*-Pr, X = O: 8% ee = 19% (*S*)
R = *t*-Bu, X = O: 20% ee = 33% (*S*)

L* =

(*R*), X = S, R = Et: 95% ee = 38% (*R*)
(*R*), X = S, R = *i*-Pr: 95% ee = 42% (*R*)
(*R*), X = S, R = Ph: 94% ee = 64% (*R*)
(*R*), X = O, R = Ph: 86% ee = 55% (*R*)
(*S*), X = S, R = *i*-Pr: 95% ee = 34% (*S*)
(*S*), X = O, R = *i*-Pr: 84% ee = 24% (*S*)
(*S*), X = S, R = *t*-Bu: 90% ee = 81% (*S*)

L* =

(*S*), X = S, R = *i*-Pr: 95% ee = 65% (*S*)
(*S*), X = O, R = *i*-Pr: 88% ee = 62% (*S*)
(*S*), X = S, R = *t*-Bu: 70% ee = 85% (*S*)
(*S*), X = O, R = *t*-Bu: 85% ee = 92% (*S*)
(*R*), X = S, R = Ph: 53% ee = 78% (*R*)
(*R*), X = O, R = Ph: 93% ee = 68% (*R*)

L* =

0%

Scheme 1.56   Test reaction with thiazoline ligands.

demonstrated a *N,N*-coordination. On the other hand, 2-pyridyl, 2-quinolyl thiazolines and a tridentate pyridine bis(thiazolines) ligand analogous to the well-known PyBox were also prepared in the same study. Both 2-pyridyl thiazolines and 2-quinolyl thiazolines gave similar results to their parent oxazolines, while the pyridine bis(thiazolines) ligands have proven to be completely inefficient (Scheme 1.56).

In 2004, another series of C_2-symmetric bis(thiazolines) ligands bridged by a dibenzo[*a,c*]-cycloheptadiene backbone was prepared by Du *et al.* through the cyclisation of the corresponding dihydroxy diamides by treatment with P_2S_5 and TEA.[100] In order to compare the efficiency of these ligands in the test reaction with that of the corresponding bis(oxazolines), the authors evaluated both families of ligands in the test reaction. In all cases, the bis(oxazolines) ligands have proven to be better ligands in terms of enantioselectivity, whereas similar yields were generally observed, as shown in Scheme 1.57. Surprisingly, it was observed that the (*S*)-bis(oxazolines) led, in most cases, to the product with the (*R*)-configuration, whereas the corresponding (*S*)-bis(thiazolines) provided the product with the reversed configuration. This was explained by the authors by the fact that both series of ligands may have a different coordination configuration in the transition states, since sulfur as a better chelating atom than oxygen may compete for chelation with nitrogen in the thiazoline structure. Due to the fact that the steric effect near the sulfur atom was less important than near the nitrogen atom, the enantioselection was much weaker.

In 2001, a new chiral bis(benzothiazines) ligand was reported by Harmata *et al.* and successfully applied to the test reaction (Scheme 1.58).[101] The authors studied, in particular, different reaction conditions to perform this reaction and proved that good yields and enantioselectivities were obtained in relatively nonpolar solvents, whereas the reaction remained racemic in dichloromethane

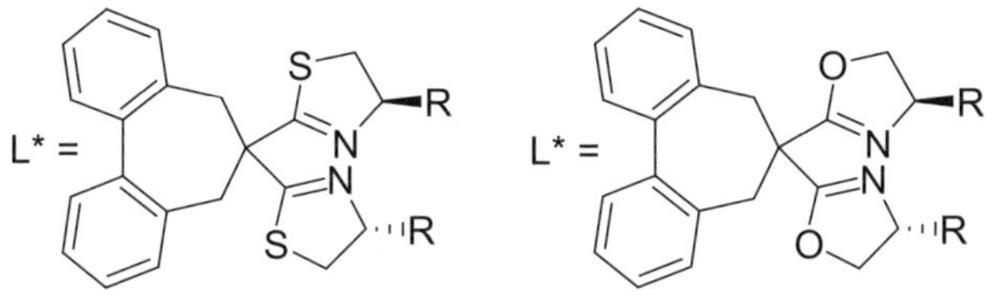

R = CH$_2$*i*-Pr: 87% ee = 16% (*S*)   R = CH$_2$*i*-Pr: 88% ee = 74% (*R*)
R = *i*-Pr: 89% ee = 56% (*S*)   R = *i*-Pr: 86% ee = 75% (*R*)
R = Bn: 84% ee = 14% (*R*)   R = Bn: 83% ee = 73% (*R*)
R = Ph: 86% ee = 19% (*S*)   R = Ph: 90% ee = 86% (*R*)

Scheme 1.57   Test reaction with bis(thiazolines) ligands bridged by a dibenzo[*a,c*]-cycloheptadiene backbone.

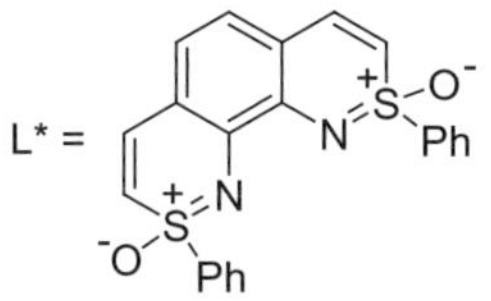

(*R,R*)-L*: 90% ee = 82% (*S*)
(*S,S*)-L*: 90% ee = 80% (*R*)

Scheme 1.58   Test reaction with bis(benzothiazines) ligand.

or acetonitrile. This ligand was also evaluated in the reaction with cyclohexenyl acetate but, in this case, the corresponding product was isolated with moderate yield and enantioselectivity (60% ee).

Moderate to high enantioselectivities of up to 79% ee have been reported for the test reaction by Kellog *et al.* by using chiral diamino sulfides derived from ephedrine (Scheme 1.59).[102] These ligands were prepared by means of an intramolecular S_N2 substitution of ephedrine by amino substituent providing the corresponding aziridine with a high stereocontrol. A subsequent stereo-selective ring opening at the benzylic centre using tetrabutylammonium fluoride and dithiols afforded the corresponding amino sulfides in excellent yields.

In 2005, Bandini *et al.* reported the synthesis of new *N,N*-palladium chelates as chiral diamino oligothiophenes that were studied by X-ray diffraction, showing a five-membered palladacycle through the nitrogen atoms but also a significant van der Waals interaction between the sulfur atom of one of inner thiophene ring and palladium.[103] As shown in Scheme 1.60, these complexes have proven to be excellent catalysts for the test reaction since the product was isolated in nearly quantitative yield and with a high enantioselectivity in almost all cases of catalysts. Moreover, the scope of the reaction could be extended to the reaction of dimethyl methylmalonate with the same carbonate, providing the corresponding product with up to 99% ee, and with 1,3-dimethylallyl carbonate, leading to the corresponding product with up to 90% ee. In order to gain insight into the role played by the thiophene ring, the authors have pre-pared analogous ligands bearing phenyl rings replacing thiophene rings, and observed that high enantioselectivities were obtained only in the cases of ligands in which a thiophene ring was maintained close to the nitrogen-chelating unit. Consequently, it was concluded that the noncovalent secondary Pd/S interaction seemed to be crucial to achieve an efficient discrimination around the metal centre. In addition, one ligand could be grafted on a soluble polymeric support, which was successfully applied to the test reaction both as homogeneous and heterogeneous catalysts (Scheme 1.60).[104] Indeed, under homogeneous conditions, this ligand in solution in dichloromethane gave the expected product with both similar yield and enantioselectivity to those obtained when using the corresponding nonsupported analogue. Similarly, when used in THF, in which the polymer was insoluble, similar results were

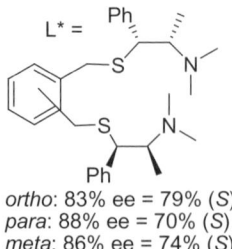

ortho: 83% ee = 79% (*S*)
para: 88% ee = 70% (*S*)
meta: 86% ee = 74% (*S*)

Scheme 1.59   Test reaction with diamino sulfides derived from ephedrine.

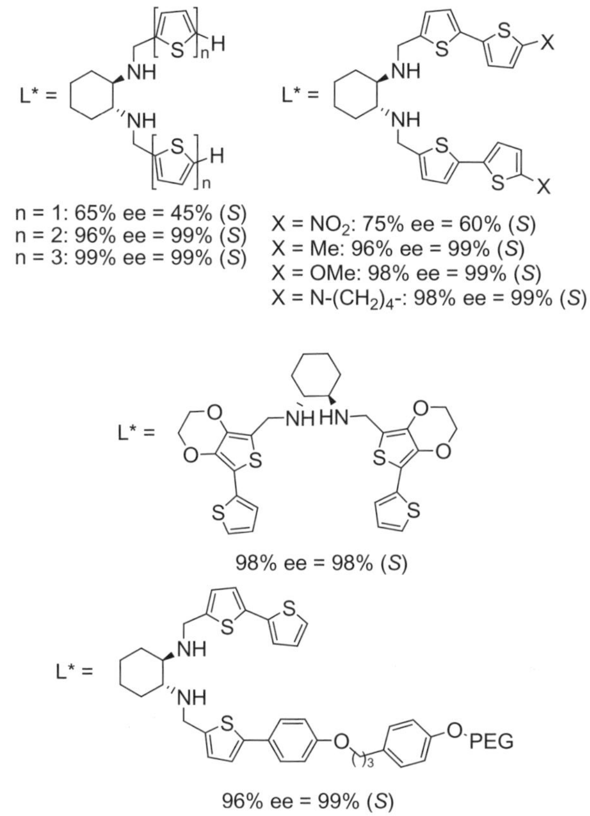

Scheme 1.60 Test reaction with diamino-oligothiophene ligands.

obtained. In the same study, a series of ligands of the same type were prepared having various electron-donating or electron-withdrawing substituents at the α-position of the bithienyl group. When applied to the test reaction, almost all the ligands allowed excellent enantioselectivities to be obtained, except with a ligand bearing a nitro group (Scheme 1.60).

Other tetradentate thiophene diamine ligands have been previously involved in the palladium-catalysed test reaction by Kim *et al.*, in 2004.[81] Prepared from (*R,R*)-1,2-diaminocyclohexane, these ligands were proved to be very efficient in terms of enantioselectivity in this reaction since an enantioselectivity of up to 98% ee was observed when the ligand/palladium ratio was 4/1 (Scheme 1.61).

Almost no attention has been paid to diphosphine sulfides employed as chiral ligands for palladium-catalysed nucleophilic substitution reactions. In this context, enantiomerically pure diphosphine sulfides derived from 2,2'-biphosphole, which combined axial chirality and phosphorus chiralities, were synthesised, in 2008, by Gouygou *et al.* through a four-step synthetic sequence.[105] Among various palladium catalytic systems derived from this type of ligands and evaluated for the test reaction, that depicted in Scheme 1.62

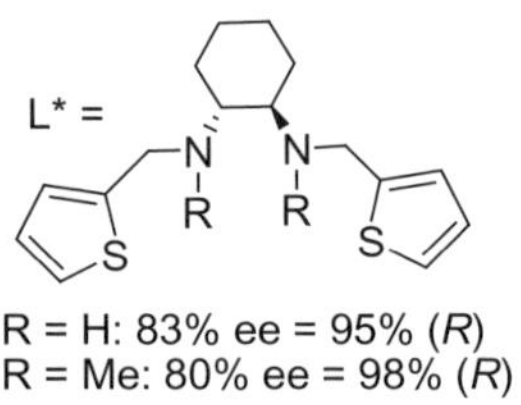

L* =

R = H: 83% ee = 95% (*R*)
R = Me: 80% ee = 98% (*R*)

Scheme 1.61   Test reaction with thiophene diamine ligands.

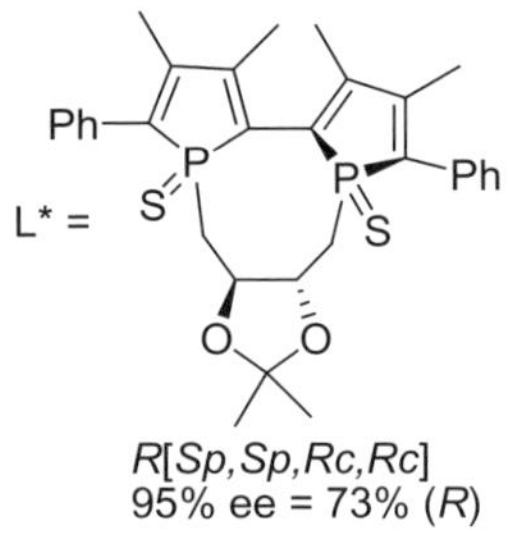

L* =

R[*Sp,Sp,Rc,Rc*]
95% ee = 73% (*R*)

Scheme 1.62   Test reaction with 2,2′-biphosphole-derived diphosphine sulfide.

allowed an excellent yield and an enantioselectivity of 73% ee to be obtained. In order to gain insight into the coordination chemistry of these ligands towards palladium, these authors have prepared the corresponding diphosphine analogues. Using these ligands, the palladium catalytic systems were more active for the same reaction than the corresponding disulfides, while the enantioselectivities were significantly lower. Consequently, the role of the sulfur atom in the coordination to the palladium was confirmed.

Palladium-catalysed asymmetric allylations of various carbonyl compounds have been studied by Hiroi *et al.* using various types of chiral sulfonamides derived from α-amino acids.[106] In particular, the chiral bidentate phosphinyl sulfonamide derived from (*S*)-proline and depicted in Scheme 1.63 was employed in the presence of palladium to catalyse the allylation of methyl aminoacetate diphenyl ketimine with allyl acetate, leading to the corresponding (*R*)-product with a moderate enantioselectivity of 62% ee. This ligand was also applied to the allylation of a series of other nucleophiles, as shown in Scheme 1.63, providing the corresponding allylated products in moderate enantioselectivities.

In the past few years, impressive results have been obtained in the development of highly efficient new copper catalytic systems capable to induce chirality in the allylic substitution.[107] In contrast to other metals (Pd, Mo, and Ir, for example), copper allows nonstabilised nucleophiles to be used. Increasing interest is being shown in catalytic systems employing Grignard,[108] organozinc, and organoaluminium reagents as the carbon nucleophiles, and diboron as the noncarbon nucleophile. In the context of copper catalysts based on sulfur-containing ligands, Gennari *et al.* have studied the copper-catalysed allylic

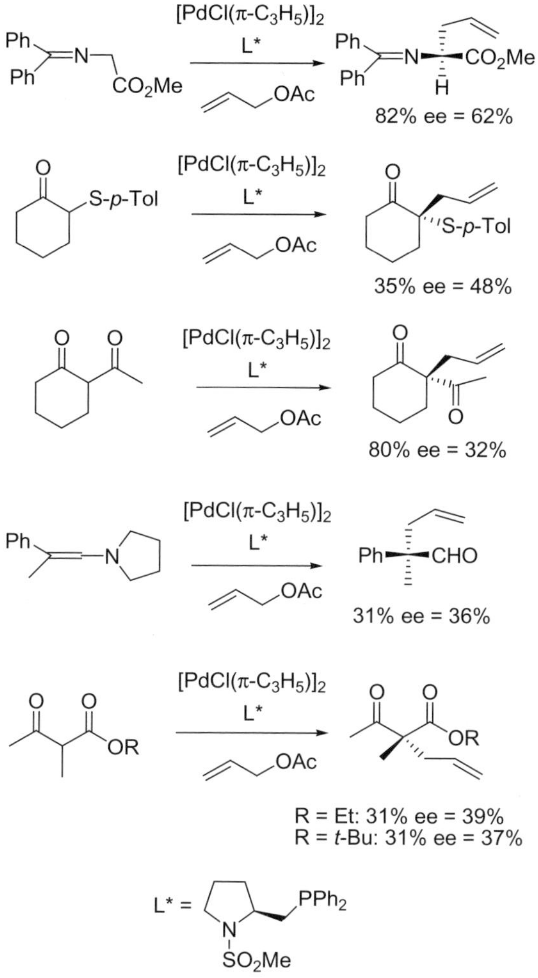

Scheme 1.63 Test reaction with (*S*)-proline-derived phosphinyl sulfonamide ligand.

substitution reaction of cinnamyl phosphate with $ZnEt_2$ as the nucleophile.[109] This reaction was performed in the presence of novel chiral β-aminosulfonamide ligands, giving excellent ratios of the corresponding S_N2'/S_N2 products combined with a moderate enantioselectivity of up to 40% ee, as shown in Scheme 1.64.

The catalytic enantioselective desymmetrization of *meso* compounds is a powerful tool for the construction of enantiomerically enriched functionalized products.[110] *Meso* cyclic allylic diol derivatives are challenging substrates for the asymmetric allylic substitution reaction owing to the potential competition of several reaction pathways. In particular, S_N2' and S_N2 substitutions can occur, and both with either retention or inversion of the stereochemistry. In the

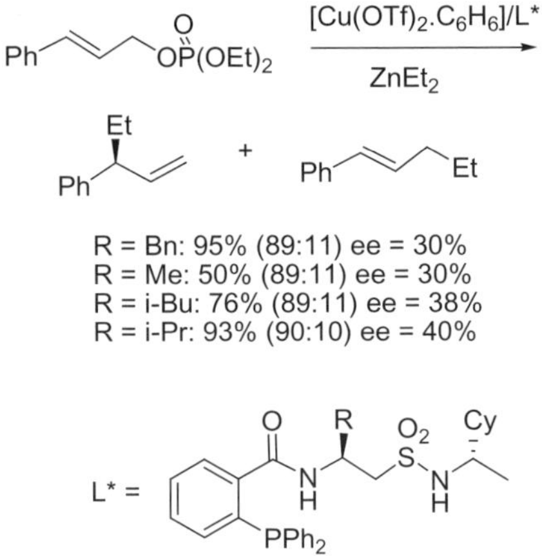

Scheme 1.64   Cu-catalysed allylic substitution of cinnamyl phosphate with ZnEt₂ in the presence of β-aminosulfonamide ligands.

case of S_N2 substitution in which an allylic alcohol derivative is obtained, a second allylic substitution might occur through the S_N2' or S_N2 mechanism with either retention or inversion of the stereochemistry. Based on this complex scenario, up to fifteen isomers (seven pairs of enantiomers and one *meso* compound) could, in principle, be obtained. In this context, Gennari *et al.* have disclosed a new highly regio-, diastereo-, and enantioselective desymmetrization of *meso* cyclic allylic bisdiethylphosphates with ZnEt₂ catalysed by copper complexes of chiral β-aminosulfonamide ligands depicted in Scheme 1.65.[111]

The catalytic enantioselective addition of vinylmetals to activated alkenes is a potentially versatile but undeveloped class of transformations. Compared to processes with arylmetals and, particularly alkylmetals, processes with the corresponding vinylic reagents are of higher synthetic utility but remain scarce, and the relatively few reported examples are Rh-catalysed conjugate additions.[112] In this context, Hoveyda *et al.* reported very recently an efficient method for catalytic asymmetric allylic alkylations with vinylaluminum reagents that were prepared and used *in situ*.[113] Thus, stereoselective reactions of commercially available DIBAL-H with readily accessible terminal alkynes efficiently delivered the vinylmetals, which reacted with an allylic phosphate in the presence of a chiral sulfur-containing *N*-heterocyclic carbene complex derived from an air stable Cu salt, such as that depicted in Scheme 1.66. The S_N2' products were obtained with an exceptional selectivity of >98% in all cases. Either substrates bearing sterically demanding groups, electron-withdrawing aryl units, or an unsubstituted phenyl underwent the catalytic asymmetric allylic alkylation in 82–94% yields and enantioselectivities of 87–98% ee in the presence of only 0.5 mol% of the sulfur-containing catalyst.

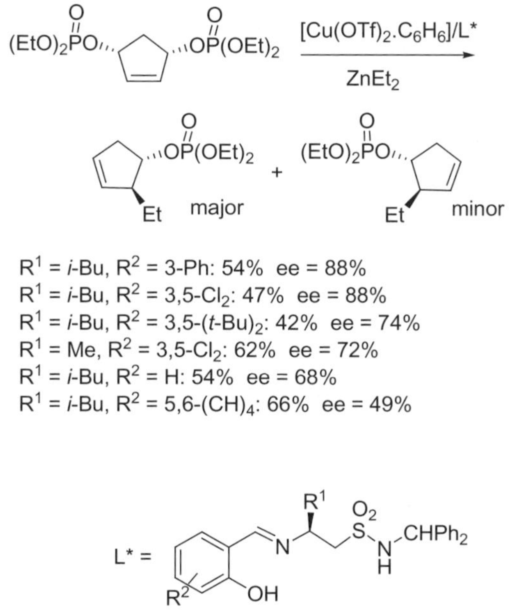

R[1] = *i*-Bu, R[2] = 3-Ph: 54% ee = 88%
R[1] = *i*-Bu, R[2] = 3,5-Cl$_2$: 47% ee = 88%
R[1] = *i*-Bu, R[2] = 3,5-(*t*-Bu)$_2$: 42% ee = 74%
R[1] = Me, R[2] = 3,5-Cl$_2$: 62% ee = 72%
R[1] = *i*-Bu, R[2] = H: 54% ee = 68%
R[1] = *i*-Bu, R[2] = 5,6-(CH)$_4$: 66% ee = 49%

Scheme 1.65 Cu-catalysed desymmetrization of *meso* cyclic allylic bisdiethylphosphate with ZnEt$_2$ in the presence of β-aminosulfonamide ligands.

1.7 Sulfur-Containing Ferrocenyl Ligands

Chiral ferrocene ligands have been widely used in asymmetric catalysis[114] and, in particular, ferrocene-based chiral phosphines.[115] Numerous reports are available on the chemistry of allylic alkylations using ferrocenyl ligands.[115,116] Planar chiral ferrocene ligands have been shown to be efficient stereocontrol elements in many metal centred asymmetric catalytic reactions due to their stability, rigid bulkiness, ease of introduction of the planar chirality, and interesting electronic properties of the ferrocene unit. Most ferrocene ligands contain both planar and central or axial chirality. In particular, numerous reports are available on the chemistry of allylic alkylations using ferrocenyl ligands.[115,116b] Indeed, the advantages of using ferrocene as a scaffold for chiral ligands are its planar chirality, rigid bulkiness, stability, inherent special electronic and stereochemical properties, and ease of derivatisation. In recent years, various sulfur-containing chiral ferrocene derivatives[117] have been successfully implicated in the test reaction, such as chiral 1-oxazolinyl-1′-(phenylthio)-ferrocenes,[118] which gave modest results, whereas high enantioselectivities of up to 98% ee were obtained by using thioether derivatives of ferrocenyloxazolines with central and planar chiralities (Scheme 1.67).[119] These authors have highlighted that the central chirality of the oxazoline ring played a major role in controlling the enantiodifferenciation.

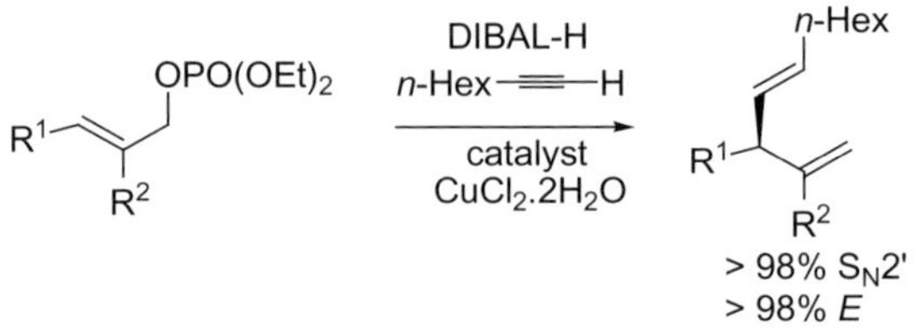

R^1 = o-Tol, R^2 = Me, R^3 = Ph: 87% ee > 98%
R^1 = o-BrC$_6$H$_4$, R^2 = Me, R^3 = Ph: 84% ee = 96%
R^1 = o-NO$_2$C$_6$H$_4$, R^2 = Me, R^3 = Ph: 94% ee = 96%
R^1 = p-NO$_2$C$_6$H$_4$, R^2 = Me, R^3 = Ph: 93% ee = 89%
R^1 = m-TsOC$_6$H$_4$, R^2 = Me, R^3 = Ph: 82% ee = 87%
R^1 = R^3 = Ph, R^2 = Me: 84% ee = 92%
R^1 = 1-Naph, R^2 = Me, R^3 = Ph: 88% ee = 91%
R^1 = BnCH$_2$, R^2 = Me, R^3 = Ph: 88% ee = 77%
R^1 = Ph, R^2 = R^3 = H: 90% ee = 79%
R^1 = Cy, R^2 = R^3 = H: 92% ee = 86%
R^1 = PhMe$_2$Si, R^2 = R^3 = H: 91% ee = 93%

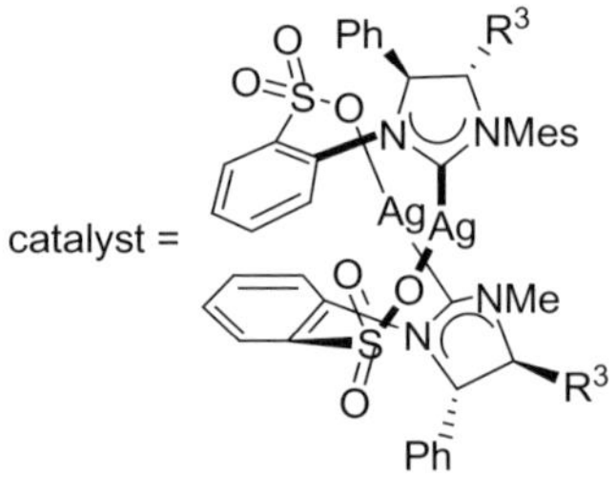

catalyst =

Scheme 1.66   Cu-catalysed allylic alkylations of allylic phosphates with vinylaluminum reagents in the presence of sulfur-containing diaminocarbene ligands.

In 2000, other S/N-ferrocenyloxazolines were prepared by Aït-Haddou *et al.* starting from chiral 2-amino-3-phenyl-1,3-propanediol.[120] The corresponding P/N-analogues were also prepared in order to compare their efficiency in the test reaction. As shown in Scheme 1.68, both ligands gave good results in terms of both activity and enantioselectivity with a better result for the S/N ligand.

In 2003, Bonini *et al.* reported a new synthesis of ferrocenyloxazolines based on an iodide-mediated ring expansion of N-ferrocenoyl-aziridine-2-carboxylic esters.[121] The thus-formed ligands were successfully employed as palladium chelates for the test reaction, since they allowed the product to be formed in quantitative yields and good to high enantioselectivities (Scheme 1.69). According to the results, it seemed that the additional chiral centre present in the oxazoline backbone of these ligands did not play a major role for the asymmetric induction and the activity of the corresponding catalysts.

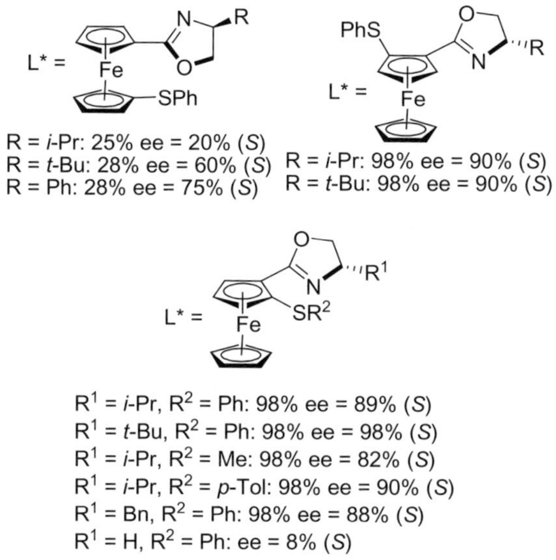
L* =
Fe
SPh
N
R
O
R = i-Pr: 25% ee = 20% (S)
R = t-Bu: 28% ee = 60% (S)
R = Ph: 28% ee = 75% (S)
PhS
O
N
R
L* =
Fe
R = i-Pr: 98% ee = 90% (S)
R = t-Bu: 98% ee = 90% (S)
O
N
R1
SR2
L* =
Fe
R1 = i-Pr, R2 = Ph: 98% ee = 89% (S)
R1 = t-Bu, R2 = Ph: 98% ee = 98% (S)
R1 = i-Pr, R2 = Me: 98% ee = 82% (S)
R1 = i-Pr, R2 = p-Tol: 98% ee = 90% (S)
R1 = Bn, R2 = Ph: 98% ee = 88% (S)
R1 = H, R2 = Ph: ee = 8% (S)

Scheme 1.67 Test reaction with sulfur-containing ferrocenyloxazoline ligands.

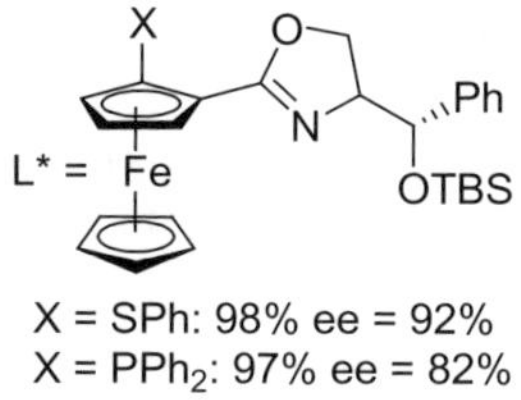
X
O
N
Ph
OTBS
L* =
Fe
X = SPh: 98% ee = 92%
X = PPh2: 97% ee = 82%

Scheme 1.68 Test reaction with S/N- and P/N-ferrocenyloxazoline ligands.

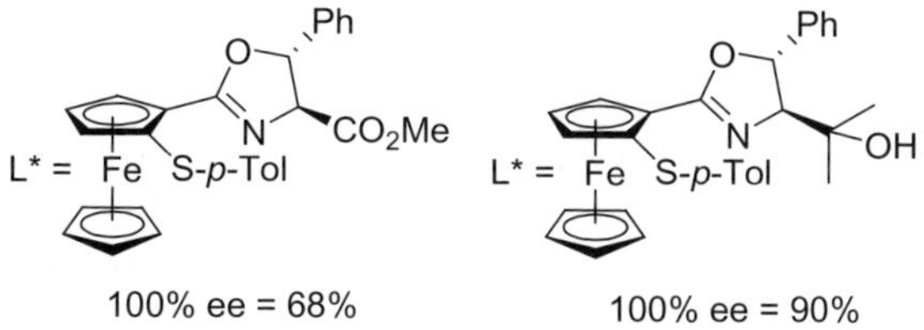
Ph
O
N
CO2Me
L* =
Fe
S-p-Tol
100% ee = 68%
Ph
O
N
OH
L* =
Fe
S-p-Tol
100% ee = 90%

Scheme 1.69 Test reaction with S/N-ferrocenyloxazoline ligands bearing two stereogenic centres.

In addition, Bryce *et al.* have studied the binding of palladium to other S/N-ferrocenyloxazoline ligands by cyclic voltammetry and proved that it was reversible.[122] These redox-active liganding systems were successfully used in the test reaction, providing the product in both high yield and enantioselectivity of up to 93% ee, as shown in Scheme 1.70.

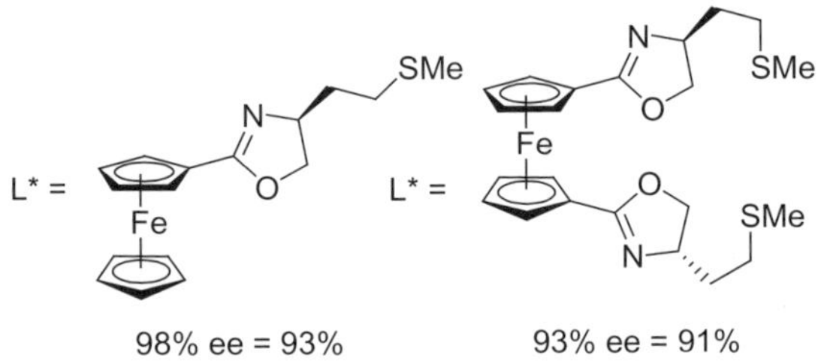

98% ee = 93%   93% ee = 91%

Scheme 1.70   Test reaction with S/N-ferrocenyloxazoline ligands.

Several groups have reported the synthesis of various chiral S/P-ferrocenyl ligands (Scheme 1.71) and have investigated them in the test reaction. As an example, Enders *et al.* have employed novel bidentate planar ferrocenyl ligands, bearing a stereogenic centre at the β-position to the metallocene backbone, which led to quantitative yields and enantioselectivities of up to 97% ee.[123] These authors observed that S/S-coordinating ligands displayed a modest enantioselectivity (20% ee) in the test reaction. In addition, the authors tested, for comparison, a similar planar chiral ferrocene S/P ligand bearing a stereogenic centre in the α-position, which proved to be less efficient in terms of enantioselectivity (34% ee) for the reaction performed under the same conditions. On the other hand, replacing one coordinating atom by phosphorus as a strong π-acceptor allowed the reaction to be performed in a quantitative yield and a high enantioselectivity of up to 97% ee (Scheme 1.71). NMR studies and X-ray analysis supported the assumption that this type of ligand allowed an electronic differentiation of the allylic carbon termini and that a preferred configuration (*exo–syn–syn*) was strongly favoured with these intermediates by intramolecular C–H/π interactions. Similar results were observed by Dai *et al.* using other planar chiral S/P-bidentate ferrocenyl ligands, prepared from the commercially available starting material, *N,N*-dimethyl-(*S*)-α-ferrocenylethylamine, providing both excellent yield (up to 93%) and enantioselectivity of up to 94% ee.[124] In addition, readily available 1-phosphino-2-sulfenylferrocenes, possessing a planar chirality as the sole source of chirality, were reported by Carretero *et al.*, providing very high enantioselectivities in the test reaction combined with excellent yields (Scheme 1.71).[125] In this study, the authors demonstrated that a sterically hindered sulfur atom (*t*-Bu-S) was essential to obtain a high asymmetric induction. The scope of these ligands was extended to their successful employment in palladium-catalysed allylic substitution with nitrogen nucleophiles. The authors proved the formation of an S/P-bidentate ligand, giving a five-membered palladacycle by X-ray analysis and NMR studies. Very recently, Manoury *et al.* have reported various new chiral ferrocenylphosphine thioethers and -thiophosphine thioethers having only a planar chirality, which proved to be successful in the test reaction (Scheme 1.71).[126] These authors have obtained enantioselectivities of up to 93% ee by using the S/S ligands, and similar results were observed with the involvement of the corresponding S/P-analogues.

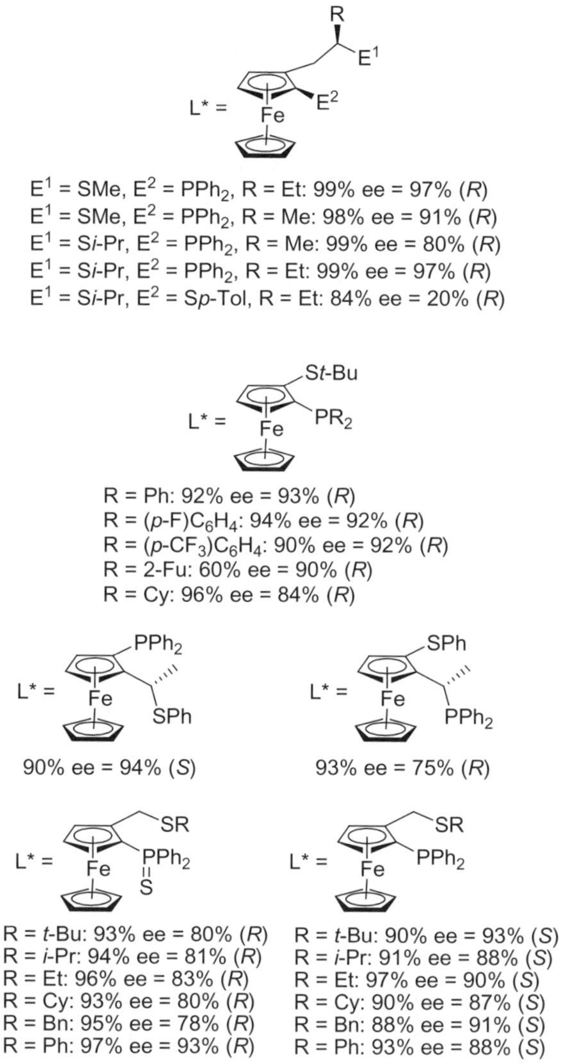
R
E1
E2
L* =
Fe
E1 = SMe, E2 = PPh2, R = Et: 99% ee = 97% (R)
E1 = SMe, E2 = PPh2, R = Me: 98% ee = 91% (R)
E1 = Si-Pr, E2 = PPh2, R = Me: 99% ee = 80% (R)
E1 = Si-Pr, E2 = PPh2, R = Et: 99% ee = 97% (R)
E1 = Si-Pr, E2 = Sp-Tol, R = Et: 84% ee = 20% (R)
St-Bu
PR2
L* =
Fe
R = Ph: 92% ee = 93% (R)
R = (p-F)C6H4: 94% ee = 92% (R)
R = (p-CF3)C6H4: 90% ee = 92% (R)
R = 2-Fu: 60% ee = 90% (R)
R = Cy: 96% ee = 84% (R)
PPh2
SPh
L* =
Fe
90% ee = 94% (S)
SPh
PPh2
L* =
Fe
93% ee = 75% (R)
SR
PPh2
S
L* =
Fe
R = t-Bu: 93% ee = 80% (R)
R = i-Pr: 94% ee = 81% (R)
R = Et: 96% ee = 83% (R)
R = Cy: 93% ee = 80% (R)
R = Bn: 95% ee = 78% (R)
R = Ph: 97% ee = 93% (R)
SR
PPh2
L* =
Fe
R = t-Bu: 90% ee = 93% (S)
R = i-Pr: 91% ee = 88% (S)
R = Et: 97% ee = 90% (S)
R = Cy: 90% ee = 87% (S)
R = Bn: 88% ee = 91% (S)
R = Ph: 93% ee = 88% (S)

Scheme 1.71 Test reaction with S/P-ferrocenyl ligands.

A sulfenylphosphinoferrocene ligand of this type could be immobilised on a polystyrene support by Carretero *et al.*, in 2007 (Scheme 1.72).[127] This polymer-supported ligand was further investigated for the test reaction, providing a higher enantioselectivity (98% ee) than that obtained when using the non-supported corresponding ligand (93% ee) whereas the yield was slightly lower (82% instead of 92% with the nonsupported ligand). Similarly, when benzylamine was used as the nucleophile, the corresponding substitution product was obtained in 79% yield and a good enantioselectivity of 91% ee, albeit somewhat lower than from the unsupported ligand (97% ee).

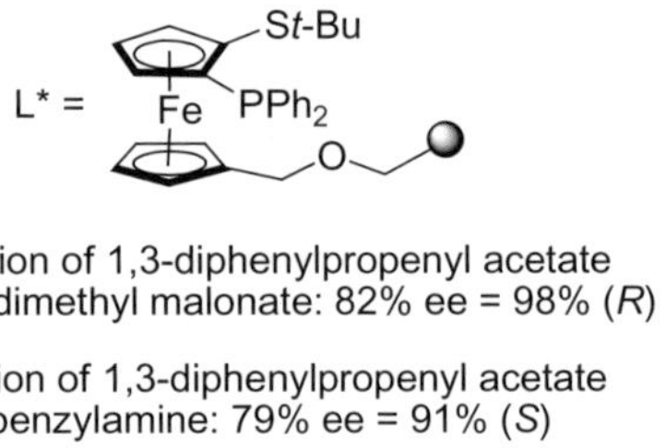

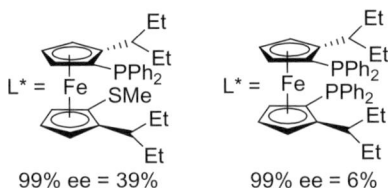

reaction of 1,3-diphenylpropenyl acetate
with dimethyl malonate: 82% ee = 98% (R)

reaction of 1,3-diphenylpropenyl acetate
with benzylamine: 79% ee = 91% (S)

Scheme 1.72   Test reaction with polymer supported S/P-ferrocenyl ligand.

99% ee = 39%   99% ee = 6%

Scheme 1.73   Test reaction with pseudo-C_2-symmetric S/P-hybrid ferrocenyl ligand
and its P/P-analogue.

In 2003, Kang *et al.* reported the preparation of a pseudo-C_2-symmetric S/P-hybrid ferrocenyl ligand (Scheme 1.73) in order to compare its efficiency in the test reaction with its C_2-symmetric biphosphine analogue.[128] This S/P-bidentate ligand containing a planar chirality was obtained in 55% yield in a multistep reaction starting from 1,1′-ferrocenedicarboaldehyde. When applied as a ligand for the test reaction, this S/P ligand led to the product in a quantitative yield and with a moderate enantioselectivity (39% ee), whereas the enantioselectivity obtained by using the corresponding P/P-analogue was much lower (6% ee).

Another class of S/P-containing ferrocenyl ligand has been developed by Pregosin *et al.*[21a] These ligands, having a planar chirality and substituted with a thioglucose functionality, allowed an enantioselectivity of 88% ee to be obtained for the test reaction product (Scheme 1.74). This good result not only arose from the planar chirality of the ligand but also from the sugar moiety, since changing the carbohydrate substituent with a cyclohexyl or an ethyl group resulted in a dramatic decrease in the reaction enantioselectivity (67 and 34% ee, respectively). Therefore, the thioglucose moiety seemed to play an essential role in the enantioselectivity of the reaction.

Moreover, a few chiral ferrocenylsulfur-imine ligands were investigated in the palladium-catalysed asymmetric allylic alkylation of 1,3-diphenylpropenyl acetate and cyclohexenyl acetate with dimethyl malonate (Scheme 1.75).[124,129,130]

The first monosubstituted ferrocene derivatives containing chiral aminosulfide backbones in the side chain were reported by Bonini *et al.*[129] These

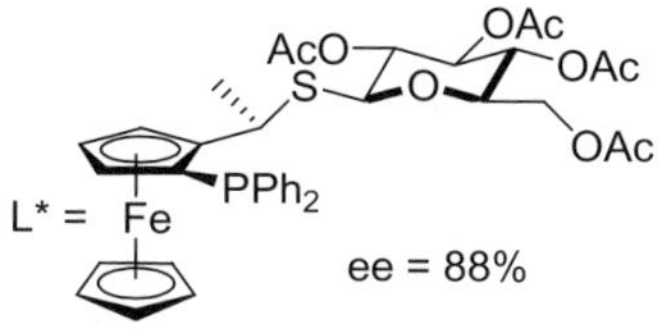
AcO
OAc
S
O
OAc
OAc
PPh2
L* = Fe
ee = 88%

Scheme 1.74 Test reaction with thioglucose-derived S/P-ferrocenyl ligand.

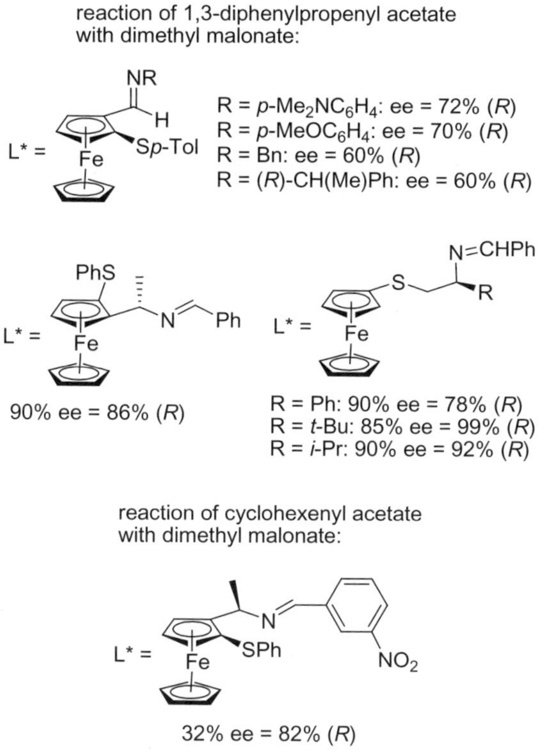
reaction of 1,3-diphenylpropenyl acetate
with dimethyl malonate:
NR
H
Sp-Tol
L* = Fe
R = p-Me2NC6H4: ee = 72% (R)
R = p-MeOC6H4: ee = 70% (R)
R = Bn: ee = 60% (R)
R = (R)-CH(Me)Ph: ee = 60% (R)
PhS
N
Ph
L* = Fe
90% ee = 86% (R)
N=CHPh
S
R
L* = Fe
R = Ph: 90% ee = 78% (R)
R = t-Bu: 85% ee = 99% (R)
R = i-Pr: 90% ee = 92% (R)
reaction of cyclohexenyl acetate
with dimethyl malonate:
N
SPh
NO2
L* = Fe
32% ee = 82% (R)

Scheme 1.75 Ferrocene-based sulfur-imine ligands for Pd-catalysed allylic alkylations.

ligands were investigated for their efficiency in the test reaction, providing an asymmetric induction of up to 99% ee, as shown in Scheme 1.76. It was shown that the β-dimethylamino derivatives were the most selective. Furthermore, increasing the steric bulk on the backbone chiral centres did not allow better selectivities to be obtained.

Although many excellent results have been observed using planar chiral ferrocenyl ligands, Toru *et al.* reported in 2004 the first chiral ferrocenyl ligand bearing a chiral sulfinyl group as the sole chirality.[131] These new chiral 1-phosphinyl-1′-sulfinylferrocene ligands were evaluated in the standard reaction, giving enantioselectivities of up to 68% ee, as shown in Scheme 1.77.

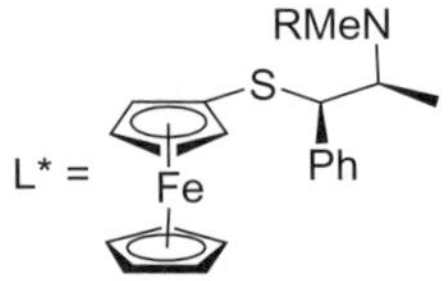

R = H: 87% ee = 43% (*S*)
R = Me: 88% ee = 99% (*S*)

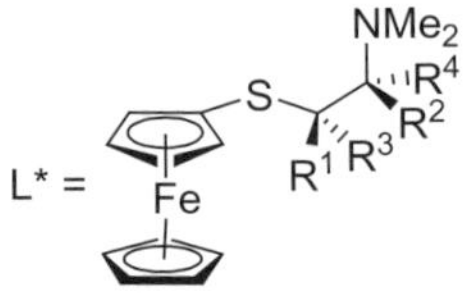

R^1 = Ph, R^2 = Me, R^3 = R^4 = H: 88% ee = 99% (*S*)
R^1 = R^2 = Ph, R^3 = R^4 = H: 99% ee = 79% (*S*)
R^1 = R^2 = H, R^3 = R^4 = Ph: 96% ee = 77% (*R*)
R^1 = R^3 = R^4 = H, R^2 = Ph: 96% ee = 99% (*S*)

Scheme 1.76   Test reaction with β-aminoalkyl ferrocenyl sulfide ligands.

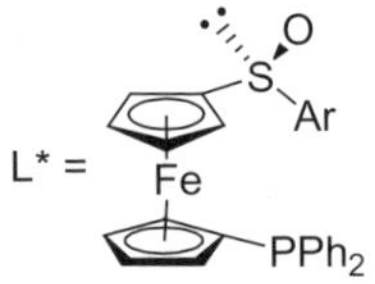

Ar = Ts: 82% ee = 58% (*R*)
Ar = *p*-MeOC$_6$H$_4$: 99% ee = 56% (*R*)
Ar = *p*-ClC$_6$H$_4$: 45% ee = 25% (*R*)
Ar = 1-Naph: 96% ee = 68% (*R*)

Scheme 1.77   Test reaction with 1-phosphinyl-1′-sulfinylferrocene ligands.

In 2005, Anderson and Osborne developed a series of new heterobidentate ferrocenyl S/N- and P/N ligands possessing only a planar chirality as the sole stereoinducing element.[132] Involved as the ligands of palladium in the test reaction, these ligands gave contrasted results. As shown in Scheme 1.78, the *ortho*-methyl sulfide ligand possessing a dimethylamino substituent gave a substantially higher enantioselectivity (59% ee) than the corresponding ligand possessing a diisopropylamino substituent (13% ee). Moreover, it was found that the ligand bearing a diisopropylamino substituent was much slower than the corresponding dimethylamine ligand. Either extra steric hindrance from the diisopropylamine group retarded the reaction or the dimethylamine ligand gave a more efficient reaction due to an enhanced ability to chelate the palladium catalyst. The highest enantioselectivity of 79% ee was obtained, however, by using an N/P-chelate ligand, as depicted in Scheme 1.78.

A new class of easily accessible 1,2-disubstituted phosphinamidite-thioether ligands derived from Ugi's amine and based on a ferrocene motif has been

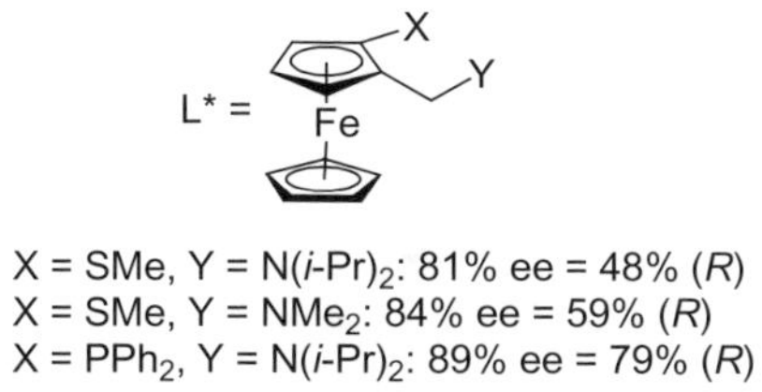
L* =
Fe
X
Y
X = SMe, Y = N(i-Pr)2: 81% ee = 48% (R)
X = SMe, Y = NMe2: 84% ee = 59% (R)
X = PPh2, Y = N(i-Pr)2: 89% ee = 79% (R)

Scheme 1.78 Test reaction with S/N- and P/N-ferrocenyl ligands.

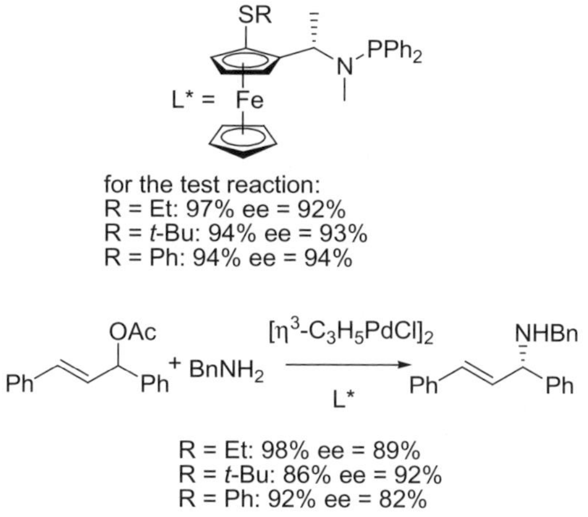
SR
PPh2
N
L* = Fe
for the test reaction:
R = Et: 97% ee = 92%
R = t-Bu: 94% ee = 93%
R = Ph: 94% ee = 94%
OAc
Ph
Ph
+ BnNH2
[η3-C3H5PdCl]2
L*
NHBn
Ph
Ph
R = Et: 98% ee = 89%
R = t-Bu: 86% ee = 92%
R = Ph: 92% ee = 82%

Scheme 1.79 Phosphinamidite-thioether ferrocenyl ligands for Pd-catalysed allylic alkylations and aminations.

developed by Chan *et al.*[133] These bidentate ligands were successfully employed for the test reaction, providing enantioselectivities of up to 93% ee and excellent yields, as shown in Scheme 1.79. Moreover, these ligands were also evaluated for analogous C–N bond formation. Thus, both excellent yield and enantioselectivity were obtained for the allyl amine product resulting from the reaction of 1,3-diphenyl-2-propenyl acetate with benzylamine (Scheme 1.79).

The stereoselective functionalisation of indoles is a subject of significance because of the common occurrence of indole moieties in many bioactive natural products and pharmaceuticals.[134] In this context, Lewis acid promoted Friedel–Crafts reactions have been explored extensively for enantioselective alkylation of indoles at the C3 position, but a high catalyst loading (10 mol%) was often required for achieving satisfactory yields and enantiopurities. In 2007, a novel chiral ferrocenyl heterobidentate N/S/P ligand was selected by Chan *et al.* through systematic screening and optimisation to induce palladium-catalysed asymmetric allylic alkylation of indoles with 1,3-diphenyl-2-propenyl acetate.[135] Thus, the ligand depicted in Scheme 1.80 and synthesised via a three-step sequence from (*S*)-Ugi's amine allowed high enantioselectivities of up to

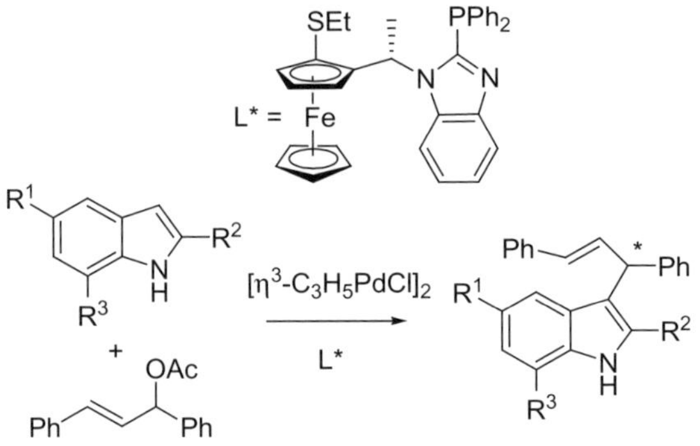

R^1 = R^2 = R^3 = H: 74% ee = 95%
R^1 = R^3 = H, R^2 = Ph: 77% ee = 92%
R^1 = R^3 = H, R^2 = Me: 78% ee = 95%
R^1 = Me, R^2 = R^3 = H: 77% ee = 96%
R^1 = R^2 = H, R^3 = Me: 75% ee = 96%
R^1 = Br, R^2 = R^3 = H: 66% ee = 94%
R^1 = Cl, R^2 = R^3 = H: 61% ee = 96%
R^1 = MeO, R^2 = R^3 = H: 85% ee = 94%
R^1 = BnO, R^2 = R^3 = H: 56% ee = 94%

Scheme 1.80   N/S/P-ferrocenyl ligand for Pd-catalysed allylic alkylations of indoles.

96% ee to be achieved irrespective of the steric or electronic nature of indoles, as shown in Scheme 1.80.

In contrast to allylic alkylations and aminations that have been well documented, the enantioselective allylic substitution of unactivated allylic acetates with relatively hard oxygen nucleophiles has only been studied sporadically. In this context, the same workers developed, in 2008, asymmetric palladium-catalysed allylic substitution of 1,3-diphenyl-2-propenyl acetate with aliphatic alcohols in the presence of a phosphinamidite-thioether ferrocenyl ligand, generating the corresponding chiral ethers in high yields and with excellent enantioselectivities (Scheme 1.81).[136] Surprisingly, these authors observed an intriguing relationship between the enantioselectivity of the reaction and the electronic nature of the substituted benzylic alcohol. A higher enantioselectivity was observed when the benzylic alcohol contained an electron-rich *para* substituent, and the selectivity diminished gradually as the substituent became more electron deficient. Such an electronic effect of a nonconjugated substrate on the enantioselectivity has never been reported previously and was complementary to the well-known electronic effect on the stereoselectivity of either a substrate or a ligand composed of conjugated aromatic systems.

Dendrimers represent a new and fascinating class of regular highly branched and well-defined macromolecules,[137] which have found numerous promising applications in materials chemistry, biology, *etc.* but also as homogeneous catalysts with the catalytic sites in the core or at the periphery of the dendrimers.[138] However, although a very large number of catalytic systems using

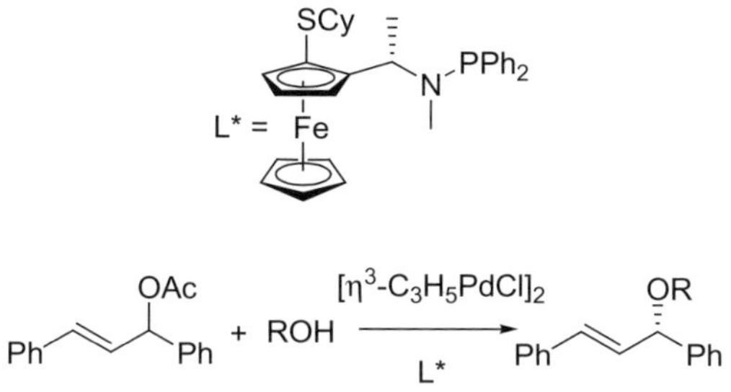

R = Bn: 98% ee = 92%
R = CH$_2$-p-(MeO)C$_6$H$_4$: 94% ee = 94%
R = CH$_2$-p-Tol: 94% ee = 93%
R = CH$_2$-p-FC$_6$H$_4$: 95% ee = 90%
R = CH$_2$-p-ClC$_6$H$_4$: 92% ee = 89%
R = CH$_2$-p-(CF$_3$)C$_6$H$_4$: 93% ee = 77%
R = CH$_2$-o-Tol: 87% ee = 95%
R = CH$_2$-m-Tol: 88% ee = 94%
R = CH$_2$-m-(MeO)C$_6$H$_4$: 94% ee = 93%
R = CH$_2$Cy: 96% ee = 93%
R = allyl: 98% ee = 93%
R = n-Bu: 98% ee = 94%
R = Cy: 58% ee = 96%

Scheme 1.81 Allylic etherifications with phosphinamidite-thioether ferrocenyl ligand.

dendrimers have already been described, far fewer reports on asymmetric catalysis involving chiral dendrimers have been disclosed.[139] In this context, Caminade *et al.* have recently succeeded to graft a ferrocenyl S/P ligand on the surface of phosphorus dendrimers having a cyclotriphosphazene core.[140] The starting material for the synthesis of dendrimers was a ferrocenyl ligand bearing a phenol, such as that depicted in Scheme 1.82, in order to covalently bind it to the surface of phosphorus-containing dendrimers. The sodium salt of this phenol ferrocenyl ligand was further readily obtained by reaction with sodium hydride in THF. The subsequent reaction of twelve equivalents of this sodium salt with the first generation dendrimer terminated with –P(S)Cl$_2$ functions proceeded smoothly at room temperature, resulting in the obtention of the chiral dendrimer in a nearly quantitative yield. Applied to the catalysis of the test reaction, this soluble polymer-supported catalyst has proved to be highly efficient since an enantioselectivity of up to 93% ee was obtained for the product (Scheme 1.82). Thus, the catalytic properties of the original monomeric ligand were left almost unaffected after grafting on the dendrimeric scaffolds.[126]

In addition, Bäckvall *et al.* have reported the application of ferrocenylthiolates as ligands in copper-catalysed substitution reactions of allylic acetates with Grignard reagents, providing a good enantioselectivity of up to 64% ee (Scheme 1.83).[141] The corresponding neutral thioether ligands were not suitable for this type of reaction, showing the importance of the anionic coordination to copper.

synthesis of L*:

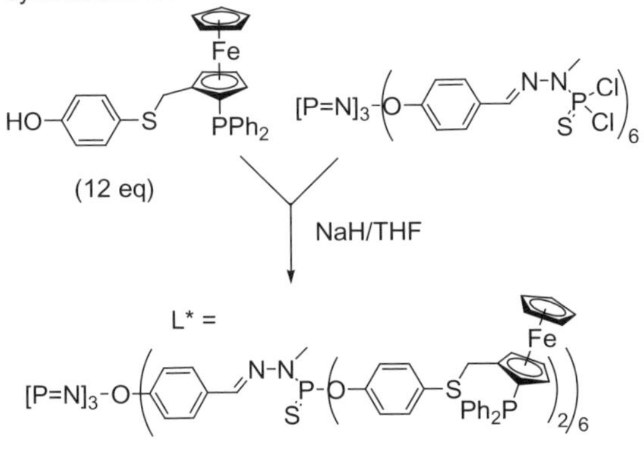

result for test reaction with L*: 89% ee = 93%

Scheme 1.82   Test reaction with dendritic ferrocenyl S/P ligand.

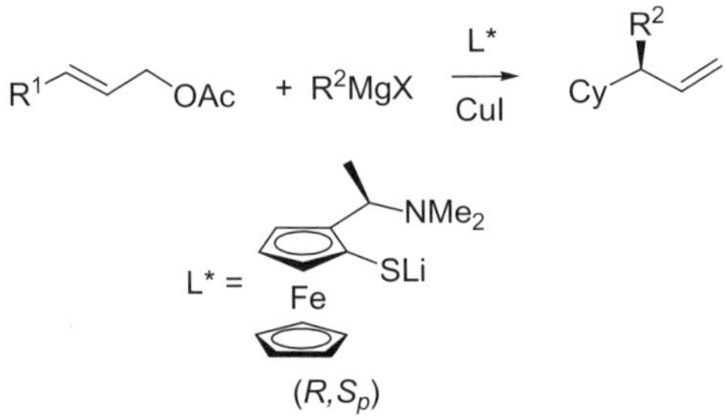

R^1 = Cy, R^2 = *n*-Bu, X = I: 88% ee = 64%
R^1 = Cy, R^2 = Et, X = I: 55% ee = 62%
R^1 = Cy, R^2 = *n*-Pr, X = I: 77% ee = 54%
R^1 = Cy, R^2 = *i*-Pr, X = Br: 51% ee = 52%
R^1 = Cy, R^2 = Me, X = I: 42% ee = 44%
R^1 = Ph, R^2 = *n*-Bu, X = I: 78% ee = 42%

Scheme 1.83   Cu-catalysed substitutions of allylic acetates by RMgX with ferrocenylthiolate ligand.

1.8   Conclusions

The enantioselective Tsuji–Trost reaction is by far the most intensively studied reaction performed in the presence of chiral sulfur-containing ligands over the past few years. This reaction has been successfully performed using various S/S, S/P, or S/N ligands. The key advantages of these new types of ligands are their easy synthesis, mostly starting from readily available commercial compounds, and their stability, which facilitates the catalytic processes. Chiral S/S ligands afforded generally the expected products with both moderate activity and enantioselectivity. Noteworthy, however, are the excellent results in terms of both activity and enantioselectivity obtained by Nakano's group by using chiral sulfideoxathiane ligands. Many more examples involving chiral S/P ligands have been developed providing excellent results, such as those reported by Nakano's group involving chiral phosphinooxathiane ligands for both Pd-catalysed alkylations and aminations, or those reported by Evans's group using chiral thioether-phosphinite ligands for similar reactions. On the other hand, S/N ligands have been by far the most studied structures, allowing excellent levels of enantioselectivity to be reached. In spite of advantages such as their stability and rapid synthesis, these ligands are, however, apart from few exceptions, generally less active and enantioselective than their analogous N/P-counterparts, which have played a dominant role among the heterodonor ligands. In particular, chiral sulfur-oxazoline ligands have been intensively investigated for the enantioselective allylic substitution reaction, providing good enantioselectivities, but the activities were generally lower than those observed by using the corresponding phosphorus-oxazoline ligands. Moreover, chiral bis(sulfoximines) developed by Bolm's group were proven to be promising ligands for this type of reactions.

In conclusion, chiral sulfur-containing ligands were not demonstrated to be more efficient in terms of both activity and enantioselectivity than their phosphorus analogous but, in some cases, they can be considered as becoming competitors of the generally most efficient nitrogen- or phosphorus-containing ligands. As particularly robust compounds, these ligands have a future in the development of asymmetric heterogeneous or homogeneous-supported C–C bond formations. Thus, several works using polymer-supported chiral sulfur-containing ligands (phosphinooxathiane, diamino oligothiophene, or sulfur-containing ferrocenyl ligands) have recently been applied to the enantioselective Tsuji–Trost reaction, providing very promising results.

References

1. (a) B. M. Trost, *Chem. Pharm. Bull.*, 2002, **50**, 1–14; (b) G. Helmchen and A. Pfaltz, *Acc. Chem. Res.*, 2000, **33**, 336–345; (c) *Catalytic Asymmetric Synthesis*, ed. I. Ojima, Wiley-VCH, New York, 2000, pp 802–856; (d) B. M. Trost and M. L. Crawley, *Chem. Rev.*, 2003, **103**, 2921–2943; (e) G. Helmchen, *J. Organomet. Chem.*, 1999, **576**, 203–214; (f) M. Johannsen and K. A. Jorgensen, *Chem. Rev.*, 1998, **98**, 1689–1708; (g) J. Tsuji, *in Palladium Reagents and Catalysis, Innovations in Organic Synthesis*, Wiley, New York, 1995; (h) T. Hayashi, *J. Organomet. Chem.*, 1999, **576**, 203–214; (i) G. Helmchen, S. Kudis, P. Sennhenn and H. Steihagen, *Pure Appl. Chem.*, 1997, **69**, 513–518; (j) J. Tsuji, *in Palladium Reagents and Catalysis*, Wiley, New York, 2004; (k) B. M. Trost and C. Lee, in *Catalytic Asymmetric Synthesis*, ed. I. Ojima, 2nd edn., Wiley-VCH, Weinheim, 2000, Chap. 8; (l) A. Heumann and M. Reglier, *Tetrahedron*, 1995, **51**, 975–1015; (m) G. Helmchen, A. Dahnz, P. Dübon,

M. Schelwies and R. Weihofen, *Chem. Commun.*, 2007, 675–691; (n) Z. Lu and S. Ma, *Angew. Chem., Int. Ed. Engl.*, 2008, **47**, 258–297.

2. (a) B. M. Trost and D. L. van Vranken, *Chem. Rev.*, 1996, **96**, 395–422; (b) B. M. Trost and C. Lee, in *Catalytic Asymmetric Synthesis*, ed. I. Ojima, 2nd edn, Wiley-VCH, New York, 2000, pp. 593–649; (c) T. Hayashi, *J. Organomet. Chem.*, 1999, **576**, 195–202.

3. E. Schulz and A. Voituriez, *Russ. Chem. Bull.*, 2003, **52**, 2588–2594.

4. E. Schulz, *Top. Organomet. Chem.*, 2005, **15**, 93–148.

5. B. L. Feringa, *Acc. Chem. Res.*, 2000, **33**, 346–353.

6. (a) P. J. Guiry and C. P. Saunders, *Adv. Synth. Catal.*, 2004, **346**, 497–537; (b) K. Selvakumar, M. Valentini, M. Worle, P. S. Pregosin and A. Albinati, *Organometallics*, 1999, **18**, 1207–1215; (c) R. Pretot and A. Pfaltz, *Angew. Chem., Int. Ed. Engl.*, 1998, **37**, 323–325; (d) S. R. Gilbertson and P. Lan, *Org. Lett.*, 2001, **3**, 2237–2240; (e) Y. Nishibayashi, K. Segawa, H. Takada, K. Ohe and S. Uemura, *Chem. Commun.*, 1996, 847–848.

7. G. Wilkinson in *Comprehensive Coordination Chemistry*, **5**, Pergamon Press, Oxford, 1987.

8. (a) A. Pfaltz, *Acc. Chem. Res.*, 1993, **26**, 339–345; (b) G. Desimoni, G. Faita and K. A. Jorgensen, *Chem. Rev.*, 2006, **106**, 3561–3651.

9. (a) B. M. Trost, *Acc. Chem. Res.*, 1996, **29**, 355–364; (b) S. G. Lee, C. W. Lim, C. E. Song, K. M. Kim and C. H. Jun, *J. Org. Chem.*, 1999, **64**, 4445–4451.

10. (a) D. A. Evans and T. A. Brandt, *Org. Lett.*, 1999, **1**, 1563–1565; (b) R. Prétôt and A. Pfaltz, *Angew. Chem., Int. Ed. Engl.*, 1998, **37**, 323–325.

11. A. Pfaltz and M. Lautens, *in Comprehensive Asymmetric Catalysis*, ed. E. N. Jacobsen, A. Pfaltz, H. Yamamoto, Vol. 2, Springer, Berlin, 1999, pp. 833–884.

12. H. Steinhagen, M. Reggelin and G. Helmchen, *Angew. Chem., Int. Ed. Engl.*, 1997, **36**, 2108–2110.

13. (a) B. M. Trost and T. R. Verhoeven, in *Comprehensive Organometallic Chemistry*, ed. G. Wilkinson, Pergamon Press, Oxford, UK, 1982, p. 799; (b) S. A. Godleski, in *Comprehensive Organic Syntheses*, ed. B. M. Trost Pergamon Press, Oxford, 1994, p. 585; (c) G. Consiglio and R. Waymouth, *Chem. Rev.*, 1989, **89**, 257–276; (d) C. G. Frost, J. Howarth and J. M. J. Williams, *Tetrahedron: Asymmetry*, 1992, **3**, 1089–1122.

14. B. M. Trost and P. E. Strege, *J. Am. Chem. Soc.*, 1977, **99**, 1649–1651.

15. (a) B. M. Trost and D. J. Murphy, *Organometallics*, 1985, **4**, 1143–1145; (b) T. Hayashi, A. Yamamoto, T. Hagihara and I. Ito, *Tetrahedron Lett.*, 1986, **27**, 191–194.

16. (a) P. B. Mackenzie, J. Whelan and B. Bosnich, *J. Am. Chem. Soc.*, 1985, **107**, 2046–2054; (b) K. L. Granberg and J. E. Backwall, *J. Am. Chem. Soc.*, 1992, **114**, 6858–6863.

17. (a) G. Chelucci and M. A. Cabras, *Tetrahedron: Asymmetry*, 1996, **7**, 965–966; (b) G. Chelucci and D. Berta, *Tetrahedron*, 1997, **53**, 3843–3848; (c) J. C. Anderson, C. James, S. Daniel and J. P. Mathias, *Tetrahedron:*

Asymmetry, 1998, **9**, 753–756; (d) C. Bolm, D. Kaufmann, M. Zehnder and M. Neuburger, *Tetrahedron Lett.*, 1996, **37**, 3985–3988.

18. (a) J. Sprinz, J. Helmchen, M. Reggelin, M. Huttner and L. Zsolnai, *Tetrahedron Lett.*, 1993, **34**, 1769–1772; (b) G. J. Dawson, C. G. Frost, J. M. J. Williams and S. J. Coote, *Tetrahedron Lett.*, 1993, **34**, 3149–3150; (c) P. Von Matt and A. Pflatz, *Angew. Chem., Int. Ed. Engl.*, 1993, **32**, 566–568; (d) A. Togni and L. Venanzi, *Angew. Chem., Int. Ed. Engl.*, 1994, **33**, 497–526; (f) A. Y. Meyers and M. Reuman, *Tetrahedron*, 1985, **41**, 837–860.

19. (a) J. V. Allen, J. F. Bower and J. M. J. Williams, *Tetrahedron: Asymmetry*, 1994, **5**, 1895–1898; (b) J. V. Allen, S. J. Coote, G. J. Dawson, C. G. Frost, C. J. Martin and J. M. J. Williams, *J. Chem. Soc., Perkin Trans. I*, 1994, 2065–2072; (c) J. V. Allen, G. J. Dawson, C. G. Frost, J. M. J. Williams and S. J. Coote, *Tetrahedron*, 1994, **50**, 799–808.

20. E. Martin and M. Diéguez, *C. R. Chimie*, 2007, **10**, 188–205.

21. (a) A. Albinati, P. S. Pregosin and K. Wick, *Organometallics*, 1996, **15**, 2419–2421; (b) J. Herrmann, P. S. Pregosin, R. Salzmann and A. Albinati, *Organometallics*, 1995, **14**, 3311–3318; (c) P. Barbaro, A. Currao, J. Herrmann, R. Nesper, P. S. Pregosin and R. Salzman, *Organometallics*, 1996, **15**, 1879–1888; (d) M. Gulea and S. Masson, *Top. Curr. Chem.*, 2003, **229**, 161–198.

22. C. G. Frost and J. M. J. Williams, *Synlett*, 1994, **7**, 551–552.

23. R. Tokunoh, M. Sodeoka, K. Aoe and M. Shibasaki, *Tetrahedron Lett.*, 1995, **36**, 8035–8038.

24. (a) B. M. Trost, K. Dogra, I. Hachiya, T. Emura, D. L. Hughes, S. Krska, R. A. Reamer, M. Palucki, N. Yasuda and P. J. Reider, *Angew. Chem., Int. Ed. Engl.*, 2002, **41**, 1929–1932; (b) B. M. Trost, P. L. Fraisse and Z. T. Ball, *Angew. Chem., Int. Ed. Engl.*, 2002, **41**, 1059–1061; (c) B. Bartels and G. Helmchen, *Chem. Commun.*, 1999, 741–742.

25. A. Albinati, J. Eckert, P. Pregosin, H. Rüegger, R. Slzmann and C. Stössel, *Organometallics*, 1997, **16**, 579–590.

26. E. W. Abel, J. Dormer, D. Ellis, K. G. Orrell, V. Sik, M. B. Hursthouse and M. A. Mazid, *J. Chem. Soc., Dalton Trans.*, 1992, 1073–1080.

27. S. Jansat, M. Gomez, G. Muller, M. Diéguez, A. Aghmiz, C. Claver, A. M. Masdeu-Bulto, L. Flores-Santos, E. Martin, M. A. Maestro and J. Mahia, *Tetrahedron: Asymmetry*, 2001, **12**, 1469–1474.

28. F. Fernandez, M. Gomez, S. Jansat, G. Muller, E. Martin, L. Flores-Santos, P. X. Garcia, A. Acosta, A. Aghmiz, M. Gimenez-Pedros, A. M. Masdeu-Bulto, M. Diéguez, C. Claver and M. A. Maestro, *Organometallics*, 2005, **24**, 3946–3956.

29. Y. Okuyama, H. Nakano, K. Takahashi, H. Hongo and C. Kabuto, *Chem. Commun.*, 2003, 524–525.

30. Y. Okuyama, H. Nakano, Y. Saito, K. Takahashi and H. Hongo, *Tetrahedron: Asymmetry*, 2005, **16**, 2551–2557.

31. N. Khiar, C. S. Araujo, E. Alvarez and I. Fernandez, *Tetrahedron Lett.*, 2003, 44, 3401–3404.

32. N. Khiar, C. S. Araujo, E. Alvarez and I. Fernandez, *Eur. J. Org. Chem.*, 2006, **1**, 1685–1700.
33. N. Khiar, C. S. Araujo, E. Alvarez and I. Fernandez, *Chem. Commun.*, 2004, 714–715.
34. E. Wojaczynska and J. Skarzewski, *Tetrahedron: Asymmetry*, 2008, **19**, 593–597.
35. I. Fernandez, C. S. Araùjo, F. Alcudia and N. Khiar, *Phosphorus, Sulfur, and Silicon*, 2005, **180**, 1509–1510.
36. W. Zhang, Q. Xu and M. Shi, *Tetrahedron: Asymmetry*, 2004, **15**, 3161–3169.
37. J. K. Whitesell, *Chem. Rev.*, 1989, **89**, 1581–1590.
38. D. A. Evans, K. R. Campos, J. S. Tedrow, F. E. Michael and M. R. Gagné, *J. Am. Chem. Soc.*, 2000, **122**, 7905–7920.
39. (a) K. Hiroi, Y. Suzuki and I. Abe, *Chem. Lett.*, 1999, 149–150; (b) Y. Suzuki, I. Abe and K. Hiroi, *Heterocycles*, 1999, **50**, 89–94; (c) K. Hiroi, Y. Suzuki and I. Abe, *Tetrahedron: Asymmetry*, 1999, **10**, 1173–1188; (d) K. Hiroi, Y. Suzuki, I. Abe and R. Kawagishi, *Tetrahedron*, 2000, **56**, 4701–4710.
40. Y.-Y. Yan and T. V. RajanBabu, *Org. Lett.*, 2000, **2**, 199–202.
41. H. Sugama, H. Saito, H. Danjo and T. Imamoto, *Synthesis*, 2001, **15**, 2348–2353.
42. J. Kang, S. H. Yu, J. I. Kim and H. G. Cho, *Bull. Korean Chem. Soc.*, 1995, **16**, 439–443.
43. S. Gladiali, S. Medici, G. Pirri, S. Pulacchini and D. Fabbri, *Can. J. Chem.*, 2001, **79**, 670–678.
44. W. Zhang and M. Shi, *Tetrahedron: Asymmetry*, 2004, **15**, 3467–3476.
45. J. W. Faller, J. C. Wilt and J. Parr, *Org. Lett.*, 2004, **6**, 1301–1304.
46. M. Diéguez, C. Claver and O. Pamies, *Eur. J. Org. Chem.*, 2007, 4621–4634.
47. O. Pàmies, G. P. F. Van Strijdonck, M. Diéguez, S. Deerenberg, G. Net, A. Ruiz, C. Claver, P. C. J. Kamer and P. W. N. M. van Leeuwen, *J. Org. Chem.*, 2001, **66**, 8867–8871.
48. E. Guimet, M. Diéguez, A. Ruiz and C. Claver, *Tetrahedron: Asymmetry*, 2005, **16**, 959–963.
49. M. Diéguez, O. Pamies and C. Claver, *J. Organomet. Chem.*, 2006, **691**, 2257–2262.
50. (a) H. Nakano, J.-i. Yokoyama, R. Fujita and H. Hongo, *Tetrahedron Lett.*, 2002, **43**, 7761–7764; (b) H. Nakano, J.-i. Yokoyama, Y. Okuyama, R. Fujita and H. Hongo, *Tetrahedron: Asymmetry*, 2003, **14**, 2361–2368.
51. (a) H. Nakano, Y. Okuyama and H. Hongo, *Tetrahedron Lett.*, 2000, **41**, 4615–4618; (b) H. Nakano, Y. Okuyama, M. Yanagida and H. Hongo, *J. Org. Chem.*, 2001, **66**, 620–625.
52. H. Nakano, K. Takahashi, Y. Suzuki and R. Fujita, *Tetrahedron: Asymmetry*, 2005, **16**, 609–614.
53. H. Nakano, Y. Okuyama, R.-s. Takahashi and R. Fujita, *Heterocycles*, 2003, **61**, 471–479.

54. N. Khiar, B. Suàrez, M. Stiller, V. Valdivia and I. Fernàndez, *Phosphorus, Sulfur, and Silicon*, 2005, **180**, 1253–1258.
55. N. Khiar, B. Suarez, V. Valdivia and I. Fernandez, *Synlett*, 2005, **19**, 2963–2967.
56. G. A. Molander, J. P. Burke and P. J. Carroll, *J. Org. Chem.*, 2004, **69**, 8062–8069.
57. K. Selvakumar, M. Valentini, P. S. Pregosin and A. Albinati, *Organometallics*, 1999, **18**, 4591–4597.
58. (a) T. Kanayama, K. Yoshida, H. Miyabe, T. Kimachi and Y. Takemoto, *J. Org. Chem.*, 2003, **68**, 6197–6201; (b) T. Kanayama, K. Yoshida, H. Miyabe and Y. Takemoto, *Angew. Chem., Int. Ed. Engl.*, 2003, **42**, 2054–2056.
59. K. Hiroi, Y. Suzuki and R. Kawagishi, *Tetrahedron Lett.*, 1999, **40**, 715–718.
60. K. Hiroi, I. Izawa, T. Takizawa and K.-i. Kawai, *Tetrahedron*, 2004, **60**, 2155–2162.
61. (a) K. Hiroi and Y. Suzuki, *Tetrahedron Lett.*, 1998, **39**, 6499–6502; (b) K. Hiroi, Y. Suzuki, Y. Kaneko, F. Kato, I. Abe and R. Kawagishi, *Polyhedron*, 2000, **19**, 525–528.
62. (a) A. Voituriez, J.-C. Fiaud and E. Schulz, *Tetrahedron Lett.*, 2002, **43**, 4907–4909; (b) A. Voituriez and E. Schulz, *Tetrahedron: Asymmetry*, 2003, **14**, 339–346.
63. Y. Imai, W. Zhang, T. Kida, Y. Nakatsuji and I. Ikeda, *Synlett*, 1999, **8**, 1319–1321.
64. S. Gladiali, G. Loriga, S. Medici and R. Taras, *J. Mol. Catal.*, 2003, **196**, 27–38.
65. X.-L. Hou, X.-W. Wu, L.-X. Dai, B.-X. Cao and J. Sun, *Chem. Commun.*, 2000, 1195–1196.
66. B.-F. Bonini, L. Giordano, M. Fochi, M. Comes-Franchini, L. Bernardi, E. Capito and A. Ricci, *Tetrahedron: Asymmetry*, 2004, **15**, 1043–1051.
67. E. Capito, L. Bernardi, M. Comes-Franchini, F. Fini, M. Fochi, S. Pollicino and A. Ricci, *Tetrahedron: Asymmetry*, 2005, **16**, 3232–3240.
68. A. Chesney and M. R. Bryce, *Tetrahedron: Asymmetry*, 1996, **7**, 3247–3254.
69. G. Chelucci, N. Culeddu, A. Saba and R. Valenti, *Tetrahedron: Asymmetry*, 1999, **10**, 3537–3546.
70. G. Chelucci, D. Muroni, G. A. Pinna, A. Saba and D. Vignola, *J. Mol. Catal.*, 2003, **191**, 1–8.
71. G. Chelucci, A. Bacchi, D. Fabbri, A. Saba and F. Ulgheri, *Tetrahedron Lett.*, 1999, **40**, 553–556.
72. B. Koning, A. Meetsma and R. M. Kellogg, *J. Org. Chem.*, 1998, **63**, 5533–5540.
73. H. Adams, J. C. Anderson, R. Cubbon, D. S. James and J. P. Mathias, *J. Org. Chem.*, 1999, **64**, 8256–8262.
74. (a) A. Saitoh, M. Misawa and T. Morimoto, *Synlett*, 1999, **4**, 483–485; (b) A. Saitoh, K. Achiwa, K. Tanaka and T. Morimoto, *J. Org. Chem.*, 2000, **65**, 4227–4240.

75. R. Tokuda, H. Matsunaga, T. Ishizuka, M. Nakajima and T. Kunieda, *Heterocycles*, 2005, **66**, 135–141.
76. G. A. Rassias, P. C. Bulman Page, S. Reignier and S. D. R. Christie, *Synlett*, 2000, **3**, 379–381.
77. P. C. Bulman Page, H. Heaney, S. Reignier and G. A. Rassias, *Synlett*, 2003, **1**, 22–28.
78. R. Siedlecka, E. Wojaczynska and J. Skarzewski, *Tetrahedron: Asymmetry*, 2004, **15**, 1437–1444.
79. T. Morimoto, K. Tachibana and K. Achiwa, *Synlett*, 1997, **1**, 783–785.
80. E.-K. Lee, S.-H. Kim, B.-H. Jung, W.-S. Ahn and G.-J. Kim, *Tetrahedron Lett.*, 2003, **44**, 1971–1974.
81. S.-H. Kim, E.-K. Lee and G.-J. Kim, *Bull. Korean Chem. Soc.*, 2004, **25**, 754–756.
82. P. H. Schneider, H. S. Schrekker, C. C. Silveira, L. A. Wessjohann and A. L. Braga, *Eur. J. Org. Chem.*, 2004, 2715–2722.
83. A. L. Braga, M. W. Paixao, P. Milani, C. Silveira and O. E. D. Rodrigues, *Synlett*, 2004, **7**, 1297–1299.
84. F. Vargas, J. A. Sehnem, F. Z. Galetto and A. L. Braga, *Tetrahedron*, 2008, **64**, 392–398.
85. (a) K. Hiroi and Y. Suzuki, *Heterocycles*, 1997, **46**, 77–81; (b) K. Hiroi, Y. Suzuki, I. Abe, Y. Hasegawa and K. Suzuki, *Tetrahedron: Asymmetry*, 1998, **9**, 3797–3817.
86. G. J. Meuzelaar, A. S. E. Karlström, M. van Klaveren, E. S. M. Persson, A. del Villar, G. van Koten and J.-E. Bäckvall, *Tetrahedron*, 2000, **56**, 2895–2903.
87. S. J. Roseblade, A. Ros, D. Monge, M. Alcarazo, E. Alvarez, J. M. Lassaletta and R. Fernandez, *Organometallics*, 2007, **26**, 2570–2578.
88. A. Ros, D. Monge, M. Alcarazo, E. Alvarez, J. M. Lassaletta and R. Fernandez, *Organometallics*, 2006, **25**, 6039–6046.
89. L. B. Schenkel and J. A. Ellman, *Org. Lett.*, 2003, **5**, 545–548.
90. (a) S. G. Pyne, *Sulfur Rep.*, 1999, **21**, 281–334; (b) M. Reggellin and C. Zur, *Synthesis*, 2000, 1–64; (c) M. Reggelin, H. Weinberger and V. Spohr, *Adv. Synth. Catal.*, 2004, **346**, 1295–1306.
91. (a) C. Bolm, O. Simic and M. Martin, *Synlett*, 2001, **12**, 1878–1880; (b) C. Bolm, M. Martin, O. Simic and M. Verrucci, *Chemtracts*, 2003, **16**, 660–666.
92. M. T. Reetz, O. G. Bondarev, H.-J. Gais and C. Bolm, *Tetrahedron Lett.*, 2005, **46**, 5643–5646.
93. (a) D. Müller, G. Umbricht, B. Weber and A. Pfaltz, *Helv. Chim. Acta*, 1991, **74**, 232–240; (b) P. von Matt, G. C. Lloyd-Jones, A. B. E. Minidis, A. Pfaltz, L. Macko, M. Neuburger, M. Zehnder, H. Rüegger and P. S. Pregosin, *Helv. Chim. Acta*, 1995, **78**, 265–284; (c) G. Chelucci, S. Deriu, G. A. Pinna, A. Saba and R. Valenti, *Tetrahedron: Asymmetry*, 1999, **10**, 3803–3809.
94. M. Gomez, S. Jansat, G. Muller, M. A. Maestro and J. Mahia, *Organometallics*, 2002, **21**, 1077–1087.

95. L. F. Tietze and J. K. Lohmann, *Synlett*, 2002, 2083–2085.

96. N. End, C. Stoessel, U. Berens, P. di Pietro and P. G. Cozzi, *Tetrahedron: Asymmetry*, 2004, **15**, 2235–2239.

97. I. Abrunhosa, M. Gulea, J. Levillain and S. Masson, *Tetrahedron: Asymmetry*, 2001, **12**, 2851–2859.

98. U. Leutenegger, G. Umbricht, C. Fahrni, P. von Matt and A. Pfaltz, *Tetrahedron*, 1992, **48**, 2143–2156.

99. I. Abrunhosa, L. Delain-Bioton, A.-C. Gaumont, M. Gulea and S. Masson, *Tetrahedron*, 2004, **60**, 9263–9272.

100. B. Fu, D.-M. Du and Q. Xia, *Synthesis*, 2004, 221–226.

101. M. Harmata and S. K. Ghosh, *Org. Lett.*, 2001, **3**, 3321–3323.

102. B. Koning, R. Hulst and R. M. Kellog, *Recl. Trav. Chim. Pays-Bas*, 1996, **115**, 49–55.

103. (a) V. G. Albano, M. Bandini, M. Melucci, M. Monari, F. Piccinelli, S. Tommasi and A. Umani-Ronchi, *Adv. Synth. Catal.*, 2005, **347**, 1507–1512; (b) V. G. Albano, M. Bandini, G. Barbarella, M. Melucci, M. Monari, F. Piccinelli, S. Tommasi and A. Umani-Ronchi, *Chem. Eur. J.*, 2006, **12**, 667–675.

104. M. Bandini, M. Benaglia, T. Quinto, S. Tommasi and A. Umani-Ronchi, *Adv. Synth. Catal.*, 2006, **348**, 1521–1527.

105. E. Robé, W. Perlikowska, C. Lemoine, L. Diab, S. Vincendeau, M. Mikolajczyk, J.-C. Daran and M. Gouygou, *J. Chem. Soc., Dalton Trans.*, 2008, 2894–2898.

106. K. Hiroi, A. Hidaka, R. Sezaki and Y. Imamura, *Chem. Pharm. Bull.*, 1997, **45**, 769–777.

107. (a) A. Alexakis, J. E. Bäckvall, N. Krause, O. Pamies and M. Diéguez, *Chem. Rev.*, 2008, **108**, 2796–2823; (b) C. A. Falciola and A. Alexakis, *Eur. J. Org. Chem.*, 2008, **22**, 3765–3780.

108. S. R. Harutyunyan, T. den Hartog, K. Geurts, A. J. Minnaard and B. L. Feringa, *Chem. Rev.*, 2008, **108**, 2824–2852.

109. S. Ongeri, U. Piarulli, M. Roux, C. Monti and C. Gennari, *Helv. Chim. Acta*, 2002, **85**, 3388–3399.

110. M. C. Willis, *J. Chem. Soc., Perkin Trans. I*, 1999, 1765–1784.

111. U. Piarulli, P. Daubos, C. Claverie, M. Roux and C. Gennari, *Angew. Chem., Int. Ed. Engl.*, 2003, **42**, 234–236.

112. T. Hayashi and K. Yamasaki, *Chem. Rev.*, 2003, **103**, 2829–2844.

113. Y. Lee, K. Akiyama, D. G. Gillingham, M. K. Brown and A. H. Hoveyda, *J. Am. Chem. Soc.*, 2008, **130**, 446–447.

114. (a) R. Gomez, J. A. Arrayas and J. C. Carretero, *Angew. Chem., Int. Ed. Engl.*, 2006, **45**, 7674–7715; (b) L.-X. Dai, T. Tu, S. L. You, W.-P. Deng and X.-L. Hou, *Acc. Chem. Res.*, 2003, **36**, 659–667.

115. (a) T. Colacot, *Chem. Rev.*, 2003, **103**, 3101–3118; (b) L. Dai, T. Tu, S. You, W. Deng and X. Hou, *Acc. Chem. Res.*, 2003, **36**, 659–667.

116. (a) T. Hayashi, in *Ferrocenes*, eds. A. Togni, T. Hayashi, VCH Weinheim, 1995, pp. 105–142; (b) A. Togni, R. L. Haltermann (ed.), in *Metallocenes*, VCH, Weinheim, Germany, 1998.

117. B. Bonini, M. Fochi and A. Ricci, *Synlett*, 2007, **3**, 360–373.

118. J. Park, Z. Quan, S. Lee, K. H. Ahn and C.-W. Cho, *J. Organomet. Chem.*, 1999, **584**, 140–146.

119. S.-L. You, Y.-G. Zhou, X.-L. Hou and L.-X. Dai, *Chem. Commun.*, 1998, 2765–2766.

120. E. Manoury, J. S. Fossey, H. Aït-Haddou, J.-C. Daran and G. G. Balavoine, *Organometallics*, 2000, **19**, 3736–3739.

121. B. F. Bonini, M. Fochi, M. Comes-Franchini, A. Ricci, L. Thijs and B. Zwanenburg, *Tetrahedron: Asymmetry*, 2003, **14**, 3321–3327.

122. A. Chesney, M. R. Bryce, R. W. J. Chubb, A. S. Batsanov and J. A. K. Howard, *Tetrahedron: Asymmetry*, 1997, **8**, 2337–2346.

123. (a) D. Enders, R. Peters, J. Runsink and J. W. Bats, *Org. Lett.*, 1999, **1**, 1863–1866; (b) D. Enders, R. Peters, R. Lochtman, G. Raabe, J. Runsink and J. W. Bats, *Eur. J. Org. Chem.*, 2000, 3399–3426.

124. T. Tu, Y.-G. Zhou, X.-L. Hou, L.-X. Dai, X.-C. Dong, Y.-H. Yu and J. Sun, *Organometallics*, 2003, **22**, 1255–1265.

125. (a) J. Priego, O. Garcia Mancheno, S. Cabrera, R. Gomez Arrayas, T. Llamas and J. C. Carretero, *Chem. Commun.*, 2002, 2512–2513; (b) O. Garcia Mancheno, J. Priego, S. Cabrera, R. Gomez Arrayas, T. Llamas and J. C. Carretero, *J. Org. Chem.*, 2003, **68**, 3679–3686.

126. L. Routaboul, S. Vincendeau, J.-C. Daran and E. Manoury, *Tetrahedron: Asymmetry*, 2005, **16**, 2685–2690.

127. B. Martin-Matute, S. I. Pereira, E. Pena-Cabrera, J. Adrio, A. M. S. Silva and J. C. Carretero, *Adv. Synth. Catal.*, 2007, **349**, 1714–1724.

128. J. Kang, J. H. Lee and K. S. Im, *J. Mol. Catal.*, 2003, **196**, 55–63.

129. (a) L. Bernardi, B. F. Bonini, M. Comes-Franchini, G. Dessole, M. Fochi and A. Ricci, *Phosphorus, Sulfur, and Silicon*, 2005, **180**, 1273–1277; (b) X. Hu, C. Bai, H. Dai, H. Chen and Z. Zheng, *J. Mol. Catal.*, 2004, **218**, 107–112.

130. L. Bernardi, B. F. Bonini, M. Comes-Franchini, M. Fochi, G. Mazzanti, A. Ricci and G. Varchi, *Eur. J. Org. Chem.*, 2002, 2776–2784.

131. S. Nakamura, T. Fukuzumi and T. Toru, *Chirality*, 2004, **16**, 10–12.

132. J. C. Anderson and J. Osborne, *Tetrahedron: Asymmetry*, 2005, **16**, 931–934.

133. F. L. Lam, T. T.-L. Au-Yeung, H. Y. Cheung, S. H. L. Kok, W. S. Lam, K. Y. Wong and A. S. C. Chan, *Tetrahedron: Asymmetry*, 2006, **17**, 497–499.

134. (a) K. W. Bentley, *Nat. Prod. Rep.*, 2004, **21**, 395–424; (b) D. J. Faulkner, *Nat. Prod. Rep.*, 2002, **19**, 1–48; (c) J. E. Saxton, *Nat. Prod. Rep.*, 1997, **14**, 559–590.

135. H. Y. Cheung, W.-Y. Yu, F. L. Lam, T. T.-L. Au-Yeung, Z. Zhou, T. H. Chan and S. C. Chan, *Org. Lett.*, 2007, **9**, 4295–4298.

136. F. L. Lam, T. T.-L. Au-Yeung, F. K. Kwong, Z. Zhou, K. Y. Wong and A. S. C. Chan, *Angew. Chem., Int. Ed. Engl.*, 2008, **47**, 1280–1283.

137. (a) G. R. Newkome C. N. Moorefield and F. Vögtle, in *Dendrimers and Dendrons: Concepts, Synthesis, Perspectives*, VCH, Weinheim,

2001; (b) J. M. J. Fréchet and D. A. Tomalia, in *Dendrimers and Other Dendritic Polymers*, Wiley, Chichester, 2001; (c) H. F. Chow, T. K. K. Mong, M. F. Nongrum and C. W. Wan, *Tetrahedron*, 1998, **54**, 8543–8660; (d) D. A. Tomalia, *High Perform. Polym.*, 2001, **13**, S1–S10.

138. D. Astruc and F. Chardac, *Chem. Rev.*, 2001, **101**, 2991–3024.
139. H. Brunner, *J. Organomet. Chem.*, 1995, **500**, 39–46.
140. L. Routaboul, S. Vincendeau, C.-O. Turrin, A.-M. Caminade, J.-P. Majoral, J.-C. Daran and E. Manoury, *J. Organomet. Chem.*, 2007, **692**, 1064–1073.
141. (a) A. S. E. Karlström, F. F. Huerta, G. J. Meuzelaar and J.-E. Bäckvall, *Synlett*, 2001, 923–926; (b) H. K. Cotton, J. Norinder and J.-E. Bäckvall, *Tetrahedron*, 2006, **62**, 5632–5640.

CHAPTER 2
Conjugate Addition

2.1 Introduction

The enantioselective metal-catalysed 1,4-conjugate addition of organometallic reagents to linear or cyclic enones is an important procedure for making C–C bonds.[1] Despite recent successes in the area of asymmetric copper-catalysed 1,4-organometallic additions to Michael acceptors, there is an ongoing quest for improved systems showing very high enantioselectivities. One problematic substrate class for this reaction is linear aliphatic enones. The high conformational mobility of these species, together with the presence of only subtle substrate–catalyst steric interactions, makes the design of effective enantioselective systems a real challenge. The accepted mechanism for this reaction involves the activation of the enone with the metal reagent through the carbonyl.[2] Since conformational exchange can take place between the *syn/anti s-cis* and *s-trans* forms of the activated enone (Scheme 2.1), linear enones usually provide lower enantioselectivities than cyclic enones.

  The use of a transition metal (traditionally copper[3] or nickel) is needed to avoid the 1,2-addition on the carbonyl group of the substrate. The sulfur-containing ligands have been generally less studied than their phosphorus counterparts. A prominent position in this rapidly expanding field is occupied by the copper-catalysed and chiral-ligand-accelerated conjugate addition of organozinc reagents.[4] This latter process has attracted considerable attention, because of the mild reaction conditions and the high functional-group tolerance that the zinc reagents offer.[5] The enantioselective version of the 1,4-addition of Grignard reagents[6] to α,β-unsaturated carbonyl compounds has been less studied in recent years, but extensively developed in the past, using different mono- and bidentate chiral ligands such as monodentate thiosugars,[7] heterobidentate S/N ligands,[8] oxazoline arenethiolates,[9] S/O ligands,[10] and S/P ligands.[11] Other chiral S/N-donor ligands derived from L-proline and

RSC Catalysis Series No. 2
Chiral Sulfur Ligands: Asymmetric Catalysis
By Hélène Pellissier
© Hélène Pellissier 2009
Published by the Royal Society of Chemistry, www.rsc.org

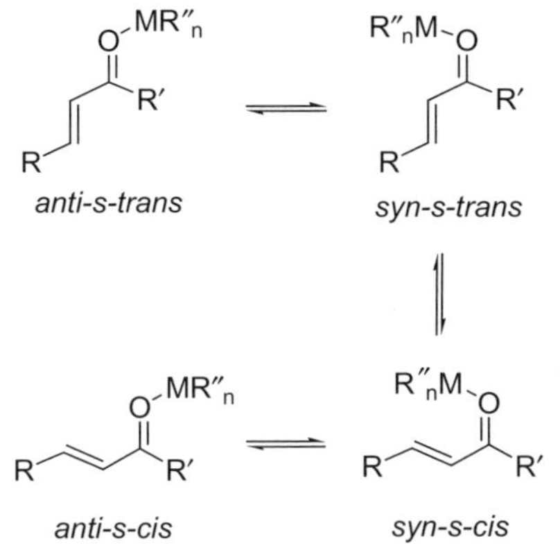

Scheme 2.1 *Syn/anti s-cis and s-trans* forms of activated enones.

(*S*)-phenylglycine have been used in the copper-catalysed conjugate addition of MeLi to enones.[12] Trialkylaluminum reagents were also involved in only a few reactions, but these represent an interesting alternative.

2.2 Copper Catalysts

Most of the successful asymmetric versions of the Cu-catalysed conjugate addition reaction have made use of organozinc reagents, especially ZnEt$_2$, a trend started by Alexakis in 1993.[13] The inherently low reactivity of organozinc reagents towards unsaturated carbonyl compounds has facilitated the development of a plethora of chiral phosphorus-based ligands capable of providing highly efficient ligand-accelerated catalysis with excellent enantioselectivities over a broad range of substrates. Therefore, viable ligand classes affording >95% enantiomeric excess for the addition of ZnR$_2$ to disubstituted cyclic or acyclic enones, lactones or lactams, nitro-olefins, amides, and malonates are now available. Trialkylaluminum reagents have recently appeared as an interesting alternative to organozinc reagents since they are also readily available. Additionally, due to their higher reactivity, they allow Cu-catalysed 1,4-addition to very challenging substrates, such as β,β′-disubstituted enones, which are inert to organozinc methodologies. Although Grignard reagents were the first species to be used, their higher reactivity leads to uncatalysed 1,2- and 1,4-additions, which limited their early applications. However, there have been recent breakthroughs that opened up the use of Grignard reagents for the highly active and enantioselective Cu-catalysed conjugate addition to a wide range of substrates. Most of the chiral sulfur-containing ligands have been employed in the Cu-catalysed conjugate addition. On the other hand, and despite the vast knowledge on sulfur–metal interactions in coordination chemistry,[14] the use of chiral BINOL ligands containing sulfur atoms in asymmetric catalysis appears to be still rather undeveloped. In this context, various chiral 1,1′-binaphthalene-derived ligands have been implicated in asymmetric conjugate additions of organometallic reagents. As an example, Woodward *et al.* have developed copper-catalysed asymmetric conjugate additions of various organometallic reagents (ZnR_2 and AlR_3) to linear enones in the presence of sulfur-containing 1,1′-binaphthyl-based ligands.[10,15] Thiourethane and thioether 1,1′-binaphthyl-based ligands were effective for the copper-catalysed 1,4-addition of $ZnEt_2$ or $AlMe_3$ to *trans*-alkyl-3-en-2-ones, yielding the corresponding products with enantioselectivities of up to 77% ee (Scheme 2.2). In comparison, the 1,4-addition of $ZnEt_2$ to a cyclic enone such as 2-2-cyclohexenone proceeded with an enantioselectivity of up to 77% ee by using the same ligand family. Ligand optimisation studies have indicated that 3,3′-dimethylthio-1,1′-binaphthalene-2,2′-diol was the most efficient with respect to enantioselectivity.[16] In addition, the variation of the enone structure has revealed that the catalyst derived from the (*S*)-binaphthalene-derived ligand caused linear enones to adopt an *s-cis* conformation with a zinc-derived Lewis acid bound to the carbonyl lone pair *anti* to the ene function (Scheme 2.2). A probable transition state is given in Scheme 2.2, taking into account that the nature of the steric block was the apparent cause of the enantioselectivity. The involvement of the corresponding unsubstituted mono- and dithio-1,1′-binaphthol ligands gave disappointing results since only a modest enantioselectivity of 36% ee combined with 66–84% yield was obtained

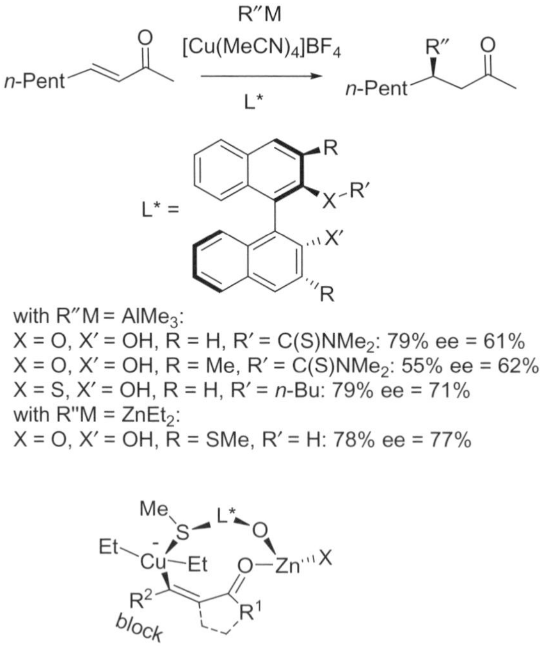
R″M
[Cu(MeCN)4]BF4
L*
n-Pent
R″
n-Pent
L* =
R
X-R′
X′
R
with R″M = AlMe3:
X = O, X′ = OH, R = H, R′ = C(S)NMe2: 79% ee = 61%
X = O, X′ = OH, R = Me, R′ = C(S)NMe2: 55% ee = 62%
X = S, X′ = OH, R = H, R′ = n-Bu: 79% ee = 71%
with R"M = ZnEt2:
X = O, X′ = OH, R = SMe, R′ = H: 78% ee = 77%
Me
L*
S
O
Et
Cu
Et
O
Zn
X
R2
R1
block

Scheme 2.2 Cu-catalysed 1,4-additions of $ZnEt_2$ and $AlMe_3$ to (*E*)-3-nonen-2-one with 1,1′-binaphthalene-derived ligands.

for the copper-catalysed addition of ZnEt$_2$ to 2-cyclohexenone by using monothiobinaphthol ligand, whereas an enantioselectivity of 55% ee combined with 30% yield was obtained in the copper-catalysed addition of *n*-BuMgCl to 2-cyclohexenone by using dithiobinaphthol as a ligand.

In 2003, the same group reported the synthesis of a series of other non-substituted 1,1'-binaphthyl thioether ligands in order to investigate these ligands in the copper-catalysed addition of AlMe$_3$ to linear aliphatic enones.[17] In this study, the authors proved that the presence of methylalumoxane, arising from hydrolysis of AlMe$_3$, reduced the enantioselectivity of the conjugate product. As shown in Scheme 2.3, the ligand optimisation studies have indicated that *n*-butylthioether ligand was the most efficient ligand with respect to enantioselectivity (86% ee) for the addition of AlMe$_3$ to (*E*)-3-nonen-2-one. These conditions were extended to many other acyclic enones with enantioselectivities of 76–93% ee (Scheme 2.3). It must be noted that the use of these ligands in the addition to benzylideneacetone or to enones bearing a conjugate vinyl function or nitroalkenes failed. Moreover, the additions of AlEt$_3$ were unsuccessful, which suggested to these authors that the structure of the active catalyst incorporated an AlR$_3$-dependent substrate binding pocket. In addition, these authors have demonstrated the need to have a directing group in the ligand backbone, such as a thioether, by preparing the corresponding *n*-butyl ether ligand that provided poor yield (<5%) and enantioselectivity (< 5% ee) for the product arising from the addition of AlMe$_3$ to (*E*)-3-nonen-2-one (Scheme 2.3). Moreover, the involvement of the corresponding octahydro-1,1'-binaphthyl thioether ligand (Scheme 2.3) in the same reaction provided the expected product in a comparable yield than its corresponding aromatic binaphthyl analogue but with a lower enantioselectivity (63% ee).[18]

In 2003, Kang *et al.* reported the synthesis of other BINOL-based thioether ligands for the enantioselective 1,4-addition of ZnEt$_2$ to 2-cyclohexenone.[19] These sulfide ligands, depicted in Scheme 2.4, were shown to give better enantioselectivities of up to 85% ee than the dithioacetals, even if the dithioacetal having a long spacer between the BINOL core and the Cu-binding site gave an enantioselectivity of 56% ee. The use of these sulfide ligands was extended to other enones to give enantioselectivities of up to 96% ee when chalcone was used as the substrate (Scheme 2.4). It was worth noting that the BINOL ligand itself gave an enantioselectivity of 7% ee with 2-cyclohexenone as the substrate, whereas the use of the monosubstituted BINOL sulfide depicted in Scheme 4 gave the 1,4-product in 74% ee.

In 2004, excellent results were reported by Shi *et al.*, dealing with the enantioselective conjugate addition of ZnEt$_2$ to enones catalysed by Cu(I) and axially chiral binaphthylthiophosphoramides as ligands, which were readily available, quite stable, recoverable and reusable (Scheme 2.5).[20] These ligands were derived from (*R,R*)-1,2-diaminocyclohexane, (*S,S*)-1,2-diphenylethylene-diamine and BINAM.[21] Their system allowed an efficient and highly enantioselective functionalisation of not only six- and seven-membered cyclic enones (97% ee), but also cyclopentenone (98% ee) and acyclic enones (up to 97% ee). The mechanism of the reaction was investigated, confirming that this series of

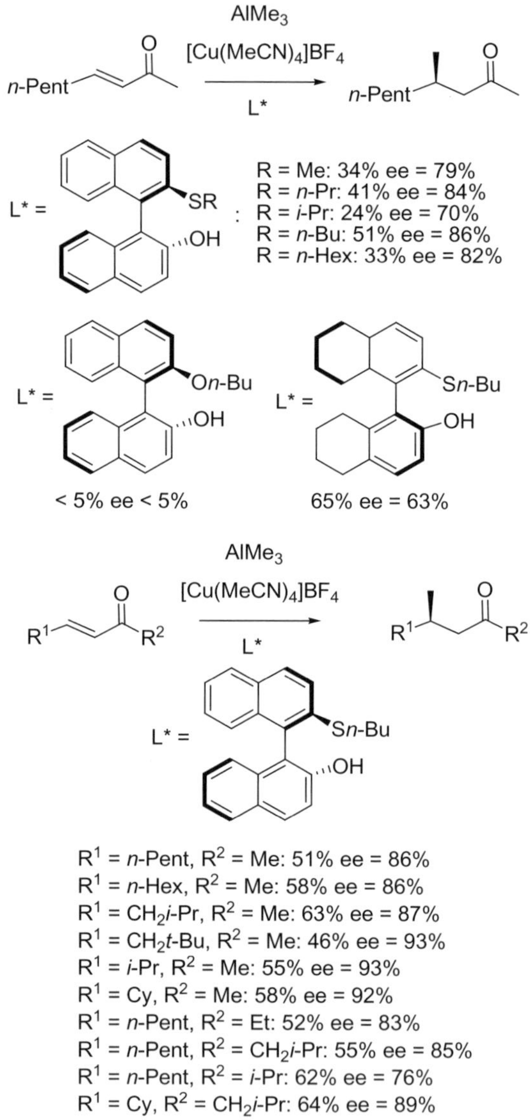

Scheme 2.3 Cu-catalysed 1,4-additions of AlMe$_3$ to linear aliphatic enones with 1,1′-binaphthalene-derived ligands.

chiral phosphoramides was a novel type of S/N-bidentate ligands through ^{31}P and ^{13}C NMR spectroscopic experiments. The scope of the process was extended to the addition of dimethyl- and diphenylzinc reagents with good enantioselectivities. The mechanism was postulated to be a bimetallic catalytic process in which the acidic proton of thiophosphoramide in the ligands played a significant role in the formation of the active species (Scheme 2.5). The linear

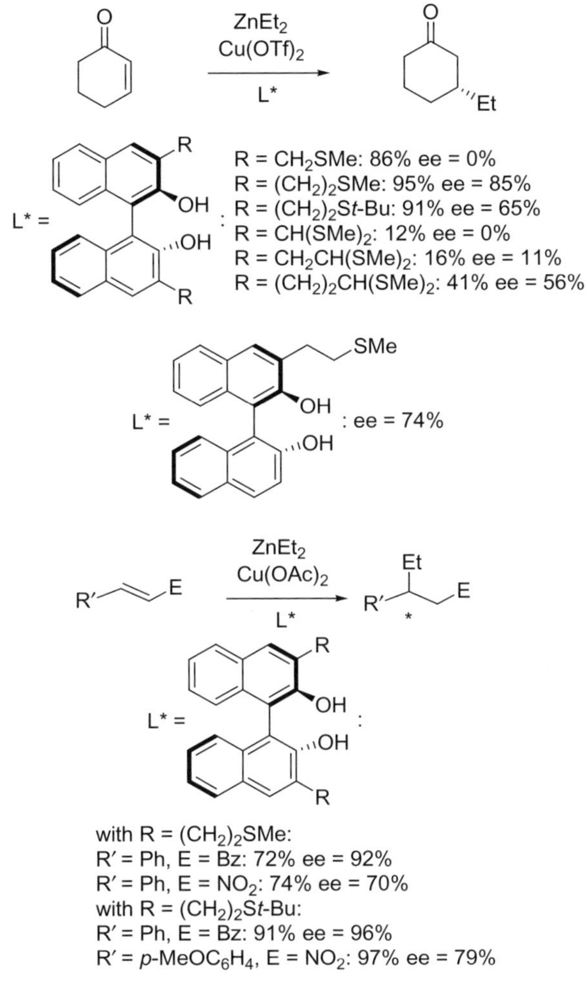

R = CH$_2$SMe: 86% ee = 0%
R = (CH$_2$)$_2$SMe: 95% ee = 85%
R = (CH$_2$)$_2$St-Bu: 91% ee = 65%
R = CH(SMe)$_2$: 12% ee = 0%
R = CH$_2$CH(SMe)$_2$: 16% ee = 11%
R = (CH$_2$)$_2$CH(SMe)$_2$: 41% ee = 56%

L* = : ee = 74%

with R = (CH$_2$)$_2$SMe:
R' = Ph, E = Bz: 72% ee = 92%
R' = Ph, E = NO$_2$: 74% ee = 70%
with R = (CH$_2$)$_2$St-Bu:
R' = Ph, E = Bz: 91% ee = 96%
R' = p-MeOC$_6$H$_4$, E = NO$_2$: 97% ee = 79%

Scheme 2.4 Cu-catalysed 1,4-additions of ZnEt$_2$ to enones and nitroalkenes with BINOL-based thioether ligands.

effect of the product enantioselectivity and the ligand enantioselectivity further revealed that the active species was a monomeric Cu(I) complex bearing a single ligand. In order to check the importance of the sulfur atom in the catalytic conjugate addition, these authors have prepared the corresponding diphenyl-phosphoramide ligand (Scheme 2.5). In the reaction of ZnEt$_2$ to 2-cyclohex-enone, the product was isolated with 64% yield but in its racemic form, whereas an enantioselectivity of 93% ee combined with 94% yield were obtained by using the corresponding sulfur-containing ligand. Due to the fact that the oxygen atom is a harder coordinating atom than the sulfur atom, it cannot smoothly coordinate to a softer metal centre, such as Cu(I), according to the

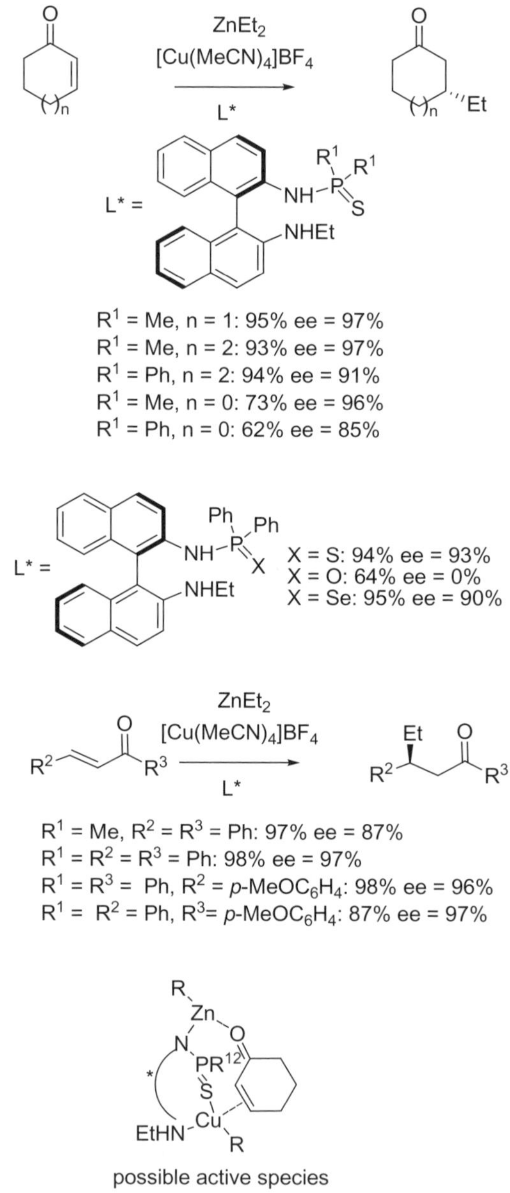
ZnEt2
[Cu(MeCN)4]BF4
L*
Et
L* =
NH
NHEt
R1 = Me, n = 1: 95% ee = 97%
R1 = Me, n = 2: 93% ee = 97%
R1 = Ph, n = 2: 94% ee = 91%
R1 = Me, n = 0: 73% ee = 96%
R1 = Ph, n = 0: 62% ee = 85%
L* =
Ph
Ph
X = S: 94% ee = 93%
X = O: 64% ee = 0%
X = Se: 95% ee = 90%
ZnEt2
[Cu(MeCN)4]BF4
L*
R1 = Me, R2 = R3 = Ph: 97% ee = 87%
R1 = R2 = R3 = Ph: 98% ee = 97%
R1 = R3 = Ph, R2 = p-MeOC6H4: 98% ee = 96%
R1 = R2 = Ph, R3= p-MeOC6H4: 87% ee = 97%
EtHN
possible active species

Scheme 2.5 Cu-catalysed 1,4-additions of $ZnEt_2$ to enones with binaphthylthiophosphoramide ligands.

hard–soft-acid–base theory.[22] In this context, the authors have prepared the corresponding selenium-containing ligand, as depicted in Scheme 2.5, and found according to this theory that this ligand afforded 95% yield combined with 90% ee for the reaction of $ZnEt_2$ to 2-cyclohexenone.

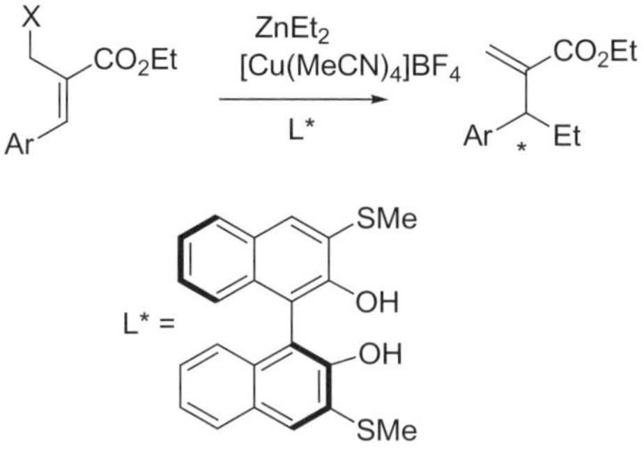

X = Br, Ar = Ph: 44% ee = 13%
X = Cl, Ar = Ph: 19% ee = 34%
X = Cl, Ar = p-$O_2NC_6H_4$: 64% ee = 64%
X = Cl, Ar = p-Tol: 19% ee = 30%
X = Cl, Ar = 1-Naph: 14% ee = 22%

Scheme 2.6 Cu-catalysed additions of $ZnEt_2$ to Baylis–Hillman-derived allylic electrophiles with BINOL-based thioether ligand.

In addition, Woodward *et al.* have described the asymmetric chemo- and regiospecific copper-catalysed addition of organozinc reagents to Baylis–Hillman-derived allylic electrophiles using sulfur-containing 1,1′-binaphthyl-based ligands, as depicted in Scheme 2.6.[23] Although this reaction giving an enantioselectivity of up to 64% ee was actually an $S_N2′$ reaction of the organometallic reagent, it was decided, however, to include it in this section. The results showed that chloride proved to be a better leaving group than bromide with respect to the enantioselectivity.

In 2008, Li *et al.* reported the synthesis of a novel class of chiral modified BINOL ligands in order to explore their catalytic ability for the enantioselective conjugate addition of $ZnEt_2$ to enones.[24] These BINOL-based ligands were bound with both sulfur-contained heterocycle (thiazole or thiadiazole) and thioether block in which the sulfur might serve as a talent anchor. As shown in Scheme 2.7, good yields but poor enantioselectivities ($\leq$ 33% ee) were obtained for the conjugate addition of $ZnEt_2$ to both cyclic and acyclic enones performed in the presence of $Cu(OTf)_2$. With chalcone derivatives, the results showed that the electron-withdrawing substituent of the chalcone obviously increased both the enantioselectivity and the yield, while the electron-donating group slightly decreased both the enantioselectivity and the yield.

On the other hand, alkoxy- and aminothiol derivatives of TADDOL have been applied as ligands by Seebach *et al.* in the enantioselective conjugate addition of *n*-butyl Grignard reagent to cycloheptenone, providing the expected 1,4-product with a good enantioselectivity of 84% ee, as shown in Scheme 2.8.[25] With the same absolute configuration of the starting ligand, the stereochemical course of the conjugate addition was reversed according to the nature of the TADDOL substituent. According to X-ray analyses of copper complexes of these ligands, the Cu-catalysed 1,4-addition proceeded via a tetranuclear

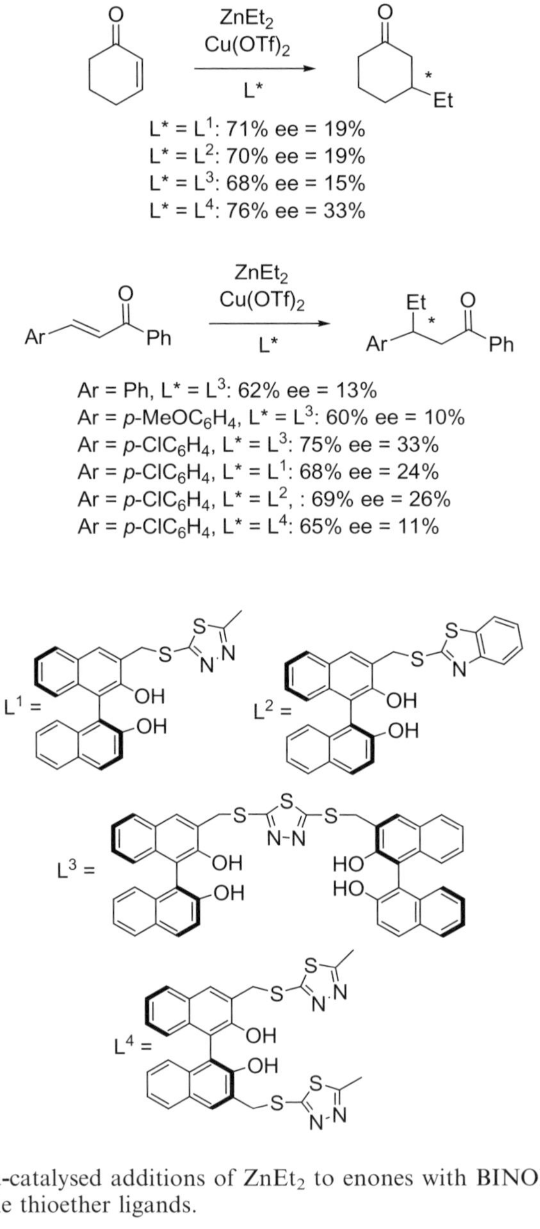

L* = L¹: 71% ee = 19%
L* = L²: 70% ee = 19%
L* = L³: 68% ee = 15%
L* = L⁴: 76% ee = 33%

Ar = Ph, L* = L³: 62% ee = 13%
Ar = p-MeOC₆H₄, L* = L³: 60% ee = 10%
Ar = p-ClC₆H₄, L* = L³: 75% ee = 33%
Ar = p-ClC₆H₄, L* = L¹: 68% ee = 24%
Ar = p-ClC₆H₄, L* = L², : 69% ee = 26%
Ar = p-ClC₆H₄, L* = L⁴: 65% ee = 11%

Scheme 2.7   Cu-catalysed additions of ZnEt₂ to enones with BINOL-based (di)thia-
zole thioether ligands.

[Cu₄S₄] species in which the TADDOL-derived ligand was an unexpected sul-
fur-monodentate and not a S/X-bidentate ligand. This hypothesis was confirmed by ¹H NOE NMR spectroscopy that suggested a different structure for
the Cu complex of the alcohol ligand relative to the methylether and the
dimethylamino ligands, which could explain the observed stereochemical
inversion.

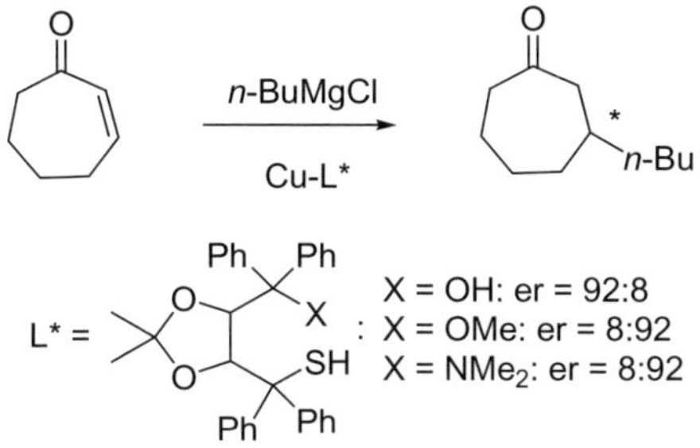

Scheme 2.8 Cu-catalysed addition of *n*-BuMgCl to cycloheptenone with TADDOL-derived ligands.

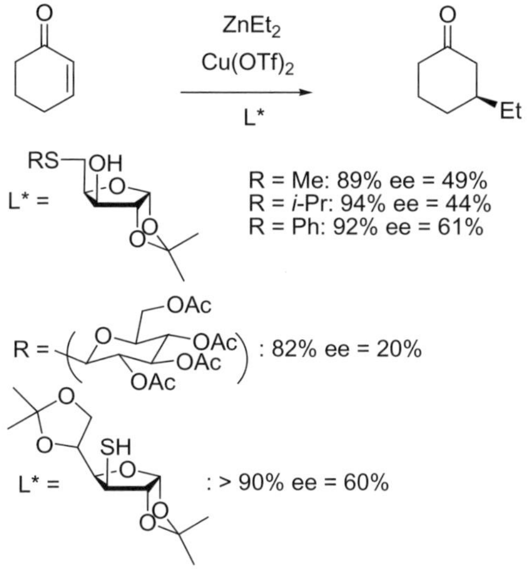

Scheme 2.9 Cu-catalysed 1,4-addition of ZnEt$_2$ to 2-cyclohexenone with thioether-alcohol ligands bearing xylofuranose backbone.

Several chiral ligands derived from sugars have been involved in asymmetric conjugate additions of organometallic reagents. Thus, the copper-catalysed conjugate addition of ZnEt$_2$ to 2-cyclohexenone could be performed by Pamies *et al.* in the presence of thioether-alcohol ligands bearing a xylofuranose backbone.[26] This was the first example of thioether-based catalysts for the copper-catalysed asymmetric conjugate addition of organometallics to enones. These ligands, readily prepared from the inexpensive (D)-xylose, provided the corresponding Michael adducts in moderate enantioselectivities (≤62% ee, Scheme 2.9). The condensation of AlMe$_3$ onto (*E*)-non-3-en-2-one was less efficient, since the enantioselectivity was of only 34% ee. In 1993, Spescha and Rihs reported the enantioselective copper-catalysed 1,4-addition of Grignard reagents to 2-cyclohexenone with another thiosugar derived ligand, depicted in

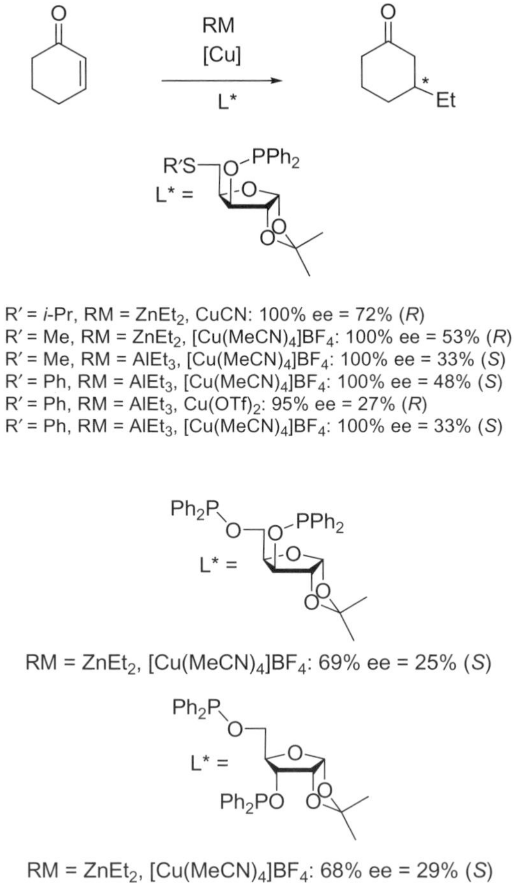
RM
[Cu]
L*
Et
R′S
PPh2
L* =
R′ = i-Pr, RM = ZnEt2, CuCN: 100% ee = 72% (R)
R′ = Me, RM = ZnEt2, [Cu(MeCN)4]BF4: 100% ee = 53% (R)
R′ = Me, RM = AlEt3, [Cu(MeCN)4]BF4: 100% ee = 33% (S)
R′ = Ph, RM = AlEt3, [Cu(MeCN)4]BF4: 100% ee = 48% (S)
R′ = Ph, RM = AlEt3, Cu(OTf)2: 95% ee = 27% (R)
R′ = Ph, RM = AlEt3, [Cu(MeCN)4]BF4: 100% ee = 33% (S)
Ph2P
PPh2
L* =
RM = ZnEt2, [Cu(MeCN)4]BF4: 69% ee = 25% (S)
Ph2P
L* =
Ph2PO
RM = ZnEt2, [Cu(MeCN)4]BF4: 68% ee = 29% (S)

Scheme 2.10 Cu-catalysed 1,4-additions of $ZnEt_2$ and $AlEt_3$ to 2-cyclohexenone with thioether-phosphinite ligands and diphosphinite ligands bearing xylofuranose backbone.

Scheme 2.9.[7] These authors observed both excellent yield and regioselectivity in the formation of the 1,4-addition product, with an enantioselectivity of up to 60% ee, strongly dependent on the reaction conditions.

In 2005, Diéguez *et al.* extended this methodology to corresponding thioether-phosphinite and -diphosphinite ligands derived from D-xylose for the copper-catalysed 1,4-addition of $ZnEt_2$ or $AlEt_3$ to 2-cyclohexenone, obtaining moderate to good enantioselectivities ($\leq$72% ee).[27] Systematically varying the functional groups at the C5 position (thioether and phosphinite) and different substituents in the thioether moieties had a strong effect on the rate and enantioselectivity (Scheme 2.10). The highest enantioselectivity was achieved with the catalyst precursor containing the thioether-phosphinite ligand that had an isopropyl substituent in the thioether moiety. The activity was, however, generally best with the diphosphinite ligands.

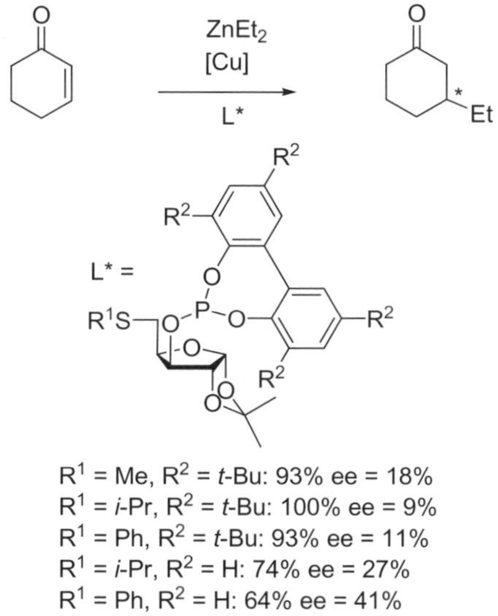

R^1 = Me, R^2 = t-Bu: 93% ee = 18%
R^1 = i-Pr, R^2 = t-Bu: 100% ee = 9%
R^1 = Ph, R^2 = t-Bu: 93% ee = 11%
R^1 = i-Pr, R^2 = H: 74% ee = 27%
R^1 = Ph, R^2 = H: 64% ee = 41%

Scheme 2.11 Cu-catalysed 1,4-addition of ZnEt$_2$ to 2-cyclohexenone with thioether-phosphite ligands bearing xylofuranose backbone.

The 1,4-addition of ZnEt$_2$ to 2-cyclohexenone was also performed by these workers in the presence of thioether-phosphite D-xylose-derived ligands, depicted in Scheme 2.11.[28] In all cases, the chemoselectivities in the 1,4-product were higher than 97% but the enantioselectivities were modest ($\leq$41% ee). It was noted that changing the substituent in the thioether moiety produced an effect on both the reactivity and the enantioselectivity, as shown in Scheme 2.11.

Oxazolines are known to have several advantages as sources of chirality, the main one being that they are readily accessible from homochiral amino alcohols, and have proved to be effective catalysts in a variety of reactions.[29,30] Furthermore, these ligands are easily modifiable and can incorporate different donor atoms in the side chains of the heterocyclic ring.[31] In this context, a series of mercaptoaryloxazoline ligands were synthesised by Pfaltz *et al.* and further investigated as ligands for the enantioselective copper-catalysed conjugate addition of Grignard reagents to cyclic enones.[9,32] These results showed that using the ligand bearing an *i*-propyl group on the oxazoline moiety allowed the enantioselectivities to be increased with the size of the cyclic enone. Indeed, enantioselectivities ranging from 16 to 37% ee were obtained for cyclopentenone, whereas enantioselectivities of 83–87% ee were obtained for cycloheptenone (Scheme 2.12). Attempts to extend the scope of this process to acyclic enones failed since lower enantioselectivities (<20% ee) were observed. In this study, a copper(I) thiolate complex was isolated in an analytically pure form and the authors proposed a trimeric Cu complex containing a six-membered chair-like (Cu$_3$S$_3$) ring.[33]

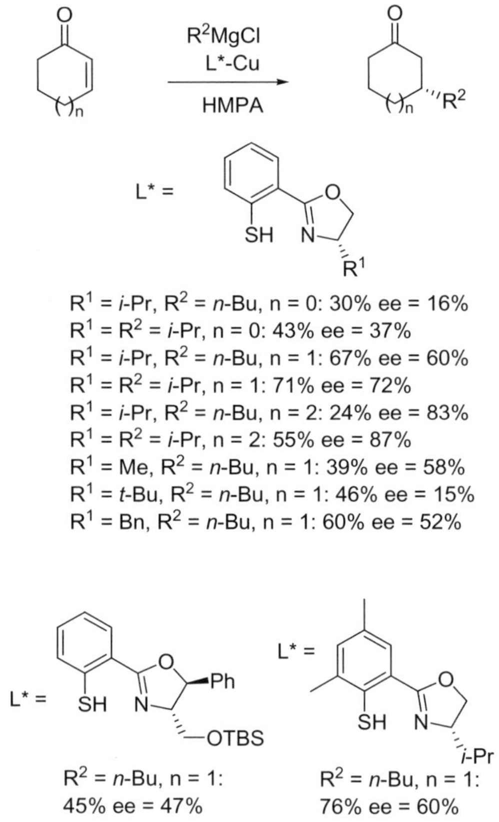
R2MgCl
L*-Cu
HMPA
R2
L* =
SH
N
O
R1
R1 = i-Pr, R2 = n-Bu, n = 0: 30% ee = 16%
R1 = R2 = i-Pr, n = 0: 43% ee = 37%
R1 = i-Pr, R2 = n-Bu, n = 1: 67% ee = 60%
R1 = R2 = i-Pr, n = 1: 71% ee = 72%
R1 = i-Pr, R2 = n-Bu, n = 2: 24% ee = 83%
R1 = R2 = i-Pr, n = 2: 55% ee = 87%
R1 = Me, R2 = n-Bu, n = 1: 39% ee = 58%
R1 = t-Bu, R2 = n-Bu, n = 1: 46% ee = 15%
R1 = Bn, R2 = n-Bu, n = 1: 60% ee = 52%
Ph
OTBS
i-Pr
R2 = n-Bu, n = 1:
45% ee = 47%
R2 = n-Bu, n = 1:
76% ee = 60%

Scheme 2.12 Cu-catalysed 1,4-additions of RMgCl to cyclic enones with mercaptoaryloxazoline ligands.

In 1996, analogous thioether ligands were studied by Iwata *et al.* as ligands for the enantioselective Cu-catalysed addition of $AlMe_3$ to a 4,4-disubstituted cyclohexa-2,5-dienone.[34] An enantioselectivity of 63% ee was obtained for the 1,4-product in each ligand, as shown in Scheme 2.13, and by using *tert*-butyldimethylsilyltriflate as an additive. No effect on the enantioselectivity was observed when replacing the thiophenyl group of the ligand by a thiomethyl group.

A new application of bis(oxazolines) ligands was reported by Reiser *et al.* who obtained some excellent results, such as that depicted in Scheme 2.14 for the 1,4-addition of $ZnEt_2$ to 2-cyclohexenone.[35] These authors used a bimetallic complex in which the substrate was locked in a two-point binding mode via a zinc atom and a copper atom.

In 2004, Ricci *et al.* reported the synthesis of several novel ligands based on an oxazoline-cyclopenta[*b*]thiophene backbone, as depicted in Scheme 2.15.[36] Moderate yields combined with enantioselectivities of up to 79% ee were obtained for the 1,4-product resulting from the copper-catalysed enantioselective

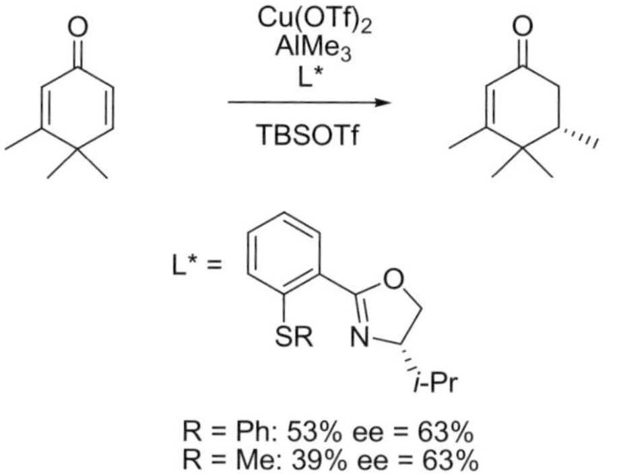

R = Ph: 53% ee = 63%
R = Me: 39% ee = 63%

Scheme 2.13 Cu-catalysed 1,4-addition of $AlMe_3$ to 4,4-disubstituted cyclohexa-2,5-dienone with thioether ligands.

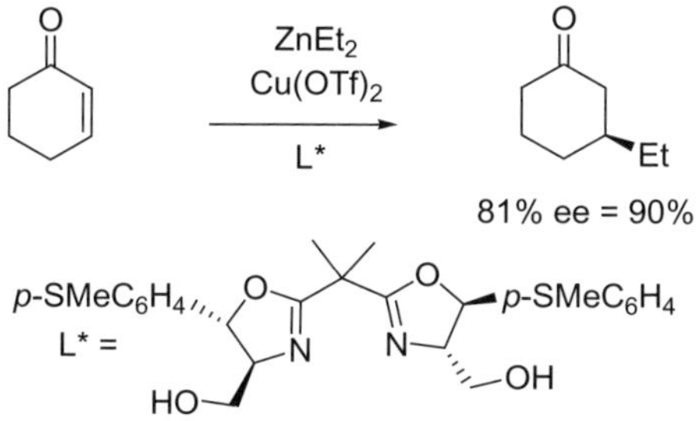

81% ee = 90%

Scheme 2.14 Cu-catalysed 1,4-addition of $ZnEt_2$ to 2-cyclohexenone with thioether-bis(oxazolines) ligand.

addition of $ZnEt_2$ to chalcone. The importance of the sulfur atom was demonstrated by the involvement in the same reaction of the ligand derived from *cis*-aminoindanol, which gave lower results in term of enantioselectivity than the corresponding cyclopenta[*b*]thiophene derived ligand (Scheme 2.15). Moreover, this result could also be explained by the fact that the electron-rich thiophene ring gave a higher electron density to the oxazoline nitrogen. In this study, the authors demonstrated that these ligands behaved as monodentate ligands by DFT-computational and ESI-mass spectral investigations. Indeed, the nitrogen atom of the oxazoline coordinated the copper, whereas the sulfur atom of the thiophene moiety did not seem to have any close contact to the metal.

Whereas the mono- and the S/S-dithioether moieties have been used to date, the 1,3-dithianyl motif was used for the first time in 2005 by Ricci *et al.* as a new hybrid ligand in asymmetric catalysis. Hence, a series of new chiral oxazoline-1,3-dithianes have been successfully applied to the copper-catalysed conjugate addition of $ZnEt_2$ to enones (Scheme 2.16).[37] The expected products were obtained in almost quantitative yields and enantioselectivities of up to 69% ee.

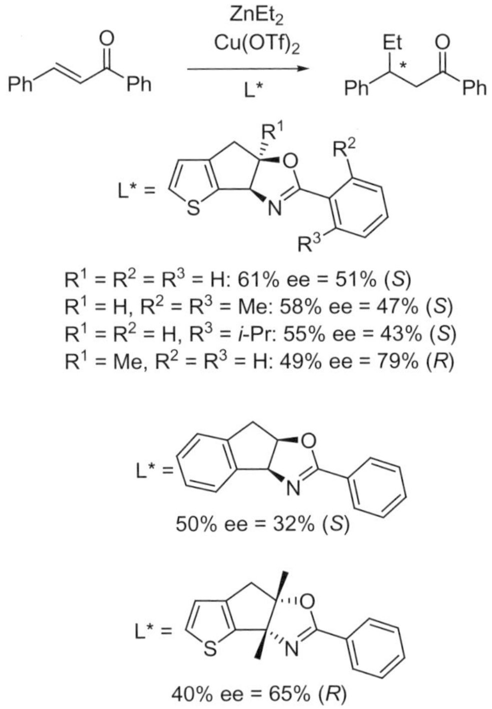
ZnEt2
Cu(OTf)2
L*
Ph
Ph
Et
O
Ph
Ph
L* =
R1
R2
R3
O
N
S
R1 = R2 = R3 = H: 61% ee = 51% (S)
R1 = H, R2 = R3 = Me: 58% ee = 47% (S)
R1 = R2 = H, R3 = i-Pr: 55% ee = 43% (S)
R1 = Me, R2 = R3 = H: 49% ee = 79% (R)
L* =
50% ee = 32% (S)
L* =
40% ee = 65% (R)

Scheme 2.15 Cu-catalysed 1,4-addition of $ZnEt_2$ to chalcone with oxazoline-cyclopenta[*b*]thiophene ligands.

The asymmetric induction appeared to be closely related to the steric hindrance exerted by the group adjacent to the oxazoline nitrogen. The conformation of the ligand has been explored using a combination of X-ray and NMR measurements, indicating the presence of a remarkable anomeric effect that accounted for the preference of the oxazoline ring for the axial location. Moreover, for a 2/1 ligand/metal ratio, the ESI-MASS analysis indicated that four of the eight ligand sites, most likely the nitrogen of each oxazoline and one of the thienyl sulfur atoms, were bound to the copper atom, indicating that these ligands behaved as bidentate S/N ligands.

Another type of sulfur-containing ligands, such as β-amino sulfide ligands, was developed by Gibson *et al.*, in 1996.[38] These ligands derived from L-proline or (*S*)-phenylglycine were used for the enantioselective conjugate addition of MeLi to 2-cyclohexenone, providing the corresponding 1,4-product with an enantioselectivity of up to 64% ee (Scheme 2.17). An X-ray analysis of a dimeric complex derived from a ligand of the same type (Scheme 2.17) and copper(I) iodide showed a C_2-symmetry and confirmed the S/N-coordination.

An arenethiolatocopper (I) complex has been used by van Koten *et al.* to catalyse the 1,4-Michael addition of Grignard reagents to acyclic enones, providing the corresponding products with excellent chemoselectivities, high

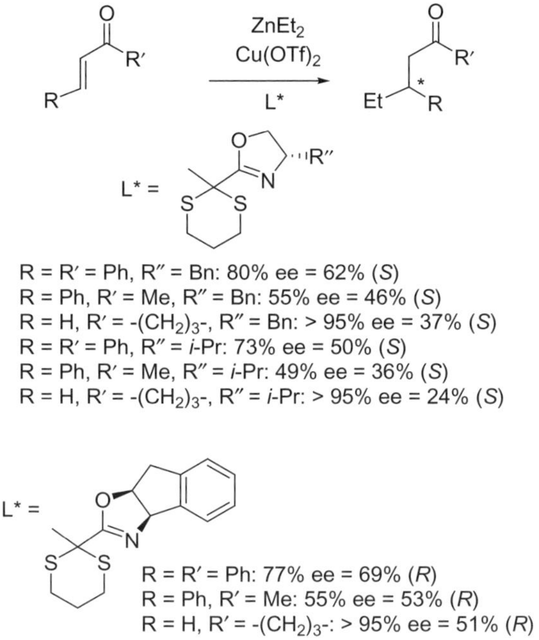
ZnEt2
Cu(OTf)2
L*
R′
R
Et
L* =
R″
R = R′ = Ph, R″ = Bn: 80% ee = 62% (S)
R = Ph, R′ = Me, R″ = Bn: 55% ee = 46% (S)
R = H, R′ = -(CH2)3-, R″ = Bn: > 95% ee = 37% (S)
R = R′ = Ph, R″ = i-Pr: 73% ee = 50% (S)
R = Ph, R′ = Me, R″ = i-Pr: 49% ee = 36% (S)
R = H, R′ = -(CH2)3-, R″ = i-Pr: > 95% ee = 24% (S)
L* =
R = R′ = Ph: 77% ee = 69% (R)
R = Ph, R′ = Me: 55% ee = 53% (R)
R = H, R′ = -(CH2)3-: > 95% ee = 51% (R)

Scheme 2.16 Cu-catalysed 1,4-additions of $ZnEt_2$ to enones with chiral oxazoline-1,3-dithiane ligands.

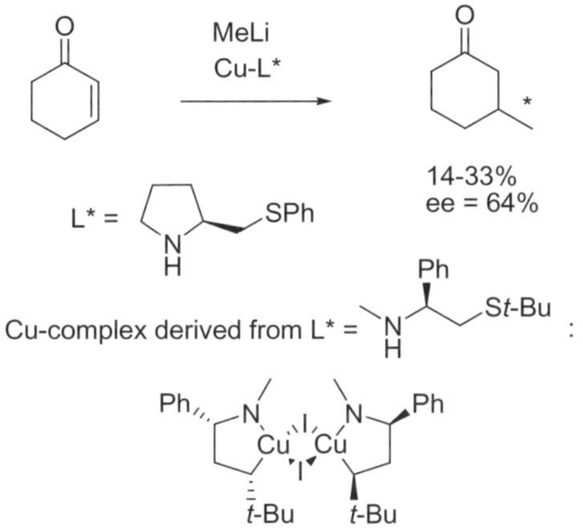
MeLi
Cu-L*
L* =
SPh
14-33%
ee = 64%
Cu-complex derived from L* =
St-Bu
Ph
t-Bu
t-Bu

Scheme 2.17 Cu-catalysed 1,4-addition of MeLi to 2-cyclohexenone with chiral β-amino sulfide ligand.

yields and good enantioselectivities of up to 76% ee (Scheme 2.18).[8a,b] These results were explained by the formation of the intermediate depicted in Scheme 2.18 in which the double bond was coordinated to copper and the oxygen atom to magnesium. In-depth NMR and crystallographic studies proved that the

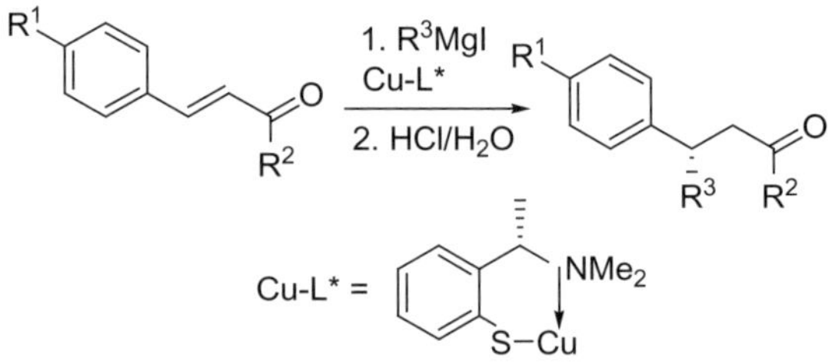

R^1 = H, R^2 = R^3 = Me: 97% ee = 76%
R^1 = H, R^2 = Me, R^3 = *n*-Bu: > 95% ee = 45%
R^1 = H, R^2 = Me, R^3 = *i*-Pr: > 85% ee = 10%
R^1 = CN, R^2 = R^3 = Me: 20% ee = 13%
R^1 = Cl, R^2 = R^3 = Me: > 99% ee = 64%
R^1 = OMe, R^2 = R^3 = Me: > 99% ee = 56%
R^1 = H, R^2 = *i*-Pr, R^3 = Me: 98% ee = 72%
R^1 = H, R^2 = *t*-Bu, R^3 = Me: > 99% ee = 45%
R^1 = H, R^2 = Ph, R^3 = Me: > 99% ee = 0%

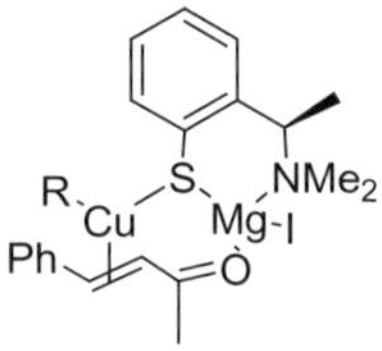

possible intermediate:

Scheme 2.18   Cu-catalysed 1,4-additions of RMgI to acyclic enones with arenethio-
          latocopper (I) complex.

arenethiolatocopper (I) complex existed both in the solid state and in solution
as a well-defined neutral trinuclear aggregate [CuSAr*]$_3$.[33,39]

   In 2004, the use of this chiral catalyst was extended by Alexakis *et al.* to
conjugate addition reactions of Grignard reagents or ZnEt$_2$ to aliphatic and
cyclic enones.[40] Thus, the diethylzinc addition to 2-cyclohexenone performed in
the presence of this catalyst led to the expected product in quantitative yield
and 83% ee. Closely related catalysts, depicted in Scheme 2.19, were also
investigated for the MeMgI addition to acyclic enones, providing the corresponding products with enantioselectivities of up to 70% ee, whereas the corresponding additions of ZnEt$_2$ to acyclic enones gave lower enantioselectivities
(≤37% ee). In contrast, the use of ZnEt$_2$ was preferred for cyclic enones,
whereas Grignard reagents led to better results in the presence of acyclic
enones.

   Another class of new ligands was prepared in quantitative yields by Feringa
et al., in 1997, by reaction between α-mercapto acids, aniline and 2-pyridine-
carboxaldehyde.[41] These pyridyl-substituted thiazolin-4-one ligands were further
involved in the copper-catalysed conjugate addition of ZnEt$_2$ to 2-cyclohexenone,

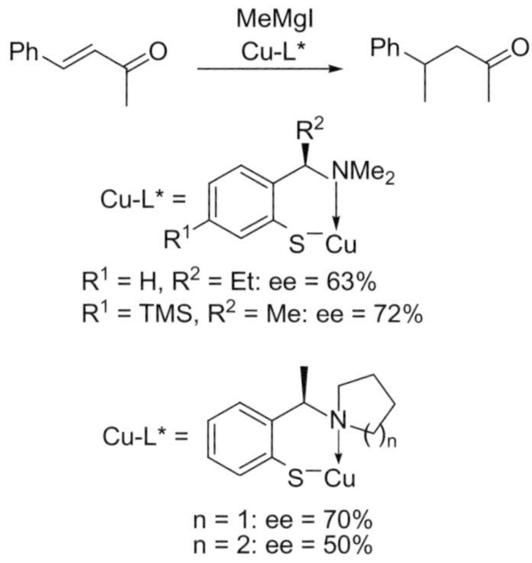

Scheme 2.19 Cu-catalysed 1,4-addition of MeMgI to acyclic enone with arenethio-latocopper (I) complexes.

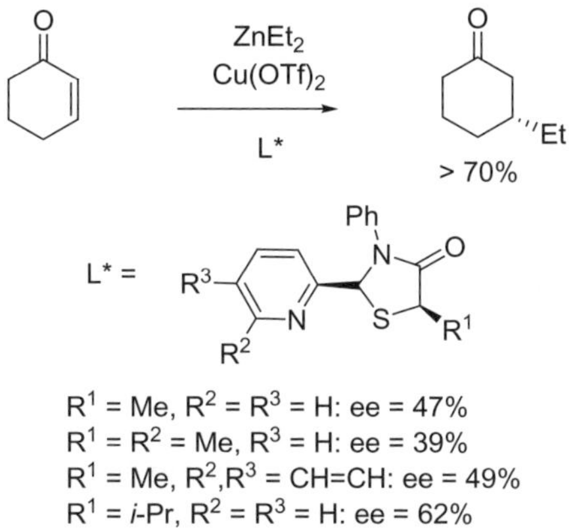

Scheme 2.20 Cu-catalysed 1,4-addition of ZnEt$_2$ to 2-cyclohexenone with pyridyl-substituted thiazolin-4-one ligands.

giving the expected product in good yields and enantioselectivities of up to 62% ee, as shown in Scheme 2.20. However, when the reaction conditions were applied to a substrate such as chalcone, the enantioselectivity of the corresponding product was low ($\leq 11\%$ ee) whereas both yield and regioselectivity remained excellent.

In order to prepare a new range of ligands incorporating both sulfoximine and phosphine moieties, Tye *et al.* reported, in 2001, the synthesis of a novel chiral phosphine-sulfoximine, which was tested as a ligand in the copper-catalysed 1,4-addition of ZnEt$_2$ to enones.[42] Whilst the reaction of acyclic enones gave the corresponding racemic 1,4-products, the best result was obtained upon

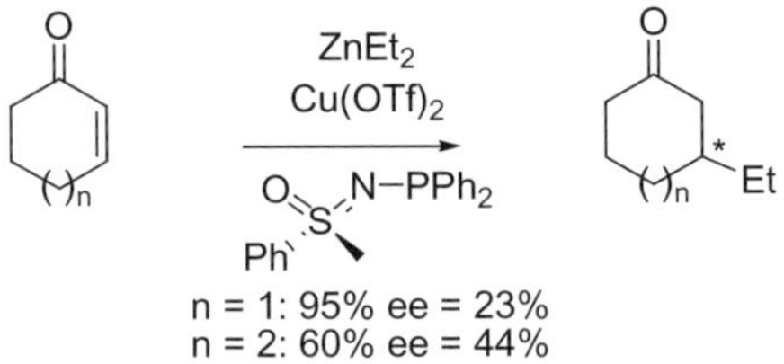
ZnEt2
Cu(OTf)2
N–PPh2
Ph
Et
n = 1: 95% ee = 23%
n = 2: 60% ee = 44%

Scheme 2.21 Cu-catalysed 1,4-additions of $ZnEt_2$ to cyclic enones with phosphine-sulfoximine ligand.

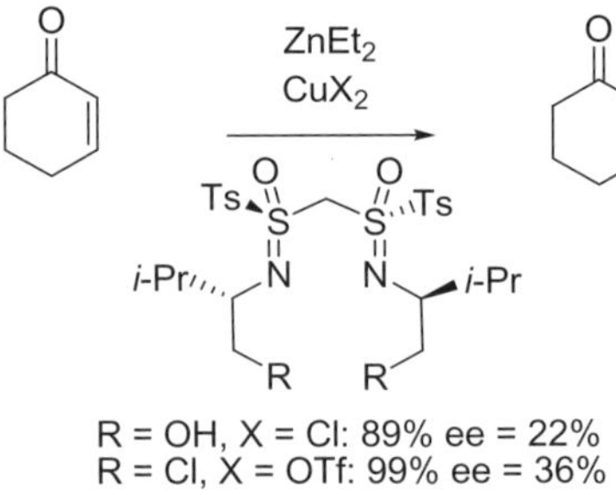
ZnEt2
CuX2
Ts
Ts
i-Pr
i-Pr
R
R
Et
R = OH, X = Cl: 89% ee = 22%
R = Cl, X = OTf: 99% ee = 36%

Scheme 2.22 Cu-catalysed 1,4-addition of $ZnEt_2$ to 2-cyclohexenone with C_2-symmetric geminal bis(sulfoximines) ligands.

reaction of cycloheptenone (44% ee), demonstrating that the ligand performed best with unhindered cyclic substrates (Scheme 2.21).

Chiral sulfoximines[43] and, in particular, chiral bidentate β-hydroxysulfoximines, are one of the most successful chiral sulfur-based ligands used in asymmetric catalysis. The X-ray analyses of complexes of ethylzinc[44] and vanadium[45] coordinated to β-hydroxysulfoximines have shown that the metal was coordinated to the hydroxy oxygen and the sulfoximine nitrogen atom. In this context, Reggelin *et al.* have developed the synthesis of a number of new C_2-symmetric geminal bis(sulfoximines), which were further investigated for the copper-catalysed 1,4-addition of $ZnEt_2$ to 2-cyclohexenone.[46] All of the ligand-copper complexes were found to be highly active catalysts but the enantioselectivities were rather disappointing with 36% ee being the best value (Scheme 2.22). Based on the observed results, it was difficult to analyse the potential role of the ligand oxygen atoms as a possible coordination site for zinc.

The first asymmetric conjugate addition of $ZnEt_2$ to aryl- and alkylidene malonates by using catalytic copper in the presence of chiral phosphorus ligands was reported by Alexakis *et al.* in 2001.[47] Among these ligands, a chiral phenylphosphorus ferrocenyl ligand bearing a benzothioether gave a better enantioselectivity than the corresponding phenylphosphorus ferrocenyl ligand without the sulfur group (57% ee, Scheme 2.23).

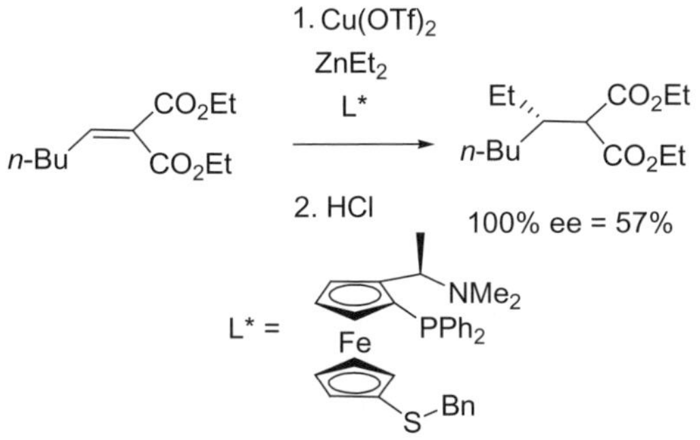

Scheme 2.23 Cu-catalysed 1,4-addition of ZnEt$_2$ to ethyl pentylidene malonate with sulfur-containing phenylphosphorus ferrocenyl ligand.

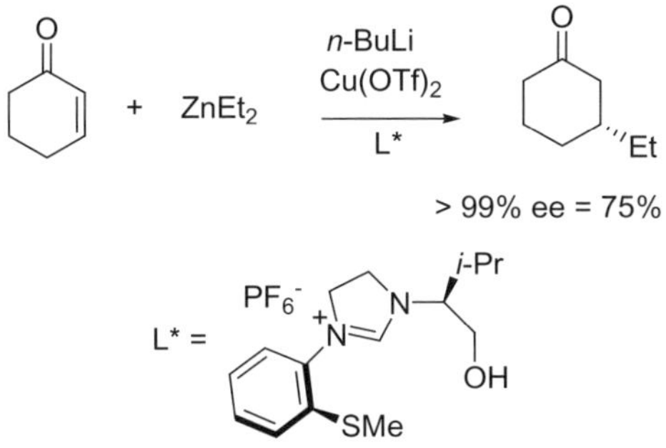

Scheme 2.24 Cu-catalysed 1,4-addition of ZnEt$_2$ to cyclohexenone with alkoxy-*N*-heterocyclic carbene ligand.

In 2005, Mauduit *et al.* developed a new class of chiral alkoxy-*N*-heterocyclic carbene ligands readily accessible in a five-step procedure from commercially available β-aminoalcohols.[48] The copper catalyst, *in situ* generated from Cu(OTf)$_2$, *n*-BuLi and the chiral alkoxy-*N*-heterocyclic carbene-imidazolinium salt, catalysed the conjugate addition of ZnEt$_2$ to cyclohexenone providing the corresponding ketone in a quantitative yield and with an enantioselectivity of 75% ee (Scheme 2.24). It was shown that the desymmetrization of the aromatic unit of the ligand with a chelating thioalkyl ether group in the *ortho* position improved the enantioselectivity of this reaction since it increased from 54% ee without the thiomethyl group to 75% ee with the thiomethyl group.

On the other hand, chiral sulfur-containing but noncoordinating ligands such as sulfonamides have been widely used in the asymmetric Michael reaction. In 1997, Sewald *et al.* reported the use of a series of chiral sulfonamides depicted in Scheme 2.25 in the Cu-catalysed conjugate addition of ZnEt$_2$ to 2-cyclohexenone.[49] Even the use of a stoichiometric amount of catalyst did not allow the enantioselectivity to be higher than 31% ee.

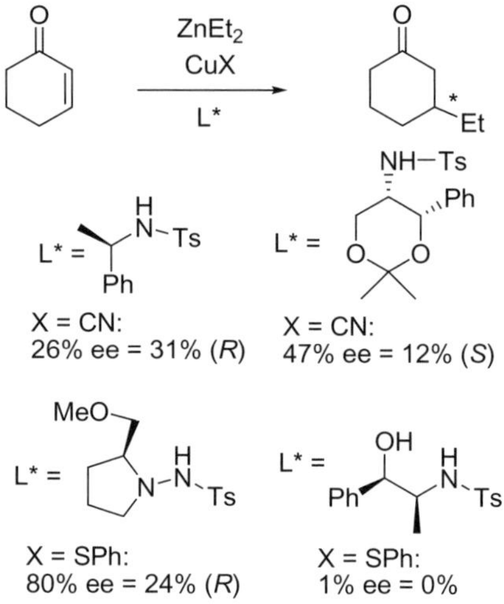

Scheme 2.25 Cu-catalysed 1,4-addition of ZnEt$_2$ to 2-cyclohexenone with sulfonamides.

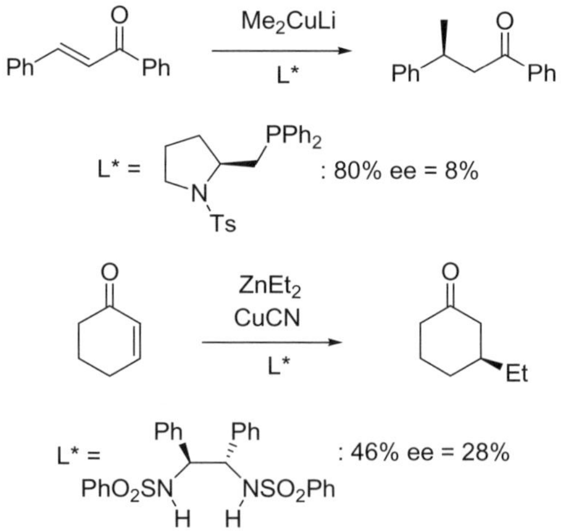

Scheme 2.26 Cu-catalysed 1,4-additions of Me$_2$CuLi and ZnEt$_2$ to enones with sulfonamide or bis(sulfonamides) ligands.

In the same context, Tomioka *et al.* have studied the addition of Me$_2$CuLi to chalcone in the presence of another chiral sulfonamide, depicted in Scheme 2.26, which provided the corresponding 1,4-product in a good yield but with a low enantioselectivity (8% ee).[50] In addition, these authors have obtained both moderate yield and enantioselectivity (≤28% ee) for the addition of ZnEt$_2$ to 2-cyclohexenone performed in the presence of a chiral bis(sulfonamides) ligand depicted in Scheme 2.26.[51]

In 2004, excellent enantioselectivities of up to 98% ee were obtained by Morimoto *et al.* by using a phosphine-sulfonamide-containing 1,1′-binaphthyl-based ligand in the enantioselective copper-catalysed conjugate addition of $ZnEt_2$ to several benzylideneacetones (Scheme 2.27).[52] Similar levels of enantioselectivity (up to 97% ee) combined with excellent yields (up to 90%) were obtained by Leighton *et al.* for the copper-catalysed enantioselective addition of various alkylzincs to cyclic enones performed in the presence of other chiral phosphine-sulfonamide ligands (Scheme 2.27).[53]

In 2000, Gennari *et al.* discovered a new family of chiral Schiff-base ligands, with the general structure, *N*-alkyl-β-(*N*′-salicylideneamino)alkanesulfonamide, depicted in Scheme 2.28.[54] These ligands were successfully implicated in the copper-catalysed conjugate addition of $ZnEt_2$ to cyclic enones (Scheme 2.28) and, less efficiently, to acyclic enones such as benzalacetone (50% ee) or chalcone

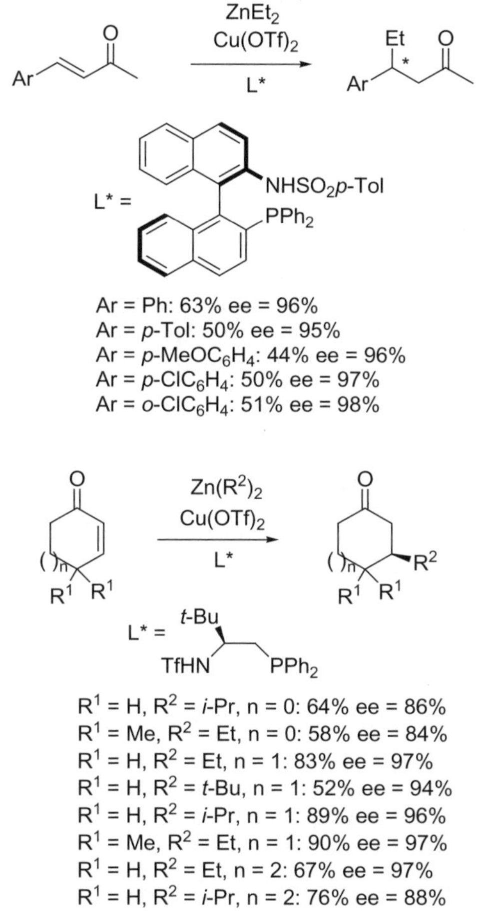
ZnEt2
Cu(OTf)2
L*
Ar
Et
O
L* =
NHSO2p-Tol
PPh2
Ar = Ph: 63% ee = 96%
Ar = p-Tol: 50% ee = 95%
Ar = p-MeOC6H4: 44% ee = 96%
Ar = p-ClC6H4: 50% ee = 97%
Ar = o-ClC6H4: 51% ee = 98%
Zn(R2)2
Cu(OTf)2
R1 R1
R2
t-Bu
TfHN
PPh2
R1 = H, R2 = i-Pr, n = 0: 64% ee = 86%
R1 = Me, R2 = Et, n = 0: 58% ee = 84%
R1 = H, R2 = Et, n = 1: 83% ee = 97%
R1 = H, R2 = t-Bu, n = 1: 52% ee = 94%
R1 = H, R2 = i-Pr, n = 1: 89% ee = 96%
R1 = Me, R2 = Et, n = 1: 90% ee = 97%
R1 = H, R2 = Et, n = 2: 67% ee = 97%
R1 = H, R2 = i-Pr, n = 2: 76% ee = 88%

Scheme 2.27 Cu-catalysed 1,4-additions of ZnR_2 to enones with phosphine-sulfonamide ligands.

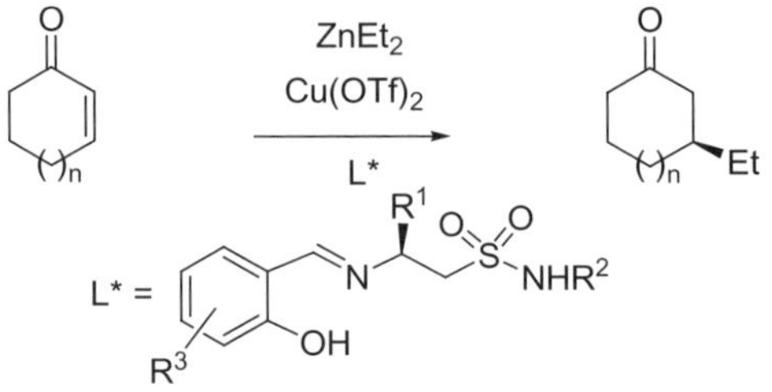

R^1 = *i*-Pr, R^2 = (*S*)-CH(Me)Cy, R^3 = 3,5-Cl$_2$, n = 1: ee = 90%
R^1 = *i*-Pr, R^2 = (*S*)-CH(Me)Cy, R^3 = 3,5-Cl$_2$, n = 2: ee = 81%
R^1 = *t*-Bu, R^2 = (*S*)-CH(Me)Cy, R^3 = 3,5-*t*-Bu$_2$, n = 1: ee = 80%
R^1 = *t*-Bu, R^2 = (*S*)-CH(Me)Cy, R^3 = 3,5-*t*-Bu$_2$, n = 2: ee = 79%
R^1 = R^2 = *i*-Pr, R^3 = 3,5-(*t*-Bu)$_2$, n = 1: ee = 76%
R^1 = R^2 = *i*-Pr, R^3 = 3,5-(*t*-Bu)$_2$, n = 2: ee = 72%
R^1 = *t*-Bu, R^2 = CHPh$_2$, R^3 = 3,5-(*t*-Bu)$_2$, n = 1: ee = 74%
R^1 = *t*-Bu, R^2 = CHPh$_2$, R^3 = 3,5-(*t*-Bu)$_2$, n = 2: ee = 75%
R^1 = *i*-Bu, R^2 = (*S*)-CH(Me)Cy, R^3 = 3,5-Cl$_2$, n = 1: ee = 73%
R^1 = *i*-Bu, R^2 = (*S*)-CH(Me)Cy, R^3 = 3,5-Cl$_2$, n = 2: ee = 74%
R^1 = *t*-Bu, R^2 = CHPh$_2$, R^3 = 3,5-(*t*-Bu)$_2$, n = 0: ee = 70%
R^1 = *i*-Bu, R^2 = (*S*)-CH(Me)Cy, R^3 = 3,5-(*t*-Bu)$_2$, n = 0: ee = 80%

Scheme 2.28   Cu-catalysed 1,4-additions of ZnEt$_2$ to cyclic enones with Schiff-base ligands.

(34% ee). These easily available ligands contained a set of different metal-binding sites such as a phenol, an imine and a secondary sulfonamide moiety. In addition, the scope of this methodology could be extended to the enantioselective addition of ZnEt$_2$ to nitroolefins, providing enantioselectivities of up to 58% ee when (*E*)-(2-nitrovinyl)benzene was used as the substrate.[55]

2.3   Other Metal Catalysts

Nickel catalysts can be used instead of copper catalysts to promote the conjugate addition reaction, providing, in some cases, better results than the corresponding copper catalysts. In 2000, Yang *et al.* discovered a series of (1*R*,2*S*,3*R*)-3-mercaptocamphan-2-ol derivatives,[56] which proved to be efficient ligands in the conjugate addition of ZnEt$_2$ to chalcone upon catalysis with Ni(acac)$_2$ (Scheme 2.29).

   This reaction has also been studied by Gibson *et al.* in the presence of a chiral β-amino disulfide, giving the 1,4-product in a good yield but with a moderate enantioselectivity (50% ee) as shown in Scheme 2.30.[8c] These authors have prepared the corresponding thiolate by treatment of this β-amino disulfide with butyllithium in THF. This thiolate was further tested for the same enantioselective conjugate addition reaction, providing a lower enantioselectivity of 37% ee (Scheme 2.30). In order to gain insight into the role played by the sulfur atom, these authors have also prepared the corresponding β-amino alcohol, which gave an enantioselectivity of only 6% ee when implicated as the ligand in

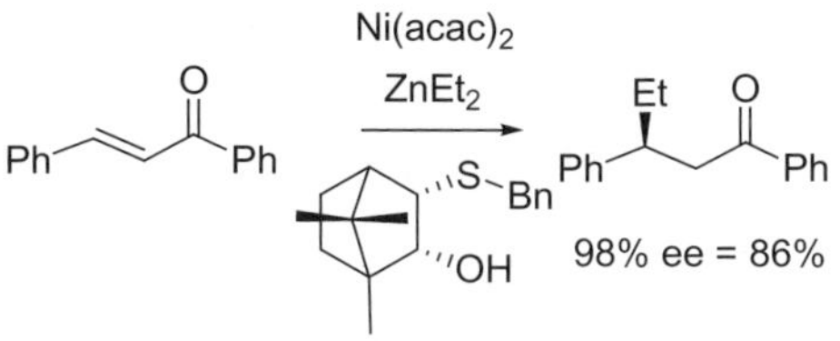

Scheme 2.29 Ni-catalysed 1,4-addition of ZnEt$_2$ to chalcone with (1R,2S,3R)-3-mercaptocamphan-2-ol-derived ligand.

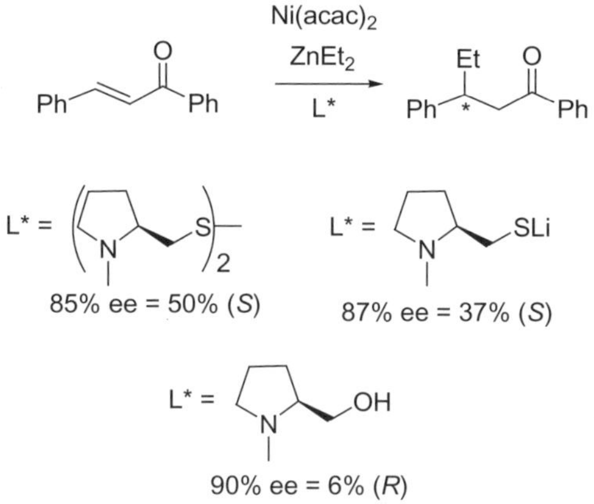

Scheme 2.30 Ni-catalysed 1,4-addition of ZnEt$_2$ to chalcone with β-amino disulfide, β-amino thiolate or β-amino alcohol ligands.

the same reaction (Scheme 2.30). This result clearly demonstrated the importance of the sulfur-chelation site for this reaction.

In the same context, Kang *et al.* have examined several bidentate thiol derivatives as ligands in the same reaction to that described above.[57] A nickel complex derived from the β-amino thiol depicted in Scheme 2.31 catalysed the reaction in a useful asymmetric level of 74% ee, whereas the use of the thiol phosphine ligand depicted in Scheme 2.31 gave a lower enantioselectivity of 27% ee. In this study, the authors found that the enantioselectivity was strongly dependent on the ligand concentration and on the nickel-to-ligand ratio.

On the other hand, the enantioselective 1,4-addition of carbanions such as enolates to linear enones is an interesting challenge, since relatively few efficient methods exist for these transformations. The Michael reaction of β-dicarbonyl compounds with α,β-unsaturated ketones can be catalysed by a number of transition-metal compounds. The asymmetric version of this reaction has been performed using chiral diol, diamine, and diphosphine ligands. In the past few years, bidentate and polydentate thioethers have begun to be considered as chiral ligands for this reaction. As an example, Christoffers *et al.* have developed the synthesis of several S/O-bidentate and S/O/S-tridentate thioether

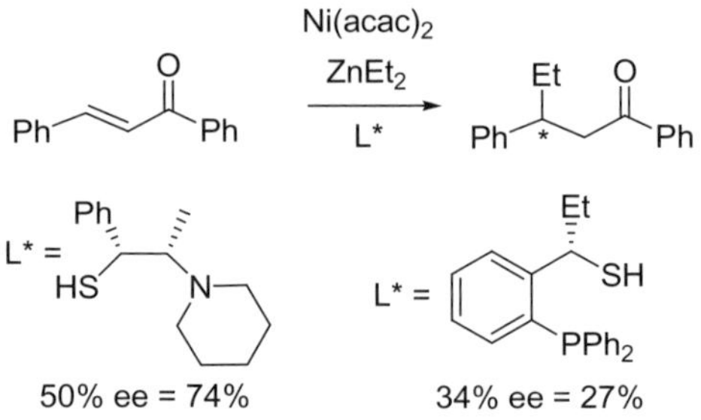

Scheme 2.31   Ni-catalysed 1,4-addition of ZnEt$_2$ to chalcone with bidentate thiol ligands.

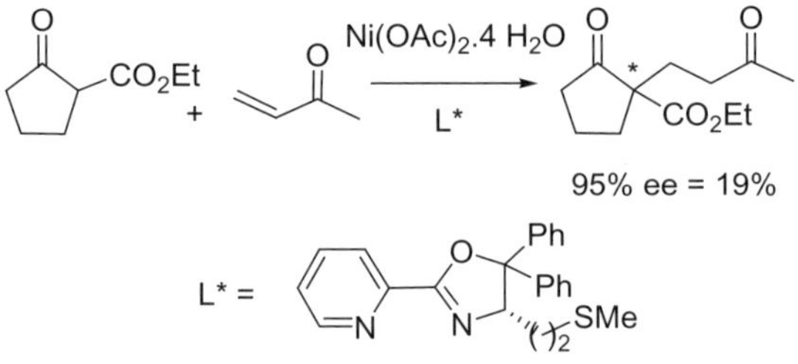

Scheme 2.32   Ni-catalysed Michael reaction of β-oxo ester with methyl vinyl ketone with tridentate oxazoline ligand.

ligands derived from chiral α-hydroxy acids.[58] These latter ligands were tested in the asymmetric catalysis version of the Michael reactions, giving, unfortunately, an enantioselectivity lower than 11% ee. As an extension of this work, these authors have prepared N/S/N-tridentate diamino thioethers and the corresponding diimino thioethers with a C_2-symmetry from chiral α-amino acids.[59] These ligands were further submitted to the Michael reaction of β-oxo esters with methyl vinyl ketone, resulting in an optimal enantiomeric excess of only 17%. In addition, various tridentate oxazoline ligands bearing adjacent thioether and heteroaryl donor groups were synthesised from *L*-cysteine and *L*-methionine by these workers.[60] All these ligands have been screened with thirteen metal salts with regard to the asymmetric catalysis version of the Michael reaction of a β-keto ester with methyl vinyl ketone to give an optimal result of 19% ee (Scheme 2.32).

   A series of chiral β-hydroxysulfoximine ligands have been synthesised by Bolm *et al.* and further investigated for the enantioselective conjugate addition of ZnEt$_2$ to various chalcone derivatives.[61] The most efficient sulfoximine, depicted in Scheme 2.33, has allowed an enantioselectivity of up to 72% ee to be obtained. These authors assumed a nonmonomeric nature of the active species in solution, as suggested by the asymmetric amplification in the catalysis with a sulfoximine of a low optical purity.

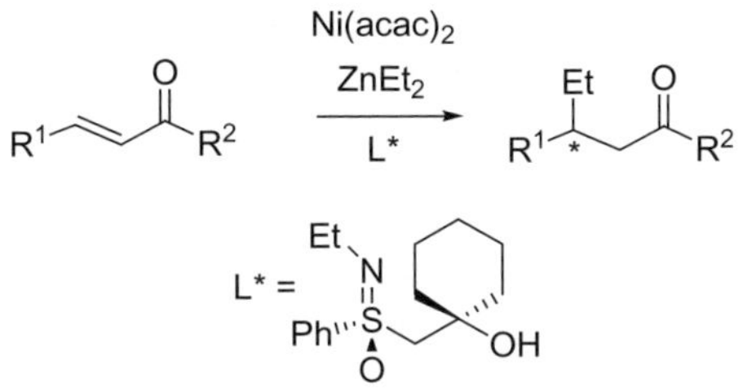

R^1 = R^2 = Ph: 71% ee = 70%
R^1 = *p*-MeOC$_6$H$_4$, R^2 = Ph: 48% ee = 70%
R^1 = Ph, R^2 = *p*-MeOC$_6$H$_4$: 45% ee = 72%
R^1 = Ph, R^2 = *p*-ClC$_6$H$_4$: 57% ee = 62%

Scheme 2.33 Ni-catalysed 1,4-additions of ZnEt$_2$ to chalcones with β-hydroxysulfoximine ligand.

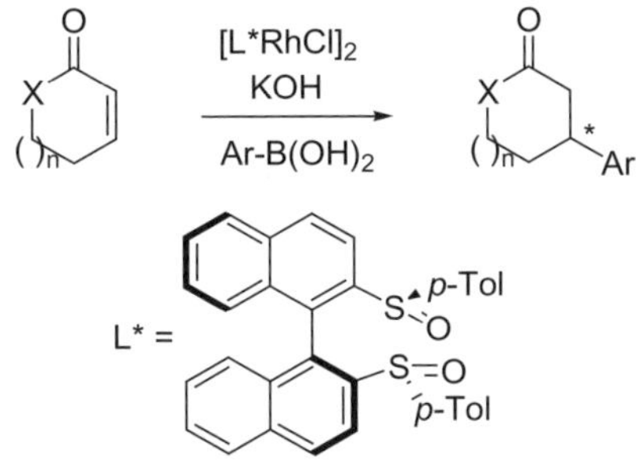

X = CH$_2$, n = 1, Ar = Ph: 99% ee = 98% (*R*)
X = CH$_2$, n = 1, Ar = *p*-Tol: 97% ee = 96% (*R*)
X = CH$_2$, n = 1, Ar = *p*-ClC$_6$H$_4$: 86% ee = 98% (*R*)
X = CH$_2$, n = 1, Ar = *p*-FC$_6$H$_4$: 90% ee = 98% (*R*)
X = CH$_2$, n = 1, Ar = *p*-MeOC$_6$H$_4$: 92% ee = 99% (*R*)
X = CH$_2$, n = 1, Ar = *m*-Tol: 93% ee = 99% (*R*)
X = CH$_2$, n = 1, Ar = *m*-CF$_3$C$_6$H$_4$: 94% ee = 97% (*R*)
X = CH$_2$, n = 1, Ar = *m*-ClC$_6$H$_4$: 93% ee = 99% (*R*)
X = CH$_2$, n = 1, Ar = *m*-FC$_6$H$_4$: 91% ee = 97% (*R*)
X = CH$_2$, n = 1, Ar = *m*-MeOC$_6$H$_4$: 55% ee = 97% (*R*)
X = CH$_2$, n = 1, Ar = *o*-Tol: 98% ee = 90% (*R*)
X = CH$_2$, n = 1, Ar = *o*-FC$_6$H$_4$: 60% ee = 99% (*R*)
X = CH$_2$, n = 1, Ar = 1-Naph: 99% ee = 90% (*R*)
X = CH$_2$, n = 1, Ar = 2-Naph: 89% ee = 96% (*R*)
X = CH$_2$, n = 0, Ar = Ph: 99% ee = 96% (*R*)
X = CH$_2$, n = 2, Ar = Ph: 98% ee = 66% (*R*)
X = O, n = 1, Ar = Ph: 98% ee = 91% (*R*)
X = CH(Me), n = 1, Ar = 2-Naph: 49% ee = 94% (*R*)

Scheme 2.34 Rh-catalysed 1,4-additions of arylboronic acids to cyclic enones with bis(sulfoxides) ligand.

Another potentially very readily available chiral ligand class with well-known coordination chemistry is represented by sulfoxides.[62] Considering that these chiral compounds play an important role as chiral auxiliaries in asymmetric synthesis,[63] it is surprising that relatively few examples exist in which this class of ligands participates in homogeneous metal catalysis. In particular, the use of rhodium bis(sulfoxides) ligands in catalysis is unknown. In this area, Dorta *et al.* reported, in 2008, the use of a chiral bis(sulfoxides) 1,1'-binaphthyl-based ligand in the rhodium-catalysed asymmetric addition of arylboronic acids to cyclic electron-deficient olefins.[64] One of the key advantages of this novel ligand over others lies in its extraordinarily easy synthesis, since it was prepared in one single step from commercially available starting materials.[65] When evaluated as a ligand in the rhodium-catalysed conjugate addition of various arylboronic acids to a series of cyclic enones, this ligand showed high reactivities and excellent enantioselectivities for the corresponding products, as shown in Scheme 2.34. In this new protocol, the authors have demonstrated that the catalyst loadings could be kept low (1.5 mol%) and an excess of boronic acid (1.1 eq) was not required for efficient catalysis. Furthermore, these new types of ligands should be more versatile than diene ligands and be able to support catalysis involving oxidative addition (H_2, $HSiR_3$, *etc.*) processes or carbon monoxide.

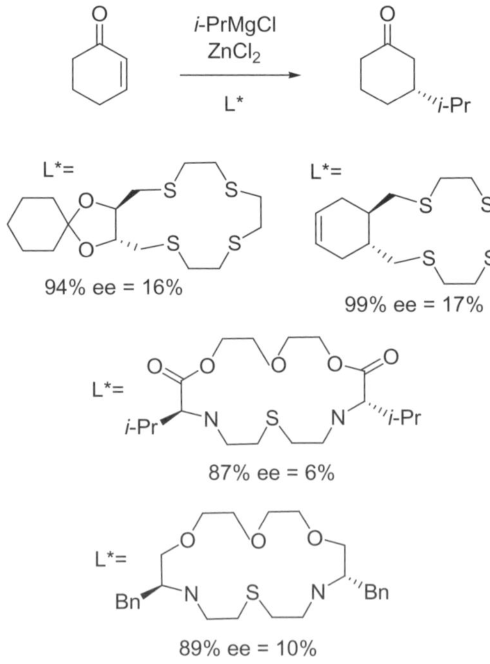

Scheme 2.35 Zn-catalysed 1,4-addition of *i*-PrMgCl to cyclohexenone with crown thioether ligands.

Modest enantioselectivities ($\leq 17\%$ ee) have been obtained by Feringa and Jansen for the enantioselective conjugate addition of Grignard reagents to enones catalysed by Zn(II) complexes of various chiral sulfur-containing ligands, such as crown thioethers depicted in Scheme 2.35.[66] Therefore, the Michael addition of *i*-PrMgCl to cyclohexenone performed in the presence of various crown thioethers did not allow good enantioselectivities to be obtained, albeit with excellent yields.

2.4 Conclusions

In conclusion, a wide variety of chiral sulfur-containing ligands have been successfully employed in the enantioselective copper-catalysed conjugate addition of various organometallic reagents to both cyclic and acyclic enones, providing results that are comparable in terms of both yield and enantio-selectivity to those of the well-known phosphorus ligands. In particular, excellent results have been obtained by using 1,1'-binaphthyl-based sulfur-containing ligands, such as 1,1'-BINOL-derived thioether ligands developed by the groups of Woodward and Kang, or 1,1'-binaphthylthiophosphoramide ligands developed by Zhang's group. Moreover, Dorta's group extended, in 2008, the use of this type of ligands (1,1'-binaphthyl-based bis(sulfoxides) ligands) to the highly enantioselective rhodium-catalysed addition of aryl-boronic acids to cyclic enones.

References

1. (a) P. Perlmutter, in *Conjugate Addition Reactions in Organic Synthesis*, Pergamon, Oxford, 1992; (b) M. P. Sibi and S. Manyem, *Tetrahedron*, 2000, **56**, 8033–8061; (c) J. Christoffers, G. Koripelly, A. Rosiak and M. Rössle, *Synthesis*, 2007, **9**, 1279–1300; (d) B. E. Rossiter and N. M. Swingle, *Chem. Rev.*, 1992, **92**, 771–806; (e) F. Lopez, A. J. Minnaard and B. L. Feringa, *Acc. Chem. Res.*, 2007, **40**, 179–188.
2. S. Woodward, *Chem. Soc. Rev.*, 2000, **29**, 393–401.
3. A. Alexakis, J. E. Bäckvall, N. Krause, O. Pamies and M. Diéguez, *Chem. Rev.*, 2008, **108**, 2796–2823.
4. (a) N. Krause, *Angew. Chem., Int. Ed. Engl.*, 1998, **37**, 283–285; (b) N. Krause and A. Hoffmann-Roder, *Synthesis*, 2001, 171–196; (c) E. Naka-mura and S. Mori, *Angew. Chem., Int. Ed. Engl.*, 2000, **39**, 3750–3771; (d) A. Alexakis and C. Benhaim, *Eur. J. Org. Chem.*, 2002, 3221–3236.
5. A. Boudier, L. O. Bromm, M. Lotz and P. Knochel, *Angew. Chem., Int. Ed. Engl.*, 2000, **39**, 4414–4435.
6. S. R. Harutyunyan, T. den Hartog, K. Geurts, A. J. Minnaard and B. L. Feringa, *Chem. Rev.*, 2008, **108**, 2824–2852.
7. M. Spescha and G. Rihs, *Helv. Chim. Acta*, 1993, **76**, 1219–1230.
8. (a) F. Lambert, D. M. Knotter, M. D. Janssen, M. van Klaveren, J. Boersma and G. van Koten, *Tetrahedron: Asymmetry*, 1991, **2**, 1097–1100; (b) M.

Van Klaveren, F. Lambert, D. J. F. M. Eijkelkamp, D. M. Grove and G. van Koten, *Tetrahedron Lett.*, 1994, **35**, 6135–6138; (c) C. L. Gibson, *Tetrahedron: Asymmetry*, 1996, **7**, 3357–3358.

9. Q. Zhou and A. Pfaltz, *Tetrahedron Lett.*, 1993, **34**, 7725–7728.
10. J. Green and S. Woodward, *Synlett*, 1995, **1**, 155–156.
11. A. Togni, G. Rihs and R. E. Blumer, *Organometallics*, 1992, **11**, 613–621.
12. G. A. Cran, C. L. Gibson, S. Handa and A. R. Kennedy, *Tetrahedron: Asymmetry*, 1996, **7**, 2511–2514.
13. A. Alexakis, J. C. Frutos and P. Mangeney, *Tetrahedron: Asymmetry*, 1993, **4**, 2427–2430.
14. S. G. Murray and F. R. Hartley, *Chem. Rev.*, 1981, **81**, 365–414.
15. (a) S. M. W. Bennett, S. M. Brown, G. Conole, M. R. Dennis, P. K. Fraser, S. Radojevic, M. McPartlin, C. M. Topping and S. Woodward, *J. Chem. Soc., Perkin Trans. I*, 1999, 3127–3132; (b) S. M. W. Bennett, S. M. Brown, A. Cunningham, M. R. Dennis, J. P. Muxworthy, M. A. Oakley and S. Woodward, *Tetrahedron*, 2000, **56**, 2847–2855; (c) S. M. W. Bennett, S. M. Brown, J. P. Muxworthy and S. Woodward, *Tetrahedron Lett.*, 1999, **40**, 1767–1770; (d) S. Woodward, *Tetrahedron*, 2002, **58**, 1017–1050; (e) S. M. Azad, S. M. W. Bennett, S. M. Brown, J. Green, E. Sinn, C. M. Topping and S. Woodward, *J. Chem. Soc., Perkin Trans. I*, 1997, 687–694.
16. (a) C. Börner, W. A. König and S. Woodward, *Tetrahedron Lett.*, 2001, **42**, 327–329; (b) C. Börner, M. R. Dennis, E. Sinn and S. Woodward, *Eur. J. Org. Chem.*, 2001, **1**, 2435–2446.
17. P. K. Fraser and S. Woodward, *Chem. Eur. J.*, 2003, **9**, 776–783.
18. V. Albrow, K. Biswas, A. Crane, N. Chaplin, T. Easun, S. Gladiali, B. Lygo and S. Woodward, *Tetrahedron: Asymmetry*, 2003, **14**, 2813–2819.
19. J. Kang, J. H. Lee and D. S. Lim, *Tetrahedron: Asymmetry*, 2003, **14**, 305–315.
20. (a) M. Shi, C.-J. Wang and W. Zhang, *Chem. Eur. J.*, 2004, **10**, 5507–5516; (b) M. Shi, W.-L. Duan and G.-B. Gong, *Chirality*, 2004, **1**, 642–651; (c) M. Shi and W. Zhang, *Adv. Synth. Catal.*, 2005, **347**, 535–540; (d) M. Shi and W. Zhang, *Tetrahedron: Asymmetry*, 2004, **15**, 167–176.
21. W. Zhang and M. Shi, *Synlett*, 2007, **1**, 19–30.
22. S. Woodward, *Tetrahedron*, 2002, **58**, 1017–1050.
23. C. Börner, J. Gimeno, S. Gladiali, P. J. Goldsmith, D. Ramazzotti and S. Woodward, *Chem. Commun.*, 2000, 2433–2434.
24. Z.-B. Dong, B. Liu, C. Fang and J.-S. Li, *J. Organomet. Chem.*, 2008, **693**, 17–22.
25. (a) D. Seebach, G. Jaeschke, A. Pichota and L. Audergon, *Helv. Chim. Acta*, 1997, **80**, 2515–2519; (b) A. Pichota, P. S. Pregosin, M. Valentini, M. Wörle and D. Seebach, *Angew. Chem., Int. Ed. Engl.*, 2000, **39**, 153–156.
26. O. Pamies, G. Net, A. Ruiz, C. Claver and S. Woodward, *Tetrahedron: Asymmetry*, 2000, **11**, 871–877.

27. E. Guimet, M. Diéguez, A. Ruiz and C. Claver, *Tetrahedron: Asymmetry*, 2005, **16**, 2161–2165.
28. M. Diéguez, O. Pamies, G. Net, A. Ruiz and C. Claver, *J. Mol. Catal.*, 2002, **185**, 11–16.
29. A. Pfaltz, *Acc. Chem. Res.*, 1993, **26**, 339–345.
30. (a) A. Togni and L. Venanzi, *Angew. Chem., Int. Ed. Engl.*, 1994, **33**, 497–526; (b) A. Y. Meyers and M. Reuman, *Tetrahedron*, 1985, **41**, 837–860.
31. (a) J. V. Allen, J. F. Bower and J. M. J. Williams, *Tetrahedron: Asymmetry*, 1994, **5**, 1895–1898; (b) J. V. Allen, S. J. Coote, G. J. Dawson, C. G. Frost, C. J. Martin and J. M. J. Williams, *J. Chem. Soc., Perkin Trans. I*, 1994, 2065–2072; (c) J. V. Allen, G. J. Dawson, C. G. Frost, J. M. J. Williams and S. J. Coote, *Tetrahedron*, 1994, **50**, 799–808.
32. Q.-L. Zhou and A. Pfaltz, *Tetrahedron*, 1994, **50**, 4467–4478.
33. D. M. Knotter, D. M. Grove, W. J. J. Smeets, A. L. Spek and G. van Koten, *J. Am. Chem. Soc.*, 1992, **114**, 3400–3410.
34. (a) Y. Takemoto, S. Kuraoka, N. Hamaue and C. Iwata, *Tetrahedron: Asymmetry*, 1996, **7**, 993–996; (b) Y. Takemoto, S. Kuraoka, N. Hamaue, K. Aoe, H. Hiramatsu and C. Iwata, *Tetrahedron*, 1996, **56**, 14177–14188.
35. M. Schinnerl, M. Seitz, A. Kaiser and O. Reiser, *Org. Lett.*, 2001, **3**, 4259–4262.
36. B. F. Bonini, E. Capito, M. Comes-Franchini, A. Ricci, A. Bottoni, F. Bernadi, G. P. Miscione, L. Giordano and A. R. Cowley, *Eur. J. Org. Chem.*, 2004, 4442–4451.
37. E. Capito, L. Bernardi, M. Comes-Franchini, F. Fini, M. Fochi, S. Pollicino and A. Ricci, *Tetrahedron: Asymmetry*, 2005, **16**, 3232–3240.
38. G. A. Cran, C. L. Gibson, S. Handa and A. R. Kennedy, *Tetrahedron: Asymmetry*, 1996, **7**, 2511–2514.
39. G. van Koten, *Pure Appl. Chem.*, 1994, **66**, 1455–1462.
40. A. M. Arink, T. W. Braam, R. Keeris, J. T. B. H. Jastrzebski, C. Benhaim, S. Rosset, A. Alexakis and G. van Koten, *Org. Lett.*, 2004, **6**, 1959–1962.
41. A. H. M. De Vries, R. P. Hof, D. Staal, R. M. Kellog and B. L. Feringa, *Tetrahedron: Asymmetry*, 1997, **8**, 1539–1543.
42. T. C. Kinahan and H. Tye, *Tetrahedron: Asymmetry*, 2001, **12**, 1255–1257.
43. H.-J. Gais, *Heteroatom. Chem.*, 2007, **18**, 472–481.
44. C. Bolm, J. Muller, G. Schlingloff, M. Zehnder and M. Neuburger, *Chem. Commun.*, 1993, **1**, 182–183.
45. C. Bolm, F. Bienewald and K. Harms, *Synlett*, 1996, 775–776.
46. M. Reggelin, H. Weinberger and V. Spohr, *Adv. Synth. Catal.*, 2004, **346**, 1295–1306.
47. A. Alexakis and C. Benhaim, *Tetrahedron: Asymmetry*, 2001, **12**, 1151–1157.
48. H. Clavier, L. Coutable, L. Toupet, J.-C. Guillemin and M. Mauduit, *J. Organomet. Chem.*, 2005, **690**, 5237–5254.
49. V. Wendisch and N. Sewald, *Tetrahedron: Asymmetry*, 1997, **8**, 1253–1257.
50. Y. Nakagawa, M. Kanai, Y. Nagaoka and K. Tomioka, *Tetrahedron Lett.*, 1996, **37**, 7805–7808.

51. Y. Nakagawa, K. Matsumoto and K. Tomioka, *Tetrahedron*, 2000, **56**, 2857–2863.
52. T. Morimoto, N. Mochizuki and M. Suzuki, *Tetrahedron Lett.*, 2004, **45**, 5717–5722.
53. I. J. Krauss and J. L. Leighton, *Org. Lett.*, 2003, **5**, 3201–3203.
54. (a) I. Chataigner, C. Gennari, U. Piarulli and S. Ceccarelli, *Angew. Chem., Int. Ed. Engl.*, 2000, **39**, 916–918; (b) I. Chataigner, C. Gennari, S. Ongeri, U. Piarulli and S. Ceccarelli, *Chem. Eur. J.*, 2001, **7**, 2628–2634.
55. S. Ongeri, U. Piarulli, R. F. W. Jackson and C. Gennari, *Eur. J. Org. Chem.*, 2001, 803–807.
56. Y. Yin, X. Li, D.-S. Lee and T.-K. Yang, *Tetrahedron: Asymmetry*, 2000, **11**, 3329–3333.
57. J. Kang, J. I. Kim, J. H. Lee, H. J. Kim and Y. H. Byun, *Bull. Korean Chem. Soc.*, 1998, **19**, 601–603.
58. J. Christoffers and U. Rößler, *Tetrahedron: Asymmetry*, 1999, **10**, 1207–1215.
59. (a) J. Christoffers and A. Mann, *Eur. J. Org. Chem.*, 1999, 1475–1479; (b) J. Christoffers and U. Rößler, *Tetrahedron: Asymmetry*, 1998, **9** 2349–2357.
60. J. Christoffers, A. Mann and J. Pickardt, *Tetrahedron*, 1999, **55**, 5377–5388.
61. C. Bolm, M. Felder and J. Müller, *Synlett*, 1992, **1**, 439–441.
62. M. Calligaris and O. Carugo, *Coord. Chem. Rev.*, 1996, **153**, 83–154.
63. (a) I. Fernandez and N. Khiar, *Chem. Rev.*, 2003, **103**, 3651–3705; (b) H. Pellissier, *Tetrahedron*, 2006, **62**, 5559–5601.
64. R. Mariz, X. Luan, M. Gatti, A. Linden and R. Dorta, *J. Am. Chem. Soc.*, 2008, **130**, 2172–2173.
65. J. Clayden, P. M. Kubinski, F. Sammiceli, M. Helliwell and L. Diorazio, *Tetrahedron*, 2004, **60**, 4387–4397.
66. J. F. G. A. Jansen and B. L. Feringa, *J. Org. Chem.*, 1990, **55**, 4168–4175.

CHAPTER 3

Addition of Organometallic Reagents to Aldehydes

3.1 Introduction

One of the major challenges in organic synthesis is the asymmetric preparation of compounds by means of forming carbon–carbon bonds. Among the different strategies, the enantioselective 1,2-addition of organometallic reagents to carbonyl compounds is probably the most straightforward and useful manner of achieving this goal. Among these approaches, the enantioselective addition of dialkylzinc reagents to aldehydes is the most widely used methodology. Moreover, the enantioselective addition of diethylzinc to aldehydes mediated by chiral ligands is one of the most studied examples of ligand-accelerated catalysis.[1] Such asymmetric reactions allow the synthesis of chiral alcohols which are ubiquitous in the structures of natural products and drug compounds. Over the past few decades and since the initial report of Oguni et al.,[2] a large number of chiral catalysts including amino alcohols, diamines and diols have been developed and high enantioselectivities have been achieved for all different types of aldehydes.[3] Indeed, approximately six hundred individual catalysts have been applied in the last decade to these types of reactions; and among these ligands, chiral β-amino alcohols have been the choice invariably for the ligand design of these reactions in particular. Several characteristics of these reactions, such as the very important nonlinear effects,[4] the autocatalysis phenomena and the derived amplification of the enantioselectivity, making them attractive from both an intellectual and industrial perspective, have ensured that sufficiently active catalytic ligands have been developed. In order to extend the diversity of the ligand structures even further, sulfur-containing chiral ligands, such as amino thiols, amino sulfides, disulfides, disulfonamides, arylthiophosphoramides and sulfinylferrocenes have recently been synthesised

RSC Catalysis Series No. 2
Chiral Sulfur Ligands: Asymmetric Catalysis
By Hélène Pellissier
© Hélène Pellissier 2009
Published by the Royal Society of Chemistry, www.rsc.org

and subsequently applied to the catalytic asymmetric dialkylzinc addition to aldehydes, allowing this process to become a mature method. Examples of enantioselective addition of arylzinc reagents performed in the presence of such ligands are still rare in the literature, as are those involving alkynylations. In addition, a few examples involving organometallic reagents other than organozinc reagents will be developed at the end of this chapter. The chapter is divided according to the nature of the organozinc reagents added to the aldehyde, dialkylzinc reagents in the first part and diaryl, alkynyl or alkenyl zinc reagents in the second part. The first and most important part is subdivided according to the nature of the chiral sulfur-containing ligands employed for the enantioselective addition of organozinc reagents to aldehydes. Thus, S/N ligands, S/N/O ligands, S/O ligands and then sulfur-containing ferrocenyl ligands are successively studied.

3.2 Addition of Organozinc Reagents to Aldehydes

3.2.1 Addition of Dialkylzinc Reagents to Aldehydes

3.2.1.1 S/N Ligands

The observation that amino sulfur catalysts offer improvements in enantioselectivity over their amino alcohol counterparts is known for a number of transition-metal-mediated C–C bond-forming processes. In particular, the zinc thiolate derivatives often led to better yields and selectivities than the corresponding zinc alcoholates. This superior catalytic activity can be explained by the following: sulfur is more polarisable than oxygen in alcohols; thiols and thiolates have a higher affinity towards metals, especially zinc; and metal thiolates are less inclined to decrease the Lewis acidity of the metal compared to metal alcoholates. This has led to the development of a wide range of chiral S/N ligands for the enantioselective 1,2-addition of dialkylzinc reagents to aldehydes. In this context, Gibson has prepared an L-proline-based β-amino tertiary thiol (Scheme 3.1), which provided (*R*)-secondary alcohols in enantioselectivities of up to 64% ee.[5] This author has also prepared a new disulfide from L-proline, depicted in Scheme 3.1, which gave better results for similar reactions in term of enantioselectivity (up to 99% ee).[6] This ligand gave, however, moderate inductions with non-α-branched aldehydes, such as dihydrocinnamaldehyde and cinnamaldehyde (69 and 70%, respectively). The decrease in enantioselectivity observed by using β-amino tertiary thiol ligand, compared to the disulfide ligand, was ascribed to the destabilisation of the major intermediate in the transition state by a steric hindrance generated by the presence of the *gem*-diphenyl group.

   Another β-amino disulfide ligand was prepared by these authors starting from (*S*)-phenylglycine or (*R*)-styrene oxide and subsequently tested for similar reactions to those described above.[7] The enantioselectivity (39–80% ee) was found to be lower than those observed for the disulfide depicted in Scheme 3.1,

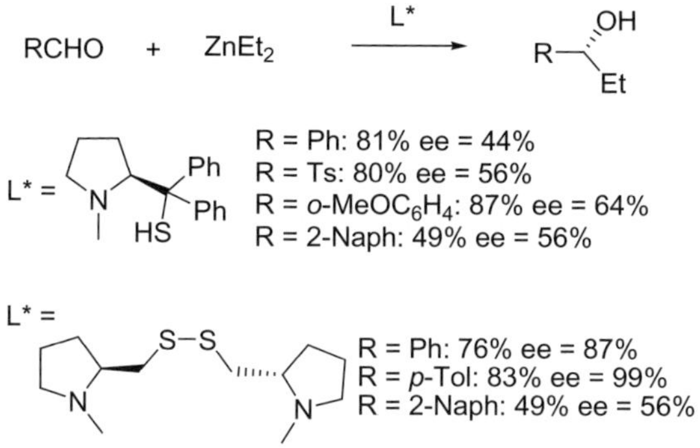

Scheme 3.1 L-Proline-derived β-amino thiol and disulfide ligands for additions of ZnEt$_2$ to aldehydes.

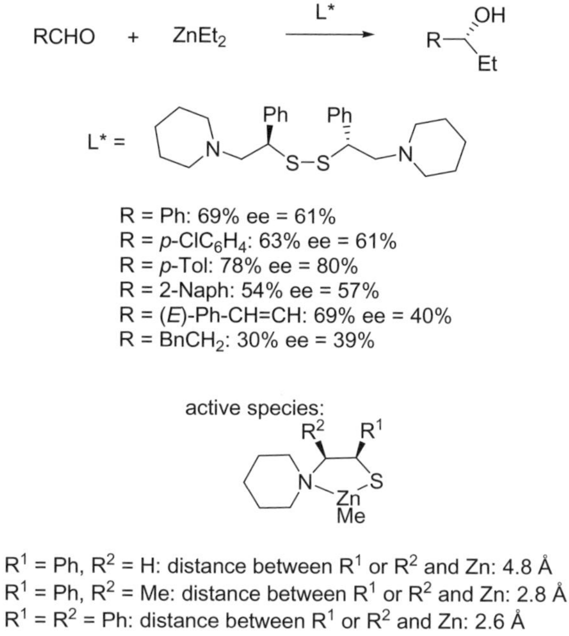

Scheme 3.2 (*S*)-Phenylglycine-derived β-amino disulfide ligand for additions of ZnEt$_2$ to aldehydes.

and this was postulated to be due to the absence of a stereodifferentiating group in the α-position to the amino group. In order to rationalize this result, a molecular mechanic analysis of the postulated active species was performed (Scheme 3.2). According to these calculations, the stereodifferentiating phenyl group in the methylzinc complex depicted in Scheme 3.2 was directed away from the zinc atom (4.8 Å). Since zinc was the key reacting centre for the catalysis, it was expected that this complex would not lead to an efficient enantiofacial discrimination, unlike the other active species (R^1=Ph, R^2=Me) in which the methyl group was close to the Zn (2.8 Å) and gave an excellent enantioselectivity.

In the same context, Aurich *et al.* have reported the synthesis of S/N ligands based on a rigid 2-azabicyclo[3.3.0]octane framework, which were tested for their general effectiveness as chiral ligands in the reaction of aldehydes with ZnEt$_2$ (Scheme 3.3).[8] It was noted that both yield and enantioselectivity were higher with these thioacetate and thio ligands than with their alcohol analogues.

In 1994, Kellogg *et al.* reported the synthesis of several sulfur derivatives of ephedrine, such as a chiral amino thiol, its corresponding phosphorylated thioephedrine, a thiazolidine and disulfides, which are depicted in Scheme 3.4.[9] These ligands were further tested for the 1,2-addition of ZnEt$_2$ to benzaldehyde, providing the product in high enantioselectivity of up to 98% ee. The enantiomeric excesses generally exceeded those obtained with the corresponding β-amino alcohols. In the case of the disulfide ligand, these authors suggested that the alkylation of the disulfide moiety occurred to generate the corresponding thioether and thiolate, with the latter being the active catalyst in the 1,2-addition reaction, as shown in Scheme 3.4.[10]

These reactions were also performed by Jin *et al.* in the presence of chiral amino thioacetate ligands derived from (+)-norephedrine.[11] As shown in Scheme 3.5, quantitative yields and excellent enantioselectivities of up to >99% ee were obtained in almost all cases of aldehydes, even when using 1.1 equivalent of ZnEt$_2$ instead of 2 equivalents as usually employed. More recently, these authors developed corresponding chiral amino thiocyanate derivatives from (−)-norephedrine and tested these new aprotic ligands for

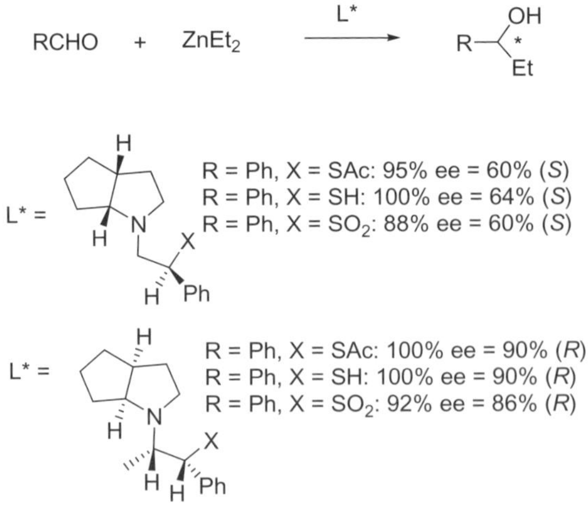

Scheme 3.3   2-Azabicyclo[3.3.0]octane-based S/N ligands for additions of ZnEt$_2$ to aldehydes.

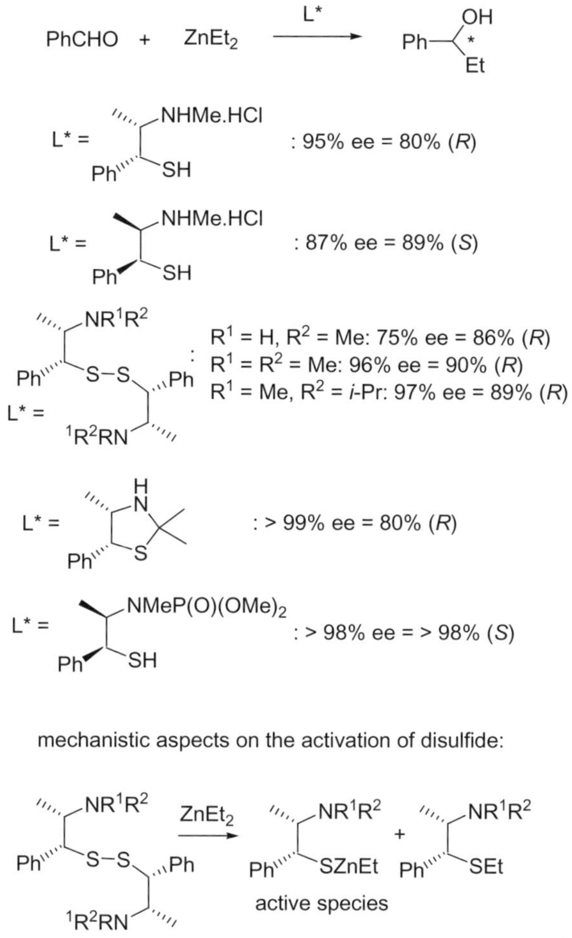

Scheme 3.4 Ephedrine-derived S/N ligands for addition of $ZnEt_2$ to benzaldehyde.

similar reactions.[12] As shown in Scheme 3.5, the products were obtained with enantioselectivities of up to 96% ee.

In 2000, novel thiazolidine derivatives were prepared from L-cysteine by Guan *et al.*, showing good enantioselectivities of up to 90% ee when used as ligands in the addition of $ZnEt_2$ to aldehydes.[13] The results collected in Scheme 3.6 show that the catalytic efficiency of the thiazolidine derivatives was influenced by the different structures of the thiazolidine rings and the bulkiness of the R^2 moiety in the ester group.

Other excellent results have been reported by Kang *et al.* for the addition of $ZnEt_2$ to aldehydes by using chiral cyclic amino thiol ligands depicted in Scheme 3.7.[14] A total enantioselectivity was obtained when (1*R*,2*S*)-1-phenyl-2-piperidinopropane-1-thiol was used as the ligand in the reactions of substituted benzaldehydes. However, *n*-hexanal and *trans*-cinnamaldehyde could only be

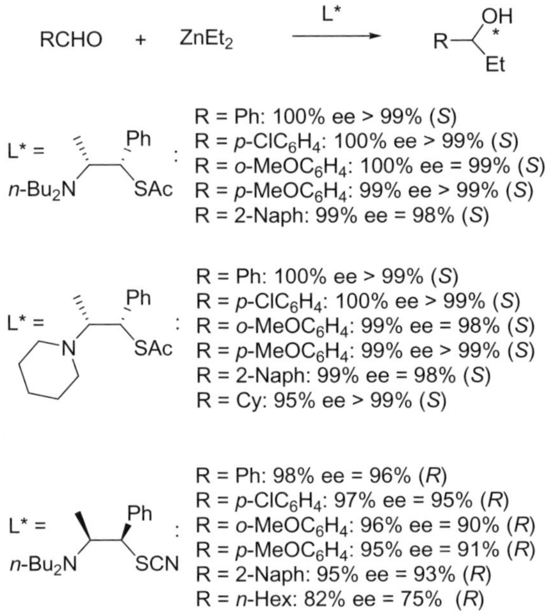

$$\text{RCHO} \quad + \quad \text{ZnEt}_2 \quad \xrightarrow{\quad L^* \quad} \quad R\underset{Et}{\overset{OH}{\diagdown}}^{*}$$

L* = (structure with *n*-Bu$_2$N, Ph, SAc)

R = Ph: 100% ee > 99% (S)
R = p-ClC$_6$H$_4$: 100% ee > 99% (S)
R = o-MeOC$_6$H$_4$: 100% ee = 99% (S)
R = p-MeOC$_6$H$_4$: 99% ee > 99% (S)
R = 2-Naph: 99% ee = 98% (S)

L* = (structure with N-piperidine ring, Ph, SAc)

R = Ph: 100% ee > 99% (S)
R = p-ClC$_6$H$_4$: 100% ee > 99% (S)
R = o-MeOC$_6$H$_4$: 99% ee = 98% (S)
R = p-MeOC$_6$H$_4$: 99% ee > 99% (S)
R = 2-Naph: 99% ee = 98% (S)
R = Cy: 95% ee > 99% (S)

L* = (structure with *n*-Bu$_2$N, Ph, SCN)

R = Ph: 98% ee = 96% (R)
R = p-ClC$_6$H$_4$: 97% ee = 95% (R)
R = o-MeOC$_6$H$_4$: 96% ee = 90% (R)
R = p-MeOC$_6$H$_4$: 95% ee = 91% (R)
R = 2-Naph: 95% ee = 93% (R)
R = n-Hex: 82% ee = 75% (R)

Scheme 3.5 Norephedrine-derived amino thioacetate and thiocyanate ligands for additions of ZnEt$_2$ to aldehydes.

ethylated in moderate enantioselectivities (62–77% ee), which showed that the present system was highly efficient in term of enantioselectivity only if aldehydes were α-branched.

In the same context, these authors have also studied the catalytic activity of the β-amino thiol and its corresponding disulfide depicted in Scheme 3.8 for similar reactions.[15] High levels of enantioselectivity were obtained for both these ligands but, however, in the case of the disulfide ligand the reaction rate was slower than that observed with the corresponding thiol, while the disulfide proved to be more chemically stable under the reaction conditions.

The effects of the sulfur substitution in the chiral amino thiols on the enantioselective addition of organozinc reagents to aldehydes have been studied by these authors.[16] The catalytic reaction in the presence of chiral amino thiols was discussed and compared to the results obtained with the corresponding amino alcohols. The quantitative and thermodynamic aspects of the monomer–dimer equilibria involved in thiazazincolidine or oxazazincolidine catalysts have also been studied on the basis of colligative properties. The results indicated that the coordination of ZnEt$_2$ prior to the coordination of the aldehyde was essential for high enantioselectivity of the thiol-catalysed reaction. On the other hand, it was proved that the thiolate-catalysed reaction was considerably faster than the corresponding alcoholate system.

In 2004, Yang and Tseng reported the synthesis of a series of new chiral amino thiols and corresponding amino thioacetate ligands derived from

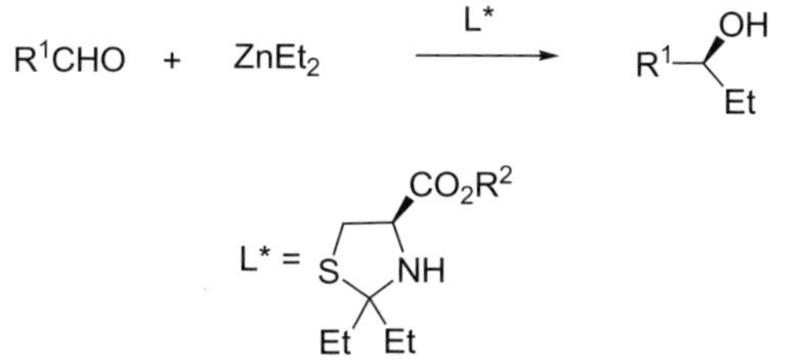

$$R^1CHO + ZnEt_2 \xrightarrow{L^*} R^1 \overset{OH}{\underset{Et}{\diagdown}}$$

$$L^* = \begin{array}{c} CO_2R^2 \\ S \diagup NH \\ Et \quad Et \end{array}$$

R^1 = Ph, R^2 = Me: 74% ee = 85%
R^1 = Ph, R^2 = Et: 70% ee = 88%
R^1 = Ph, R^2 = *i*-Pr: 98% ee = 90%
R^1 = *p*-ClC$_6$H$_4$, R^2 = *i*-Pr: 71% ee = 88%
R^1 = *p*-MeOC$_6$H$_4$, R^2 = *i*-Pr: 96% ee = 78%
R^1 = 2-Fu, R^2 = *i*-Pr: 43% ee = 78%
R^1 = (*E*)-Ph-CH=CH, R^2 = *i*-Pr: 79% ee = 75%
R^1 = *n*-Hex, R^2 = *i*-Pr: 75% ee = 52%

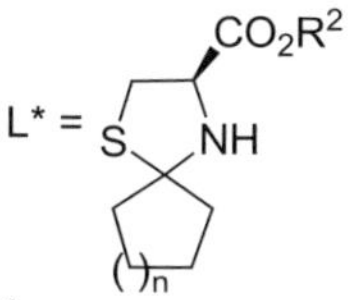

$$L^* = \begin{array}{c} CO_2R^2 \\ S \diagup NH \\ ()_n \end{array}$$

R^1 = Ph, R^2 = Me, n = 1: 65% ee = 79%
R^1 = Ph, R^2 = Et, n = 1: 67% ee = 82%
R^1 = Ph, R^2 = *i*-Pr, n = 1: 77% ee = 89%
R^1 = Ph, R^2 = Me, n = 2: 83% ee = 79%
R^1 = Ph, R^2 = Et, n = 2: 90% ee = 80%
R^1 = Ph, R^2 = *i*-Pr, n = 2: 74% ee = 81%
R^1 = Ph, R^2 = Me, n = 3: 90% ee = 86%
R^1 = Ph, R^2 = Et, n = 3: 99% ee = 88%
R^1 = Ph, R^2 = *i*-Pr, n = 3: 90% ee = 89%
R^1 = *p*-ClC$_6$H$_4$, R^2 = *i*-Pr, n = 3: 70% ee = 86%
R^1 = *p*-MeOC$_6$H$_4$, R^2 = *i*-Pr, n = 3: 99% ee = 76%
R^1 = 2-Fu, R^2 = *i*-Pr, n = 3: 45% ee = 78%
R^1 = (*E*)-Ph-CH=CH, R^2 = *i*-Pr, n = 3: 76% ee = 68%
R^1 = *n*-Hex, R^2 = *i*-Pr, n = 3: 79% ee = 56%

Scheme 3.6   L-Cysteine-derived thiazolidine ligands for additions of ZnEt$_2$ to aldehydes.

L-valine.[17] The application of these ligands to the asymmetric addition of ZnEt$_2$ to aldehydes provided the corresponding products with excellent enantioselectivities as high as 99% ee in almost all cases and with a catalytic loading as little as 0.02 mol% of the ligand depicted in Scheme 3.9.

In 1998, Anderson *et al.* developed a new series of S/N ligands derived from L-valine and analysed these in the enantioselective addition of ZnEt$_2$ to aromatic aldehydes.[18] The use of these ligands allowed enantioselectivities of up to 82% ee to be obtained (Scheme 3.10). The results showed that the ligands possessing symmetrical nitrogen substituents did not perform as well as those

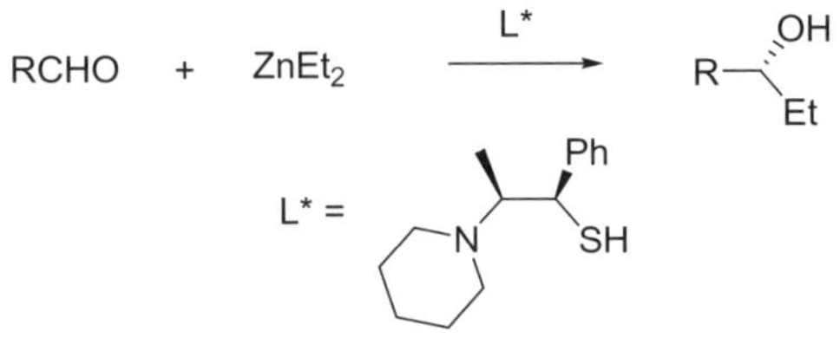

RCHO   +   ZnEt₂ $\xrightarrow{\text{L*}}$ R—$\overset{\text{OH}}{\underset{\text{Et}}{|}}$

L* =

R = Ph: 94% ee = 100%
R = o-MeOC₆H₄: 97% ee = 100%
R = p-MeOC₆H₄: 96% ee = 100%
R = p-ClC₆H₄: 99% ee = 100%
R = p-FC₆H₄: 92% ee = 100%
R = 2-Naph: 98% ee = 99%
R = 1-Naph: 100% ee = 99%
R = t-Bu: 94% ee = 100%
R = Cy: 97% ee = 100%
R = n-Pent: 96% ee = 65%
R = (E)-Ph-CH=CH: 98% ee = 77%

Scheme 3.7   Cyclic amino thiol ligand for additions of ZnEt₂ to aldehydes.

RCHO   +   ZnEt₂ $\xrightarrow{\text{L*}}$ R—$\overset{\text{OH}}{\underset{\text{Et}}{|}}$

L* =

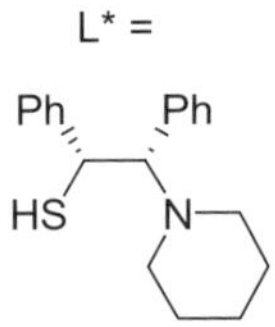

R = Ph: 90% ee = 100%
R = p-ClC₆H₄: 92% ee = 100%
R = o-MeOC₆H₄: 92% ee = 99%
R = p-MeOC₆H₄: 91% ee = 99%
R = 2-Naph: 98% ee = 100%
R = t-Bu: 83% ee = 100%
R = Cy: 87% ee = 100%
R = n-Pent: 88% ee = 79%
R = (E)-Ph-CH=CH: 88% ee = 80%

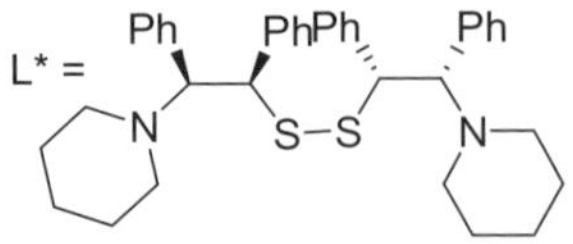

L* =

R = Ph: 92% ee = 99%
R = p-ClC₆H₄: 93% ee = 99%
R = o-MeOC₆H₄: 92% ee = 98%
R = p-MeOC₆H₄: 91% ee = 99%
R = 2-Naph: 93% ee = 98%
R = t-Bu: 63% ee = 99%
R = Cy: 92% ee = 98%
R = n-Pent: 89% ee = 87%
R = (E)-Ph-CH=CH: 93% ee = 83%

Scheme 3.8   β-Amino thiol and disulfide ligands for additions of ZnEt₂ to aldehydes.

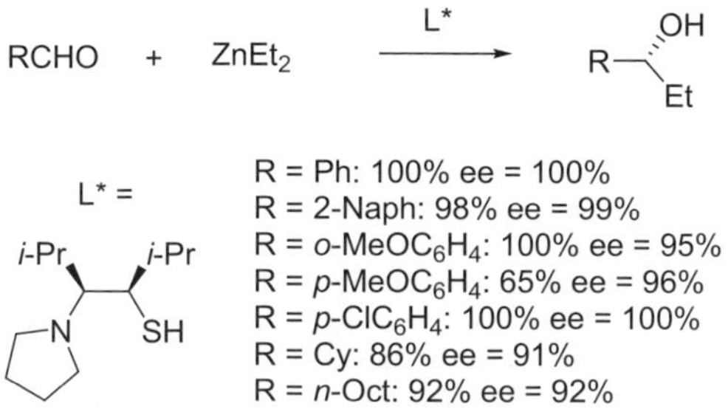

RCHO + ZnEt$_2$ $\xrightarrow{\text{L*}}$ R—CH(OH)—Et

L* =

i-Pr, *i*-Pr
N SH (pyrrolidine)

R = Ph: 100% ee = 100%
R = 2-Naph: 98% ee = 99%
R = *o*-MeOC$_6$H$_4$: 100% ee = 95%
R = *p*-MeOC$_6$H$_4$: 65% ee = 96%
R = *p*-ClC$_6$H$_4$: 100% ee = 100%
R = Cy: 86% ee = 91%
R = *n*-Oct: 92% ee = 92%

Scheme 3.9   L-Valine-derived β-amino thiol ligand for additions of ZnEt$_2$ to aldehydes.

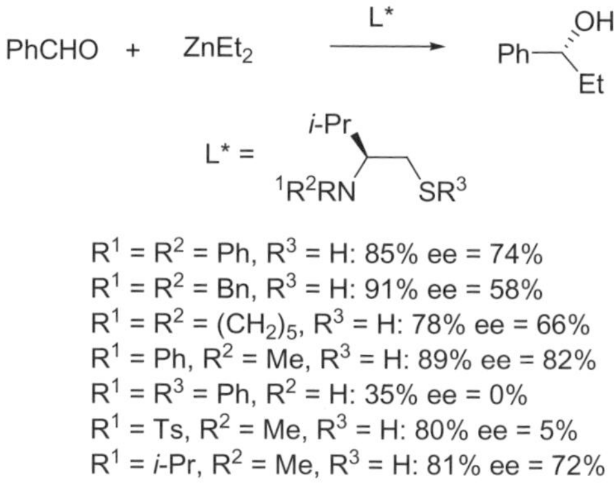

PhCHO + ZnEt$_2$ $\xrightarrow{\text{L*}}$ Ph—CH(OH)—Et

L* =

i-Pr
^{1}R^{2}RN SR^3

R^1 = R^2 = Ph, R^3 = H: 85% ee = 74%
R^1 = R^2 = Bn, R^3 = H: 91% ee = 58%
R^1 = R^2 = (CH$_2$)$_5$, R^3 = H: 78% ee = 66%
R^1 = Ph, R^2 = Me, R^3 = H: 89% ee = 82%
R^1 = R^3 = Ph, R^2 = H: 35% ee = 0%
R^1 = Ts, R^2 = Me, R^3 = H: 80% ee = 5%
R^1 = *i*-Pr, R^2 = Me, R^3 = H: 81% ee = 72%

Scheme 3.10   L-Valine-derived β-amino thio ligands for addition of ZnEt$_2$ to benzaldehyde.

having unsymmetrical nitrogen substituents. This suggested that giving the nitrogen atom the potential to become stereogenic led to enhanced enantioselection. In all cases of ligands, the best results in term of enantioselectivity were obtained with benzaldehyde. Surprisingly, the sulfur-substituted ligand led to the alcohol in the racemic form and with a low yield.

In addition, these authors have synthesised chiral β-amino disulfide ligands by oxidation of the corresponding monomers.[19] These ligands were further applied to the addition of ZnEt$_2$ to benzaldehyde, providing the product with generally less efficiency in terms of enantioselectivity (25–80% ee) than their corresponding amino thiol analogues (Scheme 3.11). These observations were consistent with the results obtained by Kang *et al.*[15] but, in contrast to those obtained by Kellogg[9b] and Gibson,[6,7] who observed better enantioselectivities by using disulfide ligands than with the corresponding amino thiol ligands.

In 2002, Braga *et al.* reported the enantioselective addition of ZnEt$_2$ to various aldehydes performed in the presence of chiral imidazolidine disulfides derived from L-cysteine.[20] In the presence of 5 mol% of ligand, the secondary alcohols were isolated in enantioselectivities of up to 91% ee. Aromatic

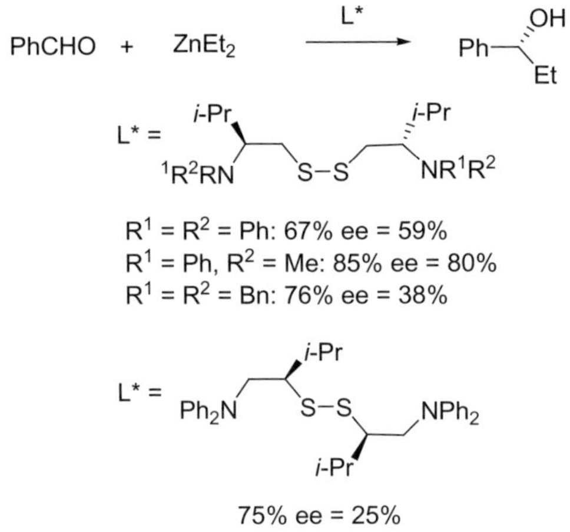

PhCHO + ZnEt₂ → Ph—CH(OH)Et

$$L^* = \text{(i-Pr, }^1R^2RN\text{, S–S, i-Pr, }NR^1R^2\text{)}$$

R¹ = R² = Ph: 67% ee = 59%
R¹ = Ph, R² = Me: 85% ee = 80%
R¹ = R² = Bn: 76% ee = 38%

$$L^* = \text{(i-Pr, }Ph_2N\text{, S–S, }NPh_2\text{, i-Pr)}$$

75% ee = 25%

Scheme 3.11 L-Valine-derived β-amino disulfide ligands for addition of ZnEt₂ to benzaldehyde.

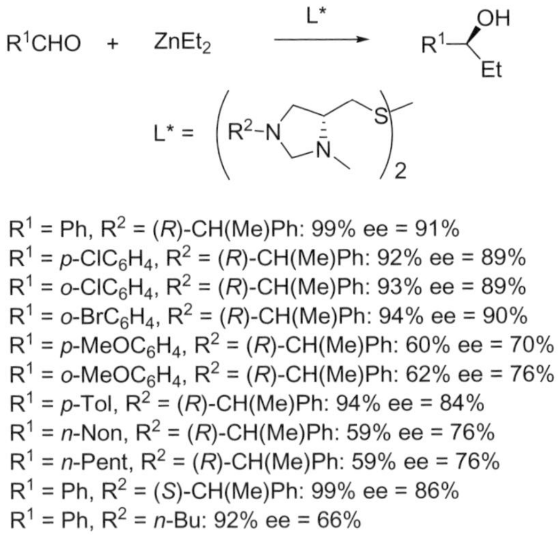

R¹CHO + ZnEt₂ → R¹—CH(OH)Et

$$L^* = \left(R^2\text{–N} \cdots \text{N} \ \text{S} \right)_2$$

R¹ = Ph, R² = (R)-CH(Me)Ph: 99% ee = 91%
R¹ = p-ClC₆H₄, R² = (R)-CH(Me)Ph: 92% ee = 89%
R¹ = o-ClC₆H₄, R² = (R)-CH(Me)Ph: 93% ee = 89%
R¹ = o-BrC₆H₄, R² = (R)-CH(Me)Ph: 94% ee = 90%
R¹ = p-MeOC₆H₄, R² = (R)-CH(Me)Ph: 60% ee = 70%
R¹ = o-MeOC₆H₄, R² = (R)-CH(Me)Ph: 62% ee = 76%
R¹ = p-Tol, R² = (R)-CH(Me)Ph: 94% ee = 84%
R¹ = n-Non, R² = (R)-CH(Me)Ph: 59% ee = 76%
R¹ = n-Pent, R² = (R)-CH(Me)Ph: 59% ee = 76%
R¹ = Ph, R² = (S)-CH(Me)Ph: 99% ee = 86%
R¹ = Ph, R² = n-Bu: 92% ee = 66%

Scheme 3.12 L-Cysteine-derived imidazolidine disulfide ligands for additions of ZnEt₂ to aldehydes.

aldehydes gave generally better enantioselectivities than aliphatic aldehydes, as shown in Scheme 3.12.

In addition, these authors have developed the synthesis of chiral aziridine sulfides and disulfides from D-cysteine through a Mitsunobu reaction.[21] The behavior of these ligands in the addition of ZnEt₂ to aldehydes was studied, showing a great catalytic potential, since the corresponding alcohols were isolated with enantioselectivities of up to 99% ee, as shown in Scheme 3.13. In

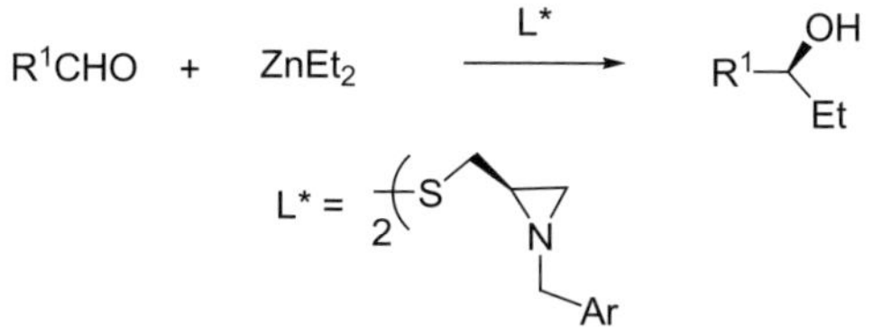

$$R^1CHO \quad + \quad ZnEt_2 \quad \xrightarrow{\text{L*}} \quad R^1\!\!-\!\!\overset{\displaystyle OH}{\underset{\displaystyle Et}{\big<}}$$

L* = $\left(S\overset{}{\underset{N}{\diagup}}\right)_2$ with N–Ar

R^1 = Ar = Ph: 61% ee > 99%
R^1 = p-ClC$_6$H$_4$, Ar = Ph: 77% ee = 94%
R^1 = p-Tol, Ar = Ph: 73% ee = 87%
R^1 = 1-Naph, Ar = Ph: 55% ee = 89%
R^1 = Py, Ar = Ph: 92% ee = 97%
R^1 = Bn, Ar = Ph: 58% ee = 89%
R^1 = n-Hex, Ar = Ph: 62% ee = 86%
R^1 = n-Dec, Ar = Ph: 61% ee = 40%
R^1 = Ph, Ar = p-Tol: 60% ee > 99%
R^1 = Ph, Ar = p-MeOC$_6$H$_4$: 58% ee = 56%

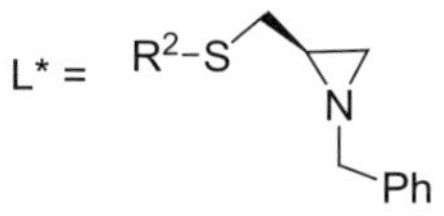

L* = R^2–S$\overset{}{\underset{N}{\diagup}}$ with N–Ph

R^1 = Ph, R^2 = Et: 42% ee = 65%
R^1 = Ph, R^2 = Bn: 57% ee = 76%

Scheme 3.13   D-Cysteine-derived aziridine disulfide ligands for additions of ZnEt$_2$ to aldehydes.

this example, the disulfide ligands proved to be more efficient than the sulfide ligands.

Air-stable bis(arenethiolates) zinc complexes, depicted in Scheme 3.14, have been prepared by van Koten *et al.* by treatment of two equivalents of the corresponding ligands with ZnCl$_2$.[22] These complexes were further treated by a large excess of ZnR$_2$, leading to the corresponding active species depicted in Scheme 3.14. These species were used for the enantioselective addition of ZnEt$_2$ to benzaldehyde, providing the product with an enantioselectivity of up to 96% ee (Scheme 3.14). Mechanistic investigation and new details of the intermediate along the reaction coordinate of the zinc-catalysed addition of diorganozinc reagents to aldehydes were proposed by these authors.[23]

On the other hand, novel diastereomeric β-amino thiol ligands possessing an isoquinuclidine skeleton have been readily prepared by Hongo *et al.* via imino-Diels–Alder reactions.[24] As shown in Scheme 3.15, when applied to the enantioselective addition of ZnEt$_2$ to various aldehydes, one of these ligands afforded the products with high enantioselectivities of up to 94% ee.

In the same area, these authors have also tested a β-amino thiol ligand containing a bicyclo[2.2.1] ring system, the 2-azanorbornylmethanethiol, which gave

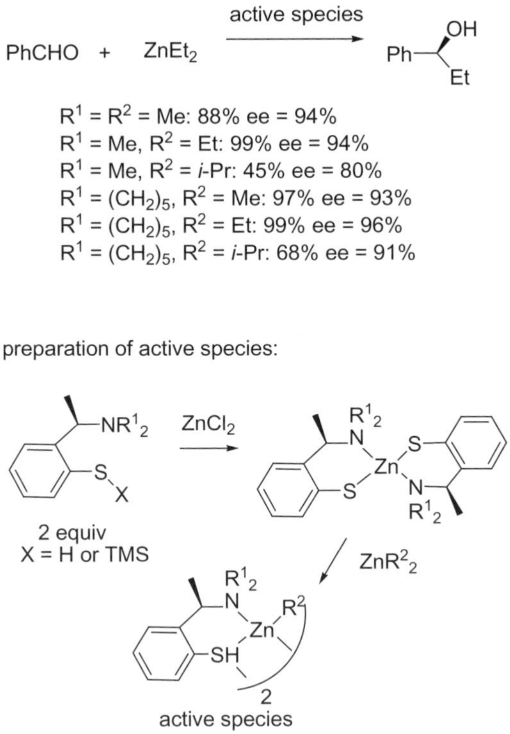
active species
PhCHO + ZnEt2
Ph
OH
Et
R1 = R2 = Me: 88% ee = 94%
R1 = Me, R2 = Et: 99% ee = 94%
R1 = Me, R2 = i-Pr: 45% ee = 80%
R1 = (CH2)5, R2 = Me: 97% ee = 93%
R1 = (CH2)5, R2 = Et: 99% ee = 96%
R1 = (CH2)5, R2 = i-Pr: 68% ee = 91%
preparation of active species:
NR12
S
X
2 equiv
X = H or TMS
ZnCl2
Zn
ZnR22
SH
R2
2
active species

Scheme 3.14 Bis(arenethiolates) zinc complexes for addition of $ZnEt_2$ to benzaldehyde.

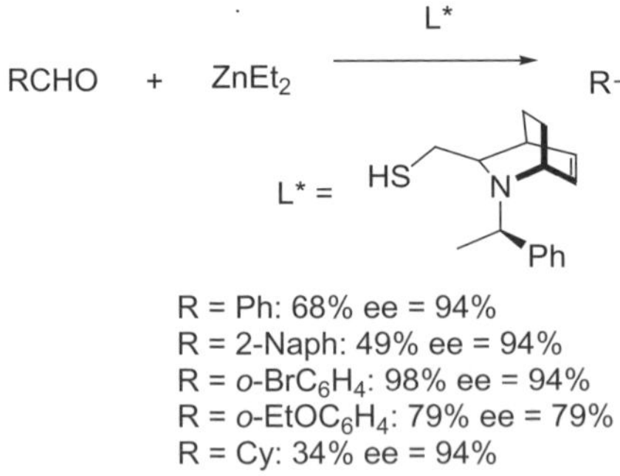
L*
RCHO + ZnEt2
R
OH
Et
L* =
HS
N
Ph
R = Ph: 68% ee = 94%
R = 2-Naph: 49% ee = 94%
R = o-BrC6H4: 98% ee = 94%
R = o-EtOC6H4: 79% ee = 79%
R = Cy: 34% ee = 94%

Scheme 3.15 β-Amino thiol ligand based on isoquinuclidine skeleton for additions of $ZnEt_2$ to aldehydes.

even better results in the same reactions to those described above, since enantioselectivities of up to >99% ee were attained, as shown in Scheme 3.16.[25] The role of the sulfur in this ligand was clearly demonstrated by the investigation in the reaction of $ZnEt_2$ to benzaldehyde of the corresponding β-amino alcohol ligand that gave an enantioselectivity of only 22% ee (Scheme 3.16).

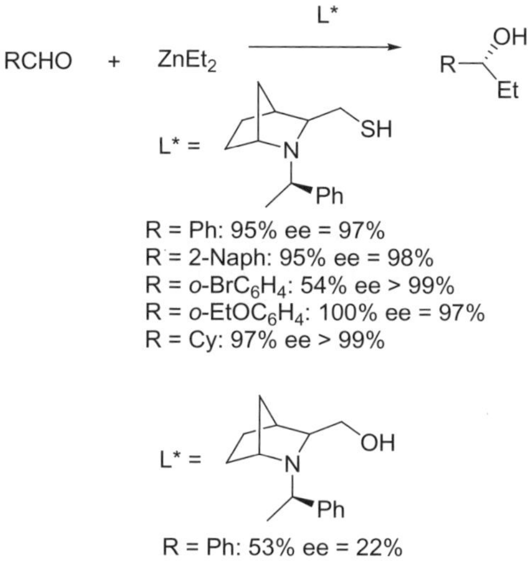
L*
RCHO
+
ZnEt2
OH
R
Et
L* =
SH
N
Ph
R = Ph: 95% ee = 97%
R = 2-Naph: 95% ee = 98%
R = o-BrC6H4: 54% ee > 99%
R = o-EtOC6H4: 100% ee = 97%
R = Cy: 97% ee > 99%
L* =
OH
N
Ph
R = Ph: 53% ee = 22%

Scheme 3.16 β-Amino thiol ligand based on bicycle[2.2.1] ring system and its β-amino alcohol analogue for additions of $ZnEt_2$ to aldehydes.

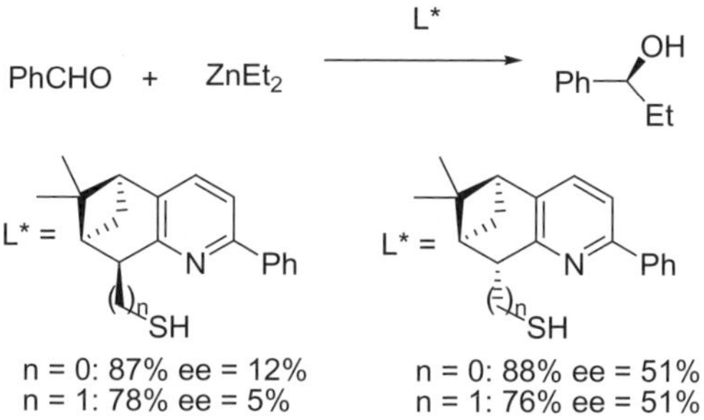
L*
PhCHO
+
ZnEt2
OH
Ph
Et
L* =
N
Ph
SH
n = 0: 87% ee = 12%
n = 1: 78% ee = 5%
L* =
N
Ph
SH
n = 0: 88% ee = 51%
n = 1: 76% ee = 51%

Scheme 3.17 Sulfur-containing pyridine ligands for addition of $ZnEt_2$ to benzaldehyde.

On the other hand, sulfur-containing pyridine ligands have been prepared by Chelucci *et al.*[26] These ligands gave only moderate enantioselectivities ($\leq$51% ee) in the enantioselective addition of $ZnEt_2$ to benzaldehyde, as shown in Scheme 3.17.

Chiral dialkyl thiophosphoramidates have scarcely been used as chiral ligands in asymmetric synthesis. In 1993, Soai *et al.* demonstrated that chiral dialkyl thiophosphoramidates derived from norephedrine were highly enantioselective catalysts for the addition of organozinc reagents to aldehydes, providing enantioselectivities of up to 98% ee, as shown in Scheme 3.18.[27] It was shown that the presence of Ti(O*i*-Pr)$_4$ increased the enantioselectivity. In addition, the enantioselectivities obtained with the thiophosphoramidate ligands were higher than those obtained by using the corresponding oxygen analogues.[28]

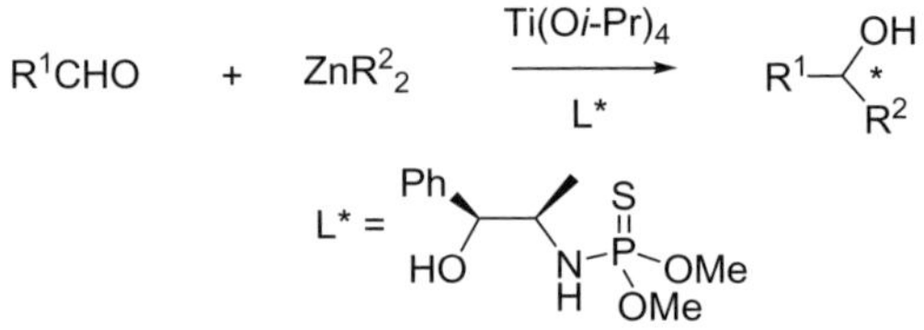

$$R^1CHO \quad + \quad ZnR^2_2 \quad \xrightarrow[\text{L*}]{\text{Ti(O}i\text{-Pr)}_4} \quad R^1\underset{R^2}{\overset{OH}{\underset{*}{<}}}$$

L* =

Ph
HO — / — N(H)—P(=S)(OMe)OMe

R^1 = Ph, R^2 = Et: 90% ee = 96% (*S*)
R^1 = Ph, R^2 = Me: 78% ee = 97% (*S*)
R^1 = *m*-MeOC$_6$H$_4$, R^2 = Et: 85% ee = 98% (*S*)
R^1 = (*E*)-Ph-CH=CH, R^2 = Et: 87% ee = 91% (*S*)
R^1 = *n*-Oct, R^2 = Et: 45% ee = 79% (*S*)

Scheme 3.18   Norephedrine-derived thiophosphoramidate ligand for Ti-catalysed additions of ZnR$_2$ to aldehydes.

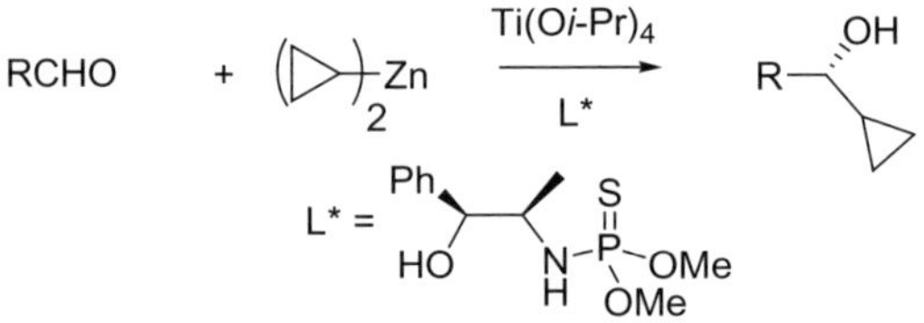

$$RCHO \quad + \quad (\triangleright)_2 Zn \quad \xrightarrow[\text{L*}]{\text{Ti(O}i\text{-Pr)}_4} \quad R\overset{OH}{\underset{\triangle}{<}}$$

L* =

Ph
HO — / — N(H)—P(=S)(OMe)OMe

R = Ph: 93% ee = 96%
R = *p*-MeOC$_6$H$_4$: 99% ee = 97%
R = *p*-ClC$_6$H$_4$: 85% ee = 91%
R = 1-Naph: 89% ee = 95%
R = (*E*)-Ph-CH=CH: 94% ee = 67%
R = Cy: 87% ee = 71%

Scheme 3.19   Norephedrine-derived thiophosphoramidate ligand for Ti-catalysed additions of ZnCy$_2$ to aldehydes.

The scope of this methodology was extended to the enantioselective cyclo-propylation of various aldehydes using dicyclopropylzinc in the presence of this norephedrine-derived thiophosphoramidate ligand combined with Ti(O*i*-Pr)$_4$ to achieve the corresponding cyclopropyl alkanols with enantioselectivities of up to 97% ee (Scheme 3.19).[29]

In 1999, Shi *et al.* showed that a diphenylthiophosphoramide derived from (1*R*,2*R*)-1,2-diaminocyclohexane could be used as a ligand in the catalytic asymmetric addition of ZnEt$_2$ to aldehydes in the presence of Ti(O*i*-Pr)$_4$, providing the corresponding alcohols in enantioselectivities of 40–50% ee (Scheme 3.20).[30] Another class of new ligands such as the phenylthio-phosphoramide of (*R*)-1,1′-binaphthyl-2,2′-diamine was developed by the same group, and further tested as a ligand in the same conditions (Scheme 3.20).[31]

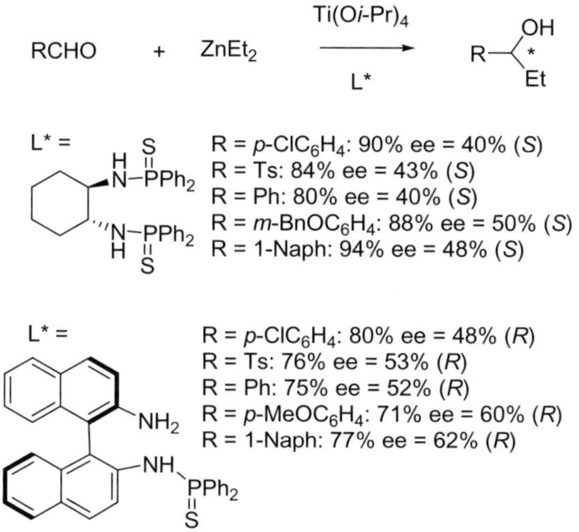

Ti(O*i*-Pr)4
RCHO + ZnEt2
L*
OH
R
*
Et
L* =
S
H
N–PPh2
N–PPh2
H
S
R = *p*-ClC6H4: 90% ee = 40% (*S*)
R = Ts: 84% ee = 43% (*S*)
R = Ph: 80% ee = 40% (*S*)
R = *m*-BnOC6H4: 88% ee = 50% (*S*)
R = 1-Naph: 94% ee = 48% (*S*)
L* =
NH2
NH
PPh2
S
R = *p*-ClC6H4: 80% ee = 48% (*R*)
R = Ts: 76% ee = 53% (*R*)
R = Ph: 75% ee = 52% (*R*)
R = *p*-MeOC6H4: 71% ee = 60% (*R*)
R = 1-Naph: 77% ee = 62% (*R*)

Scheme 3.20 (Di)phenylthiophosphoramide ligands for Ti-catalysed additions of $ZnEt_2$ to aldehydes.

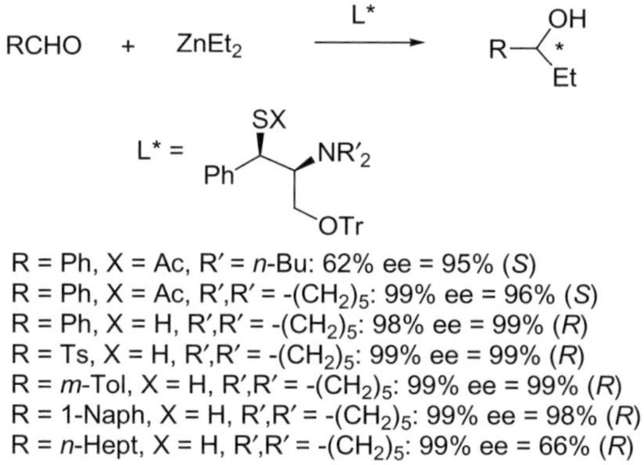

RCHO + ZnEt2
L*
OH
R
*
Et
L* =
SX
Ph
NR′2
OTr
R = Ph, X = Ac, R′ = *n*-Bu: 62% ee = 95% (*S*)
R = Ph, X = Ac, R′,R′ = -(CH2)5: 99% ee = 96% (*S*)
R = Ph, X = H, R′,R′ = -(CH2)5: 98% ee = 99% (*R*)
R = Ts, X = H, R′,R′ = -(CH2)5: 99% ee = 99% (*R*)
R = *m*-Tol, X = H, R′,R′ = -(CH2)5: 99% ee = 99% (*R*)
R = 1-Naph, X = H, R′,R′ = -(CH2)5: 99% ee = 98% (*R*)
R = *n*-Hept, X = H, R′,R′ = -(CH2)5: 99% ee = 66% (*R*)

Scheme 3.21 Norephedrine thiol and thioacetate ligands for additions of $ZnEt_2$ to aldehydes.

3.2.1.2 S/N/O Ligands

On the other hand, numerous ligands possessing three heterodonor atoms such as N, S and O have been developed and further applied to the enantioselective addition of organozinc reagents to aldehydes. As an example, a high enantioselectivity (up to 99% ee) has been obtained by Pericas *et al.*, using a new family of amino thiols, the norephedrine thiols depicted in Scheme 3.21.[32] The corresponding acetyl derivatives also showed very high enantioselectivities (up to 96% ee) for similar reactions. In contrast, for aliphatic aldehydes such as *n*-heptanal, the enantioselectivity remained moderate (66% ee). The authors have

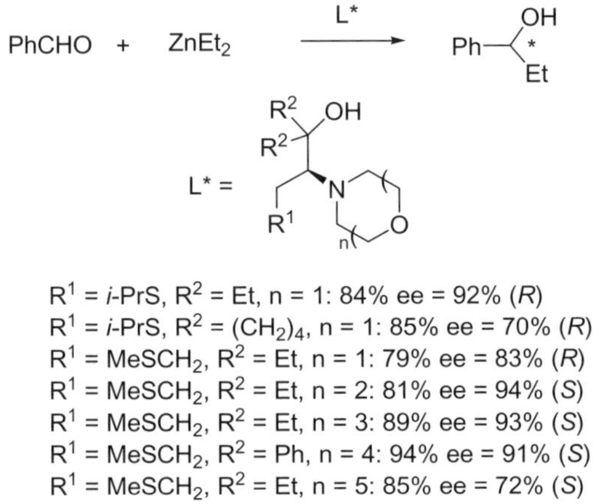

$$\text{PhCHO} + \text{ZnEt}_2 \xrightarrow{\text{L*}} \text{Ph}\overset{\text{OH}}{\underset{\text{Et}}{\overset{*}{-\!\!\!<}}}$$

L* =

R¹ = *i*-PrS, R² = Et, n = 1: 84% ee = 92% (*R*)
R¹ = *i*-PrS, R² = (CH₂)₄, n = 1: 85% ee = 70% (*R*)
R¹ = MeSCH₂, R² = Et, n = 1: 79% ee = 83% (*R*)
R¹ = MeSCH₂, R² = Et, n = 2: 81% ee = 94% (*S*)
R¹ = MeSCH₂, R² = Et, n = 3: 89% ee = 93% (*S*)
R¹ = MeSCH₂, R² = Ph, n = 4: 94% ee = 91% (*S*)
R¹ = MeSCH₂, R² = Et, n = 5: 85% ee = 72% (*S*)

Scheme 3.22   Sulfur-containing N,O-heterocycle ligands for addition of ZnEt_2 to benzaldehyde.

proposed a positive influence of the steric effect generated by the primary hydroxyl-protecting group on the catalytic activity.

The enantioselectivity exerted by a series of new chiral sulfur-containing catalysts containing N,O-heterocycles derived from natural chiral amino acids has been checked in the addition of ZnEt_2 to benzaldehyde (Scheme 3.22).[33] Molecular mechanic calculations suggested that the production of the (*R*)-alcohol might be explained by a mechanism similar to that described by Noyori, in which ZnEt_2 interacted solely with the N–C–C–OH fragment, whereas the formation of the (*S*)-enantiomer needed the direct participation of the lateral chain of the parent amino acid and the N,O-heterocycle. In the same context, Reiser *et al.* have successfully used bis(oxazolines) ligands bearing a thioether group for the addition of ZnEt_2 to benzaldehyde.[34]

In 2001, Braga *et al.* reported the synthesis of new chiral C_2-symmetric oxazolidine disulfide ligands from (*R*)-cysteine and successfully applied them as catalysts in the asymmetric addition of ZnEt_2 to various aldehydes (Scheme 3.23).[35] In the presence of 2 mol% of ligand, excellent enantioselectivities of up to >99% ee were obtained even with aliphatic aldehydes such as *n*-decanal or *n*-hexanal. These authors proposed that the active catalyst did not maintain its C_2-symmetry during the reaction. The disulfide bond was probably cleaved *in situ* by ZnEt_2.

In order to gain insight into the relative importance of the various donor atoms (N, S, O) available in free or alkylated form resulting in covalent or dative bonds to the metal, respectively, Braga *et al.* tested other chiral sulfides and disulfides derived from D-cysteine.[36] Their application in the addition of ZnEt_2 to aldehydes provided the corresponding alcohols with enantioselec-tivities of up to 99% ee, when catalytic amounts of the disulfide ligand depicted in Scheme 3.24 with the ability to form an S–Zn bond were used. According to these results, these authors demonstrated that, for the addition of ZnEt_2 to

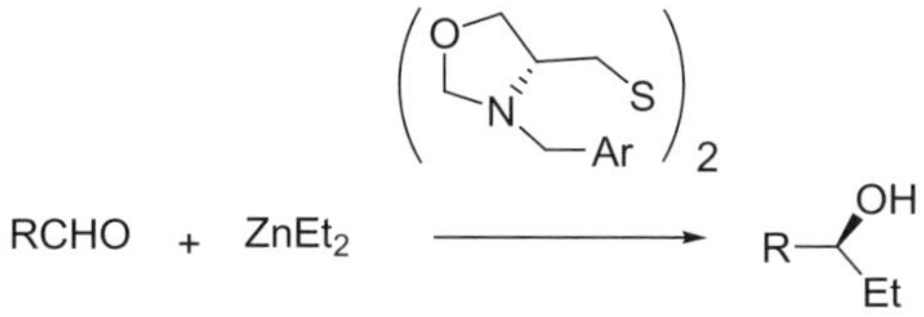

RCHO + ZnEt$_2$ $\longrightarrow$ R$-$

R = Ar = Ph: 81% ee > 99%
R = *n*-Hex, Ar = Ph: 58% ee > 99%
R = Ph, Ar = Ts: 100% ee = 92%
R = *n*-Dec, Ar = Ts: 64% ee > 99%
R = Ph, Ar = *p*-MeOC$_6$H$_4$: 83% ee = 84%
R = *n*-Hex, Ar = *p*-MeOC$_6$H$_4$: 81% ee > 99%
R = *n*-Dec, Ar = *p*-MeOC$_6$H$_4$: 86% ee = 94%

Scheme 3.23 Oxazolidine disulfide ligands for additions of ZnEt$_2$ to aldehydes.

RCHO + ZnEt$_2$ $\longrightarrow$ R$-$

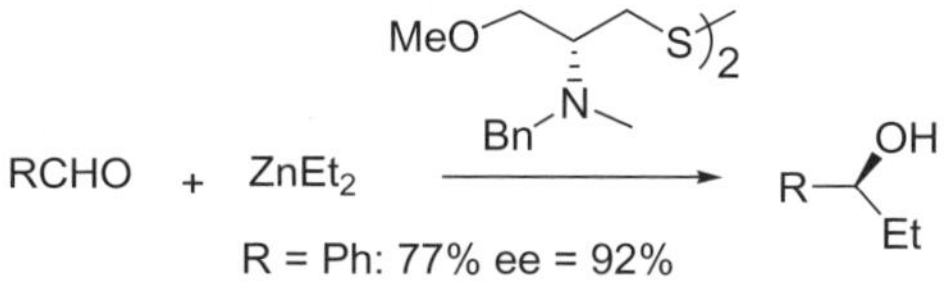

R = Ph: 77% ee = 92%
R = *p*-Tol: 67% ee = 81%
R = *p*-MeOC$_6$H$_4$: 43% ee > 99%
R = Bn: 51% ee = 91%
R = *n*-Hex: 92% ee = 88%
R = *n*-Dec: 86% ee = 80%

Scheme 3.24 D-Cysteine-derived disulfide ligand for additions of ZnEt$_2$ to aldehydes.

aldehydes, a covalent donor–Zn bond proved to be crucial and a thiolate–Zn bond was mandatory for good enantioselectivity.

In addition, other chiral sulfide ligands containing oxazolidines have been tested for the enantioselective addition of ZnEt$_2$ to aldehydes, providing moderate enantioselectivities, as shown in Scheme 3.25.[37]

In 2000, Cho *et al.* reported the synthesis of a β-amino thiol and its corresponding acetate derived from D-mannitol.[38] These ligands were applied to the enantioselective addition of ZnEt$_2$ to benzaldehyde and *n*-heptanal, providing the corresponding products in moderate to good enantioselectivities (Scheme 3.26). In both cases, the best ligand was the β-amino thiol ligand. The authors have compared this ligand with its corresponding β-amino alcohol ligand, obtaining a better enantioselectivity (92% ee instead of 82% ee) for the addition of ZnEt$_2$ to benzaldehyde. This result constituted one of the rare examples in the literature in which the amino alcohol ligand proved better than the corresponding thiol in term of enantioselectivity, in the enantioselective addition of ZnEt$_2$ to aldehydes.

Enantioselectivities ranging from 18 to 94% ee were obtained by Martens *et al.* by using C_2-symmetric bis-β-amino alcohols derived from D-cysteine for the enantioselective addition of ZnEt$_2$ to benzaldehyde (Scheme 3.27).[39]

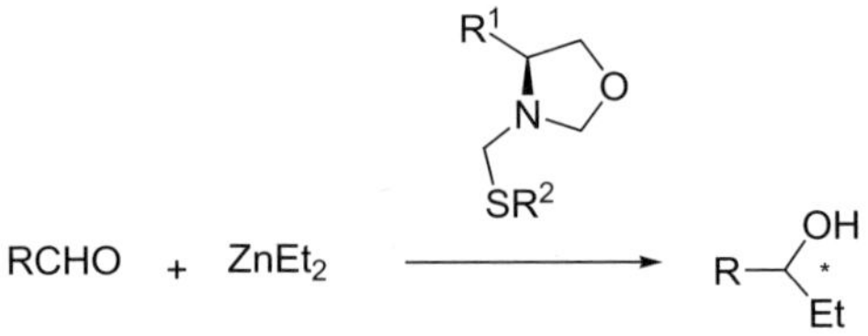

R^1 = Bn, R^2 = Ph: 97% ee = 30%
R^1 = R^2 = Bn: 95% ee = 13%
R^1 = *i*-Pr, R^2 = Bn: 75% ee = 37%

Scheme 3.25 Oxazolidine sulfide ligands for additions of ZnEt$_2$ to aldehydes.

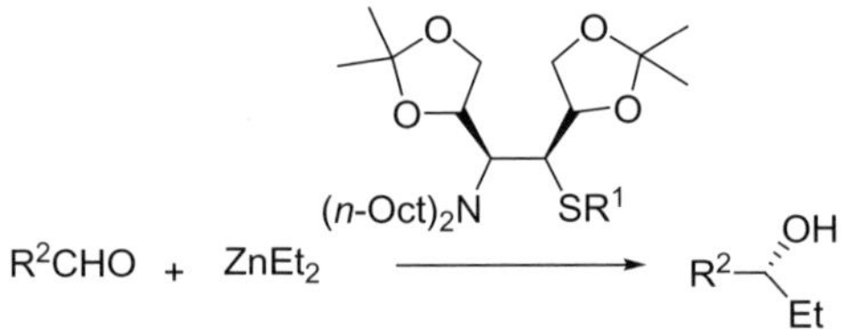

R^1 = H, R^2 = Ph: 97% ee = 82%
R^1 = Ac, R^2 = Ph: 76% ee = 43%
R^1 = H, R^2 = *n*-Hept: 80% ee = 73%
R^1 = Ac, R^2 = *n*-Hept: 76% ee = 45%

Scheme 3.26 D-Mannitol β-amino thio ligands for additions of ZnEt$_2$ to aldehydes.

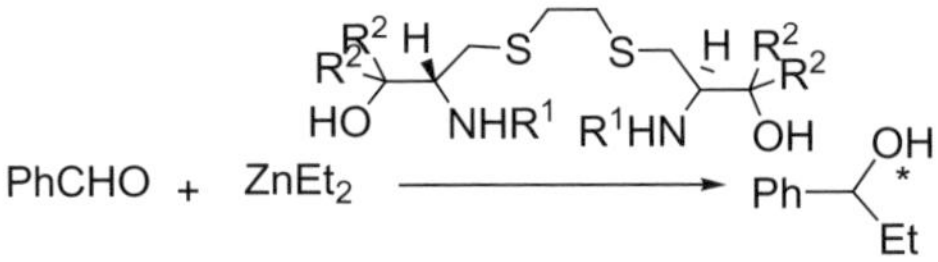

R^1 = H, R^2 = Et: ee = 18%
R^1 = H, R^2 = (CH$_2$)$_4$: ee = 29%
R^1 = H, R^2 = Ph: ee = 38%
R^1 = Me, R^2 = Et: ee = 21%
R^1 = Me, R^2 = (CH$_2$)$_4$: ee = 33%
R^1 = Me, R^2 = Ph: ee = 94%

Scheme 3.27 D-Cysteine-derived C_2-symmetric bis-β-amino alcohol ligands for addition of ZnEt$_2$ to benzaldehyde.

An examination of the catalytic abilities of C_2-symmetric chiral *N*-(β-mercaptoethyl)pyrrolidines in the enantioselective addition of ZnEt$_2$ to a variety of aldehydes was reported by Masaki *et al.*, in 1996.[40] The best enantioselectivity of up to 99% ee was obtained by using a C_2-symmetric chiral *N*-(β-mercaptoethyl)pyrrolidine bearing six-membered benzylidene acetal functionalities fused at the 2,3- and 4,5-positions as a ligand, as shown in Scheme 3.28.

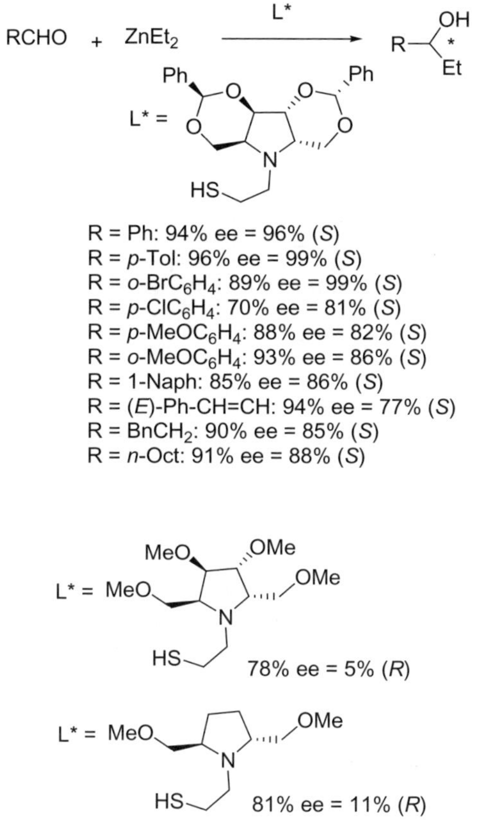

RCHO + ZnEt$_2$ $\xrightarrow{\text{L*}}$ R$-$CH(OH)(Et)*

R = Ph: 94% ee = 96% (S)
R = p-Tol: 96% ee = 99% (S)
R = o-BrC$_6$H$_4$: 89% ee = 99% (S)
R = p-ClC$_6$H$_4$: 70% ee = 81% (S)
R = p-MeOC$_6$H$_4$: 88% ee = 82% (S)
R = o-MeOC$_6$H$_4$: 93% ee = 86% (S)
R = 1-Naph: 85% ee = 86% (S)
R = (E)-Ph-CH=CH: 94% ee = 77% (S)
R = BnCH$_2$: 90% ee = 85% (S)
R = n-Oct: 91% ee = 88% (S)

78% ee = 5% (R)

81% ee = 11% (R)

Scheme 3.28 C_2-Symmetric N-(β-mercaptoethyl)pyrrolidine ligands for additions of ZnEt$_2$ to aldehydes.

The enantioselective addition of ZnEt$_2$ to benzaldehyde was also performed in the presence of chiral 2,2-disubstituted thiaprolinol derivatives as ligands by Liu *et al.*, providing the product with an enantioselectivity of up to 81% ee (Scheme 3.29).[41]

Various sulfur-containing β-amino alcohols, depicted in Scheme 3.30, have been investigated as ligands for the enantioselective addition of ZnEt$_2$ to benzaldehyde by Martens *et al.*, providing enantioselectivities of up to 91% ee.[42] It was shown that the use of the lithiated analogues of these ligands led to better enantioselectivities, as shown in Scheme 3.30. In order to explain these results, the authors have proposed that the lithium cation might coordinate more easily with the oxygen atom of the approaching aldehyde than zinc. This coordination should restrict the number of possible stereochemical courses of the reaction, allowing high optical purities to be reached.

In 2008, Juaristi *et al.* developed the synthesis of a series of novel chiral thioureas that were further examined as possible ligands for the enantioselective addition of ZnEt$_2$ to benzaldehyde.[43] The expected carbinol was isolated in

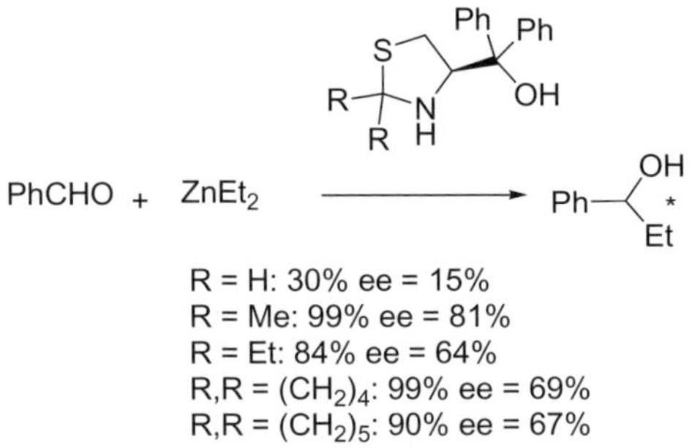

PhCHO + ZnEt$_2$ ⟶ Ph—$\overset{OH}{\underset{Et}{\overset{|}{C}}}$*

R = H: 30% ee = 15%
R = Me: 99% ee = 81%
R = Et: 84% ee = 64%
R,R = (CH$_2$)$_4$: 99% ee = 69%
R,R = (CH$_2$)$_5$: 90% ee = 67%

Scheme 3.29 2,2-Disubstituted thiaprolinol ligands for addition of ZnEt$_2$ to benzaldehyde.

PhCHO + ZnEt$_2$ $\xrightarrow{\text{L*}}$ Ph—$\overset{OH}{\underset{Et}{\overset{|}{C}}}$*

L* = MeS, R^{1}HN, OX, Ph, Ph

R^1 = X = H: ee = 5% (R)
R^1 = H, X = Li: ee = 48% (S)
R^1 = Et, X = H: ee = 76% (S)
R^1 = Et, X = Li: ee = 79% (S)
R^1 = n-Pr, X = H: ee = 91% (S)
R^1 = n-Pr, X = Li: ee = 89% (S)
R^1 = n-Bu, X = H: ee = 63% (S)
R^1 = n-Bu, X = Li: ee = 78% (S)

L* = i-PrS, R^{1}HN, OH, Ph, Ph

R^1 = X = H : ee = 45% (R)
R^1 = H, X = Li: ee = 62% (R)
R^1 = Et, X = H: ee = 74% (R)
R^1 = Et, X = Li: ee = 82% (S)
R^1 = n-Pr, X = H: ee = 79% (S)
R^1 = n-Pr, X = Li: ee = 94% (S)

L* = Ph, Ph, OX, R^2, R^2, S, R^3, R^3, N, H

R^2 = R^3 = X = H: 93% (S)
R^2 = R^3 = H, X = Li: 6% (S)
R^2,R^2 = (CH$_2$)$_4$, R^3 = X = H: 84% (S)
R^2,R^2 = (CH$_2$)$_5$, R^3 = X = H: 80% (S)

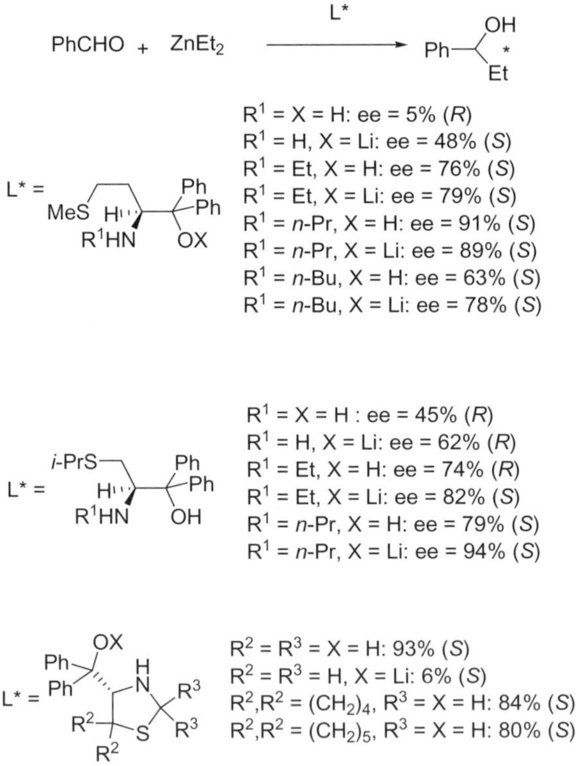

Scheme 3.30 Sulfur-containing β-amino alcohol ligands for addition of ZnEt$_2$ to benzaldehyde.

nearly quantitative yield in almost all cases of ligands but with a low enantioselectivity ($\leq$26% ee), as shown in Scheme 3.31.

In 2008, the titanium complex of a novel BINOL-based thiazole thioether ligand was found by Li *et al.* to be an efficient catalyst in the enantioselective

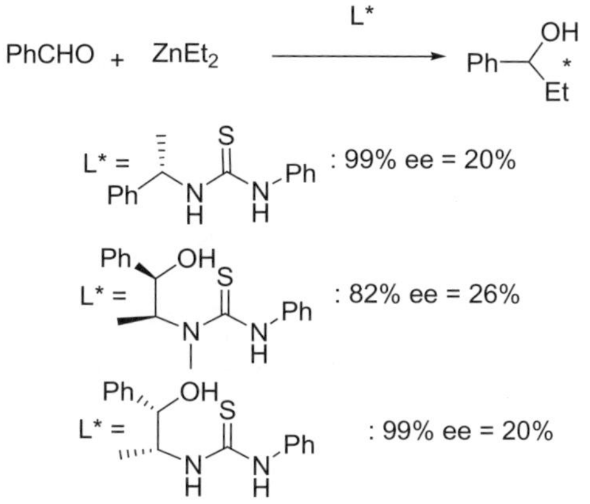
PhCHO + ZnEt2
L*
Ph
OH
Et
L* =
S
Ph
N
H
: 99% ee = 20%
: 82% ee = 26%
: 99% ee = 20%

Scheme 3.31 Thiourea ligands for addition of $ZnEt_2$ to benzaldehyde.

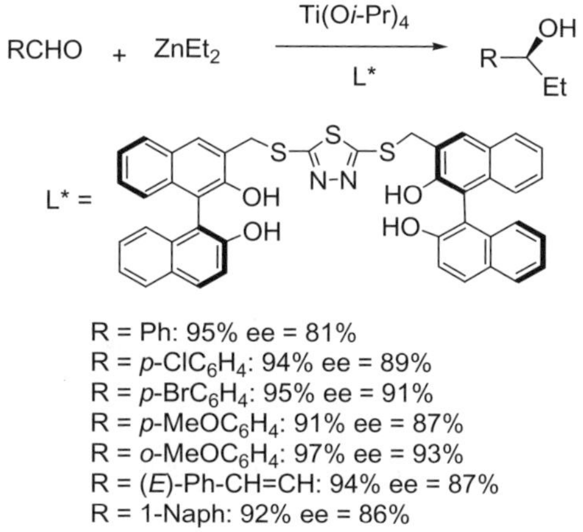
RCHO + ZnEt2
Ti(Oi-Pr)4
L*
R
OH
Et
L* =
S
N-N
OH
HO
R = Ph: 95% ee = 81%
R = p-ClC6H4: 94% ee = 89%
R = p-BrC6H4: 95% ee = 91%
R = p-MeOC6H4: 91% ee = 87%
R = o-MeOC6H4: 97% ee = 93%
R = (E)-Ph-CH=CH: 94% ee = 87%
R = 1-Naph: 92% ee = 86%

Scheme 3.32 BINOL-based thiazole thioether ligand for additions of $ZnEt_2$ to aldehydes.

addition of $ZnEt_2$ to aldehydes.[44] The use of this new type of ligand, bound with both sulfur-contained heterocycle (thiazole) and thioether block in which the sulfur might serve as a talent anchor, allowed enantioselectivities of up to 93% ee to be obtained, as shown in Scheme 3.32.

In 1997, Chelucci *et al.* developed new chiral 2-(1-*p*-tolylsulfinyl)alkylpyridines and assessed these ligands in the enantioselective addition of $ZnEt_2$ to benzaldehyde.[45] This produced in all cases catalysts in good yields but with low enantioselectivities ($\leq$19% ee). An examination of the results collected in Scheme 3.33 indicates that the enantioselectivity and the configuration of the resulting carbinol were independent of both the substituent and the configuration at the carbon centre. Therefore, this suggests that the stereochemical

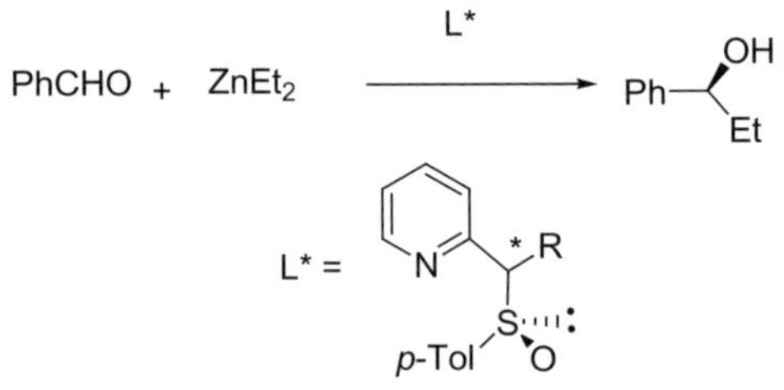

PhCHO + ZnEt₂ →(L*) Ph—〈OH / Et〉

L* = 〈pyridyl〉—CH(*R)—S(p-Tol)(⁝O)

(-)-L*, R = *i*-Pr: 87% ee = 13%
(+)-L*, R = *i*-Pr: 91% ee = 11%
(+)-L*, R = Ph: 86% ee = 14%
(-)-L*, R = Ph: 92% ee = 3%
(-)-L*, R = *t*-Bu: 85% ee = 19%
(+)-L*, R = *t*-Bu: 89% ee = 15%

Scheme 3.33   β-Pyridylsulfoxide ligands for addition of ZnEt₂ to benzaldehyde.

PhCHO + ZnEt₂ →(L*) Ph—〈OH / Et〉*

L* = 〈bipyridine with S-Me and S=O groups〉 : 40% ee = 14%

L* = 〈bipyridine with O=S-R groups〉 : R = *i*-Pr: 41% ee = 14%
R = CH₂CO₂Et: 35% ee = 11%

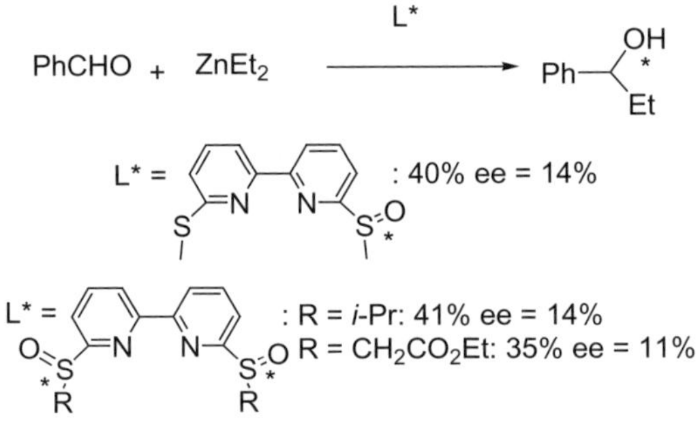

Scheme 3.34   2,2′-Bipyridine alkyl sulfoxide ligands for addition of ZnEt₂ to benzaldehyde.

R¹CHO + ZnEt₂ →(Ph–S(=O)(=NR²)–CH₂–C(OH)(R³)(R³)) R¹—〈OH / Et〉*

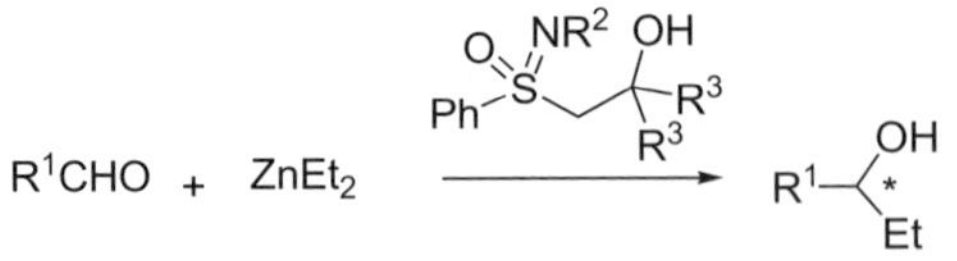

R¹ = Ph, R² = Me, R³ = (CH₂)₃: 55% ee = 84%
R¹ = Ph, R² = Me, R³ = (CH₂)₉: 50% ee = 88%
R¹ = Ph, R² = R³ = Me: 73% ee = 85%
R¹ = *p*-ClC₆H₄, R² = R³ = Me: 79% ee = 79%
R¹ = *p*-Tol, R² = R³ = Me: 48% ee = 75%
R¹ = BnCH₂, R² = R³ = Me: 50% ee = 61%
R¹ = Ph, R² = Et, R³ = Me: 94% ee = 68%

Scheme 3.35   β-Hydroxy sulfoximine ligands for additions of ZnEt₂ to aldehydes.

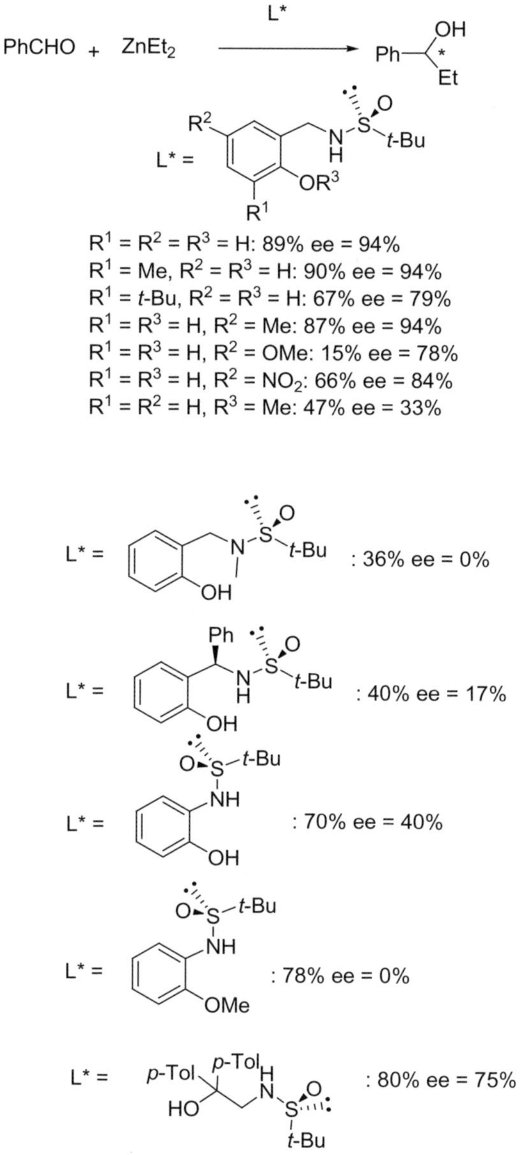

PhCHO + ZnEt$_2$ $\xrightarrow{\text{L*}}$ Ph—C*(OH)(Et)

L* = (structure with R^2, N-H-S(=O)-t-Bu, OR^3, R^1)

R^1 = R^2 = R^3 = H: 89% ee = 94%
R^1 = Me, R^2 = R^3 = H: 90% ee = 94%
R^1 = t-Bu, R^2 = R^3 = H: 67% ee = 79%
R^1 = R^3 = H, R^2 = Me: 87% ee = 94%
R^1 = R^3 = H, R^2 = OMe: 15% ee = 78%
R^1 = R^3 = H, R^2 = NO$_2$: 66% ee = 84%
R^1 = R^2 = H, R^3 = Me: 47% ee = 33%

L* = (structure) : 36% ee = 0%

L* = (structure) : 40% ee = 17%

L* = (structure) : 70% ee = 40%

L* = (structure) : 78% ee = 0%

L* = (structure) : 80% ee = 75%

Scheme 3.36 Sulfinamido ligands for addition of ZnEt$_2$ to benzaldehyde.

outcome of the reaction was determined by the stereochemistry of the sulfoxide unit.

In 2008, Rykowski *et al.* reported the synthesis of optically active 2,2′-bipyridine alkyl sulfoxides by asymmetric oxidation of their corresponding readily accessible 2,2′-bipyridine alkyl sulfides.[46] These sulfoxides were further evaluated as ligands for the enantioselective addition of ZnEt$_2$ to benzaldehyde, providing only low enantioselectivities of up to 14% ee (Scheme 3.34).

Another class of ligands containing a chiral sulfur atom is constituted by sulfoximines. In this area, Bolm *et al.* have synthesised β-hydroxy sulfoximines depicted in Scheme 3.35, and further studied them as ligands in the enantioselective addition of ZnEt$_2$ to aldehydes.[47] In spite of the fact that these ligands were not S-coordinating, they allowed the corresponding products to be obtained in high enantioselectivities of up to 88% ee. According to the results, it seemed that the asymmetric induction was dependent on the catalyst structure. The best enantioselectivities were obtained with two alkyl groups at the hydroxyl-bearing β-carbon. The authors have studied the reaction of ZnEt$_2$ with benzaldehyde by NMR spectroscopy.

In 2007, Qin *et al.* reported the synthesis of new chiral sulfinamido ligands that were tested in the enantioselective addition of ZnEt$_2$ to benzaldehyde.[48] As shown in Scheme 3.36, the results were strongly dependent on the ligand structure with enantioselectivities ranging from 0 to 94% ee. These authors have suggested that both the nitrogen atom and the oxygen atom of the sulfinamido group, as well as the oxygen atom of the phenol group, participated in the coordination with zinc to form a *O,N,O*-chelating transition state.

Numerous chiral non-S-coordinating bis(sulfonamides) have been successfully involved in the enantioselective addition of various organozinc reagents to aldehydes since the first use of *trans*-1(R),2(R)-bis(trifluoromethanesulfonamido)cyclohexane reported in 1989 by Ohno *et al.*[49] These authors demonstrated the usefulness of this ligand in the Ti-catalysed enantioselective addition of ZnR$_2$ to a variety of aldehydes, allowing enantioselectivities of up to 99% ee to be obtained (Scheme 3.37).

Knochel *et al.* have widely developed the scope of this process by employing a large variety of functionalised aldehydes and dialkylzinc reagents to provide a library of elaborate secondary alcohols with high optical purities.[50] Thus, variously functionalised dialkylzinc reagents have been added to a wide variety of functionalised aldehydes, including α,β-unsaturated aldehydes. Indeed, a wide range of functions were tolerated for both organozinc reagents and aldehydes, as

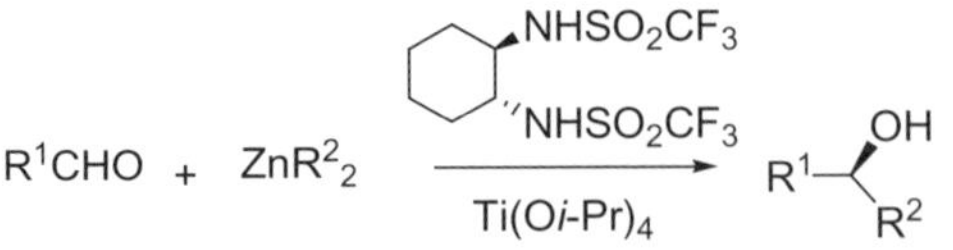

R^1 = Ph, R^2 = Et: 97% ee = 98%
R^1 = (*E*)-Ph-CH=CH, R^2 = Et: 85% ee = 99%
R^1 = BnCH$_2$, R^2 = Et: 100% ee = 92%
R^1 = *n*-Pent, R^2 = Et: 78% ee = 99%
R^1 = Ph, R^2 = *n*-Bu: 99% ee = 98%
R^1 = Ph, R^2 = Me: 99% ee = 73%
R^1 = Ph, R^2 = *n*-Pent: 99% ee = 99%

Scheme 3.37   *Trans*-1(R),2(R)-bis(trifluoromethanesulfonamido)cyclohexane ligand for additions of ZnR$_2$ to aldehydes.

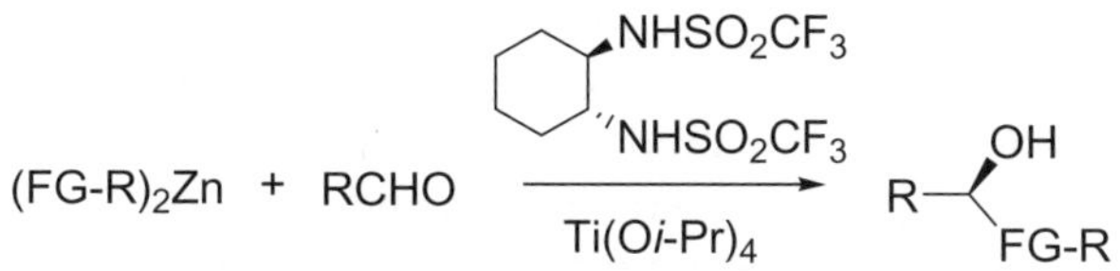

$$(FG\text{-}R)_2Zn \ + \ RCHO \quad \xrightarrow[\text{Ti}(Oi\text{-Pr})_4]{\text{NHSO}_2\text{CF}_3 \ \ \text{NHSO}_2\text{CF}_3} \quad R\text{---}\overset{OH}{\underset{FG\text{-}R}{}}$$

FG-R = AcO(CH$_2$)$_5$, R = Ph: 83% ee = 97%
FG-R = AcO(CH$_2$)$_4$, R = Ph: 72% ee = 92%
FG-R = *t*-BuCO$_2$(CH$_2$)$_3$, R = Ph: 90% ee = 92%
FG-R = Cl(CH$_2$)$_4$, R = Ph: 95% ee = 93%
FG-R = EtCO$_2$(CH$_2$)$_3$, R = Ph: 75% ee = 60%
FG-R = *n*-Dec, R = Ph: 88% ee = 96%
FG-R = Cy, R = Ph: 67% ee = 80%
FG-R = PivO(CH$_2$)$_3$, R = Ph: 58% ee = 93%
FG-R = I(CH$_2$)$_6$, R = Ph: 58% ee = 86%
FG-R = *n*-Oct, R = Me$_3$Sn(CH$_2$)$_2$: 80% ee = 97%
FG-R = Et, R = Me$_3$Sn(CH$_2$)$_2$: 89% ee = 93%
FG-R = *n*-Oct, R = (*n*-Bu)$_3$Sn(CH$_2$)$_2$: 82% ee = 93%
FG-R = AcO(CH$_2$)$_4$, R = (*n*-Bu)$_3$Sn(CH$_2$)$_2$: 85% ee = 93%
FG-R = AcO(CH$_2$)$_5$, R = (*n*-Bu)$_3$Sn(CH$_2$)$_2$: 81% ee = 92%
FG-R = Cl(CH$_2$)$_4$, R = (*n*-Bu)$_3$Sn(CH$_2$)$_2$: 79% ee = 95%
FG-R = *n*-Pent, R = (*n*-Bu)$_3$Sn-CH=CH: 88% ee = 92%
FG-R = AcO(CH$_2$)$_5$, R = (*n*-Bu)$_3$Sn-CH=CH: 75% ee = 91%

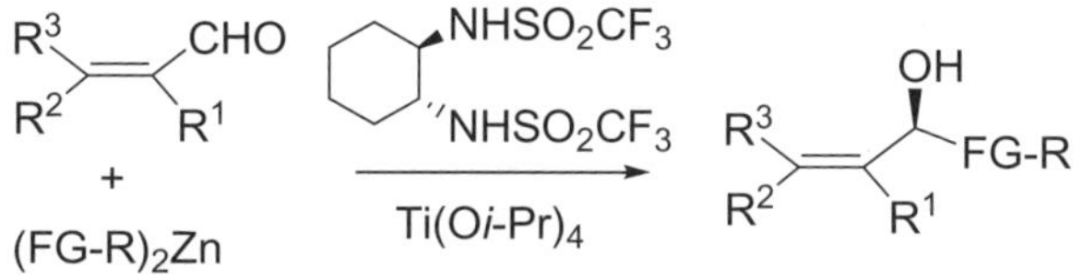

$$\underset{R^2}{\overset{R^3}{}}\!\!=\!\!\underset{R^1}{\overset{CHO}{}} \ + \ (FG\text{-}R)_2Zn \quad \xrightarrow[\text{Ti}(Oi\text{-Pr})_4]{\text{NHSO}_2\text{CF}_3 \ \ \text{NHSO}_2\text{CF}_3} \quad \underset{R^2}{\overset{R^3}{}}\!\!=\!\!\underset{R^1}{\overset{}{}}\overset{OH}{\underset{FG\text{-}R}{}}$$

FG-R = AcO(CH$_2$)$_5$, R^1 = R^2 = Me, R^3 = H: 70% ee = 98%
FG-R = *n*-Oct, R^1 = R^2 = Me, R^3 = H: 65% ee = 98%
FG-R = AcO(CH$_2$)$_5$, R^1 = Br, R^2 = *n*-Pr, R^3 = H: 95% ee = 94%
FG-R = Cl(CH$_2$)$_4$, R^1 = Br, R^2 = *n*-Pr, R^3 = H: 68% ee = 95%
FG-R = PivO(CH$_2$)$_3$, R^1 = Br, R^2 = *n*-Pr, R^3 = H: 68% ee = 95%
FG-R = Et, R^1 = Br, R^2 = *n*-Pr, R^3 = H: 94% ee = 95%
FG-R = Et, R^1 = I, R^2 = Sn(*n*-Bu)$_3$, R^3 = H: 75% ee = 94%

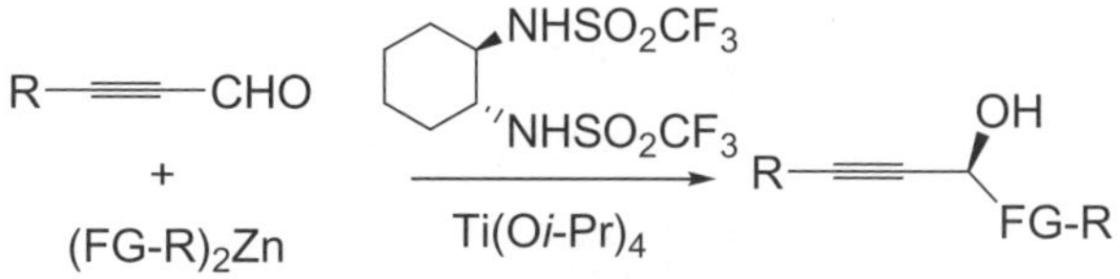

$$R\text{---}\!\!\equiv\!\!\text{---}CHO \ + \ (FG\text{-}R)_2Zn \quad \xrightarrow[\text{Ti}(Oi\text{-Pr})_4]{\text{NHSO}_2\text{CF}_3 \ \ \text{NHSO}_2\text{CF}_3} \quad R\text{---}\!\!\equiv\!\!\text{---}\overset{OH}{\underset{FG\text{-}R}{}}$$

FG-R = *n*-Pent, R = TIPS: 92% ee = 96%
FG-R = *n*-Pent, R = *n*-Bu: 81% ee = 91%
FG-R = *n*-Pent, R = Ph: 87% ee = 92%
FG-R = *n*-Pent, R = *n*-Pr: 89% ee > 96%
FG-R = PivO(CH$_2$)$_4$, R = *n*-Pr: 88% ee > 96%

Scheme 3.38 *Trans*-1(*R*),2(*R*)-bis(trifluoromethanesulfonamido)cyclohexane ligand for additions of Zn(FG-R)$_2$ to functionalised aldehydes.

shown in Scheme 3.38. The proposed catalyst for this process was the bis(sulfonamido) titanium bis(isopropoxides) complex depicted in Scheme 3.38.

A number of other bis(sulfonamides) derived from *trans*-1,2-diaminocyclohexane have been developed by several groups and successfully applied to these reactions. Thus, Walsh *et al.* have reported the use of a series of these ligands in the enantioselective addition of ZnEt$_2$ to benzaldehyde giving excellent enantioselectivities, as shown in Scheme 3.39.[51]

In the same context, *trans*-1,2-dicamphorsulfonamidocyclohexane was found by Uang and Hwang to promote the Ti-catalysed enantioselective addition of ZnEt$_2$ to various aldehydes with high yields and moderate to high enantioselectivities (Scheme 3.40).[52] Moreover, several chiral tetradentate ligands derived from *trans*-1,2-diaminocyclohexane were also applied to these reactions by the groups of Walsh[53] and Zhang,[54] providing the corresponding products in good to excellent enantioselectivities (Scheme 3.40).

The isoborneol-10-sulfonamide unit represents a new type of chiral structure that has been successfully used in different reactions of organometallic addition to aldehydes with excellent results in some cases. In particular, this unique type of ligand has been investigated by Yus and Ramon for the enantioselective

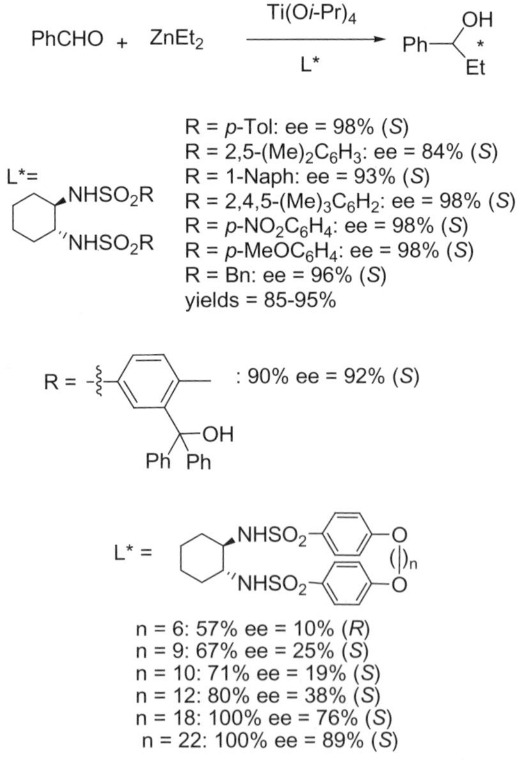

$$PhCHO + ZnEt_2 \xrightarrow[L^*]{Ti(Oi\text{-}Pr)_4} Ph\text{---}\overset{OH}{\underset{Et}{\overset{*}{\text{<}}}}$$

L* = (structure: cyclohexane with NHSO$_2$R, NHSO$_2$R)

R = *p*-Tol: ee = 98% (*S*)
R = 2,5-(Me)$_2$C$_6$H$_3$: ee = 84% (*S*)
R = 1-Naph: ee = 93% (*S*)
R = 2,4,5-(Me)$_3$C$_6$H$_2$: ee = 98% (*S*)
R = *p*-NO$_2$C$_6$H$_4$: ee = 98% (*S*)
R = *p*-MeOC$_6$H$_4$: ee = 98% (*S*)
R = Bn: ee = 96% (*S*)
yields = 85-95%

R = (aryl structure with OH, Ph Ph) : 90% ee = 92% (*S*)

L* = (structure: cyclohexane with NHSO$_2$-aryl-O, NHSO$_2$-aryl-O, (CH$_2$)$_n$)

n = 6: 57% ee = 10% (*R*)
n = 9: 67% ee = 25% (*S*)
n = 10: 71% ee = 19% (*S*)
n = 12: 80% ee = 38% (*S*)
n = 18: 100% ee = 76% (*S*)
n = 22: 100% ee = 89% (*S*)

Scheme 3.39 *Trans*-1,2-diaminocyclohexane-derived bis(sulfonamides) ligands for addition of ZnEt$_2$ to benzaldehyde.

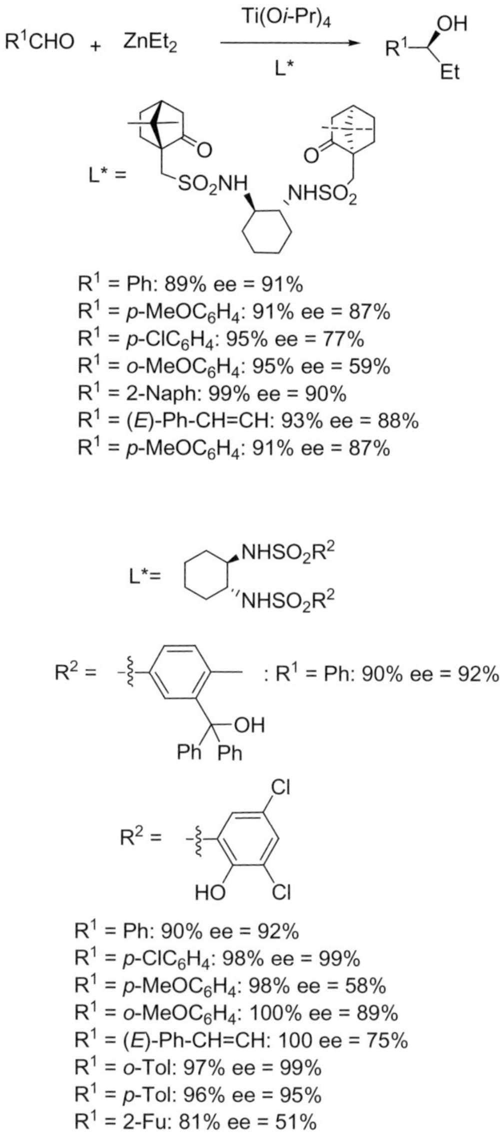
R1CHO + ZnEt2
Ti(Oi-Pr)4
L*
OH
R1
Et
L* =
O
SO2NH
O
NHSO2
R1 = Ph: 89% ee = 91%
R1 = p-MeOC6H4: 91% ee = 87%
R1 = p-ClC6H4: 95% ee = 77%
R1 = o-MeOC6H4: 95% ee = 59%
R1 = 2-Naph: 99% ee = 90%
R1 = (E)-Ph-CH=CH: 93% ee = 88%
R1 = p-MeOC6H4: 91% ee = 87%
L*=
NHSO2R2
NHSO2R2
R2 =
: R1 = Ph: 90% ee = 92%
OH
Ph
Ph
Cl
R2 =
HO
Cl
R1 = Ph: 90% ee = 92%
R1 = p-ClC6H4: 98% ee = 99%
R1 = p-MeOC6H4: 98% ee = 58%
R1 = o-MeOC6H4: 100% ee = 89%
R1 = (E)-Ph-CH=CH: 100 ee = 75%
R1 = o-Tol: 97% ee = 99%
R1 = p-Tol: 96% ee = 95%
R1 = 2-Fu: 81% ee = 51%

Scheme 3.40 *Trans*-1,2-diaminocyclohexane-derived bis(sulfonamides) ligands for additions of $ZnEt_2$ to aldehydes.

addition of organozinc reagents to aldehydes.[55] These authors have introduced hydroxycamphorsulfonamide derivatives as chiral ligands for this type of reaction, obtaining modest enantioselectivities (Scheme 3.41).[56]

In 2000, these workers described disulfonamide ligands derived from camphor, which could be easily prepared from chiral camphorsulfonyl chloride, and further successfully used in the enantioselective addition of $ZnEt_2$ to

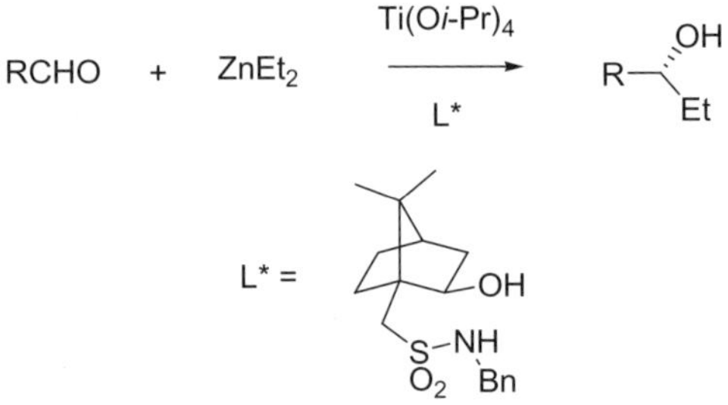

$$\text{RCHO} \quad + \quad \text{ZnEt}_2 \quad \xrightarrow[\substack{\text{L*}}]{\text{Ti(O}i\text{-Pr)}_4} \quad R{-}\overset{\text{OH}}{\underset{\text{Et}}{|}}$$

L* =

R = *n*-Hex: > 95% ee = 68%
R = *c*-Hex: > 95% ee = 92%
R = BnCH$_2$: > 95% ee = 94%
R = *p*-ClC$_6$H$_4$: > 95% ee = 68%
R = *p*-CNC$_6$H$_4$: > 95% ee = 56%
R = *p*-MeOC$_6$H$_4$: > 95% ee = 54%
R = *p*-(Me$_2$N)C$_6$H$_4$: > 95% ee = 14%
R = Ph: > 95% ee = 65%

Scheme 3.41   Hydroxycamphorsulfonamide ligand for additions of ZnEt$_2$ to aldehydes.

$$\text{RCHO} \quad + \quad \text{ZnEt}_2 \quad \xrightarrow[\substack{\text{L*}}]{\text{Ti(O}i\text{-Pr)}_4} \quad R{-}\overset{\text{OH}}{\underset{\text{Et}}{|}}$$

L* =

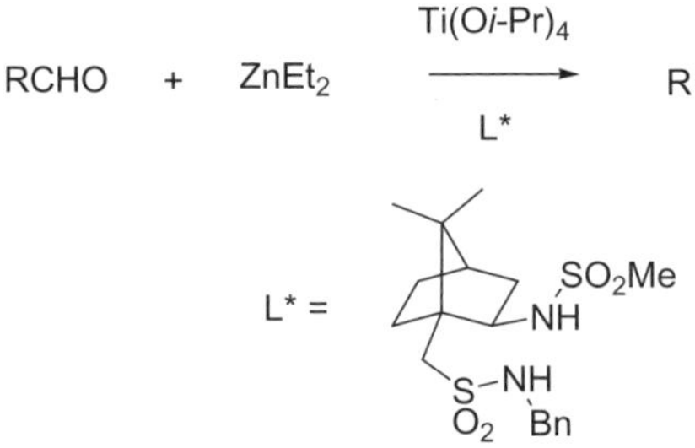

R = *n*-Hex: > 95% ee = 96%
R = *c*-Hex: > 95% ee = 92%
R = BnCH$_2$: > 95% ee = 94%

Scheme 3.42   Camphor-derived disulfonamide ligand for additions of ZnEt$_2$ to aldehydes.

aldehydes in good to high enantioselectivities of up to 96% ee (Scheme 3.42).[57] It is worth noting that the enantioselectivity was higher for aliphatic aldehydes than for aromatic aldehydes, this behaviour being unusual for this type of reaction.

In 2006, Wang *et al.* reported the synthesis of a new camphor-derived di-sulfonamide ligand based on L-tartaric acid that was employed in similar reactions to those described above, giving rise to enantioselectivities of up to 83% ee by using 5 mol% of catalyst loading (Scheme 3.43).[58]

In the same area, Moreau *et al.* have developed new hydrophobic ionic liquids containing chiral camphorsulfonamide units and showed that they could be used as ligands for the Ti-catalysed addition of ZnEt$_2$ to benzaldehyde.[59] The ionic

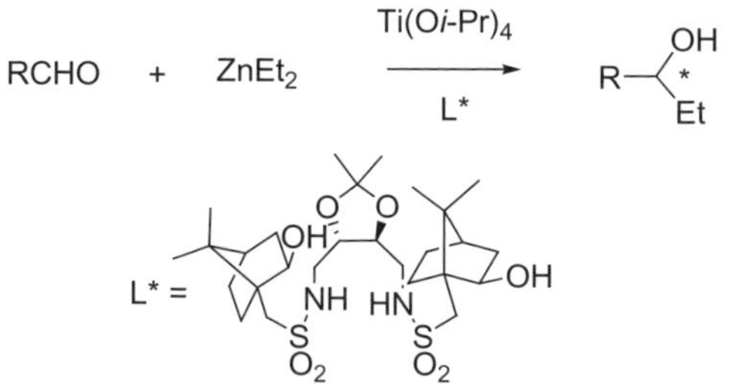

R = Ph: 90% ee = 65%
R = p-FC$_6$H$_4$: 93% ee = 75%
R = p-ClC$_6$H$_4$: 95% ee = 70%
R = o-Tol: 96% ee = 81%
R = o-MeOC$_6$H$_4$: 95% ee = 63%
R = 1-Naph: 85% ee = 83%

Scheme 3.43 L-Tartaric acid-derived disulfonamide ligand for additions of ZnEt$_2$ to aldehydes.

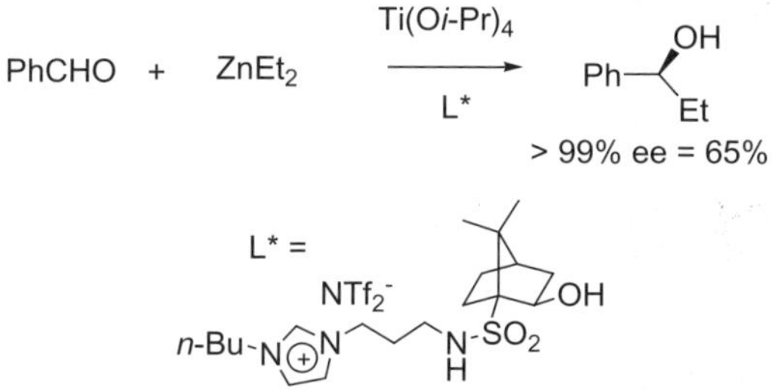

Scheme 3.44 Ionic liquids containing camphor-derived sulfonamide ligand for addition of ZnEt$_2$ to benzaldehyde.

catalytic system showed catalytic properties similar to those of related nonionic counterparts in terms of both activity and enantioselectivity. Interestingly, the ionic ligands could be easily recycled and reused without loss of activity or selectivity. This novel approach offered advantages in combining homogeneous catalysis and easy separation of the chiral ionic ligand. The best result is presented in Scheme 3.44.

The use of chiral C_2-symmetric trifluoromethanesulfonamides derived from (R)-1,1'-binaphthyl-2,2'-diamine in similar reactions to those described above has led to the formation of the expected alcohols with enantioselectivities of 43–54% ees.[31] Better enantioselectivities were observed by Paquette *et al.*, resulting from the use of chiral C_2-symmetric VERDI (verbenone dimers) disulfonamides derived from the dimerisation of (+)-verbenone. Stereoselectivity levels ranging from 72 to 98% ee were observed, depending on the structural characteristics of the aldehyde (Scheme 3.45).[60]

In 1997, Moberg *et al.* showed that chiral bis(ethylsulfonamides)amines, obtained from N-triflyl-2-alkylaziridines via reaction with benzylamine,

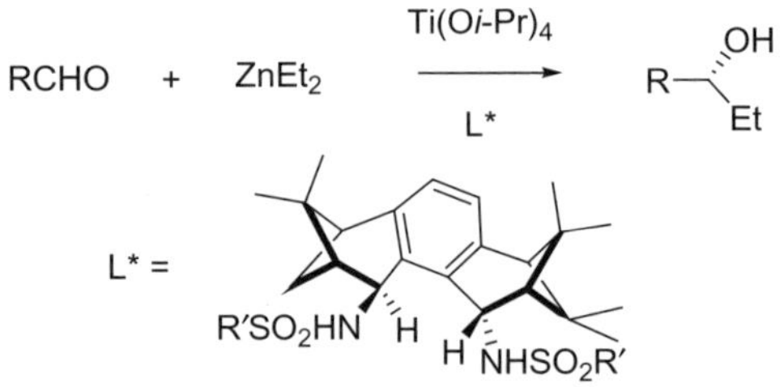

RCHO + ZnEt$_2$ $\xrightarrow[\text{L*}]{\text{Ti(O}i\text{-Pr)}_4}$ R—CH(OH)Et

L* =

R'SO$_2$HN H   H NHSO$_2$R'

R = Ph, R' = Me: 93% ee = 98%
R = Ph, R' = Ts: 93% ee = 89%
R = Ph, R' = CF$_3$: 90% ee = 95%
R = p-FC$_6$H$_4$, R' = Me: 97% ee = 81%
R = p-FC$_6$H$_4$, R' = Ts: 83% ee = 81%
R = p-FC$_6$H$_4$, R' = CF$_3$: 83% ee = 86%
R = Ph(CH$_2$)$_2$, R' = Me: 86% ee = 97%
R = Ph(CH$_2$)$_2$, R' = Ts: 74% ee = 96%
R = n-Pent, R' = Me: 87% ee = 98%
R = n-Pent, R' = Ts: 77% ee = 92%

Scheme 3.45 VERDI disulfonamide ligands for additions of ZnEt$_2$ to aldehydes.

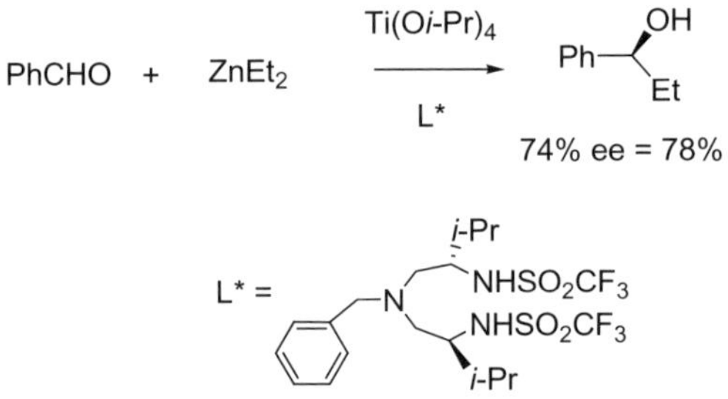

PhCHO + ZnEt$_2$ $\xrightarrow[\text{L*}]{\text{Ti(O}i\text{-Pr)}_4}$ Ph—CH(OH)Et

74% ee = 78%

L* =

i-Pr
NHSO$_2$CF$_3$
NHSO$_2$CF$_3$
i-Pr

Scheme 3.46 Bis(ethylsulfonamides)amine ligand for addition of ZnEt$_2$ to benzaldehyde.

catalysed the Ti-mediated addition of ZnEt$_2$ to benzaldehyde to give the product in an enantioselectivity of up to 78% ee (Scheme 3.46).[61]

Among various amino alcohols used as ligands, derivatives of simple and easily available carbohydrates are only rarely reported and remain underestimated. In this context, α-hydroxy sulfonamides derived from D-glucosamine have been used by Bauer *et al.* for addition of ZnEt$_2$ to aldehydes, providing the corresponding products in high enantioselectivities of up to 97% ee, as shown in Scheme 3.47.[62]

In 2003, these authors developed the Ti-catalysed enantioselective addition of ZnEt$_2$ to benzaldehyde in the presence of C_2-symmetric bis(camphorsulfonamides) ligands derived from achiral 1,2- and 1,3-diamines and (+)-camphorsulfonic acid.[63] These authors showed that the stereochemical outcome of the reaction was highly influenced both by the structure of the

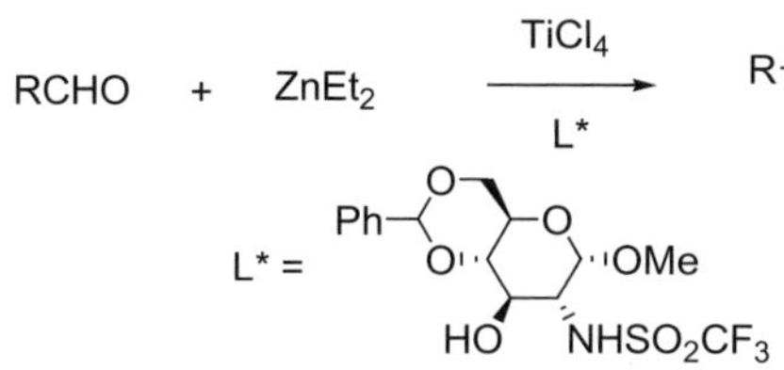

R = Ph: 99% ee = 97%
R = *m*-ClC$_6$H$_4$: 85% ee = 90%
R = *p*-MeOC$_6$H$_4$: 95% ee = 89%
R = (*E*)-Ph-CH=CH: 94% ee = 86%
R = Cy: 73% ee = 39%
R = *n*-Hex: 80% ee = 88%

Scheme 3.47 D-Glucosamine-derived α-hydroxy sulfonamide ligand for additions of ZnEt$_2$ to aldehydes.

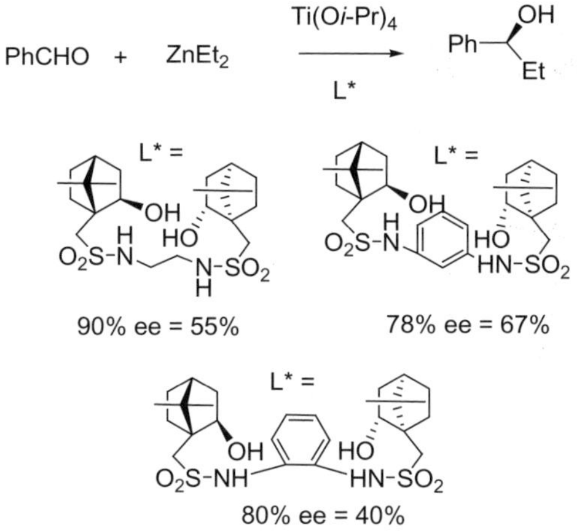

90% ee = 55%

78% ee = 67%

80% ee = 40%

Scheme 3.48 *C$_2$*-Symmetric bis(camphorsulfonamides) ligands for addition of ZnEt$_2$ to benzaldehyde.

ligand and the reaction parameters. The best enantioselectivity in this study was of 67% ee (Scheme 3.48). Comparison of these results with results obtained for phenyl hydroxycamphorsulfonamide by Yus and Ramon[56a] suggested that the *C$_2$*-symmetry was not always advantageous and, in some cases, it seemed that both sides of the *C$_2$*-symmetric molecules could behave independently.

On the other hand, excellent enantioselectivities of up to 98% ee have been obtained by Gau *et al.* for the Ti-catalysed enantioselective addition of ZnEt$_2$ to aldehydes in the presence of *N*-sulfonylated amino alcohols as ligands (Scheme 3.49).[64]

In addition, these authors have reported the synthesis of other *N*-sulfonylated amino alcohols derived from camphor that have been examined as ligands

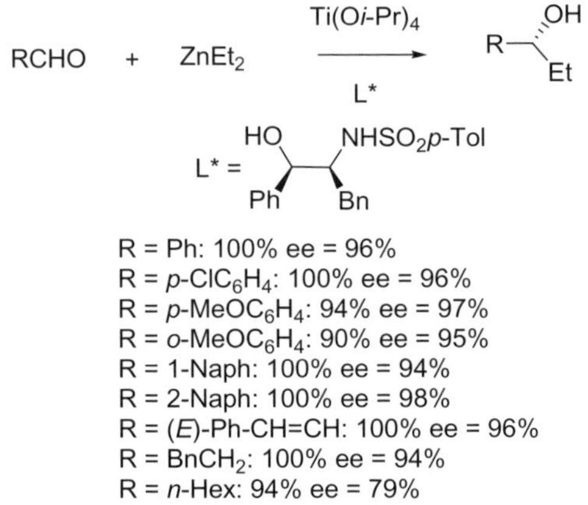

$$RCHO \quad + \quad ZnEt_2 \quad \xrightarrow[\text{L*}]{\text{Ti(O}i\text{-Pr)}_4} \quad R\overset{OH}{\underset{Et}{\diagdown}}$$

L* = HO NHSO₂p-Tol structure with Ph and Bn

R = Ph: 100% ee = 96%
R = p-ClC₆H₄: 100% ee = 96%
R = p-MeOC₆H₄: 94% ee = 97%
R = o-MeOC₆H₄: 90% ee = 95%
R = 1-Naph: 100% ee = 94%
R = 2-Naph: 100% ee = 98%
R = (E)-Ph-CH=CH: 100% ee = 96%
R = BnCH₂: 100% ee = 94%
R = n-Hex: 94% ee = 79%

Scheme 3.49   *N*-Sulfonylated amino alcohol ligand for additions of ZnEt₂ to aldehydes.

$$RCHO \quad + \quad ZnEt_2 \quad \xrightarrow[\text{L*}]{\text{Ti(O}i\text{-Pr)}_4} \quad R\overset{OH}{\underset{Et}{\diagdown}}$$

L* = camphor-derived structure with O₂S, HO, NH, Ph, Bn

R = Ph: 91% ee = 91%
R = p-ClC₆H₄: 90% ee = 94%
R = o-ClC₆H₄: 90% ee = 79%
R = p-MeOC₆H₄: 83% ee = 93%
R = o-MeOC₆H₄: 88% ee = 83%
R = m-MeOC₆H₄: 90% ee = 86%
R = 1-Naph: 87% ee = 87%
R = 2-Naph: 87% ee = 91%
R = Bn: 66% ee = 39%
R = BnCH₂: 69% ee = 58%

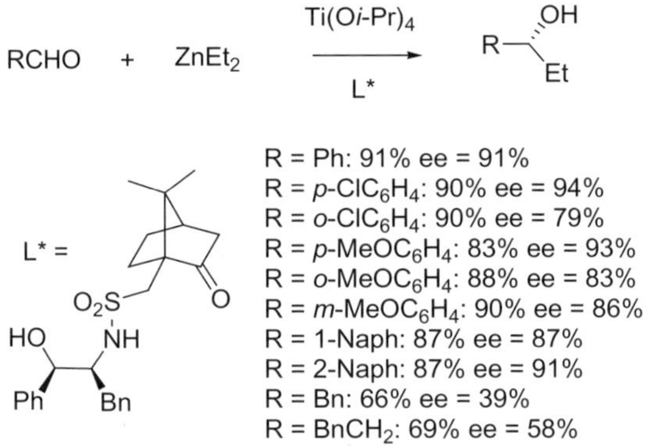

Scheme 3.50   Camphor-derived *N*-sulfonylated amino alcohol ligand for additions of ZnEt₂ to aldehydes.

in similar reactions to those described above.[65] The best ligand depicted in Scheme 3.50 allowed enantioselectivities of up to 94% ee to be obtained. This study revealed that the steric bulk of the camphor group did not affect the stereocontrol except for the requirement of a higher excess of Ti(O*i*-Pr)₄. Although this ligand contained a carbonyl group on the camphor moiety, which was a potential donor for coordination of the metal centre, this study strongly suggested that the ligand precursor behaved as a bidentate ligand. In contrast, the hydroxyl group on the camphor moiety of the other ligands tested had a negative effect on stereocontrol.

The preparation and the use of several C_2-symmetric disulfonamides derived from 1,2-amino alcohols in the Ti-catalysed enantioselective addition of di-alkylzinc reagents to aldehydes was described by Yus *et al.*, in 2002.[66] The best

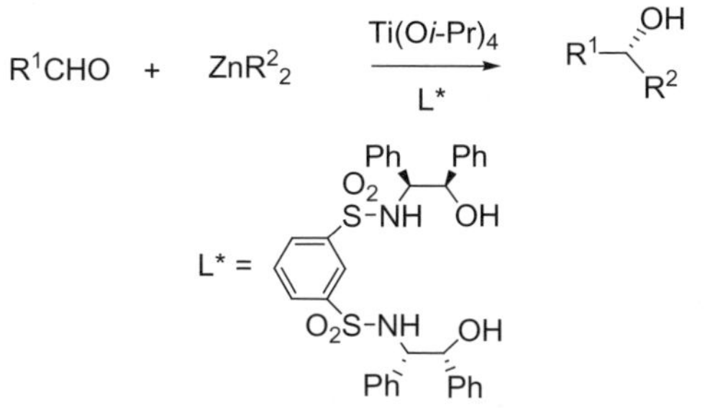

R^1 = Ph, R^2 = Me: > 95% ee = 78%
R^1 = p-MeOC$_6$H$_4$, R^2 = Et: > 95% ee = 92%
R^1 = p-ClC$_6$H$_4$, R^2 = Et: > 95% ee = 88%
R^1 = p-CF$_3$C$_6$H$_4$, R^2 = Et: > 95% ee = 78%
R^1 = p-CNC$_6$H$_4$, R^2 = Et: > 95% ee = 74%
R^1 = p-MeOC$_6$H$_4$, R^2 = Et: > 95% ee = 92%
R^1 = n-Hex, R^2 = Et: > 95% ee = 60%
R^1 = Cy, R^2 = Et: > 95% ee = 54%
R^1 = (E)-Ph-CH=CH, R^2 = Et: > 95% ee = 74%
R^1 = PhC≡C, R^2 = Et: > 95% ee = 72%

Scheme 3.51 C_2-Symmetric *N*-sulfonylated amino alcohol ligand for additions of ZnR$_2$ to aldehydes.

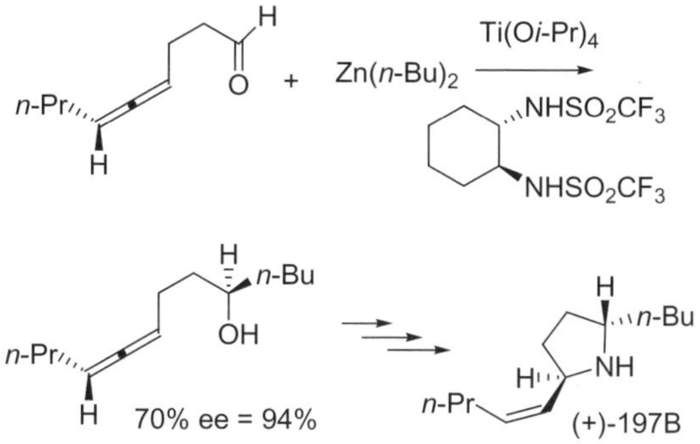

Scheme 3.52 Synthesis of alkaloid (+)-197B.

enantioselectivity of up to 92% ee was obtained with the ligand depicted in Scheme 3.51. The enantioselectivities were generally higher for aromatic aldehydes than for either aliphatic or α,β-unsaturated ones.

In order to synthesise the pyrrolidine alkaloid, (+)-197B, bis-(*R,R*)-trifluoromethanesulfonamide ligand was employed in the enantioselective addition of Zn(*n*-Bu)$_2$ to an allene-aldehyde, affording the corresponding (*R*)-alcohol in 70% yield and 94% ee (Scheme 3.52).[67]

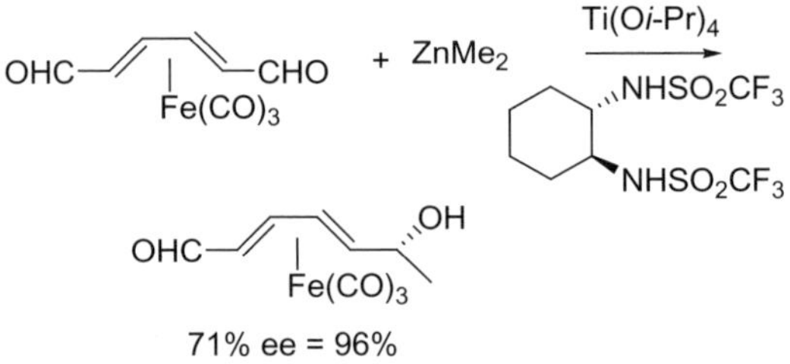

Scheme 3.53   Bis-(R,R)-trifluoromethanesulfonamide ligand for addition of ZnMe$_2$ to a dialdehyde-Fe(CO)$_3$ complex.

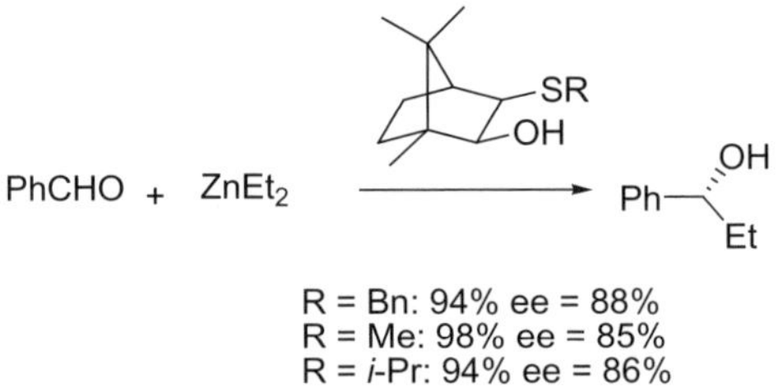

Scheme 3.54   Isoborneol-derived β-hydroxy-sulfide ligand for addition of ZnEt$_2$ to benzaldehyde.

In addition, this ligand was applied to the Ti-catalysed enantioselective reaction of ZnMe$_2$ with a dialdehyde-Fe(CO)$_3$ complex, giving rise to the corresponding monomethylated complex with 96% ee (Scheme 3.53).[68]

3.2.1.3 S/O Ligands

On the other hand, several S/O ligands have been successfully applied to the enantioselective addition of ZnEt$_2$ to aldehydes. As an example, Arai *et al.* have developed isoborneol-derived β-hydroxy-sulfide ligands and employed them in the enantioselective addition of ZnEt$_2$ to benzaldehyde, providing enantio-selectivities of up to 88% ee (Scheme 3.54).[69] These authors showed that the enantioselectivity of the reaction did not depend on the substituent of the sulfur atom.

In 2002, Shiina *et al.* reported the synthesis of ((R)-thiolan-2-yl)diphenylmethanol and its application as a ligand to the enantioselective addition of ZnEt$_2$ to benzaldehyde (Scheme 3.55).[70] A low enantioselectivity (22% ee) was obtained but the authors could increase it up to 92% ee by adding 10 mol% of B(Oi-Pr)$_3$. A stable four-membered ring structure formed by two oxygens and two metallic species was proposed as a transition state for the addition of ZnMe$_2$ to formaldehyde in which the boron atom was coordinated with the oxygen atom of the zinc alkoxide of the ligand (Scheme 3.55).

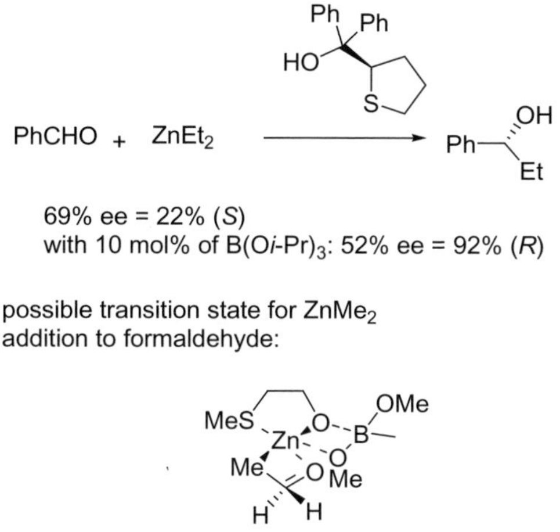
Ph
Ph
HO
S
PhCHO + ZnEt2
OH
Ph
Et
69% ee = 22% (S)
with 10 mol% of B(Oi-Pr)3: 52% ee = 92% (R)
possible transition state for ZnMe2
addition to formaldehyde:
MeS
O
OMe
Zn
B
Me
O
O
Me
H
H

Scheme 3.55 ((*R*)-Thiolan-2-yl)diphenylmethanol ligand for addition of $ZnEt_2$ to benzaldehyde.

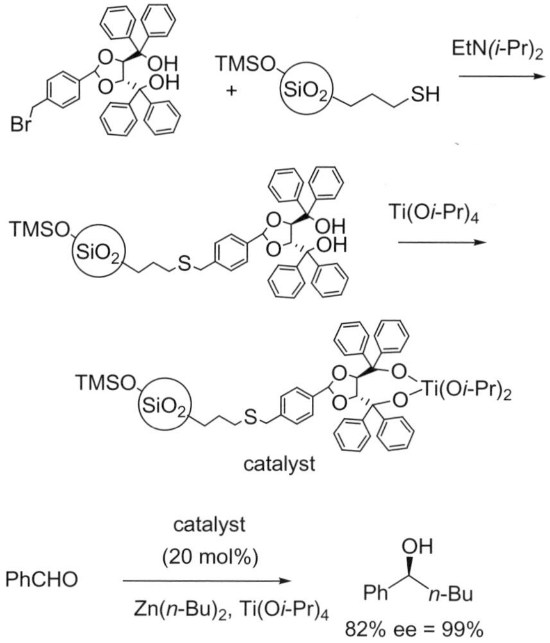
OH
OH
TMSO
+
SiO2
SH
EtN(i-Pr)2
Br
TMSO
SiO2
S
OH
OH
Ti(Oi-Pr)4
TMSO
SiO2
S
Ti(Oi-Pr)2
catalyst
catalyst
(20 mol%)
PhCHO
Zn(n-Bu)2, Ti(Oi-Pr)4
OH
Ph
n-Bu
82% ee = 99%

Scheme 3.56 CPG-Immobilised sulfur-containing Ti-TADDOLate for butylation of benzaldehyde.

On the other hand, Seebach and Heckel have demonstrated that sulfur-containing TADDOL derivatives could be immobilised on hydrophobic controlled-pore glass silica gel.[71] Indeed, controlled-pore glass (CPG) is a rigid support that offers an openly accessible pore structure in all possible solvents

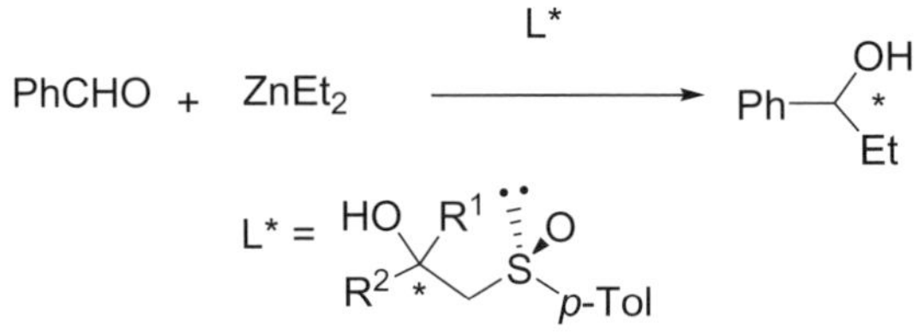

L*

PhCHO + ZnEt$_2$ $\xrightarrow{\hspace{3cm}}$ Ph$-\overset{\text{OH}}{\underset{\text{Et}}{\overset{|}{\text{C}}}}$*

L* = HO R^1 :O
$\underset{R^2}{\overset{}{\diagup}}$ ... S$\diagdown$p-Tol

from [2S,(S)R]-L*: R^1 = t-Bu, R^2 = H: ee = 9% (R)
from [2R,(S)R]-L*: R^1 = t-Bu, R^2 = H: ee = 22% (S)
from [2R,(S)R]-L*: R^1 = t-Bu, R^2 = Me: ee = 35% (S)
from [1S,(S)R]-L*: R^1 = Ph, R^2 = H: ee = 18% (R)
from [1R,(S)R]-L*: R^1 = Ph, R^2 = H: ee = 16% (S)
from [2R,(S)R]-L*: R^1 = Ph, R^2 = Me: ee = 25% (S)
from [5R,(S)R]-L*: R^1 = CO$_2$Me-(CH$_2$)$_3$, R^2 = H: ee = 7% (S)

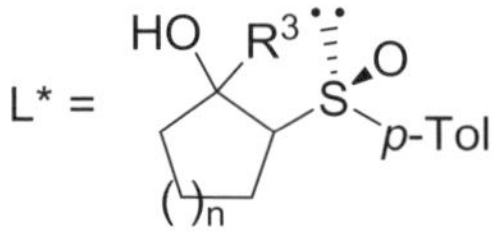

L* = HO R^3 :O
S$\diagup$p-Tol
($\,$)$_n$

from [1S,2S, (S)R]-L*: n = 1, R^3 = H: ee = 2% (S)
from [1R,2S, (S)R]-L*: n = 1, R^3 = H: ee = 11% (S)
from [1S,2S, (S)R]-L*: n = 2, R^3 = H: ee = 10% (S)
from [1R,2S, (S)R]-L*: n = 2, R^3 = H: ee = 23% (S)
from [1R,2S, (S)R]-L*: n = 2, R^3 = Me: ee = 45% (S)
with AlMe$_3$:
from [1R,2S, (S)R]-L*: n = 2, R^3 = Me: ee = 55% (S)

Scheme 3.57 β-Hydroxysulfoxide ligands for addition of ZnEt$_2$ to benzaldehyde.

and in a wide temperature and pressure range. These novel CPG-immobilised catalysts were prepared, as summarised in Scheme 3.56, and then applied to the condensation of Zn(n-Bu)$_2$ onto benzaldehyde. Both excellent yield and enantioselectivity were observed, thus demonstrating that these catalysts were in many ways superior to polymer-bound or polymer-incorporated analogues.

There are relatively few studies dealing with the use of sulfoxides as chiral ligands in this type of reaction. In this area, various chiral β-hydroxysulfoxide ligands have been investigated by Carreno *et al.* in the enantioselective addition of ZnEt$_2$ to benzaldehyde, providing low to moderate enantioselectivities.[72] The best enantioselectivity of 45% ee was obtained by using a β-disubstituted cyclohexylsulfoxide as a ligand (Scheme 3.57). This enantioselectivity could be increased up to 55% ee when the active species was formed by addition of this ligand to AlMe$_3$.

3.2.1.4 Sulfur-Containing Ferrocenyl Ligands

Chiral ferrocenyl ligands have been widely used in asymmetric catalysis.[73] In particular, numerous reports are available on the chemistry of allylic alkylations using ferrocenyl ligands.[74] The advantages of using ferrocene as a scaffold for chiral ligands are its planar chirality, rigid bulkiness, stability, inherent special electronic and stereochemical properties, and ease of derivation. In recent years, various sulfur-containing chiral ferrocenyl derivatives have been successfully implicated in the enantioselective addition of organozinc reagents to aldehydes. In this context, Carretero *et al.* demonstrated, in 2001, the utility of a new family of these chiral ligands, 2-amino-substituted *tert*-butylsulfinylferrocenes, for the asymmetric addition of ZnEt$_2$ to aromatic aldehydes (Scheme 3.58).[75] Therefore, the corresponding products were isolated in good to high enantioselectivities of up to 96% ee. The results showed that the planar chirality of the ferrocene was more important than the stereogenic sulfur atom to perform an efficient asymmetric induction. They concluded that these ligands behaved as *N*-mono-coordinating ligands rather than bidentate *N,O*- or *S,N*-chelating ligands.

In addition, Bonini *et al.* have shown that a planar chiral sulfur-containing ferrocenyl-oxazoline carbinol ligand, depicted in Scheme 3.59, could also be used to catalyse the addition of ZnEt$_2$ to benzaldehyde with a moderate enantioselectivity (46% ee).[76]

The immobilisation of organic compounds on highly porous SiO$_2$ instead of organic polymers for solid-phase synthesis and catalysis has several advantages, such as the rigidity of the structure, the fact that it does not swell in solvents, and the stability at low and high temperature and pressure. In 2000, Seebach *et al.* succeeded in grafting a TADDOL sulfur-containing derivative onto a commercially available controlled-pore glass, and then loaded the TADDOL

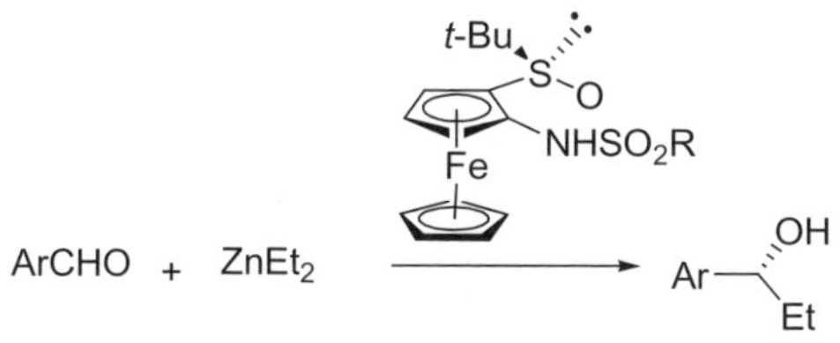

$$\text{ArCHO} \; + \; \text{ZnEt}_2 \longrightarrow \text{Ar}\overset{\text{OH}}{\underset{\text{Et}}{\diagup}}$$

Ar = Ts, R = Ph: 80% ee = 88%
Ar = *p*-MeOC$_6$H$_4$, R = Ph: 76% ee = 86%
Ar = Ts, R = *p*-MeOC$_6$H$_4$: 65% ee = 66%
Ar = Ts, R = *p*-MeO$_2$CC$_6$H$_4$: 71% ee = 88%
Ar = Ts, R = 2-Naph: 73% ee = 82%
Ar = Ts, R = *m*-FC$_6$H$_4$: 76% ee = 80%
Ar = *p*-MeOC$_6$H$_4$, R = Ts: 72% ee = 66%
Ar = *p*-MeOC$_6$H$_4$, R = *p*-(MeO$_2$C)C$_6$H$_4$: 78% ee = 96%
Ar = *p*-MeOC$_6$H$_4$, R = 2-Naph: 79% ee = 92%
Ar = *p*-MeOC$_6$H$_4$, R = *m*-FC$_6$H$_4$: 74% ee = 72%

Scheme 3.58 2-Amino-substituted *tert*-butylsulfinylferrocene ligands for additions of ZnEt$_2$ to aromatic aldehydes.

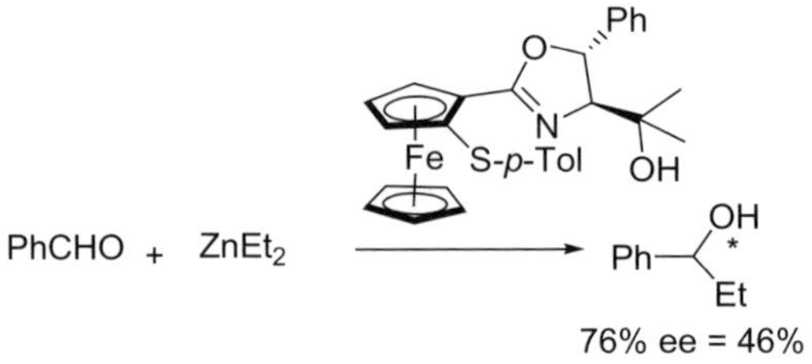

Scheme 3.59 Sulfur-containing ferrocenyl-oxazoline carbinol ligand for addition of ZnEt$_2$ to benzaldehyde.

moieties with titanates. The resulting material was tested in the enantioselective addition of ZnEt$_2$ to benzaldehyde, giving a quantitative yield and an enantioselectivity of 98% ee.[77]

3.2.2 Addition of Aryl-, Alkenyl- and Alkynylzinc Reagents to Aldehydes

On the other hand, diarylmethanols are key intermediates in the synthesis of antihistamic agents, such as neobenodine, orphenadrine, carbinoxamine, antimuscarinics, antidepressants and endothelin antagonists. Consequently, the asymmetric synthesis of optically active arylcarbinols has attracted much attention due to their high availability as starting materials. Among the available methods for the synthesis of chiral diarylmethanols, the asymmetric addition of arylzinc showed the advantages of a wide substituent tolerance, mild reaction conditions and the use of relatively nontoxic zinc metal.[78] The initial approach was centered upon the use of the expensive ZnPh$_2$ as the aryl source.[79] However, its enantioselective addition to aldehydes is a challenging endeavour, since this reagent is much more reactive than the well-known ZnEt$_2$. The uncatalysed background addition thus competes with the enantioselective pathway, leading to the formation of the racemic product. In order to circumvent this problem, ZnEt$_2$ was found to reduce the reactivity of the arylzinc reagent, by forming PhZnEt, which is less reactive than ZnPh$_2$ itself. This strategy improves the overall system performance, since the aryl transfer reaction proceeds slowly when compared to the reaction with ZnPh$_2$ alone. Additionally, it accounts for a higher selectivity for the phenyl transfer and allows the use of a reduced amount of the expensive ZnPh$_2$. These two protocols, however, have a serious drawback. The scope of the aryl group to be transferred is limited to the phenyl ring, since only ZnPh$_2$ is a commercially available diarylzinc reagent. Thus, there is a growing interest in the development of methods that allow the asymmetric transfer of a broader range of substituted aryl groups, starting from inexpensive and readily accessible sources. In this context, an interesting protocol was recently introduced by Bolm and Rudolph, which took advantage of the use of boronic acids as the source of the nucleophilic aryl species, by a boron-to-zinc exchange reaction with

ZnEt$_2$.[80] This modified methodology broadened the scope of such an addition reaction and enabled synthetic chemists to elaborate functionalised arylzinc reagents as nucleophiles in salt-free conditions. Unfortunately, only a few elegant and efficient catalysts, such as chiral β-aminoalcohols[81] have been developed for this purpose. In 2006, Braga *et al.* reported the synthesis of novel chiral thiazolidine derivatives in a single step from inexpensive and commercially available starting materials such as L-cysteine ester.[82] These ligands catalysed enantioselective arylation of various aldehydes using aryl boronic acids as a source of transferable aryl groups. The expected diarylmethanols were obtained in good yields and enantioselectivities of up to 81% ee, as shown in Scheme 3.60. The importance of these results lies in the fact that this was the first study involving ester groups in asymmetric catalytic aryl transfer to aldehydes with PhB(OH)$_2$/Et$_2$Zn systems.

Similar reactions have been developed more recently by Jin *et al.* using chiral amino thioacetate ligands derived from the corresponding amino alcohols.[83] Low catalyst loadings of only 1–2.5 mol% were sufficient to achieve excellent enantioselectivities of up to 98% ee as well as high conversions in short times (Scheme 3.61). These authors have shown that the thioacetoxy moiety of the amino thioacetates has a surprisingly beneficial effect in enhancing the asymmetric induction.

The direct generation of organometallic reagents from alkenes and alkynes is a useful strategy for efficient C–C bond formation, since unsaturated substrates are readily available. Hydrozirconation of alkynes with Cp$_2$ZrHCl (Schwartz reagent)[84] provides organozirconocene reagents that can readily be added to aldehydes.[85] In 1994, Wipf and Xu reported a high-yielding protocol for the *in situ* transmetalation of alkenylzirconocenes to alkylzinc species.[86] In the presence of an aldehyde, the corresponding allylic alcohol product is generated in a one-pot reaction in high yield. An asymmetric version of this methodology

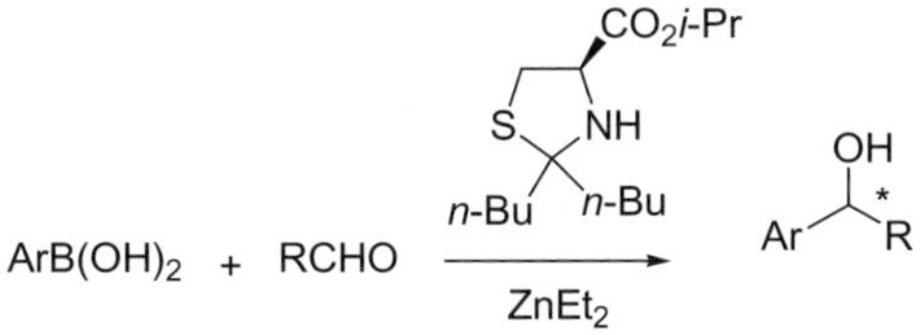

$$ArB(OH)_2 + RCHO \xrightarrow{ZnEt_2} $$

Ar = Ph, R = *p*-Tol: 97% ee = 81% (*S*)
Ar = Ph, R = *p*-MeOC$_6$H$_4$: 91% ee = 75% (*S*)
Ar = Ph, R = *p*-ClC$_6$H$_4$: 98% ee = 79% (*S*)
Ar = Ph, R = *o*-Tol: 97% ee = 73% (*S*)
Ar = Ph, R = *o*-MeOC$_6$H$_4$: 93% ee = 56% (*S*)
Ar = Ph, R = *o*-ClC$_6$H$_4$: 97% ee = 42% (*S*)
Ar = Ph, R = 2-Thio: 89% ee = 33% (*S*)
Ar = Ph, R = *n*-Hex: 63% ee = 6% (*S*)
Ar = Ph, R = *t*-Bu: 59% ee = 20% (*S*)
Ar = *p*-AcC$_6$H$_4$, R = Ph: 90% ee = 80% (*S*)
Ar = *p*-ClC$_6$H$_4$, R = Ph: 99% ee = 57% (*S*)

Scheme 3.60 Thiazolidine ligand for arylations of aldehydes with aryl boronic acids.

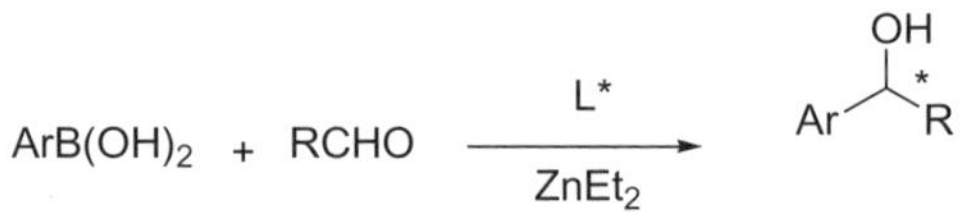

$$\text{ArB(OH)}_2 \ + \ \text{RCHO} \quad \xrightarrow[\text{ZnEt}_2]{\text{L*}} \quad \underset{\text{Ar}}{\overset{\text{OH}}{\bigwedge}} \text{R}$$

L* = L²*, Ar = Ph, R = o-ClC₆H₄: 97% ee = 98%
L* = L²*, Ar = Ph, R = p-MeOC₆H₄: 98% ee = 93% (R)
L* = L²*, Ar = Ph, R = o-MeOC₆H₄: 93% ee = 91% (R)
L* = L²*, Ar = Ph, R = p-Tol: 94% ee = 94% (R)
L* = L²*, Ar = Ph, R = 2-C₁₀H₈: 91% ee = 94% (R)
L* = L¹*, Ar = p-MeOC₆H₄, R = Ph: 93% ee = 92% (S)
L* = L³*, Ar = p-ClC₆H₄, R = Ph: 92% ee = 95% (S)
L* = L¹*, Ar = p-Tol, R = Ph: 90% ee = 92% (S)
L* = L²*, Ar = p-Tol, R = Ph: 90% ee = 90% (S)
L* = L²*, Ar = m-ClC₆H₄, R = Ph: 91% ee = 88% (S)
L* = L²*, Ar = o-ClC₆H₄, R = Ph: 86% ee = 80% (S)

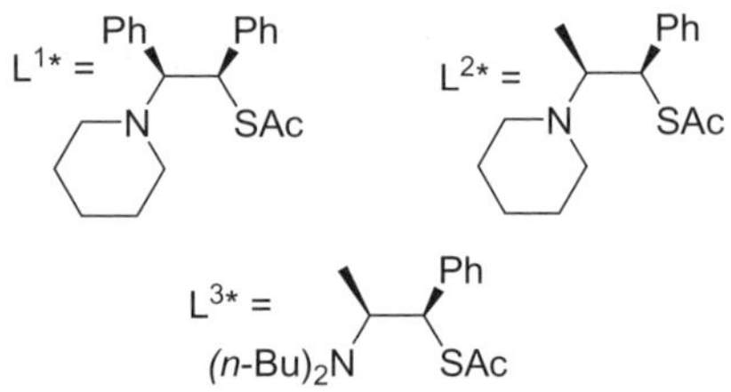

Scheme 3.61 Aminothioacetate ligands for arylations of aldehydes with aryl boronic acids.

was developed by these authors by performing the reaction in the presence of chiral sulfur-containing ligands such as that depicted in Scheme 3.62.[87] Therefore, the *in situ* hydrozirconation of alkynes, transmetalation to dimethylzinc, and chiral amino thiol-catalysed addition to aldehydes provided an efficient protocol for the asymmetric preparation of (*E*)-allylic alcohols with high yields and enantioselectivities of up to 99% ee (Scheme 3.62).

In 2008, Wipf and Jayasuriya extended this methodology to the use of another new β-amino thiol scaffold as a ligand (Scheme 3.63).[88] This ligand showed comparable enantioselectivity at a lower ligand loading of 5 mol% compared to the ligand of Scheme 3.62 that was used at 10 mol% of loading. Moreover, this new ligand could potentially increase the asymmetric induction in substrates, such as aliphatic aldehydes, for which the first ligand has been shown to be mediocre.

In 2002, Braga *et al.* employed a chiral C₂-symmetric oxazolidine disulfide as a ligand for the enantioselective synthesis of propargylic alcohols[89] by direct addition of alkynes to aldehydes (Scheme 3.64).[90] Good yields but moderate enantioselectivities (≤58% ee) were obtained for the enantioselective alkynylation of aldehydes in the presence of ZnEt₂.

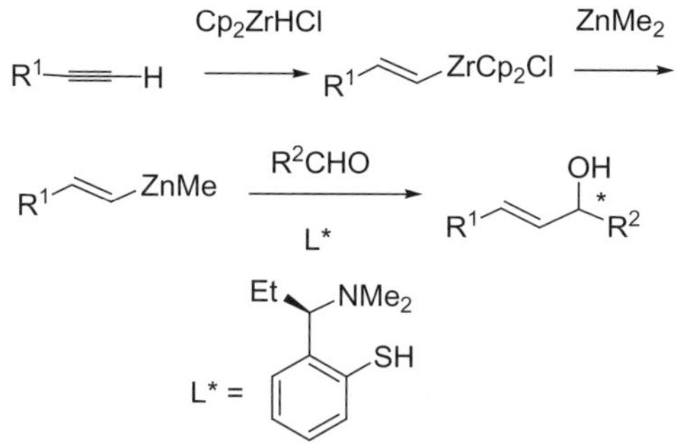

$$R^1 \!\!-\!\!\!\equiv\!\!\!-\! H \xrightarrow{\text{Cp}_2\text{ZrHCl}} R^1 \diagdown\!\!\!\diagup ZrCp_2Cl \xrightarrow{\text{ZnMe}_2}$$

$$R^1 \diagdown\!\!\!\diagup ZnMe \xrightarrow[\text{L*}]{R^2CHO} R^1 \diagdown\!\!\!\diagup \overset{OH}{\underset{*}{\diagup}} R^2$$

$$L^* = \quad \begin{array}{c} Et \diagdown \diagup NMe_2 \\ \diagup SH \end{array}$$

R^1 = *n*-Bu, R^2 = *p*-ClC$_6$H$_4$: 83% ee = 97%
R^1 = *n*-Bu, R^2 = *p*-CF$_3$C$_6$H$_4$: 71% ee = 93%
R^1 = *n*-Bu, R^2 = *p*-MeOC$_6$H$_4$: 75% ee = 63%
R^1 = *n*-Bu, R^2 = *m*-MeOC$_6$H$_4$: 79% ee = 99%
R^1 = *n*-Bu, R^2 = Cy: 63% ee = 74%
R^1 = *n*-Bu, R^2 = BnCH$_2$: 71% ee = 64%
R^1 = *t*-Bu, R^2 = Ph: 73% ee = 83%
R^1 = TIPSOC(O)(CH$_2$)$_2$, R^2 = Ph: 67% ee = 92%

Scheme 3.62 Zirconocene–zinc transmetalations and *in situ* additions to aldehydes in the presence of amino thiol ligand.

$$R^1 \!\!-\!\!\!\equiv\!\!\!-\! R^2 \xrightarrow[\substack{3.\ L^* \\ 4.\ R^3CHO}]{\substack{1.\ Cp_2ZrHCl \\ 2.\ ZnMe_2}} R^1 \diagdown\!\!\!\overset{R^2}{\diagup}\!\!\!\overset{}{\underset{\overset{|}{OH}}{\diagup}} R^3$$

$$L^* = \quad \begin{array}{c} i\text{-}Pr \diagdown \quad \diagup i\text{-}Pr \\ \text{N} \quad SH \end{array}$$

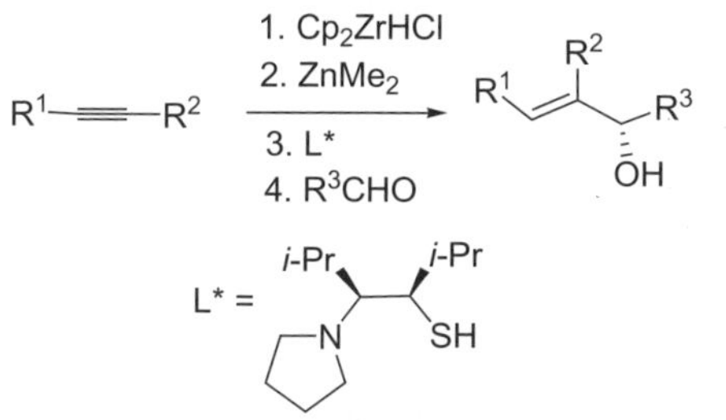

R^1 = *n*-Bu, R^2 = H, R^3 = Ph: 81% ee = 94%
R^1 = *n*-Bu, R^2 = H, R^3 = BnCH$_2$: 78% ee = 79%
R^1 = *n*-Bu, R^2 = H, R^3 = *p*-MeOC$_6$H$_4$: 65% ee = 42%
R^1 = *n*-Bu, R^2 = H, R^3 = Cy: 70% ee = 73%
R^1 = R^2 = Et, R^3 = Ph: 89% ee = 90%
R^1 = R^2 = Et, R^3 = BnCH$_2$: 65% ee = 53%
R^1 = R^2 = Et, R^3 = *p*-MeOC$_6$H$_4$: 63% ee = 87%
R^1 = TIPSOC(O)(CH$_2$)$_2$, R^2 = H, R^3 = Ph: 72% ee = 71%

Scheme 3.63 Zirconocene–zinc transmetalations and *in situ* additions to aldehydes in the presence of amino thiol ligand.

In 2005, a novel C_2-symmetric bis(sulfonamides) ligand was easily prepared in three steps by Wang *et al.* and further applied to the enantioselective addition of alkynylzinc reagents to aldehydes performed in the presence of Ti(O*i*-Pr)$_4$.[91] When a catalyst loading of 4 mol% was used, a high enantioselectivity of up to 97% ee was achieved, as shown in Scheme 3.65. When the amount of the

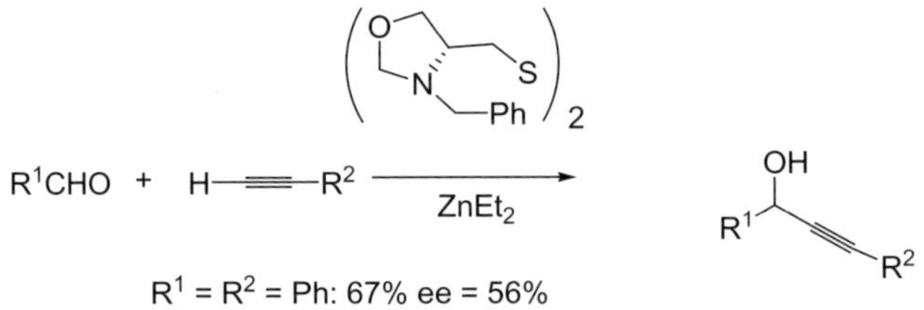

R^{1}CHO + H$\equiv$R^2 $\xrightarrow{\text{ZnEt}_2}$

R^1 = R^2 = Ph: 67% ee = 56%
R^1 = Ph, R^2 = n-Bu: 55% ee = 36%
R^1 = p-Tol, R^2 = Ph: 76% ee = 58%
R^1 = o-ClC$_6$H$_4$, R^2 = Ph: 80% ee = 50%
R^1 = p-MeOC$_6$H$_4$, R^2 = n-Bu: 81% ee = 52%
R^1 = o-ClC$_6$H$_4$, R^2 = n-Bu: 72% ee = 58%
R^1 = n-Pent, R^2 = n-Bu: 73% ee = 43%
R^1 = n-Pent, R^2 = Ph: 82% ee = 51%

Scheme 3.64   Oxazolidine disulfide ligand for alkynylations of aldehydes.

RCHO + Ph$\equiv$H $\xrightarrow[\text{L* (4 mol%)}]{\substack{\text{Ti(O}i\text{-Pr)}_4 \\ \text{ZnEt}_2}}$

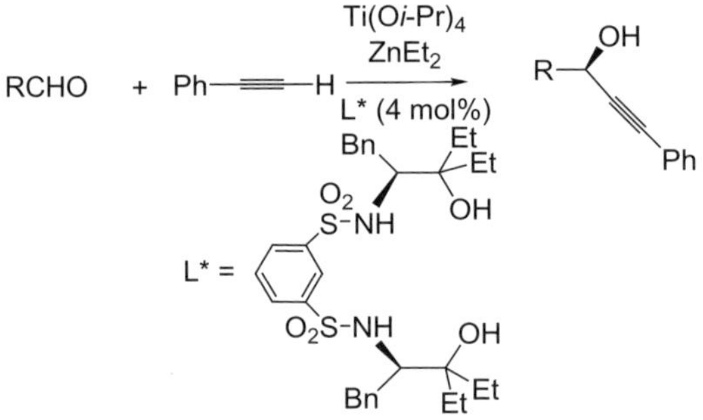

R = Ph: 92% ee = 95%
R = m-Tol: 86% ee = 94%
R = p-Tol: 90% ee = 97%
R = p-FC$_6$H$_4$: 86% ee = 90%
R = o-MeOC$_6$H$_4$: 85% ee = 84%
R = m-MeOC$_6$H$_4$: 89% ee = 91%
R = p-MeOC$_6$H$_4$: 89% ee = 91%
R = m-BrC$_6$H$_4$: 83% ee = 84%
R = p-BrC$_6$H$_4$: 86% ee = 93%
R = p-ClC$_6$H$_4$: 89% ee = 91%
R = 2-Naph: 68% ee = 91%
R = 1-Naph: 65% ee = 83%
R = (E)-Ph-CH=CH: 63% ee = 71%
R = i-Bu: 89% ee = 57%

Scheme 3.65 C_2-Symmetric bis(sulfonamides) ligand for alkynylations of aldehydes.

ligand was lowered to 2 mol%, excellent levels of enantioselectivity of up to 95% ee were still achieved, for example, 95% ee for reaction of benzaldehyde instead of 97% ee by using 4 mol% of the ligand.

A few vinylzinc reagents have been added to aldehydes in the presence of chiral sulfur-containing ligands. As an example, Seto *et al.* reported, in 2005, the synthesis of dipeptide *N*-acylethylenediamine-based ligands by parallel

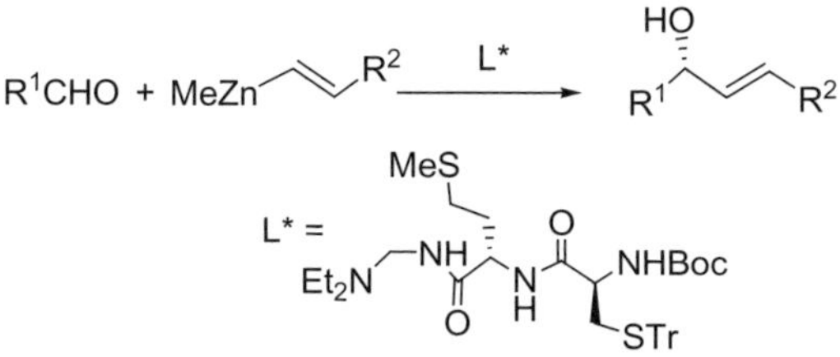

R^1 = Ph, R^2 = *t*-Bu: 80% ee = 93%
R^1 = Ph, R^2 = *c*-Pr: 55% ee = 92%
R^1 = Ph, R^2 = *n*-Hex: 63% ee = 91%
R^1 = Ph, R^2 = *n*-Bu: 53% ee = 90%
R^1 = Ph, R^2 = Cy: 66% ee = 89%
R^1 = *p*-ClC$_6$H$_4$, R^2 = *t*-Bu: 80% ee = 93%
R^1 = *p*-BrC$_6$H$_4$, R^2 = *t*-Bu: 74% ee = 85%
R^1 = Ph, R^2 = *t*-Bu: 80% ee = 93%
R^1 = Cy, R^2 = *t*-Bu: 56% ee = 93%
R^1 = *p*-MeOC$_6$H$_4$, R^2 = *t*-Bu: 60% ee = 91%
R^1 = 2-Naph, R^2 = *t*-Bu: 83% ee = 91%
R^1 = BnCH$_2$, R^2 = *t*-Bu: 62% ee = 83%
R^1 = (*E*)-Ph-CH=CH, R^2 = *t*-Bu: 67% ee = 79%
R^1 = 1-Naph, R^2 = *t*-Bu: 72% ee = 62%

Scheme 3.66 Dipeptide *N*-acylethylenediamine-based ligand for additions of alkenylzinc reagents to aldehydes.

solid-phase methods.[92] These ligands were screened in crude form as catalysts for the enantioselective addition of alkenylzinc reagents to aldehydes to give the corresponding allylic alcohols. Three sites of diversity on the ligands were optimised using a positional scanning approach. The optimised structure from the library, depicted in Scheme 3.66, was found to catalyse the formation of a wide variety of (*E*)-allylic alcohols with enantioselectivities ranging from 90 to 95% ee. This ligand was effective for both aromatic and α-branched aldehydes, and vinylzinc reagents derived from both bulky and straight chain terminal alkynes.

In 2004, Yang and Tseng reported the synthesis of a series of new chiral amino thiol ligands derived from L-valine, which were further employed (1 mol%) in the enantioselective alkenylzinc addition to aldehydes, providing an efficient route for chiral (*E*)-allylic alcohols with enantioselectivities of up to >99% ee, as shown in Scheme 3.67.[93]

3.3 Addition of Organometallic Reagents other than Organozinc Reagents

A second class of organometallic derivatives, namely alkyllithium reagents, has been condensed onto aldehydes in the presence of chiral sulfur-containing

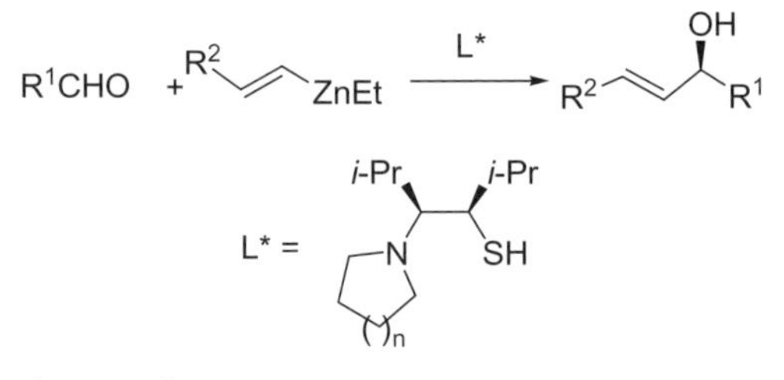

Scheme 3.67 Amino thiol ligand for additions of alkenylzinc reagents to aldehydes.

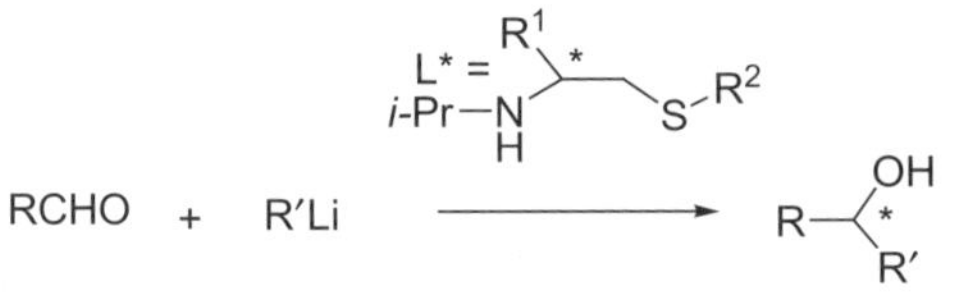

R^1 = Ph, R^2 = *n*-Hex, n = 1: 90% ee = 99%
R^1 = Ph, R^2 = *t*-Bu, n = 1: 86% ee = 99%
R^1 = Cy, R^2 = Ph, n = 1: 72% ee = 66%
R^1 = (*E*)-Ph-CH=CH, R^2 = *n*-Hex, n = 1: 81% ee = 70%
R^1 = Ph, R^2 = *n*-Hex, n = 2: 94% ee = 99%
R^1 = *p*-MeOC$_6$H$_4$, R^2 = *n*-Hex, n = 2: 90% ee = 98%
R^1 = *o*-ClC$_6$H$_4$, R^2 = *n*-Hex, n = 2: 86% ee = 93%
R^1 = Ph, R^2 = *n*-Bu, n = 2: 94% ee = 99%
R^1 = *o*-ClC$_6$H$_4$, R^2 = *n*-Bu, n = 2: 90% ee = 98%

(*R*)-L*, R = R^1 = R^2 = Ph, R′ = *n*-Bu: 82% ee > 98% (*S*)
(*S*)-L*, R = R^1 = Ph, R^2 = Et, R′ = *n*-Bu: 84% ee = 94% (*R*)
(*S*)-L*, R = R^2 = Ph, R^1 = R′ = *n*-Bu: 87% ee = 68% (*R*)
(*S*)-L*, R = Ph, R^1 = R′ = *n*-Bu, R^2 = Et: 92% ee = 81% (*R*)
(*S*)-L*, R = R^2 = Ph, R^1 = *i*-Pr, R′ = *n*-Bu, R^2 = Et: 80% ee = 97% (*R*)
(*R*)-L*, R = R^1 = R^2 = Ph, R′ = Me: 67% ee = 95% (*S*)
(*S*)-L*, R = R^1 = R^2 = Ph, R′ = *n*-Bu: 79% ee = 90% (*R*)
(*S*)-L*, R = Ts, R^1 = *n*-Bn, R^2 = Ph, R′ = Me: 69% ee = 91% (*R*)

Scheme 3.68 Amino sulfide ligands for additions of alkyllithium reagents to aldehydes.

ligands. Hence, several chiral amino sulfides have been synthesised by Hilmersson *et al.* from amino acids, such as phenylalanine, phenylglycine and valine. These amino sulfides were used as chiral ligands in the asymmetric addition of *n*-butyllithium and methyllithium to various aldehydes. The highest stereoselectivities were obtained with benzaldehyde, which gave rise to 1-phenyl-1-pentanol and 1-phenyl-1-ethanol in >98 and 95% ee, respectively (Scheme 3.68).[94] It was shown that these ligands were superior, with respect to the enantioselectivity, to their oxygen analogues in these reactions. The authors assumed that the Li–S bond (~2.5 Å), which is longer than the Li–O bond (~2.0 Å), could affect the geometry of the chelate and partially explain the difference in enantioselectivity.

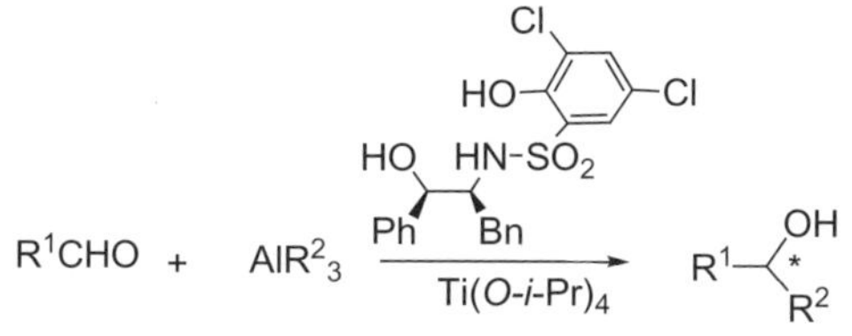

R[1]CHO + AlR[2]$_3$ $\xrightarrow{\text{Ti(O-}i\text{-Pr)}_4}$ R[1]$-\overset{\text{OH}}{\underset{\text{R}^2}{\overset{*}{\text{C}}}}$

R[1] = Ph, AlR[2]$_3$ = AlEt$_3$: 98% ee = 96% (*R*)
R[1] = *p*-ClC$_6$H$_4$, AlR[2]$_3$ = AlEt$_3$: 100% ee = 94% (*R*)
R[1] = 1-Naph, AlR[2]$_3$ = AlEt$_3$: 94% ee = 92% (*R*)
R[1] = 2-Naph, AlR[2]$_3$ = AlEt$_3$: 100% ee = 92% (*R*)
R[1] = (*E*)-Ph-CH=CH, AlR[2]$_3$ = AlEt$_3$: 100% ee = 88% (*R*)
R[1] = Cy, AlR[2]$_3$ = AlEt$_3$: 54% ee = 91% (*R*)
R[1] = Ph, AlR[2]$_3$ = AlMe$_3$: 100% ee = 98% (*R*)
R[1] = 1-Naph, AlR[2]$_3$ = AlMe$_3$: 95% ee = 96% (*R*)
R[1] = Cy, AlR[2]$_3$ = AlMe$_3$: 100% ee = 91% (*R*)
R[1] = (*E*)-Ph-CH=CH, AlR[2]$_3$ = AlMe$_3$: 99% ee > 99% (*R*)
R[1] = Ph, AlR[2]$_3$ = (allyl)AlEt$_2$: 100% ee = 90% (*R*)
R[1] = 2-Naph, AlR[2]$_3$ = (allyl)AlEt$_2$: 100% ee = 96% (*R*)

Scheme 3.69 *N*-Sulfonylated amino alcohol ligand for additions of AlR$_3$ to aldehydes.

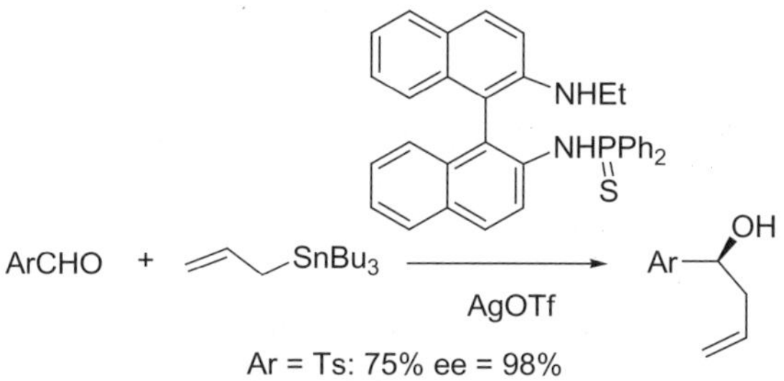

ArCHO + ⟋⟍SnBu$_3$ $\longrightarrow$ Ar$-\overset{\text{OH}}{\underset{}{\overset{*}{\text{C}}}}$⟍⟍
$\quad\quad\quad\quad\quad\quad$ AgOTf

Ar = Ts: 75% ee = 98%
Ar = *p*-ClC$_6$H$_4$: 74% ee = 96%
Ar = *p*-MeOC$_6$H$_4$: 56% ee = 80%
Ar = (*E*)-Ph-CH=CH: 80% ee = 68%
Ar = Ph: 84% ee = 94%

Scheme 3.70 Ag-promoted allylations of aldehydes with binaphthylthiopho-sphoramide ligand.

As alkylation reagents, trialkylaluminum reagents are more interesting than organozinc reagents since they are economically obtained in industrial scale.[95] Moreover, unlike the diethylzinc addition to aldehydes, which is extremely slow in the absence of a catalyst, trialkylaluminum reagents themselves are known to add aldehydes at room temperature within hours. Due to competing reactions, the development of enantioselective catalysts for trialkylaluminum additions has become more challenging. In this context, chiral titanium complexes of *N*-sulfonylated amino alcohols have been used by Gau *et al.* to catalyse the asymmetric trialkylaluminum addition to aldehydes.[96] Therefore, asymmetric methylation, ethylation and allylation reactions of various aldehydes using trialkylaluminum reagents in the presence of the chiral sulfur ligand depicted in Scheme 3.69 gave excellent yields and enantioselectivities of up to 99% ee.

An extension of the asymmetric condensation of organometallics onto aldehydes is the enantioselective silver-promoted allylation reaction of aldehydes with allyltributyltin, which has recently been performed by Shi *et al.* in the presence of chiral diphenylthiophosphoramide ligands and more efficiently, with binaphthylthiophosphoramide ligands.[97] According to the nature of the ligand substituents, the corresponding allylation products were obtained in enantioselectivities of up to 98% ee, as depicted in Scheme 3.70.

3.4 Conclusions

In conclusion, a wide variety of sulfur-containing ligands have been successfully applied to the enantioselective addition of organozinc reagents to aldehydes. The best results have been obtained by using S/N ligands, such as amino thiol ligands, which gave in most cases better enantioselectivities than their amino alcohol ligand counterparts. The enantioselectivities were generally excellent with aromatic aldehydes; however, some improvement in the enantioselective addition of organozinc reagents to aliphatic aldehydes performed in the presence of chiral sulfur-containing ligands has still to be achieved.

Even if hundreds of chiral catalysts have been developed to promote the enantioselective addition of alkylzinc reagents to aldehydes with enantioselectivities over 90% ee, the addition of organozinc reagents to aldehydes is not a solved problem. For example, only very few studies on the addition of vinyl groups or acetylides and even arylzinc reagents to aldehydes have been published, in spite of the fact that the products of these reactions, chiral allylic, propargylic and aryl alcohols, are valuable chiral building blocks.

References

1. (a) L. Pu and H.-B. Yu, *Chem. Rev.*, 2001, **101**, 757–824; (b) D. J. Berrisford, C. Bolm and K. B. Sharpless, *Angew. Chem., Int. Ed. Engl.*, 1995, **34**, 1059–1070; (c) P. Knochel, J. J. Almera Perea and P. Jones, *Tetrahedron*, 1998, **54**, 8275–8319.
2. N. Oguni, T. Omi, Y. Yamamoto and A. Nakamura, *Chem. Lett.*, 1983, **6**, 841–842.
3. (a) D. A. Evans, *Science*, 1988, **240**, 420–426; (b) K. Soai and S. Niwa, *Chem. Rev.*, 1992, **92**, 833–856; (c) P. Knochel and R. D. Singer, *Chem. Rev.*, 1993, **93**, 2117–2188.
4. (a) C. Girard and H. B. Kagan, *Angew. Chem., Int. Ed. Engl.*, 1998, **37**, 2923–2959; (b) R. Noyori and M. Kitamura, *Angew. Chem., Int. Ed. Engl.*, 1991, **30**, 34–48.
5. C. L. Gibson, *Tetrahedron: Asymmetry*, 1999, **10**, 1551–1561.
6. C. L. Gibson, *J. Chem. Soc., Chem. Commun.*, 1996, 645–646.
7. D. A. Fulton and C. L. Gibson, *Tetrahedron Lett.*, 1997, **38**, 2019–2022.

8. (a) M. Kossenjans, M. Soeberdt, S. Wallbaum, K. Harms, J. Martens and H. G. Aurich, *J. Chem. Soc., Perkin Trans. I*, 1999, 2353–2365; (b) H. G. Aurich and M. Soeberdt, *Tetrahedron, Lett.*, 1998, **39**, 2553–2554.

9. (a) M. A. Poelert, R. P. Hof, N. C. M. W. Peper and R. M. Kellogg, *Heterocycles*, 1994, **37**, 461–475; (b) R. P. Hof, M. A. Poelert, N. C. M. W. Peper and R. M. Kellogg, *Tetrahedron: Asymmetry*, 1994, **1**, 31–34; (c) R. Hulst, H. Heres, N. C. M. W. Peper and R. M. Kellogg, *Tetrahedron: Asymmetry*, 1996, **7**, 2755–2760.

10. K. Fitzpatrick, R. Hulst and R. M. Kellogg, *Tetrahedron: Asymmetry*, 1995, **6**, 1861–1864.

11. M.-J. Jin, S.-J. Ahn and K.-S. Lee, *Tetrahedron Lett.*, 1996, **37**, 8767–8770.

12. M.-J. Jin, Y.-M. Kim and K.-S. Lee, *Tetrahedron Lett.*, 2005, **46**, 2695–2696.

13. Q. Meng, Y. Li, Y. He and Y. Guan, *Tetrahedron: Asymmetry*, 2000, **11**, 4255–4261.

14. (a) J. Kang, J. W. Lee and J. I. Kim, *J. Chem. Soc., Chem. Commun.*, 1994, 2009–2010; (b) J. Kang, J. W. Kim, J. W. Lee, D. S. Kim and J. I. Kim, *Bull. Korean Chem. Soc.*, 1996, **17**, 1135–1142.

15. J. Kang, D. S. Kim and J. I. Kim, *Synlett*, 1994, **1**, 842–844.

16. (a) J. Kang, J. B. Kim, J. W. Kim and D. Lee, *J. Chem. Soc., Perkin Trans. 2*, 1997, 189–194; (b) J. Kang, J. B. Kim, J. W. Kim and D. Lee, *Bull. Korean Chem. Soc.*, 1998, **19**, 475–481.

17. S.-L. Tseng and T.-K. Yang, *Tetrahedron: Asymmetry*, 2004, **15**, 3375–3380.

18. J. C. Anderson, R. Cubbon, M. Harding and D. S. James, *Tetrahedron: Asymmetry*, 1998, **9**, 3461–3490.

19. J. C. Anderson and M. Harding, *J. Chem. Soc., Chem. Commun.*, 1998, 393–394.

20. A. L. Braga, F. Vargas, C. C. Silveira and L. H. de Andrade, *Tetrahedron Lett.*, 2002, **43**, 2335–2337.

21. A. L. Braga, P. Milani, M. W. Paixao, G. Zeni, O. E. D. Rodrigues and E. F. Alves, *Chem. Commun.*, 2004, 2488–2489.

22. E. Rijnberg, J. T. B. H. Jastrzebski, M. D. Janssen, J. Boersma and G. van Koten, *Tetrahedron Lett.*, 1994, **35**, 6521–6524.

23. E. Rijnberg, N. J. Hovestad, W. K. Kleij, Arjan, J. T. B. H. Jadtrzebski, J. Boersma, M. D. Janssen, A. L. Spek and G. van Koten, *Organometallics*, 1997, **16**, 2847–2857.

24. H. Nakano, K. Iwasa and H. Hongo, *Heterocycles*, 1997, **46**, 267–274.

25. H. Nakano, N. Kumagai, H. Matsuzaki, C. Kabuto and H. Hongo, *Tetrahedron: Asymmetry*, 1997, **9**, 1391–1401.

26. G. Chelucci, D. Berta, D. Fabbri, G. A. Pinna, A. Saba and F. Ulgheri, *Tetrahedron: Asymmetry*, 1998, **9**, 1933–1940.

27. K. Soai, Y. Hirose and Y. Ohno, *Tetrahedron: Asymmetry*, 1993, **4**, 1473–1474.

28. K. Soai, Y. Ohno, Y. Inoue, T. Tsuruoka and Y. Hirose, *Recl. Trav. Chim. Pays-Bas*, 1995, **114**, 145–152.

29. T. Shibata, H. Tabira and K. Soai, *J. Chem. Soc., Perkin Trans. 1*, 1998, 177–178.
30. (a) M. Shi and W.-S. Sui, *Tetrahedron: Asymmetry*, 1999, **10**, 3319–3325; (b) M. Shi, X.-F. Wu and G. Rong, *Chirality*, 2002, **14**, 90–95.
31. M. Shi and W.-S. Sui, *Chirality*, 2000, **12**, 574–580.
32. C. Jimeno, A. Moyano, M. A. Pericas and A. Riera, *Synlett*, 2001, **7**, 1155–1157.
33. (a) O. Juanes, J. C. Rodriguez-Ubis, E. Brunet, H. Pennemann, M. Kossenjans and J. Martens, *Eur. J. Org. Chem.*, 1999, 3323–3333; (b) M. Kossenjans, H. Pennemann, J. Martens, O. Juanes, J. C. Rodriguez-Ubis and E. Brunet, *Tetrahedron: Asymmetry*, 1998, **9**, 4123–4125.
34. M. Schinnerl, M. Seitz, A. Kaiser and O. Reiser, *Org. Lett.*, 2001, **3**, 4259–4262.
35. (a) A. L. Braga, H. R. Appelt, P. H. Schneider, C. C. Silveira and L. A. Wessjohann, *Tetrahedron: Asymmetry*, 1999, **10**, 1733–1738; (b) A. L. Braga, H. R. Appelt, P. H. Schneider, O. E. D. Rodrigues, C. C. Silveira and L. A. Wessjohann, *Tetrahedron*, 2001, **57**, 3291–3295.
36. A. L. Braga, E. F. Alves, C. C. Silveira, G. Zeni, H. R. Appelt and L. A. Wessjohann, *Synthesis*, 2005, **4**, 588–594.
37. A. L. Braga, O. E. D. Rodrigues, M. W. Paixao, H. R. Appelt, C. C. Silveira and D. P. Bottega, *Synthesis*, 2002, **16**, 2338–2340.
38. B. T. Cho, Y. S. Chun and W. K. Yang, *Tetrahedron: Asymmetry*, 2000, **11**, 2149–2157.
39. M. Kossenjans and J. Martens, *Tetrahedron: Asymmetry*, 1998, **9**, 1409–1417.
40. Y. Masaki, Y. Satoh, T. Makihara and M. Shi, *Chem. Pharm. Bull.*, 1996, **44**, 454–456.
41. H.-L. Huang, Y.-C. Lin, S.-F. Chen, C.-L. J. Wang and L. T. Liu, *Tetrahedron: Asymmetry*, 1996, **7**, 3067–3070.
42. (a) T. Mehler and J. Martens, *Tetrahedron: Asymmetry*, 1994, **5**, 207–210; (b) W. Trentmann, T. Mehler and J. Martens, *Tetrahedron: Asymmetry*, 1997, **8**, 2033–2043.
43. M. Hernandez-Rodriguez, C. G. Avila-Ortiz, J. M. del Campo, D. Hernandez-Romero, M. Rosales-Hoz and E. Juaristi, *Aust. J. Chem.*, 2008, **61**, 364–375.
44. Z.-B. Dong, B. Liu, C. Fang and J.-S. Li, *J. Organomet. Chem.*, 2008, **693**, 17–22.
45. G. Chelucci, D. Berta and A. Saba, *Tetrahedron*, 1997, **53**, 3843–3848.
46. J. Lawecka, B. Bujnicki, J. Drabowicz and A. Rykowski, *Tetrahedron Lett.*, 2008, **49**, 719–722.
47. (a) C. Bolm, J. Müller, G. Schlingloff, M. Zehnder and M. Neuburger, *J. Chem. Soc., Chem. Commun.*, 1993, 182–183; (b) C. Bolm and J. Müller, *Tetrahedron*, 1994, **50**, 4355–4362.
48. Z. Huang, H. Lai and Y. Qin, *J. Org. Chem.*, 2007, **72**, 1373–1378.
49. (a) H. Takahashi, T. Kawakita, M. Yoshioka, S. Kobayashi and M. Ohno, *Tetrahedron Lett.*, 1989, **30**, 7095–7098; (b) M. Yoshioka, T. Kawakita and M. Ohno, *Tetrahedron Lett.*, 1989, **30**, 1657–1660; (c) H. Takahashi,

T. Kawakita, M. Ohno, M. Yoshioka and S. Kobayashi, *Tetrahedron*, 1992, **48**, 5691–5700.

50. (a) M. J. Rozema, A. Sidduri and P. Knochel, *J. Org. Chem.*, 1992, **57**, 1956–1958; (b) W. Brieden, R. Ostwald and P. Knochel, *Angew. Chem., Int. Ed. Engl.*, 1993, **32**, 582–584; (c) M. J. Rozema, C. Eisenberg, H. Lütjens, R. Ostwald, K. Belyk and P. Knochel, *Tetrahedron Lett.*, 1993, **34**, 3115–3118; (d) P. Knochel, W. Brieden, M. J. Rozema and C. Eisenberg, *Tetrahedron Lett.*, 1993, **34**, 5881–5884; (e) F. Langer, A. Devasagayaraj, P.-Y. Chavant and P. Knochel, *Synlett*, 1994, **1**, 410–412; (f) C. Eisenberg and P. Knochel, *J. Org. Chem.*, 1994, **59**, 3760–3761; (g) R. Ostwald, P.-Y. Chavant, H. Stadtmüller and P. Knochel, *J. Org. Chem.*, 1994, **59**, 4143–4153; (h) L. Schwink and P. Knochel, *Tetrahedron Lett.*, 1994, **35**, 9007–9010; (i) S. Nowotny, S. Vettel and P. Knochel, *Tetrahedron Lett.*, 1994, **35**, 4539–4540; (j) S. Vettel and P. Knochel, *Tetrahedron Lett.*, 1994, **35**, 5849–5852; (k) H. Lütjens and P. Knochel, *Tetrahedron: Asymmetry*, 1994, **5**, 1161–1162; (l) H. Lütjens, S. Nowotny and P. Knochel, *Tetrahedron: Asymmetry*, 1995, **6**, 2675–2678; (m) S. Vettel, C. Lutz and P. Knochel, *Synlett*, 1996, **1**, 731–733; (n) F. Langer, L. Schwink, A. Devasagayaraj, P.-Y. Chavant and P. Knochel, *J. Org. Chem.*, 1996, **61**, 8229–8243; (o) S. Vettel, C. Lutz, A. Diefenbach, G. Haderlein, S. Hammerschmidt, K. Kühling, M.-R. Mofid, T. Zimmermann and P. Knochel, *Tetrahedron: Asymmetry*, 1997, **8**, 779–800; (p) C. Lutz and P. Knochel, *J. Org. Chem.*, 1997, **62**, 7895–7898.

51. (a) S. Pritchett, D. H. Woodmansee, T. J. Davis and P. J. Walsh, *Tetrahedron Lett.*, 1998, **39**, 5941–5942; (b) S. Pritchett, P. Gantzel and P. J. Walsh, *Organometallics*, 1997, **16**, 5130–5132; (c) E. Royo, J. M. Betancort, T. J. Davis, P. Carroll and P. J. Walsh, *Organometallics*, 2000, **19**, 4840–4851; (d) J. Balsells and P. J. Walsh, *J. Am. Chem. Soc.*, 2000, **122**, 3250–3251; (e) S. Pritchett, D. H. Woodmansee, P. Gantzel and P. J. Walsh, *J. Am. Chem. Soc.*, 1998, **120**, 6423–6424; (f) P. J. Walsh, *Acc. Chem. Res.*, 2003, **36**, 739–749; (g) J. Balsells and P. J. Walsh, *J. Am. Chem. Soc.*, 2000, **122**, 1802–1803; (h) J. Balsells, J. M. Betancort and P. J. Walsh, *Angew. Chem., Int. Ed. Engl.*, 2000, **39**, 3428–3430.

52. C.-D. Hwang and B.-J. Uang, *Tetrahedron: Asymmetry*, 1998, **9**, 3979–3984.

53. D. E. Ho, J. M. Betancort, M. L. L. Woodmansee and P. J. Walsh, *Tetrahedron Lett.*, 1997, **38**, 3867–3870.

54. (a) X. Zhang and C. Guo, *Tetrahedron Lett.*, 1995, **36**, 4947–4950; (b) J. Qiu, C. Guo and X. Zhang, *J. Org. Chem.*, 1997, **62**, 2665–2668; (c) C. Guo, J. Qiu and X. Zhang, *Tetrahedron*, 1997, **53**, 4145–4158.

55. D. J. Ramon and M. Yus, *Synlett*, 2007, **15**, 2309–2320.

56. (a) D. J. Ramon and M. Yus, *Tetrahedron: Asymmetry*, 1997, **8**, 2479–2496; (b) M. Yus, D. J. Ramon and O. Prieto, *Tetrahedron: Asymmetry*, 2003, **14**, 1103–1114.

57. O. Prieto, D. J. Ramon and M. Yus, *Tetrahedron: Asymmetry*, 2000, **11**, 1629–1644.

58. A. Hui, J. Zhang, J. Fan and Z. Wang, *Tetrahedron: Asymmetry*, 2006, **17**, 2101–2107.
59. (a) B. Gadenne, P. Hesemann and J. J. E. Moreau, *Tetrahedron Lett.*, 2004, **45**, 8157–8160; (b) B. Gadenne, P. Hesemann, V. Polshettiwar and J. J. E. Moreau, *Eur. J. Inorg. Chem.*, 2006, 3697–3702.
60. L. A Paquette and R. Zhou, *J. Org. Chem.*, 1999, **64**, 7929–7934.
61. M. Cernerud, A. Skrinning, I. Bérgère and C. Moberg, *Tetrahedron: Asymmetry*, 1997, **8**, 3437–3441.
62. T. Bauer, J. Tarasiuk and K. Pasniczek, *Tetrahedron: Asymmetry*, 2002, **13**, 77–82.
63. T. Bauer and J. Gajewiak, *Tetrahedron*, 2003, **59**, 10009–10012.
64. (a) J.-S. You, M.-Y. Shao and H.-M. Gau, *Tetrahedron: Asymmetry*, 2001, **12**, 2971–2975; (b) K.-H. Wu and H.-M. Gau, *Organometallics*, 2003, **22**, 5193–5200; (c) K.-H. Wu and H.-M. Gau, *Organometallics*, 2004, **23**, 580–588.
65. X.-P. Hui, C.-A. Chen and H.-M. Gau, *Chirality*, 2005, **17**, 51–56.
66. M. Yus, D. J. Ramon and O. Prieto, *Tetrahedron: Asymmetry*, 2002, **13**, 1573–1579.
67. V. M. Arredondo, S. Tian, F. E. McDonald and T. J. Marks, *J. Am. Chem. Soc.*, 1999, **121**, 3633–3639.
68. Y. Takemoto, N. Yoshikawa, Y. Baba, C. Iwata, T. Tanaka, T. Ibuka and H. Ohishi, *J. Am. Chem. Soc.*, 1999, **121**, 9143–9154.
69. Y. Arai, N. Nagata and Y. Mazaki, *Chem. Pharm. Bull.*, 1995, **12**, 2243–2245.
70. I. Shiina, K. Konishi and Y.-S. Kuramoto, *Chem. Lett.*, 2002, **1**, 164–165.
71. A. Heckel and D. Seebach, *Chem. Eur. J.*, 2002, **8**, 560–572.
72. M. C. Carreno, J. L. Ruano, M. C. Maestro and L. M. M. Cabrejas, *Tetrahedron: Asymmetry*, 1993, **4**, 727–734.
73. (a) R. Gomez, J. A. Arrayas and J. C. Carretero, *Angew. Chem., Int. Ed. Engl.*, 2006, **45**, 7674–7715; (b) L.-X. Dai, T. Tu, S. L. You, W.-P. Deng and X.-L. Hou, *Acc. Chem. Res.*, 2003, **36**, 659–667.
74. (a) T. Hayashi, *Ferrocenes*, ed. A. Togni, T. Hayashi, VCH, Weinheim, 1995, pp. 105–142; (b) A. Togni, R. L. Haltermann, (eds.), *Metallocenes*, VCH: Weinheim, Germany, 1998; (c) T. J. Colacot, *Chem. Rev.*, 2003, **103**, 3101–3118; (d) L.-X. Dai, T. Tu, S.-L. You, W.-P. Deng and X.-L. Hou, *Acc. Chem. Res.*, 2003, **36**, 659–667.
75. (a) J. Priego, O. G. Mancheno, S. Cabrera and J. C. Carretero, *Chem. Commun.*, 2001, 2026–2027; (b) J. Priego, O. G. Mancheno, S. Cabrera and J. C. Carretero, *J. Org. Chem.*, 2002, **67**, 1346–1353.
76. B. F. Bonini, M. Fochi, M. Comes-Franchini, A. Ricci, L. Thijs and B. Zwanenburg, *Tetrahedron: Asymmetry*, 2003, **14**, 3321–3327.
77. A. Heckel and D. Seebach, *Angew. Chem., Int. Ed. Engl.*, 2000, **39**, 163–165.
78. C. Bolm, N. Hermanns, J. P. Hildebrand and K. Muniz, *Angew. Chem., Int. Ed. Engl.*, 2001, **40**, 3284–3308.
79. P. I. Dosa, J. C. Ruble and J. C. Fu, *J. Org. Chem.*, 1997, **62**, 444–445.
80. C. Bolm and J. Rudolph, *J. Am. Chem. Soc.*, 2002, **124**, 14850–14851.

81. (a) F. Schmidt, R. T. Stemmler, J. Rudolph and C. Bolm, *Chem. Soc. Rev.*, 2006, **35**, 454–470; (b) J. G. Kim and P. J. Walsh, *Angew. Chem., Int. Ed. Engl.*, 2006, **45**, 4175–4178.

82. A. L. Braga, P. Milani, F. Vargas, M. W. Paixao and J. A. Sehnem, *Tetrahedron: Asymmetry*, 2006, **17**, 2793–2797.

83. M.-J. Jin, S. M. Sarkar, D.-H. Lee and H. Qiu, *Org. Lett.*, 2008, **10**, 1235–1237.

84. J. Schwartz and J. A. Labinger, *Angew. Chem., Int. Ed. Engl.*, 1976, **15**, 333–334.

85. P. Wipf and W. Xu, *Org. Synth.*, 1997, **74**, 205–211.

86. P. Wipf and W. Xu, *Tetrahedron Lett.*, 1994, **35**, 5197–5200.

87. P. Wipf and S. Ribe, *J. Org. Chem.*, 1998, **63**, 6454–6455.

88. P. Wipf and N. Jayasuriya, *Chirality*, 2008, **20**, 425–430.

89. L. Pu, *Tetrahedron*, 2003, **59**, 9873–9886.

90. A. L. Braga, H. R. Appelt, C. C. Silveira and L. H. de Andrade, *Tetrahedron*, 2002, **58**, 10413–10416.

91. M. Ni, R. Wang, Z.-j. Han, B. Mao, C.-s. Da, L. Liu and C. Chen, *Adv. Synth. Catal.*, 2005, **347**, 1659–1665.

92. C. M. Sprout, M. L. Richmond and C. T. Seto, *J. Org. Chem.*, 2005, **70**, 7408–74.

93. S.-L. Tseng and T.-K. Yang, *Tetrahedron: Asymmetry*, 2005, **16**, 773–782.

94. J. Granander, R. Sott and G. Hilmersson, *Tetrahedron: Asymmetry*, 2003, **14**, 439–447.

95. F. A. Cotton and G. Wilkinson, *Adv. Inorg. Chem. 4th edn*, Wiley, New York, 1980, p. 342.

96. J.-S. You, S.-H. Hsieh and H.-M. Gau, *Chem. Commun.*, 2001, **1**, 1546–1547.

97. (a) M. Shi and W.-S. Sui, *Tetrahedron: Asymmetry*, 2000, **11**, 773–779; (b) C.-J. Wang and M. Shi, *Eur. J. Org. Chem.*, 2003, **1**, 2823–2828.

CHAPTER 4

Addition of Organozinc Reagents to Ketones

Nowadays, the generation of tertiary carbon-atom stereocentres can be achieved easily in most cases by using the appropriate chiral auxiliary, reagent or catalyst. However, the approach to complex compounds bearing quaternary stereocentres, such as tertiary alcohols and related systems, which are important building blocks of naturally occurring and artificial biologically active molecules, is still a challenge for synthetic organic chemists, and every enantioselective procedure for the construction of a fully substituted carbon centre is of great value.[1] Among the different approaches for the synthesis of these types of compounds, such as kinetic resolution, enantioselective desymmetrization, oxidation, cyclisation, electrophilic alkylation and nucleophilic alkylation, those that involve stereoselective C–C bond formation are of particular interest, since one C–C bond and one stereoelement are created in only one synthetic step. The simplest approach for the preparation of chiral tertiary alcohols is the enantioselective addition of organometallic reagents to a ketone. Indeed, the preparation of chiral molecules bearing a quaternary stereocentre is still a very important challenge in synthetic organic chemistry, mainly as a result of steric factors, with the diastereoselective approach being the most commonly used.[2] However, the enantioselective version of these reactions have, in general, some advantages over the diastereoselective ones, such as for the repeated preparation of a compound, the need for the attachment and detachment of a chiral auxiliary at the beginning and end of the synthetic process is avoided. Indeed, every novel enantioselective procedure for the construction of a fully substituted carbon centre is of great value,[3] the simplest approach for the preparation of chiral tertiary alcohols being the enantioselective 1,2-addition of organometallic reagents to ketones.[2] Indeed, one of the most common methods to extend the carbon skeleton is through addition of organometallic reagents to

RSC Catalysis Series No. 2
Chiral Sulfur Ligands: Asymmetric Catalysis
By Hélène Pellissier
© Hélène Pellissier 2009
Published by the Royal Society of Chemistry, www.rsc.org

carbonyl compounds.[4] Among nonstabilised organometallic compounds,[5] dialkylzinc reagents are ideal alkyl nucleophile synthons as it is possible to prepare many different functionalised organometallic reagents due to their low reactivity.[6]

Despite the monumental success in the enantioselective addition of organozinc reagents to aldehydes with literally hundreds of catalysts capable of promoting these reactions with very high enantioselectivities, catalytic asymmetric additions of these low reactive reagents to ketones have received much less attention with only a few systems that are able to catalyse these reactions.[7] Thus, a long-standing problem in asymmetric catalysis has been the absence of efficient and highly enantioselective methods to prepare tertiary alcohols with high levels of enantioselectivity.[3b,7b] This discrepancy is due to the greater reactivity of aldehydes over ketones and to the reduced propensity of ketones to coordinate to Lewis acids, relative to aldehydes. Furthermore, to achieve high enantioselectivities, catalysts must readily distinguish between the lone pairs of the carbonyl oxygen. Differentiation of aldehyde lone pairs is simple since little steric hindrance is encountered on binding *syn* to the aldehyde hydrogen relative to binding *syn* to the alkyl substituent. With ketones, however, discrimination of the lone pairs is difficult due to their similar environments. Although several examples involving organolithium[8] and Grignard[9] reagents have been reported, these usually require greater than stoichiometric amounts of chiral ligands. In addition, the high reactivity of these reagents precludes the presence of many functional groups and, therefore, reduces the attractiveness of such strongly basic reagents. Demands for increased functional group tolerance and enantioselective variants have reduced the attractiveness of such strongly basic and reactive reagents, making the search for milder organometallic reagents capable of addition to ketones essential. In contrast to organolithium and Grignard reagents, organozinc reagents are mild and exhibit a very low nucleophilicity and excellent functional group compatibility,[10] making them ideal for use in the synthesis of complex, highly functionalised targets. Because of the low reactivity of organozinc reagents, their reaction with ketones does not yield the addition product,[11] giving instead the corresponding secondary alcohol, arising from a reduction process.[11,12] Only in 1997 did Walsh *et al.* report the use of novel bis(sulfonamides) diol ligands, which were successfully applied to catalyse the first enantioselective addition of alkylzinc reagents to ketones in the presence of $Ti(Oi\text{-}Pr)_4$, providing the expected tertiary alcohols in moderate to good yields and with enantioselectivities of up to 88% ee, as shown in Scheme 4.1.[7a,13] However, the sluggish nature of these catalysts was particularly troubling, because even if high enantioselectivities into the 90% ee range could be reached, a catalyst that required 20 mol% loading and several days to consume the starting material was of limited use.

In 1998, two other examples of chiral ligands that enabled the asymmetric addition of organozinc reagents to ketones were reported by two groups independently. Thus, Dosa and Fu employed the nonsulfur-containing Noyori's DAIB[14] ligand in the asymmetric addition of $ZnPh_2$ to ketones with

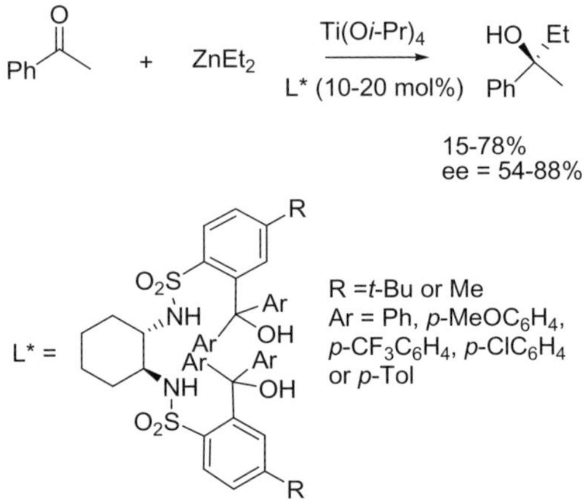

Scheme 4.1   Addition of ZnEt$_2$ to acetophenone with bis(sulfonamides) diol ligands.

enantioselectivities of up to 91% ee.[15] A simple mechanistic study showed that the catalytic cycle, as well as active species, should be similar to those described for the corresponding aldehyde reaction. A phenyl group was exchanged for a methoxide group on the zinc atom of the chiral Lewis acid, and the chiral system activated the ketone and the zinc reagent simultaneously. The success of this reaction was attributed, on the one hand, to the increase of the Lewis acidity of the zinc atom due to the presence of a methoxide moiety, and, on the other hand, to the higher reactivity of ZnPh$_2$. This reagent did not have hydrogen atoms in α-position, and consequently, the formation of zinc hydride and subsequent reduction processes were limited or prevented. In the same context, Ramon and Yus reported, also in 1998, the second example of enantioselective addition of organozinc reagents to ketones performed in the presence of a chiral sulfur-containing ligand, for the first time, such as isoborneol-derived sulfonamide ligand and Ti(Oi-Pr)$_4$.[16] The reaction yielded the expected tertiary alcohols with modest to good yields and enantioselectivities of up to 89% ee, as shown in Scheme 4.2. Although enantioselectivities approaching 90% ee were observed for substrates such as alkyl aryl ketones, others such as heteroaryl, α,β-unsaturated or dialkyl ketones resulted in moderate or low enantioselectivities. The reaction seemed to be quite sensitive to changes in the conditions, such as solvents, reagent amounts, temperature *etc.* Furthermore, the catalysts exhibited a low turnover frequency, requiring 4–14 days for completion of the reaction at 20 mol% catalyst loading. A preliminary mechanistic study showed a small nonlinear effect when a stoichiometric amount of the chiral ligand was used, which disappeared when the reaction was performed with substoichiometric amounts of the chiral ligand. Moreover, it was shown that the enantioselectivity was independent of the chemical yield but highly dependent on the steric bulkiness of the ligand. No autoinductive process was observed, with the

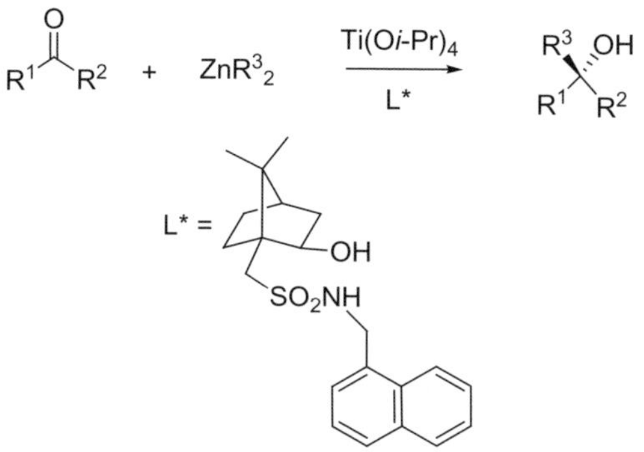

R^1 = Ph, R^2 = R^3 = Et: 63% ee = 84%
R^1 = *(E)*-Ph-CH=CH, R^2 = Me, R^3 = Et: 72% ee = 31%
R^1 = Ph, R^2 = CH$_2$Br, R^3 = Et: 79% ee = 76%
R^1 = Ph, R^2 = Me, R^3 = Et: 85% ee = 86%
R^1 = Ph, R^2 = *n*-Bu, R^3 = Et: 78% ee = 86%
R^1 = Me, R^2 = 2-Naph, R^3 = Et: 75% ee = 42%
R^1 = *p*-BrC$_6$H$_4$, R^2 = Me, R^3 = Et: 83% ee = 81%
R^1 = Ph, R^2 = Et, R^3 = Me: 89% ee = 89%
R^1 = Cy, R^2 = *n*-Bu, R^3 = Me: 26% ee = 0%

catalytic active species:

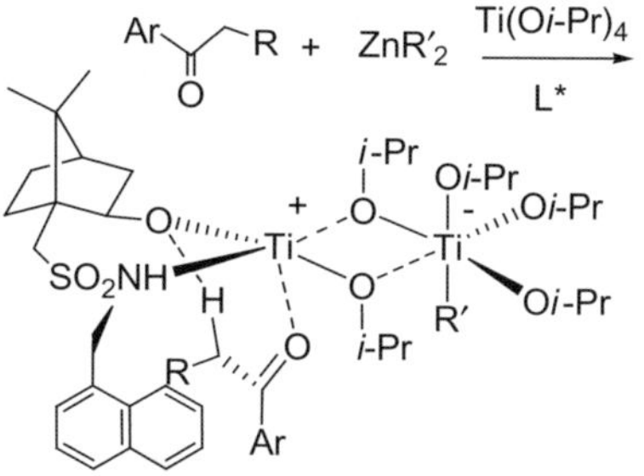

Scheme 4.2 Additions of ZnR$_2$ to ketones with hydroxycamphorsulfonamide ligand.

enantioselectivity depending only on the steric bulkiness of the ketone substituents and not on their electronic character. All these facts were similar to those previously described for the known corresponding reaction with aldehydes. Consequently, a similar catalytic cycle and species involved were proposed for this new reaction. In other words, the catalytic species was postulated to be a dinuclear titanium complex, in which one titanium atom bore the chiral ligand and the ketone and the other titanium atom bore the alkyl moiety, the two metal centres being connected by two isopropoxy bridges (Scheme 4.2).

The catalytic enantioselective alkynylation of ketones is a powerful method to prepare chiral tertiary propargylic alcohols whose building block is often present in numerous attractive pharmaceuticals.[17] In this context, the hydroxycamphorsulfonamide ligand has been more recently used by Lu *et al.* in the alkynylation of ketones, yielding the corresponding chiral propargylic alcohols with enantioselectivities of up to 97% ee, as shown in Scheme 4.3.[18] According to these authors, the results were quite sensitive to both the solvent and the source of copper used, with dichloromethane and copper(II) triflate giving the best enantioselectivities. It should be pointed out that the scope of the reaction was not very broad since changing either the alkyl aryl ketone to dialkyl derivatives, or phenylacetylene to aliphatic substituted acetylene derivatives,

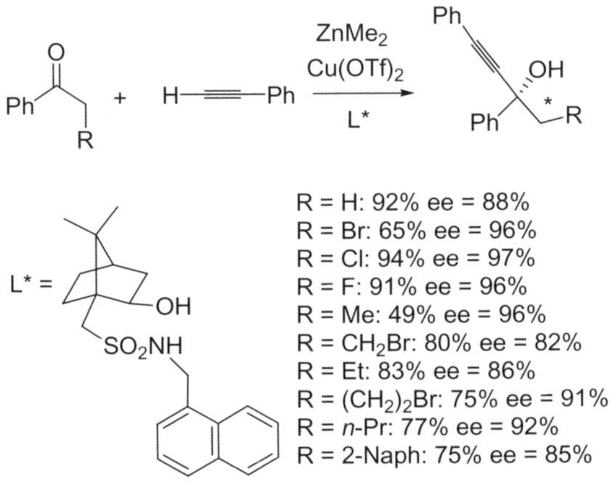
ZnMe2
Cu(OTf)2
L*
Ph
OH
R
H—Ph
L* =
OH
SO2NH
R = H: 92% ee = 88%
R = Br: 65% ee = 96%
R = Cl: 94% ee = 97%
R = F: 91% ee = 96%
R = Me: 49% ee = 96%
R = CH2Br: 80% ee = 82%
R = Et: 83% ee = 86%
R = (CH2)2Br: 75% ee = 91%
R = n-Pr: 77% ee = 92%
R = 2-Naph: 75% ee = 85%

Scheme 4.3 Alkynylations of ketones with hydroxycamphorsulfonamide ligand.

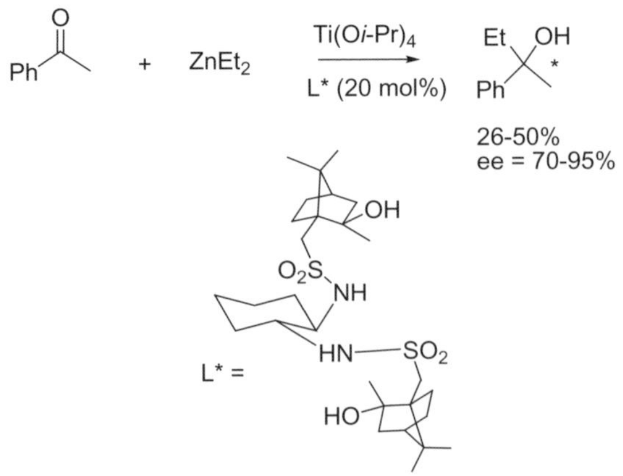
O
Ph
+ ZnEt2
Ti(Oi-Pr)4
L* (20 mol%)
Et OH
Ph
26-50%
ee = 70-95%
OH
O2S
NH
HN—SO2
L* =
HO

Scheme 4.4 Addition of $ZnEt_2$ to acetophenone with bis(hydroxycamphorsulfonamides) ligand.

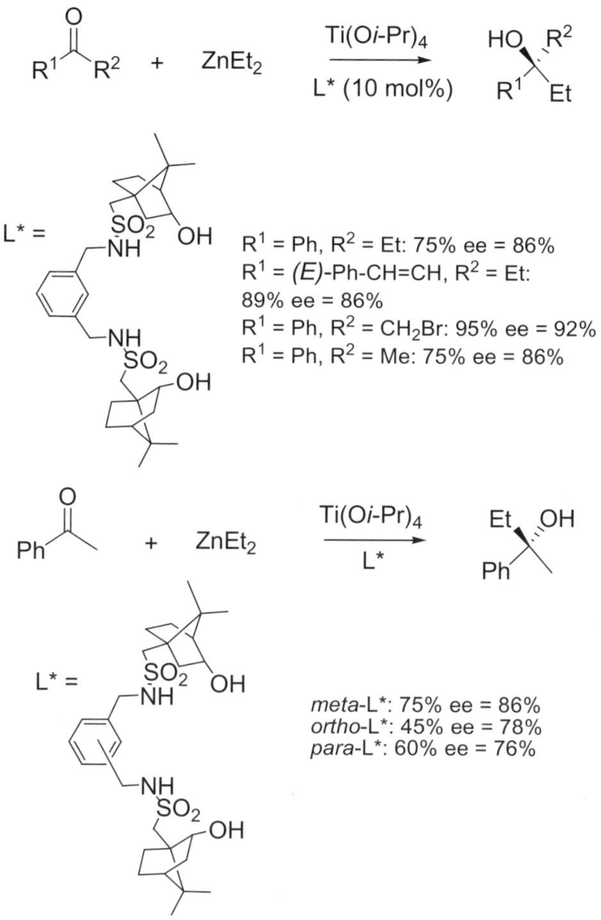

Scheme 4.5 Additions of ZnEt$_2$ to ketones with bis(hydroxycamphorsulfonamides) ligands bearing xylenediamine as linker.

decreased the enantioselectivity, while the relative positions (*ortho-, meta-*) of these substituents did not. Thus, the best results were obtained when *ortho*-substituted aryl ketones were used as the substrates. This was rationalized as being due to possible restrictions on the orientation of the substrates due to a higher steric hindrance.

In 2000, Walsh *et al.* reported the synthesis of a novel chiral bis(sulfonamides) ligand derived from camphor, which was subsequently tested as a ligand for the enantioselective addition of ZnEt$_2$ to acetophenone (Scheme 4.4).[7a,19] In order to obtain the corresponding tertiary alcohol in a high enantioselectivity of up to 95% ee, the reaction had to be performed at a high catalyst loading (20 mol%) and frequently did not go to completion after several days. Furthermore, the product was isolated in low to moderate yields ($\leq$50%).

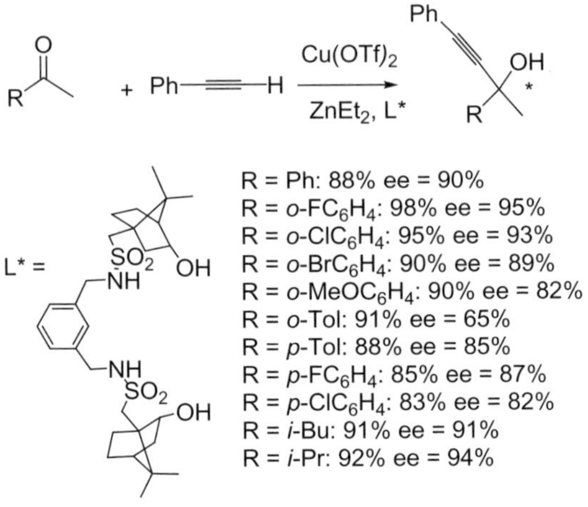

Scheme 4.6 Alkynylations of ketones with bis(hydroxycamphorsulfonamides) ligand bearing xylenediamine as linker.

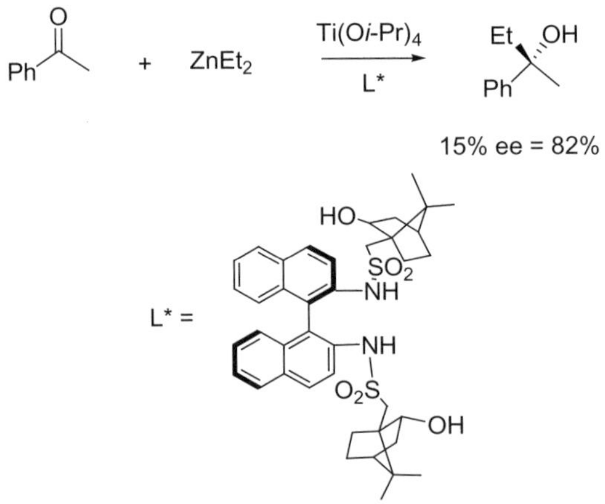

Scheme 4.7 Addition of ZnEt$_2$ to acetophenone with bis(hydroxycamphor-sulfonamides) ligand bearing binaphthyldiamine as linker.

The assumption that the dinuclear catalytic species depicted in Scheme 4.2 was the common intermediate for the alkylation of either aldehydes or ketones has brought Yus *et al.* to the possibility of improving the ligand by replacing the flexible alkoxide bridge ligands between both titanium atoms in the catalytic species with the proper ligand. In this way, playing with the length and the angle between both guest titanium atoms could control the synergistic effect of both metallic centres and, therefore, control the enantioselectivity. The novel ligand would be able to bind to two titanium atoms at

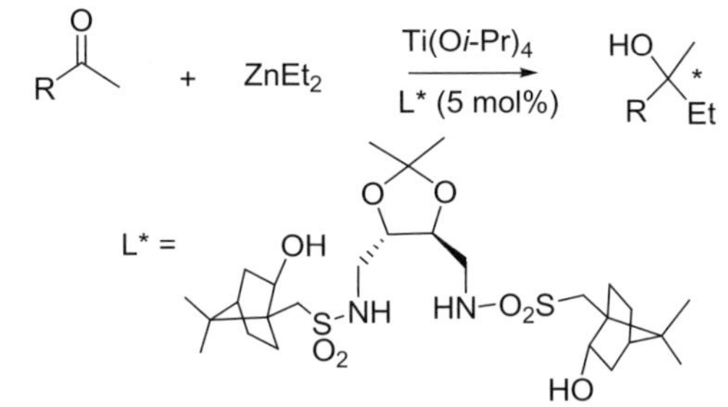

R = p-ClC$_6$H$_4$: 75% ee = 81%
R = p-FC$_6$H$_4$: 78% ee = 83%
R = p-BrC$_6$H$_4$: 78% ee = 80%
R = Ph: 70% ee = 99%
R = p-MeOC$_6$H$_4$: 68% ee = 35%
R = p-Tol: 70% ee = 70%
R = p-PhC$_6$H$_4$: 80% ee = 78%
R = 2-Fu: 88% ee = 10%
R = 2-Naph: 80% ee = 82%

Scheme 4.8 Additions of ZnEt$_2$ to ketones with bis(hydroxycamphorsulfonamide) ligand based on tartaric acid.

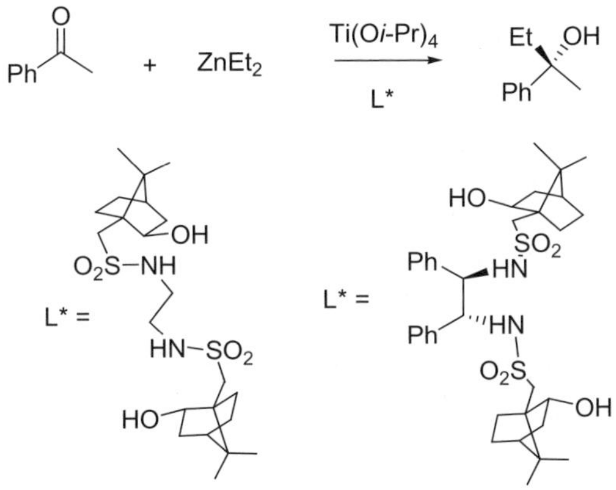

60% ee = 92%   25% ee = 33%

Scheme 4.9 Addition of ZnEt$_2$ to acetophenone with ethanediamine ligand and its phenyl derivative.

the same time. In this context, and inspired by the first successful study developed by Walsh *et al.* (Scheme 4.4),[19] Yus *et al.* have prepared several C_2-symmetric disulfonamides derived from chiral camphorsulfonyl chloride and different diamines.[20] In the screening of diamines as the covalent linker,

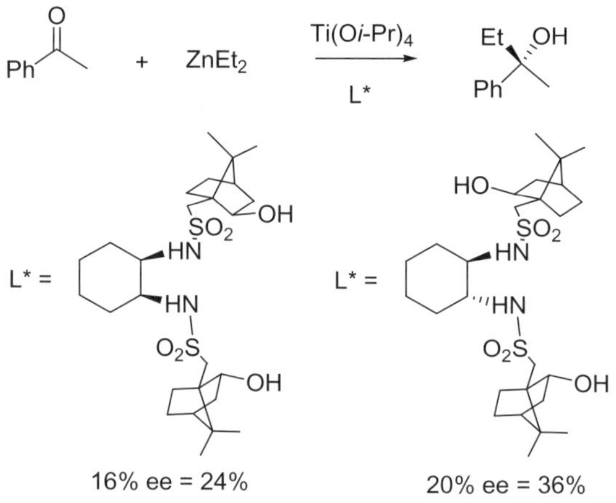

Scheme 4.10   Addition of ZnEt₂ to acetophenone with 1,2-cyclohexanediamine-derived ligands.

xylenediamine emerged as the best one. Indeed, using xylenediamine derivatives as connectors for the isoborneol-10-sulfonamide moieties, these authors found that their relative position had an important effect not only on the enantioselectivity, but also on the chemical yield and reaction rates, with the *meta*-derivative depicted in Scheme 4.5 giving the best results with enantioselectivities of up to 92% ee. Moreover, a reasonable 10 mol% catalyst loading was applied to achieve this process.

In addition, the most efficient *meta*-ligand depicted above was successfully applied, in 2006, to the alkynylation of ketones.[21] Thus, Liu *et al.* showed that this ligand was able to catalyse the enantioselective addition of phenylacetylene to various ketones, using Cu(OTf)₂ as the starting base in toluene. The results were excellent and homogeneous not only for substituted aryl alkyl ketones, but also for aliphatic methyl ketones (Scheme 4.6).

The involvement by Yus *et al.* of other diamines, such as the binaphthyl-diamine derivative depicted in Scheme 4.7, did not produce any improvement for the enantioselective addition of ZnEt₂ to acetophenone, since the corresponding alcohol was produced with an enantioselectivity of 82% ee instead of 86% ee obtained with the corresponding *meta*-xylenediamine derivative.[20]

In the same area, Wang *et al.* reported, in 2006, the synthesis of another diamine derivative based on L-tartaric acid and employed this novel di-sulfonamide ligand at a low catalyst loading of 5 mol% for the enantioselective addition of ZnEt₂ to acetophenone.[22] This provided an excellent enantioselectivity of up to 99% ee for the formation of the corresponding tertiary alcohol but, however, the enantioselectivity of this addition was not constant and, for other ketones, never reached 84% ee, as shown in Scheme 4.8. These authors showed that the electronic properties of the substituents had a great influence on both the reaction yield and enantioselectivity. Thus, the presence of an electron-withdrawing group on the aromatic ring of the

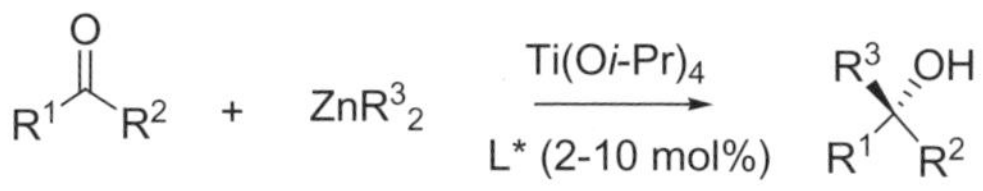

$$R^1 \overset{O}{\underset{}{\underset{}{\bigvee}} } R^2 \; + \; ZnR^3_2 \quad \xrightarrow[L^* \,(2\text{-}10\text{ mol}\%)]{Ti(Oi\text{-}Pr)_4} \quad R^1 \overset{R^3 \;\; OH}{\underset{}{\underset{}{\bigvee}}} R^2$$

R^1 = Ph, R^2 = R^3 = Et: 80% ee = 98%
R^1 = (E)-Ph-CH=CH, R^2 = R^3 = Et: 90% ee > 99%
R^1 = Ph, R^2 = CH$_2$Br, R^3 = Et: 70% ee = 50%
R^1 = Ph, R^2 = Me, R^3 = Et: 90% ee > 99%
R^1 = Ph, R^2 = Et, R^3 = Me: > 95% ee = 98%
R^1 = Cy, R^2 = n-Bu, R^3 = Me: 70% ee = 65%
R^1 = Ph, R^2 = n-Bu, R^3 = Et: 83% ee = 87%
R^1 = p-Tol, R^2 = Me, R^3 = Et: 90% ee = 95%
R^1 = m-Tol, R^2 = Me, R^3 = Et: 82% ee = 99%
R^1 = o-Tol, R^2 = Me, R^3 = Et: 24% ee = 96%
R^1 = Ph, R^2 = (CH$_2$)$_2$Cl, R^3 = Et: 82% ee = 89%
R^1 = p-CF$_3$C$_6$H$_4$, R^2 = Me, R^3 = Et: 90% ee = 93%
R^1 = m-CF$_3$C$_6$H$_4$, R^2 = Me, R^3 = Et: 56% ee = 98%
R^1 = p-MeOC$_6$H$_4$, R^2 = Me, R^3 = Et: 85% ee = 94%
R^1 = m-ClC$_6$H$_4$, R^2 = Et, R^3 = Me: 90% ee = 96%
R^1 = Ph-C≡C, R^2 = Me, R^3 = Et: > 95% ee > 99%

L* =

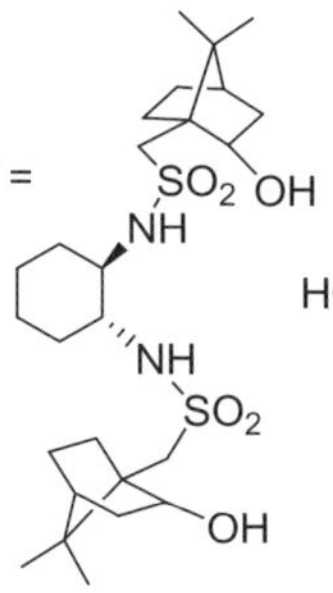

HOCSAC ligand

Scheme 4.11 Additions of ZnR$_2$ to ketones with HOCSAC ligand.

substrate increased the enantioselectivity. In addition, heteroaromatic ketone derivatives, such as 2-acetyl furan and 2-acetylthiophene, also gave good yields but with a poor enantioselectivity.

In 2006, Yus *et al.* investigated the ethanediamine ligand depicted in Scheme 4.9 as a new ligand for the enantioselective addition of ZnEt$_2$ to acetophenone, providing the corresponding tertiary alcohol in a good yield and with a high enantioselectivity of 92% ee.[23] When the same reaction was performed in the presence of the corresponding phenyl derivative as the ligand, it yielded very modest results (25%, 33% ee) as shown in Scheme 4.9.

Another idea developed by Yus *et al.* was to use other linkers with a low conformational freedom, such as 1,2-cyclohexanediamine derivatives.[24]

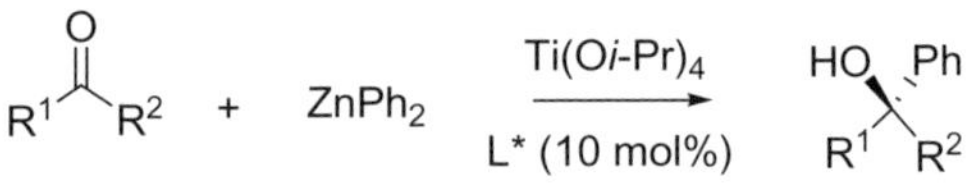

$R^1 = m\text{-}MeOC_6H_4$, $R^2 = Et$: 99% ee = 92%
$R^1 = p\text{-}MeOC_6H_4$, $R^2 = Et$: 99% ee = 88%
$R^1 = m\text{-}CF_3C_6H_4$, $R^2 = Me$: 93% ee = 95%
$R^1 = o\text{-}BrC_6H_4$, $R^2 = Me$: 76% ee = 95%
$R^1 = 2\text{-}Naph$, $R^2 = Me$: 99% ee = 96%
$R^1 = (E)\text{-}Ph\text{-}CH=CH$, $R^2 = Me$: 58% ee = 91%
$R^1 = Cy$, $R^2 = Me$: 81% ee = 87%
$R^1 = i\text{-}Pr$, $R^2 = Me$: 55% ee = 75%
$R^1 = p\text{-}Tol$, $R^2 = Me$: 98% ee = 92%
$R^1 = p\text{-}BrC_6H_4$, $R^2 = Me$: 98% ee = 96%
$R^1 = p\text{-}CF_3C_6H_4$, $R^2 = Me$: 97% ee = 91%
$R^1 = p\text{-}BrC_6H_4$, $R^2 = Et$: 95% ee = 80%

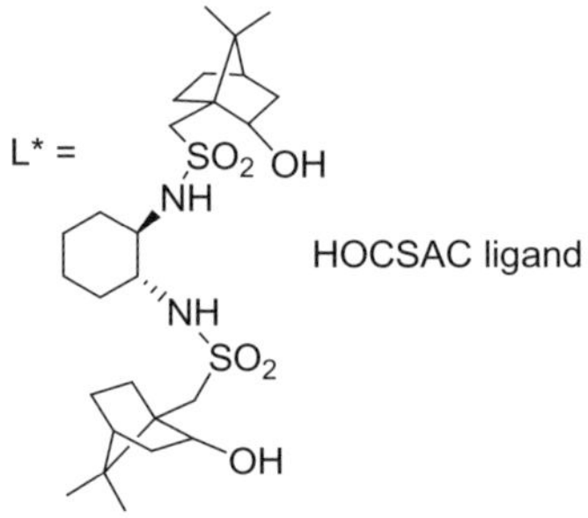

HOCSAC ligand

Scheme 4.12 Additions of ZnPh$_2$ to ketones with HOCSAC ligand.

Thus, several cyclohexane-1,2-diamine derivatives, such as those depicted in Scheme 4.10, were tested as the ligands for the enantioselective addition of ZnEt$_2$ to acetophenone, giving only low enantioselectivities ($\leq$36% ee).

It was independently found by two groups[23–25] that the *exo*-diol derived from bis(camphorsulfonyl)-substituted *trans*-cyclohexane-1,2-diamine ligand (HOCSAC) was an excellent promoter for the enantioselective addition of dialkylzinc reagents to any type of ketones, even dialkyl ketones, in the presence of Ti(O*i*-Pr)$_4$. As shown in Scheme 4.11, excellent enantioselectivities of up to 99% ee were obtained in these conditions in combination with high yields and with a low catalyst loading of 2–10 mol%.

These results were comparable to those obtained for the related additions to aldehydes. Furthermore, it was shown that this ligand was also efficient for the addition of arylzinc reagents, providing enantioselectivities of up to 96% ee, as shown in Scheme 4.12.[23,26]

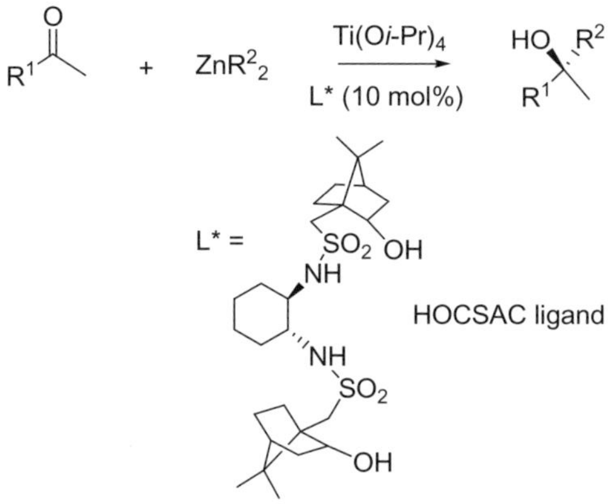

$$R^1 = \text{m-Tol}, R^2 = \text{n-Oct: 91\% ee = 98\%}$$

R^1 = m-Tol, R^2 = n-Oct: 91% ee = 98%
R^1 = m-Tol, R^2 = (CH$_2$)$_4$Cl: 95% ee = 99%
R^1 = m-Tol, R^2 = (CH$_2$)$_4$OPiv: 58% ee = 96%
R^1 = m-Tol, R^2 = (CH$_2$)$_3$i-Pr: 77% ee = 96%
R^1 = m-Tol, R^2 = (CH$_2$)$_4$OTBS: 89% ee = 98%
R^1 = m-Tol, R^2 = (CH$_2$)$_5$Br: 89% ee = 96%
R^1 = 2-Naph, R^2 = n-Oct: 70% ee = 99%
R^1 = 2-Naph, R^2 = (CH$_2$)$_4$Cl: 83% ee = 99%
R^1 = 2-Naph, R^2 = (CH$_2$)$_3$i-Pr: 75% ee = 90%
R^1 = m-CF$_3$C$_6$H$_4$, R^2 = (CH$_2$)$_4$OPiv: 53% ee = 99%
R^1 = m-CF$_3$C$_6$H$_4$, R^2 = (CH$_2$)$_3$i-Pr: 54% ee = 92%
R^1 = m-ClC$_6$H$_4$, R^2 = (CH$_2$)$_3$i-Pr: 69% ee = 94%
R^1 = m-ClC$_6$H$_4$, R^2 = (CH$_2$)$_4$OTBS: 51% ee = 93%
R^1 = (E)-Ph-CH=CH, R^2 = (CH$_2$)$_3$i-Pr: 86% ee = 93%

Scheme 4.13 Additions of functionalised dialkylzinc reagents to ketones with HOCSAC ligand.

In addition, the success of the HOCSAC ligand was further confirmed by Walsh *et al.* by the use of functionalised dialkylzinc reagents,[25b] as well as under very highly concentrated reaction conditions allowing catalyst loadings to be reduced by up to 40-fold (0.25 mol%) while maintaining the enantioselectivity over 90% ee.[25c] Thus, various functionalised dialkylzinc reagents could be successfully added to a broad range of saturated ketones and enones, providing a broad range of the corresponding tertiary alcohols in good yields and enantioselectivities of up to 91% ee, as shown in Scheme 4.13.

In 2007, the high efficiency of this ligand was applied by Yus *et al.* to the total synthesis of biologically active products, such as the cocaine-abuse therapeutic agent depicted in Scheme 4.14.[27] The key step of this synthesis was the enantioselective addition of ZnMe$_2$ to α-halo-substituted acetophenones leading to the corresponding chiral halohydrins with moderate to good results, as shown in Scheme 4.14. The subsequent reaction of this alcohol with a piperazine, depicted in Scheme 4.14 and prepared by reaction of 4-fluoro-phenylmagnesium bromide with ethyl formate followed by nucleophilic

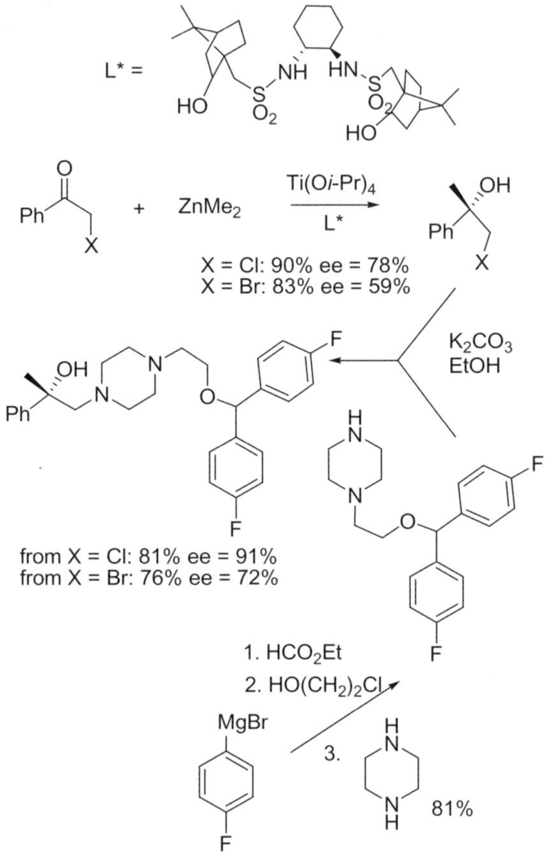

Scheme 4.14 Synthesis of cocaine-abuse therapeutic agent.

substitution with 2-chloroethanol and piperazine, gave the final tertiary alcohol. This product was a promising candidate as a cocaine-abuse therapeutic agent. It was noteworthy that although the initial enantiomeric excess of the tertiary alcohol derived from the addition of $ZnMe_2$ to α-halo-substituted acetophenone was moderate to good, the enantiomeric excess increased after a simple column chromatography of the final product, probably due to the formation of strong dimers of the final chiral product. These enantiomeric excesses could be further improved simply by recrystallisation of the corresponding dimaleate salt from *N,N*-dimethylformamide–methanol.

This ligand has also been used by the same authors to promote the addition of $ZnMe_2$ to a functionalised α,β-unsaturated ketone in the asymmetric key step of the first enantioselective synthesis of (–)-frontalin.[28] This synthesis started with the naphthalene-catalysed lithiation of a chlorinated ketal (Scheme 4.15) that, after several transmetalation processes, was trapped by reaction

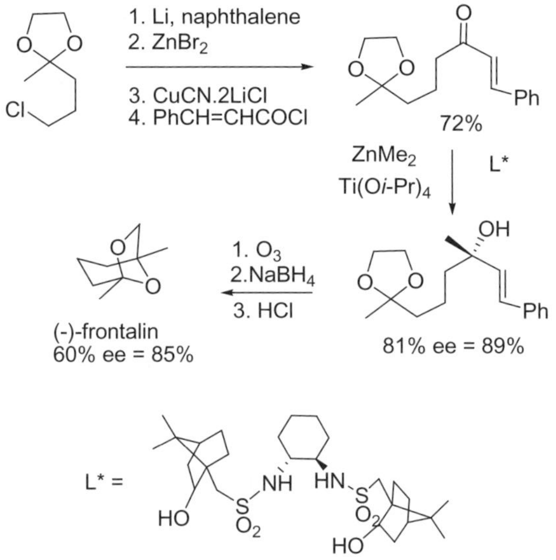

Scheme 4.15 Synthesis of (–)-frontalin.

with cinnamoyl chloride to give the corresponding functionalised ketone. The enantioselective addition of $ZnMe_2$ to this ketone gave the corresponding tertiary alcohol with a high enantioselectivity of up to 89% ee. The ozonolysis of the carbon–carbon double bond, followed by the reduction of the *in situ* formed carbonylic compound gave a primary alcohol, and a final acidic hydrolysis gave the expected (–)-frontalin. This product is one of the semi-chemicals (aggregation pheromones) secreted from pine beetles of the *Dendroctonus* family, and since these beetles destroy large areas of pine forest, frontalin has been used to control the progression of harmful insect infestations.

The use of cyclic α,β-unsaturated ketones as starting materials in the enantioselective addition of dimethyl- and diethylzinc reagents catalysed by the HOCSAC ligand was introduced by Walsh and Jeon, in 2003.[25b,29] As shown in Scheme 4.16, the corresponding cyclic tertiary alcohols were formed in high enantioselectivities of up to 99% ee.

The subsequent epoxidation of these *in situ* formed allylic tertiary alcohols yielded the corresponding *syn*-epoxy alcohols with high levels of diastereo- and enantioselectivity, thus providing a novel one-pot asymmetric synthesis of acyclic chiral epoxyalcohols via a domino vinylation epoxidation reaction (Scheme 4.17).[30]

Not only dialkylzinc reagents were suitable nucleophiles for this addition, but also highly reactive $ZnPh_2$ could be used in the enantioselective addition to cyclic α,β-unsaturated ketones catalysed by the HOCSAC ligand, providing excellent results.[26] As shown in Scheme 4.18, the addition of $ZnPh_2$ to α-substituted α,β-unsaturated ketones catalysed by the HOCSAC ligand gave the

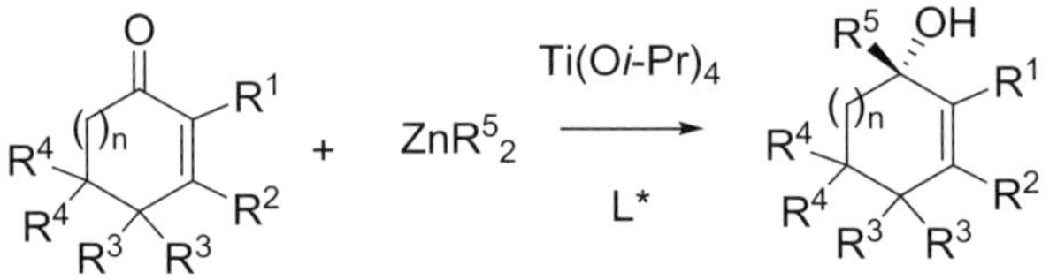

R^1 = R^3 = Me, R^2 = R^4 = H, R^5 = Me, n = 1: 84% ee = 99%
R^1 = R^3 = Me, R^2 = R^4 = H, R^5 = Et, n = 1: 76% ee = 98%
R^1 = R^5 = Me, R^2 = R^3 = R^4 = H, n = 0: 55% ee = 98%
R^1 = Me, R^2 = R^3 = R^4 = H, R^5 = Et, n = 0: 65% ee = 96%
R^1 = n-Pent, R^2 = R^3 = R^4 = H, R^5 = Me, n = 0: 62% ee = 99%
R^1 = n-Pent, R^2 = R^3 = R^4 = H, R^5 = Et, n = 0: 50% ee = 99%
R^1 = Ph, R^2 = R^3 = R^4 = H, R^5 = Me, n = 1: 54% ee = 95%
R^1 = Ph, R^2 = R^3 = R^4 = H, R^5 = Et, n = 1: 40% ee = 95%
R^1 = R^2 = R^3 = R^4 = H, R^5 = Et, n = 1: 75% ee = 52%
R^1 = R^3 = H, R^2 = R^4 = Me, R^5 = Et, n = 1: 50% ee = 61%
R^1 = CH$_2$OTBS, R^2 = R^3 = R^4 = H, R^5 = Me, n = 1: 81% ee > 99%
R^1 = CH$_2$OTBS, R^2 = R^3 = R^4 = H, R^5 = Et, n = 1: 81% ee > 99%
R^1,R^2 = CH=CH-O, R^3 = R^4 = H, R^5 = Me, n = 1: 79% ee = 89%
R^1,R^2 = CH=CH-O, R^3 = R^4 = H, R^5 = Et, n = 1: 88% ee = 89%

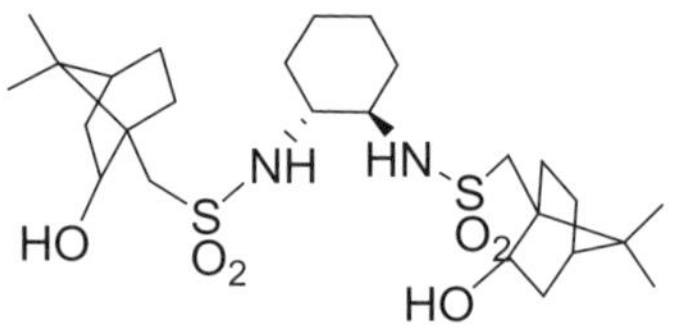

Scheme 4.16   Additions of ZnR$_2$ to cyclic α,β-unsaturated ketones with HOCSAC ligand.

corresponding allylic tertiary alcohols with excellent enantioselectivities of up to 97% ee.[31]

The use of expensive and unstable ZnPh$_2$ in the preparation of chiral diarylmethanol derivatives, with electronically and sterically similar aryl rings, made this approach less attractive for the enantioselective synthesis. In order to avoid this inconvenience, other alternative preparations of arylzinc reagents were evaluated.[11b,32] As a first choice, Yus *et al.* proposed the use of arylboronic acids as a viable source of phenyl (Scheme 4.19). Thus, the reaction of various boronic acids with an excess of ZnEt$_2$ at 70 °C gave the corresponding arylzinc intermediates (probably aryl(ethyl)zincs), which were trapped by reaction with different ketones.[23,26a] In fact, the reaction of phenylboronic acid with 4-bromophenyl methyl ketone produced the expected corresponding alcohol with similar results to those using ZnPh$_2$ as the initial substrate. The results were very good for the reactions with alkyl aryl ketones, decreasing as the alkyl group

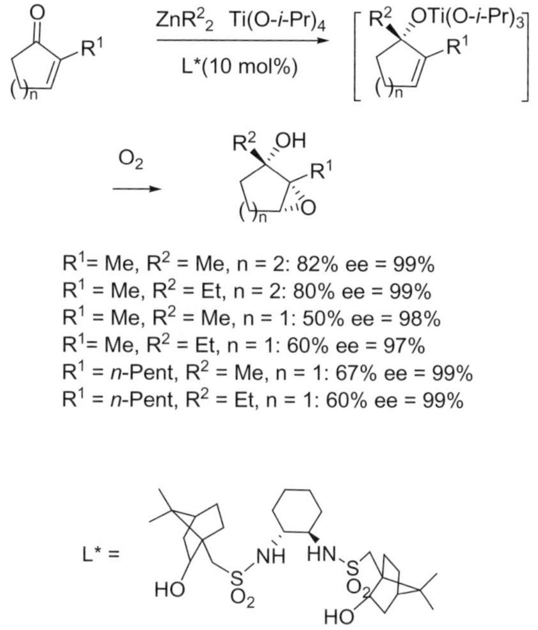

Scheme 4.17 Asymmetric domino alkylation epoxidation reactions.

became more hindered. However, the reaction with simple unsymmetrical dialkyl ketones gave low enantioselectivities. The substitution at the *para*-position of the arylboronic acid seemed to have a very little effect on the results.

One of the drawbacks of the aforementioned arylation protocol was the use of a large excess of ZnEt$_2$. The difficulty in adjusting this excess meant that one of the byproducts was the corresponding ethylated tertiary alcohol. In order to overcome this problem, Yus *et al.* investigated triarylboranes as an initial source of nucleophiles. Therefore, the reaction of ZnEt$_2$ with BPh$_3$, and then the reaction of the *in situ* formed zinc intermediate with ketones led to the expected alcohols, as shown in Scheme 4.20.[23] The obtained enantio-selectivities were similar, if no better, than those obtained from previous protocols, with the size increase of the alkyl moiety of the ketone decreasing the enantioselectivity.

The scope of this methodology could be extended to the use of arylmagne-sium derivatives as the initial source of nucleophiles. The successive transme-talations of the arylmagnesium derivative to a triarylboron intermediate and then to an arylzinc, followed by reaction with acetophenone gave the expected tertiary alcohols, as shown in Scheme 4.21.[23] The results were as good as the previous ones. However, the presence of magnesium salts after the first trans-metalation process had an important negative effect not only on the enan-tioselectivity, but also on the chemical yield. Thus, the complete precipitation of the magnesium halide was compulsory.

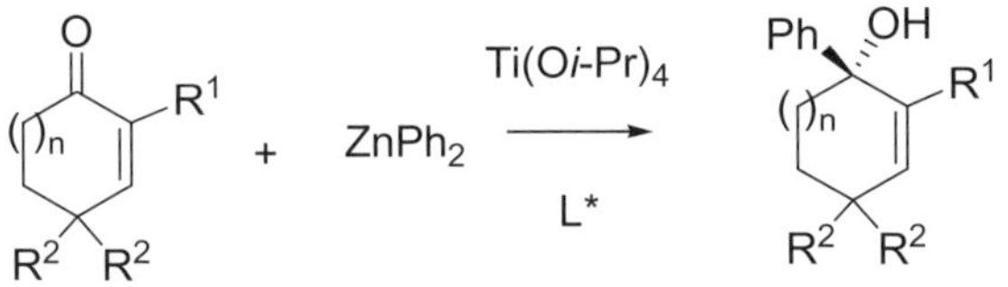

R¹ = R² = H, n = 1: 81% ee = 1%
R¹ = R² = Me, n = 1: 92% ee = 97%
R¹ = CH₂OTBS, R² = H, n = 1: 64% ee = 80%
R¹ = Me, R² = H, n = 0: 60% ee = 84%
R¹ = n-Pent, R² = H, n = 0: 74% ee = 97%
R¹,R² = CH=CH-O, R² = H, n = 1: 81% ee = 71%
R¹ = I, R² = H, n = 0: 46% ee = 93%
R¹ = Br, R² = H, n = 0: 66% ee = 94%
R¹ = I, R² = H, n = 1: 77% ee = 93%

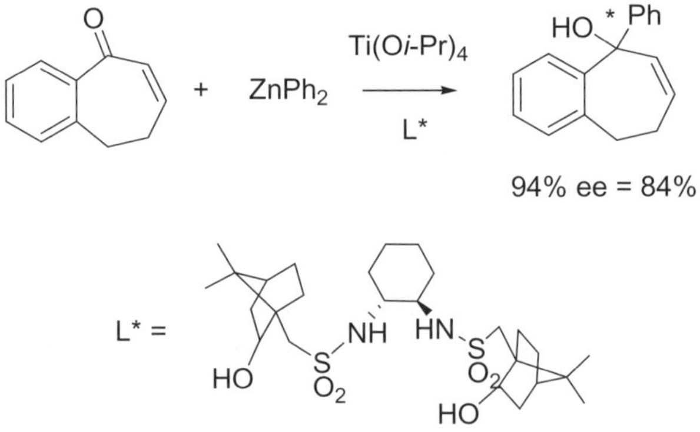

94% ee = 84%

Scheme 4.18 Additions of ZnPh₂ to cyclic α,β-unsaturated ketones with HOCSAC ligand.

Despite the great success of the transmetalation process in the enantioselective arylation of ketones, its extension to allylation or alkynylation reactions failed, providing the corresponding tertiary alcohols with enantiomeric excesses never higher than 50% ee.[23] On the other hand, more success has been found in the alkenylation of ketones.[23,33] The process started with the hydrozirconation of terminal alkynes to give the corresponding alkenylzirconium intermediates, which were transmetalated by reaction, in this case, with various ketones in the presence of the HOCSAC ligand. This protocol tolerated the presence of other carbon–carbon multiple bonds on the alkyne, as well as different functionalities and achieved excellent results for alkyl ketones, α,β-unsaturated ketones and even dialkylketones, as shown in Scheme 4.22.

^{1}H NMR spectroscopic studies of the titanium tetraisopropoxide–HOCSAC mixture seemed to prove that the bimetallic species was formed and was very

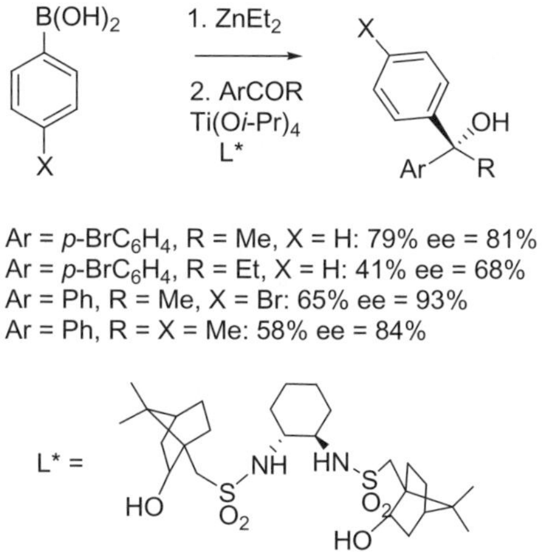
B(OH)2
X
1. ZnEt2
2. ArCOR
Ti(Oi-Pr)4
L*
X
OH
Ar
R
Ar = p-BrC6H4, R = Me, X = H: 79% ee = 81%
Ar = p-BrC6H4, R = Et, X = H: 41% ee = 68%
Ar = Ph, R = Me, X = Br: 65% ee = 93%
Ar = Ph, R = X = Me: 58% ee = 84%
L* =
NH
HN
S
O2
HO
HO

Scheme 4.19 Arylations of ketones with boronic acids and HOCSAC ligand.

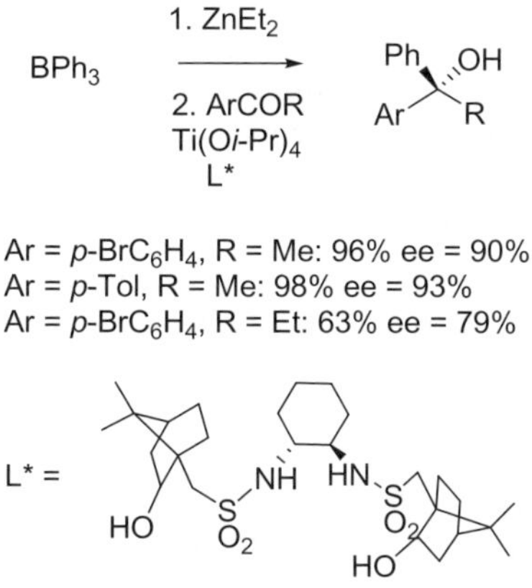
BPh3
1. ZnEt2
2. ArCOR
Ti(Oi-Pr)4
L*
Ph
OH
Ar
R
Ar = p-BrC6H4, R = Me: 96% ee = 90%
Ar = p-Tol, R = Me: 98% ee = 93%
Ar = p-BrC6H4, R = Et: 63% ee = 79%
L* =
NH
HN
S
O2
HO
HO

Scheme 4.20 Additions of BPh_3 to ketones with HOCSAC ligand.

stable compared with other possibilities. Thus, the spectra of the 1:1 molar mixture showed the presence of a mixture of the free HOCSAC ligand and a C_2-symmetric complex.[22] Once the versatility of the HOCSAC ligand[34] was demonstrated, another isoborneol-10-sulfonamide derivative connected, in this case, with a cyclopentane-1,2-diamine linker was introduced as a ligand by Walsh and de Parrodi (Scheme 4.23).[35] Even if the use of this new ligand for the enantioselective addition of ethyl-, phenyl-, and 1-hexenyl groups to a range of ketones provided good to excellent enantioselectivities, however, the previously obtained level of enantioselectivity (with HOCSAC ligand) could not be reached, not only for the alkylation process, but also for the arylation and the

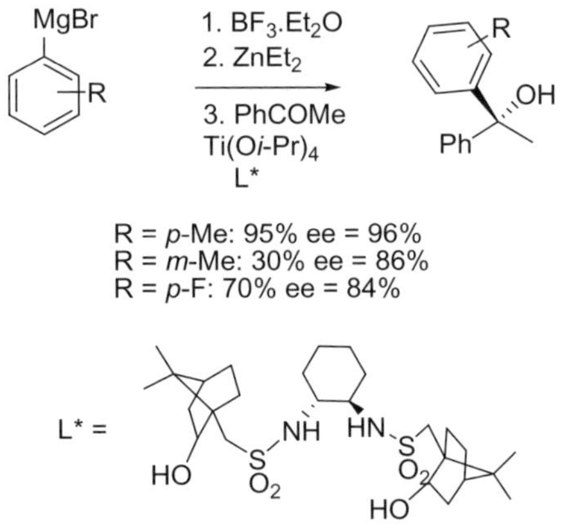

R = *p*-Me: 95% ee = 96%
R = *m*-Me: 30% ee = 86%
R = *p*-F: 70% ee = 84%

Scheme 4.21   Additions of ArMgBr to acetophenone with HOCSAC ligand.

alkenylation processes. The authors have speculated that the reason for the lower enantioselectivity obtained with this ligand compared to its diaminocyclohexane analogue might be due to the greater conformational freedom of *trans*-1,2-disubstituted five-membered rings.[36] Moreover, this catalyst required longer reaction times than its diaminocyclohexane analogue, the HOCSAC ligand, and resulted in diminished yields.

After finding the right combination for the diamine linkers, Yus *et al.* tried to determine whether it was compulsory to use two isoborneol-10-sulfonamide moieties. In this context, these authors have prepared the ligand depicted in Scheme 4.24 by reaction of the best amine linker, *trans*-cyclohexane-1,2-diamine, with camphorsulfonyl chloride and then with methanesulfonyl chloride, followed by reduction with AlH(*i*-Bu)$_2$ and then hydrolysis.[23] When this new ligand was involved in the enantioselective addition of ZnEt$_2$ to acetophenone, the expected tertiary alcohol was obtained in excellent yield and enantioselectivity of 96% ee, as shown in Scheme 4.24. According to this result, the authors concluded that the second isoborneol unit seemed not to be necessary to obtain a high enantioselectivity.

In addition, corresponding arenesulfonyl derivatives of this ligand were prepared by the same group.[27,37] Their preparation was a more challenging task as they could only be formed by changing the order of the sulfonyl chloride additions. Therefore, the first step of their synthesis was the reaction with arenesulfonyl chloride under biphasic conditions, followed by the reaction with camphorsulfonyl chloride using the previous protocol, followed by reduction to give the final ligands depicted in Scheme 4.25 in moderate yields. When these novel ligands were investigated as catalysts for the enantioselective addition of ZnEt$_2$ to acetophenone, the corresponding tertiary alcohol was obtained as only one enantiomer (Scheme 4.25). These unbeatable results were also reached in the case of the reaction of BPh$_3$ with 4-bromoacetophenone to give the corresponding chiral tertiary diarylmethanol in yields greater than 90%.[37]

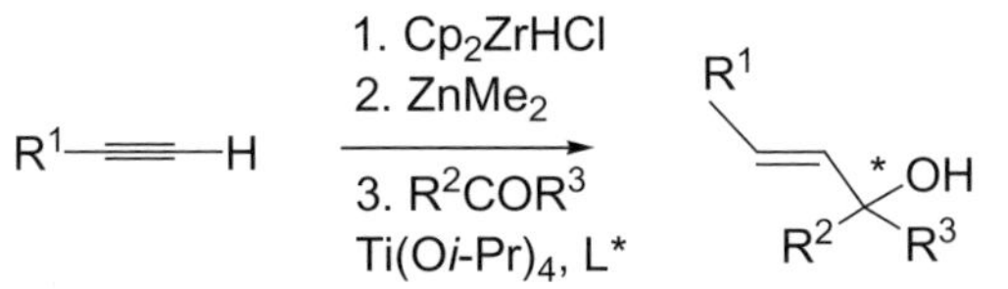
1. Cp2ZrHCl
2. ZnMe2
3. R2COR3
Ti(Oi-Pr)4, L*
R1
H
OH
R2
R3

R^1 = *n*-Bu, R^2 = *(E)*-Ph-CH=CH, R^3 = Me: 87% ee = 92%
R^1 = *n*-Bu, R^2 = Ph, R^3 = Me: 85% ee = 93%
R^1 = *n*-Bu, R^2 = Ph, R^3 = Et: 90% ee = 94%
R^1 = *n*-Bu, R^2 = *m*-ClC_6H_4, R^3 = Et: 93% ee = 93%
R^1 = *n*-Bu, R^2 = *i*-Bu, R^3 = Me: 85% ee = 79%
R^1 = *t*-Bu, R^2 = *m*-Tol, R^3 = Me: 93% ee = 92%
R^1 = Cy, R^2 = Ph, R^3 = Me: 90% ee = 95%
R^1 = $CH_2OTBDPS$, R^2 = Ph, R^3 = Me: 92% ee = 89%
R^1 = Ph, R^2 = *m*-$CF_3C_6H_4$, R^3 = Me: 84% ee = 88%
R^1 = $(CH_2)_2OTBDPS$, R^2 = *m*-$CF_3C_6H_4$, R^3 = Me: 94% ee = 90%
R^1 = $(CH_2)_4Cl$, R^2 = *m*-Tol, R^3 = Me: 98% ee = 90%

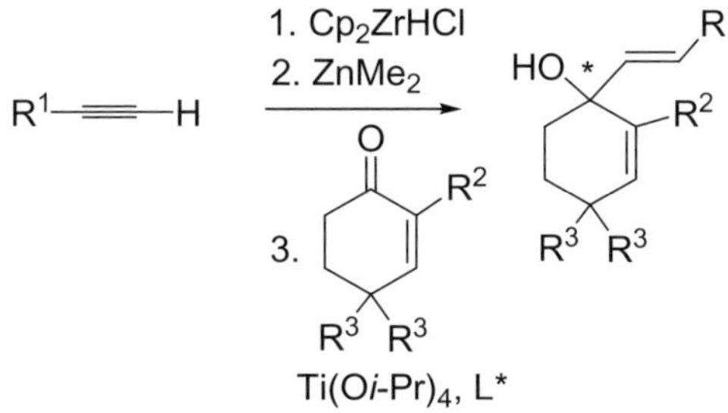
1. Cp2ZrHCl
2. ZnMe2
3.
Ti(Oi-Pr)4, L*
R1
H
O
R2
R3
HO
*

R^1 = *n*-Bu, R^2 = R^3 = Me: 94% ee = 97%
R^1 = $(CH_2)_2OTBDPS$, R^2 = R^3 = Me: 85% ee = 94%

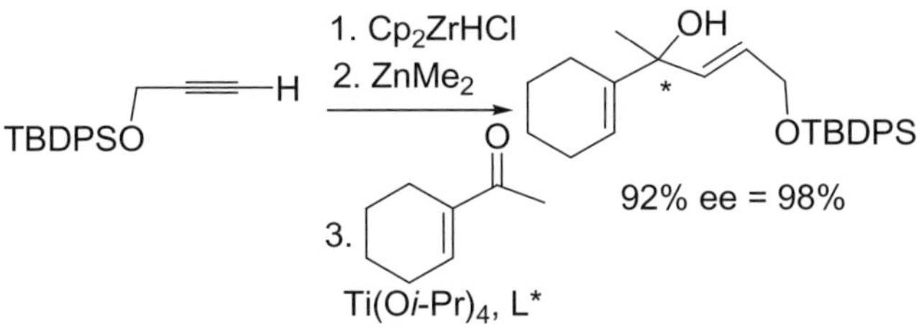
TBDPSO
H
1. Cp2ZrHCl
2. ZnMe2
3.
O
Ti(Oi-Pr)4, L*
OH
*
OTBDPS
92% ee = 98%

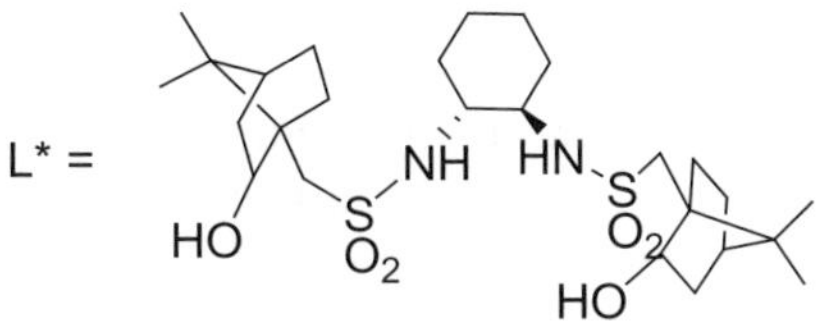
L* =
HO
S
O2
NH
HN
S
O2
HO

Scheme 4.22 Vinylations of ketones with HOCSAC ligand.

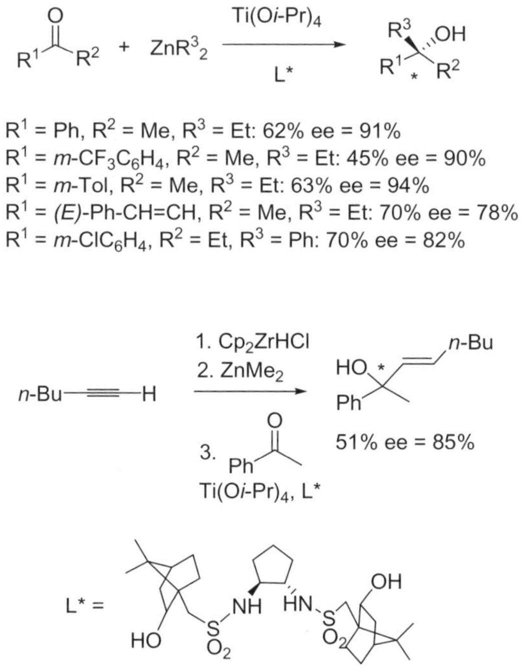
Ti(Oi-Pr)4
L*
R1 = Ph, R2 = Me, R3 = Et: 62% ee = 91%
R1 = m-CF3C6H4, R2 = Me, R3 = Et: 45% ee = 90%
R1 = m-Tol, R2 = Me, R3 = Et: 63% ee = 94%
R1 = (E)-Ph-CH=CH, R2 = Me, R3 = Et: 70% ee = 78%
R1 = m-ClC6H4, R2 = Et, R3 = Ph: 70% ee = 82%
1. Cp2ZrHCl
2. ZnMe2
3.
Ti(Oi-Pr)4, L*
51% ee = 85%
L* =

Scheme 4.23 Alkylations, arylation and vinylation of ketones with *trans*-1,2-diaminocyclopentane-based ligand.

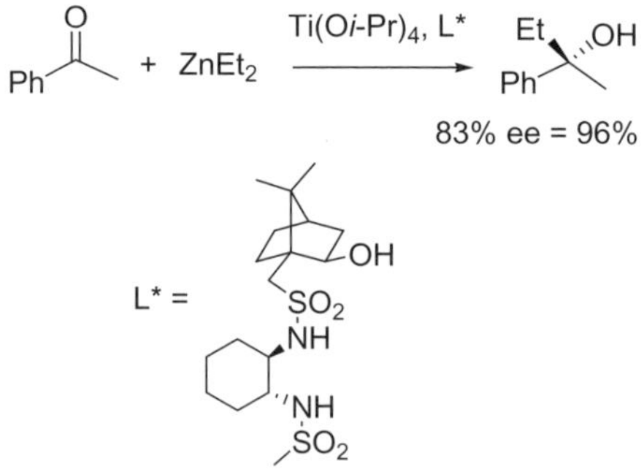
Ti(Oi-Pr)4, L*
83% ee = 96%
L* =

Scheme 4.24 Addition of $ZnEt_2$ to acetophenone with camphor-derived mesylamide ligand.

The high economic and environmental impact of a single-use chiral ligand has forced the development of new alternatives, such as the anchoring of a chiral ligand onto inert polymers.[38] In this context, it should be pointed out that a ligand of the type of those depicted above ($X{=}CH{=}CH_2$) has permitted its heterogenization by the radical preparation of polymers using styrene

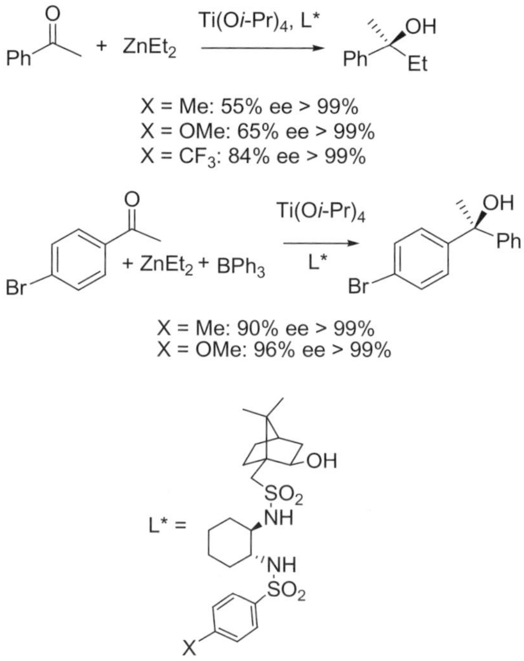
O
Ph
+ ZnEt2
Ti(Oi-Pr)4, L*
OH
Ph
Et
X = Me: 55% ee > 99%
X = OMe: 65% ee > 99%
X = CF3: 84% ee > 99%
O
Br
+ ZnEt2 + BPh3
Ti(Oi-Pr)4
L*
OH
Ph
Br
X = Me: 90% ee > 99%
X = OMe: 96% ee > 99%
OH
SO2
NH
L* =
NH
SO2
X

Scheme 4.25 Alkylation and arylation of ketones with *trans*-1-arenesulfonylamino-2-isoborneolsulfonylaminocyclohexane ligands.

and divinylbenzene (Scheme 4.26).[39] Its application as a ligand in the enantioselective addition of $ZnEt_2$ to acetophenone allowed the corresponding tertiary alcohol to be formed in an excellent enantioselectivity of up to 99% ee but in a moderate yield (56%). The reaction rates were dependent on the amount of styrene added in the preparation of the polymer, and the higher the amount of styrene used, the lower the chemical yield. On the other hand, the yields of the reaction of BPh_3 with alkyl aryl ketones were in the range of those obtained using the corresponding homogeneous ligands, but the products had slightly lower enantiomeric excesses. The polymer-supported ligand could be easily recovered in 75–90% yield by precipitation from the crude mixture, using methanol after hydrolysis. Unfortunately, the activity of the recovered polymer decreased rapidly after a few reuses. The reasons for this decrease were not clear since the transmission electron microscopy experiments on the twofold recovered polymer did not show the expected presence of titanium occluded in the polymer.

In 2008, these authors reported a new strategy to attach chiral *trans*-1-arenesulfonylamino-2-isoborneolsulfonylaminocyclohexane to an achiral Fréchet dendron (polyether having a repeated 3,5-dioxybenzyl structure) by a radical approach.[40] The dendrimers obtained were successfully used in the enantioselective nucleophilic alkylation and arylation of ketones, providing

preparation of ligands:

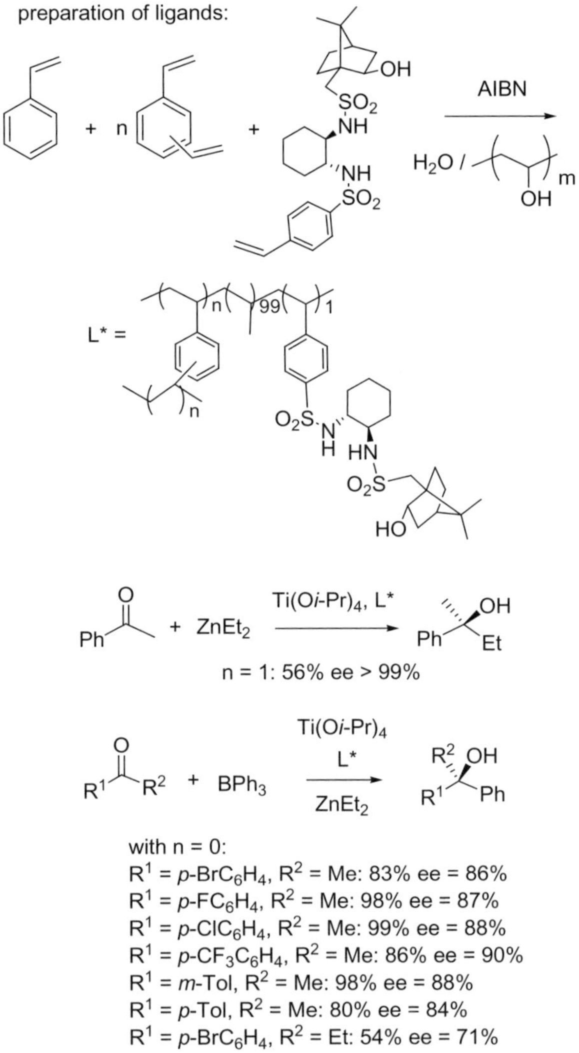

Scheme 4.26   Polymer-supported *trans*-1-arenesulfonylamino-2-isoborneolsulfony-
laminocyclohexane ligand for alkylation and arylations of ketones.

the corresponding tertiary alcohols with enantioselectivities of up to > 99%
ee, as shown in Scheme 4.27. These complexes had diameters in the nanoscale
range, which could permit their use in a continuous-flow membrane reactor,
although so far the reaction rates made this probability difficult to
accomplish.

In addition, the enantioselective addition of akynylzinc reagents to aromatic
ketones has been developed by Wang *et al.* using novel C_2-symmetric

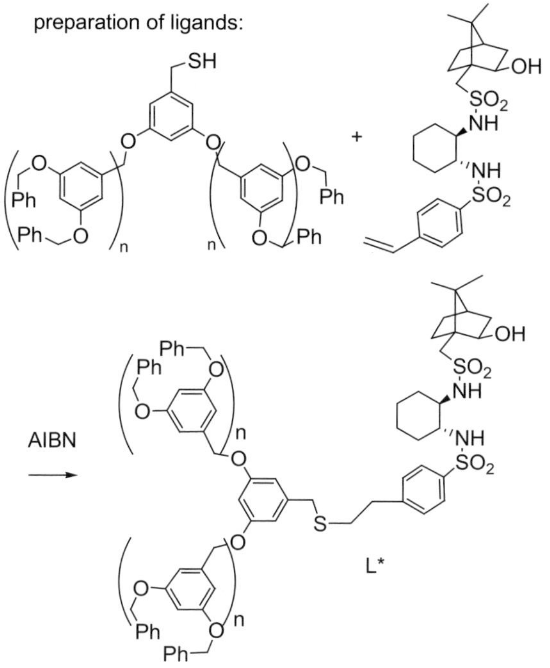

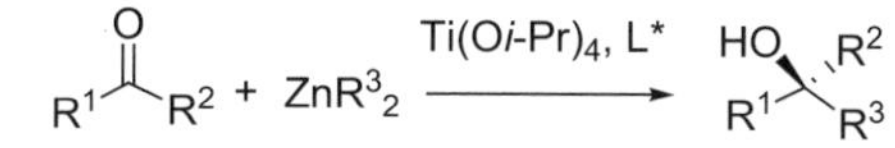

preparation of ligands:

$R^1 = Ph, R^2 = Et, R^3 = Me, n = 0$: 21% ee > 99%
$R^1 = Ph, R^2 = Me, R^3 = Et, n = 0$: 53% ee = 97%
$R^1 = p\text{-}Tol, R^2 = Me, R^3 = Et, n = 0$: 29% ee = 96%
$R^1 = p\text{-}FC_6H_4, R^2 = Me, R^3 = Et, n = 0$: 92% ee = 85%
$R^1 = p\text{-}BrC_6H_4, R^2 = Me, R^3 = Et, n = 0$: 98% ee = 90%
$R^1 = 2\text{-}Naph, R^2 = Me, R^3 = Et, n = 0$: 98% ee = 57%
$R^1 = Ph, R^2 = Me, R^3 = Et, n = 1$: 59% ee = 99%
$R^1 = p\text{-}Tol, R^2 = Me, R^3 = Et, n = 1$: 34% ee = 98%

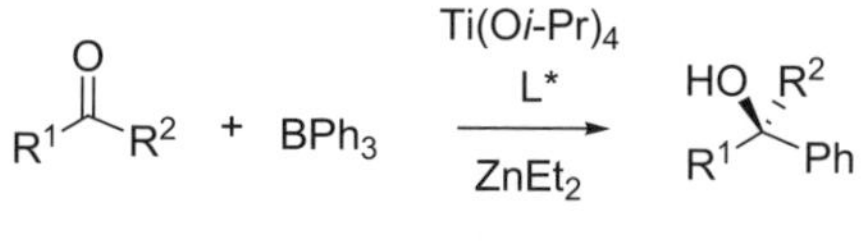

$R^1 = p\text{-}Tol, R^2 = Me, n = 0$: 65% ee = 80%
$R^1 = m\text{-}Tol, R^2 = Me, n = 0$: 67% ee = 84%
$R^1 = p\text{-}CF_3C_6H_4, R^2 = Me, n = 0$: 88% ee = 83%
$R^1 = p\text{-}FC_6H_4, R^2 = Me, n = 0$: 90% ee = 84%
$R^1 = p\text{-}ClC_6H_4, R^2 = Me, n = 0$: 91% ee = 64%
$R^1 = p\text{-}BrC_6H_4, R^2 = Me, n = 0$: 89% ee = 78%
$R^1 = p\text{-}BrC_6H_4, R^2 = Et, n = 0$: 51% ee = 78%
$R^1 = m\text{-}Tol, R^2 = Me, n = 1$: 95% ee = 84%
$R^1 = p\text{-}CF_3C_6H_4, R^2 = Me, n = 1$: 97% ee = 81%
$R^1 = p\text{-}BrC_6H_4, R^2 = Me, n = 1$: 84% ee = 75%

Scheme 4.27 Fréchet dendrimeric *trans*-1-arenesulfonylamino-2-isoborneolsulfonyl-aminocyclohexane ligand for alkylations and arylations of ketones.

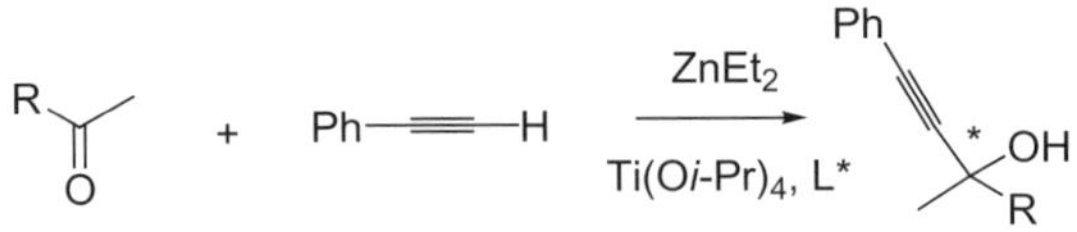

R = Ph: 64% ee = 71%
R = o-FC$_6$H$_4$: 62% ee = 62%
R = p-FC$_6$H$_4$: 46% ee = 67%
R = m-BrC$_6$H$_4$: 74% ee = 77%
R = p-ClC$_6$H$_4$: 51% ee = 70%
R = 1-Naph: 36% ee = 81%
R = m-Tol: 60% ee = 74%
R = p-Tol: 42% ee = 68%
R = m-MeOC$_6$H$_4$: 58% ee = 74%
R = Bn: 70% ee = 55%

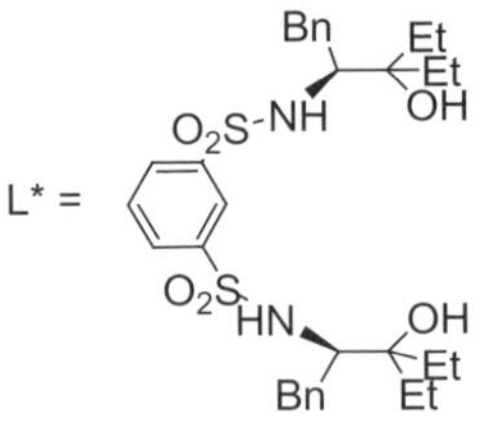

Scheme 4.28   Alkynylations of ketones with C_2-symmetric bis(sulfonamides) ligand.

bis(sulfonamides) ligands, such as that depicted in Scheme 4.28, in combination with Ti(Oi-Pr)$_4$.[41] Good enantioselectivities of up to 81% ee were obtained with a range of ketones combined with yields of up to 74%.

In conclusion, the enantioselective addition of organozinc reagents to ketones performed in the presence of sulfonamide ligands has become possible and constitutes a very useful and promising reaction for the synthesis of complicated chiral structures. Since 1998, the use of ligands bearing an isoborneol-10-sulfonamide motif, mainly developed independently by the groups of Yus[7b,42] and Walsh,[7] has allowed the enantioselective addition of various organozinc reagents to be added with high enantioselectivities to a broad range of ketones. Today, this type of ligand remains the only one able to efficiently catalyse these reactions, allowing a highly efficient route to a broad range of chiral tertiary alcohols. In addition, the easy modular introduction of the isoborneol-10-sulfonamide motif into different structures permits prediction of an important increase in the application of these ligands in the future, in spite of the fact that the great success of the enantioselective addition of organozinc reagents to ketones using this type of ligand has so far eclipsed pertinent efforts on the application of such ligands in other reactions. Moreover, the possible attachment of this motif to different polymers will open up new possibilities, not only in its use in the now classical catalysed alkylation of ketones, but also in other enantioselective reactions.

References

1. (a) J. Christoffers and A. Mann, *Angew. Chem., Int. Ed. Engl*, 2001, **40**, 4591–4597; (b) J. Christoffers and A. Baro, *Angew. Chem., Int. Ed. Engl.*, 2003, **42**, 1688–1690.
2. *Quaternary Stereocenters–Challenges and Solutions for Organic Synthesis*, J. Christoffers, A. Baro, ed., Wiley-VCH, Weinheim, 2005.
3. (a) K. Fuji, *Chem. Rev.*, 1993, **93**, 2037–2066; (b) E. J. Corey and A. Guzman-Perez, *Angew. Chem., Int. Ed. Engl.*, 1998, **37**, 388–401; (c) D. J. Ramon and M. Yus, *Curr. Org. Chem.*, 2004, **8**, 149–183; (d) J. Christoffers and A. Baro, *Adv. Synth. Catal.*, 2005, **347**, 1473–1482; (e) O. Riant and J. Hannedouche, *Org. Biomol. Chem.*, 2007, **5**, 873–888; (f) P. G. Cozzi, R. Hilgraf and N. Zimmermann, *Eur. J. Org. Chem.*, 2007, 5969–5994; (g) M. Hatano and K. Ishihara, *Synthesis*, 2008, 1647–1675.
4. (a) R. Noyori and M. Kitamura, *Angew. Chem., Int. Ed. Engl.*, 1991, **30**, 34–48; (b) K. Soai and S. Niwa, *Chem. Rev.*, 1992, **92**, 833–856; (c) L. Pu and H.-B. Yu, *Chem. Rev.*, 2001, **101**, 757–824; (d) L. Pu, *Tetrahedron*, 2003, **59**, 9873–9886; (e) M. Yus and D. J. Ramon, *Rec. Res. Dev. Org. Chem.*, 2002, **6**, 297–378; (f) M. Hatano, T. Miyamoto and K. Ishihara, *Curr. Org. Chem.*, 2007, **11**, 127–157; (g) C. Garcia and V. S. Martin, *Curr. Org. Chem.*, 2006, **10**, 1849–1889.
5. (a) M. Yus and D. J. Ramon, *Latv. Kim. Z.*, 2002, 79–92; (b) A. Boudier, L. O. Bromm, M. Lotz and P. Knochel, *Angew. Chem., Int. Ed. Engl.*, 2000, **39**, 4414–4435; (c) C. Najera and M. Yus, *Trends Org. Chem.*, 1991, **2**, 155–181.
6. (a) P. Knochel, J. J. Almena Perea and P. Jones, *Tetrahedron*, 1998, **54**, 8275–8319; (b) P. Knochel, in *Metal-Catalysed Cross-Coupling Reactions*, ed. F. Diederich, P. J. Stang, Wiley-VCH, Weinheim, 1998, pp. 387–419; (c) E. Erlik, in *Organozinc Reagents in Organic Synthesis*, CRC Press, Boca Raton, 1996; (d) P. Knochel, *Synlett*, 1995, 393–403; (e) P. Knochel and R. D. Singer, *Chem. Rev.*, 1993, **93**, 2117–2188.
7. (a) J. M. Betancort, C. Garcia and P. J. Walsh, *Synlett*, 2004, 749–760; (b) D. J. Ramon and M. Yus, *Angew. Chem., Int. Ed. Engl.*, 2004, **43**, 284–287.
8. (a) A. S. Thompson, E. G. Corley, M. F. Huntington and E. J. J. Grabowski, *Tetrahedron Lett.*, 1995, **36**, 8937–8940; (b) A. S. Thompson, E. G. Corley, M. F. Huntington, E. J. J. Grabowski, J. F. Remenar and D. B. Collum, *J. Am. Chem. Soc.*, 1998, **120**, 2028–2038; (c) L. Tan, C.-y. Chen, R. D. Tillyer, E. J. J. Grabowski and P. J. Reider, *Angew. Chem., Int. Ed. Engl.*, 1999, **38**, 711–712; (d) M. A. Huffman, N. Yasuda, A. E. DeCamp and E. J. J. Grabowski, *J. Am. Chem. Soc.*, 1995, **60**, 1590–1594; (e) G. S. Kauffman, G. D. Harris, R. L. Dorow, B. R. P. Stone, R. L. Parsons, J. A. Pesti, N. A. Magnus, J. M. Fortunak, P. N. Confalone and W. A. Nugent, *Org. Lett.*, 2000, **2**, 3119–3121.
9. (a) B. Weber and D. Seebach, *Angew. Chem., Int. Ed. Engl.*, 1992, **31**, 84–86; (b) B. Weber and D. Seebach, *Tetrahedron*, 1994, **50**, 6117–6128.

10. E. Erdik, in *Organozinc Reagents in Organic Synthesis*, CRC Press, Boca Raton, 1996.

11. (a) M. Watanabe and K. Soai, *J. Chem. Soc., Perkin Trans. I*, 1994, 3125–3128; (b) R. Noyori, S. Suga, K. Kawai, S. Okada, M. Kitamura, N. Oguni, T. Kaneko and Y. Matsuda, *J. Organomet. Chem.*, 1990, **382**, 19–37.

12. (a) P. Knochel, in *Encyclopedia of Reagents for Organic Synthesis*, ed. L. A. Paquette, John Wiley and Sons, Chichester, 1995, **Vol. 3**, pp. 1861–1866; (b) J. Boersma, in *Comprehensive Organometallic Chemistry*, ed. G. Wilkinson, Pergamon Press, Oxford, 1982, **Vol. 2**, pp. 823–862; (c) M. B. Marx, E. Henry-Basch and M. P. Freon, *C.R. Acad. Sci. Ser. C*, 1967, **264**, 527–530; (d) G. E. Coates and D. Ridley, *J. Chem. Soc. (A)*, 1966, 1064–1069.

13. (a) D. E. Ho, J. M. Betancort, D. H. Woodmansee, M. L. Larter and P. J. Walsh, *Tetrahedron Lett.*, 1997, **38**, 3867–3870; (b) S. Pritchett, D. H. Woodmansee, P. Gantzel and P. J. Walsh, *J. Am. Chem. Soc.*, 1998, **120**, 6423–6424.

14. (a) M. Kitamura, S. Suga, K. Kawai and R. Noyori, *J. Am. Chem. Soc.*, 1986, **108**, 6071–6072; (b) R. Noyori, S. Suga, K. Kawai, S. Okada and M. Kitamura, *Pure Appl. Chem.*, 1988, **60**, 1597–1606; (c) R. Noyori, in *Asymmetric Catalysis in Organic Synthesis*, Wiley, New York, 1994.

15. P. I. Dosa and G. C. Fu, *J. Am. Chem. Soc.*, 1998, **120**, 445–446.

16. (a) D. J. Ramon and M. Yus, *Tetrahedron Lett.*, 1998, **39**, 1239–1242; (b) D. J. Ramon and M. Yus, *Tetrahedron*, 1998, **54**, 5651–5666.

17. L. Pu, *Tetrahedron*, 2003, **59**, 9873–9886.

18. (a) G. Lu, X. Li, X. Jia, W. L. Chan and A. S. C. Chan, *Angew. Chem., Int. Ed. Engl.*, 2003, **42**, 5057–5058; (b) G. Lu, X. Li, Y.-M. Li, F. Y. Kwong and A. S. C. Chan, *Adv. Synth. Catal.*, 2006, **348**, 1926–1933.

19. J. Balsells and P. J. Walsh, *J. Am. Chem. Soc.*, 2000, **122**, 3250–3251.

20. M. Yus, D. J. Ramon and O. Prieto, *Tetrahedron: Asymmetry*, 2003, **14**, 1103–1114.

21. L. Liu, R. Wang, Y.-F. Kang and C. Cai Chen, *Synlett*, 2006, 1245–1249.

22. A. Hui, J. Zhang, J. Fan and Z. Wang, *Tetrahedron: Asymmetry*, 2006, **17**, 2101–2107.

23. V. J. Forrat, O. Prieto, D. J. Ramon and M. Yus, *Chem. Eur. J.*, 2006, **12**, 4431–4445.

24. M. Yus, D. J. Ramon and O. Prieto, *Tetrahedron: Asymmetry*, 2002, **13**, 2291–2293.

25. (a) C. Garcia, L. K. LaRochelle and P. J. Walsh, *J. Am. Chem. Soc.*, 2002, **124**, 10970–10971; (b) S.-J. Jeon, H. Li, C. Garcia, L. K. LaRochelle and P. J. Walsh, *J. Org. Chem.*, 2005, **70**, 448–455; (c) S.-J. Jeon, H. Li and P. J. Walsh, *J. Am. Chem. Soc.*, 2005, **127**, 16416–16425.

26. (a) O. Prieto, M. Yus and D. J. Ramon, *Tetrahedron: Asymmetry*, 2003, **14**, 1955–1957; (b) C. Garcia and P. J. Walsh, *Org. Lett.*, 2003, **5**, 3641–3644.

27. V. J. Forrat, D. J. Ramon and M. Yus, *Tetrahedron: Asymmetry*, 2007, **18**, 400–405.

28. M. Yus, D. J. Ramon and O. Prieto, *Eur. J. Org. Chem.*, 2003, 2745–2748.
29. S.-J. Jeon and P. J. Walsh, *J. Am. Chem. Soc.*, 2003, **125**, 9544–9545.
30. A. E. Lurain, P. J. Carroll and P. J. Walsh, *J. Org. Chem.*, 2005, **70**, 1262–1268.
31. H. Li, C. Garcia and P. J. Walsh, *Proc. Natl. Acad. Sci. U.S.A.*, 2004, **101**, 5425–5427.
32. (a) P. Knochel, H. Leuser, L.-Z. Gong, S. Perrone and F. K. Kneissel, in *Handbook of Functionalized Organometallics: Applications in Synthesis*, ed. P. Knochel, **Vol. 1**, Wiley-VCH, Weinheim, 2005, 251–346; (b) F. Schmidt, R. T. Stemmler, J. Rudolph and C. Bolm, *Chem. Soc. Rev.*, 2006, **35**, 454–470.
33. (a) H. Li and P. J. Walsh, *J. Am. Chem. Soc.*, 2004, **126**, 6538–6539; (b) H. Li and P. J. Walsh, *J. Am. Chem. Soc.*, 2005, **127**, 8355–8361.
34. P. J. Walsh, US Patent 6660884, 2003.
35. C. A. De Parrodi and P. J. Walsh, *Synlett*, 2004, 2417–2420.
36. E. L. Eliel and S. H. Wilen, in *Stereochemistry of Organic Compounds*, Wiley and Sons, New York, 1994.
37. V. J. Forrat, D. J. Ramon and M. Yus, *Tetrahedron: Asymmetry*, 2005, **16**, 3341–3344.
38. (a) J. A. Gladysz, *Chem. Rev.*, 2002, **102**, 3215–3216; (b) A. Corma and H. Garcia, *Chem. Rev.*, 2003, **103**, 4307–4365; (c) L.-X. Dai, *Angew. Chem., Int. Ed. Engl.*, 2004, **43**, 5726–5729.
39. V. J. Forrat, D. J. Ramon and M. Yus, *Tetrahedron: Asymmetry*, 2006, **17**, 2054–2058.
40. V. J. Forrat, D. J. Ramon and M. Yus, *Tetrahedron: Asymmetry*, 2008, **19**, 537–541.
41. M. Ni, R. Wang, Z.-j. Han, B. Mao, C.-s. Da and L. Liu, *Adv. Synth. Catal.*, 2005, **347**, 1659–1665.
42. D. J. Ramon and M. Yus, *Synlett*, 2007, **15**, 2309–2320.

CHAPTER 5
Diels–Alder Reaction

Catalytic asymmetric cycloaddition reactions have been intensively developed as powerful atom-economic ways to prepare chiral carbocyclic and heterocyclic compounds. The Diels–Alder reaction is one of the rare C–C bond-forming reactions that permit the rapid development of molecular complexity.[1] It allows the stereoselective formation of as many as four stereogenic centres, and as many as three carbocyclic rings in the intramolecular and transannular variation. For this reason, the recent development of highly enantioselective catalytic Diels–Alder reactions represents a great advance in synthetic chemistry. A large number of metals, ligands and dienophiles have been studied. Although the chiral auxiliary-based reactions retain a position of central importance, catalytic variants are developing rapidly. These catalytic processes are normally promoted by Lewis-acid catalysts[2] in the presence of oxygen-chelating ligands or phosphorus-containing ligands,[3] which have been the most successful catalysts until recently. More recently, a number of chiral N-containing ligands, such as oxazoline ligands,[4] have been introduced to catalyse these reactions on the basis of the excellent results reported by Evans et al.,[5] using chiral bis(oxazolines) ligands in combination with copper salts. Finally, a large amount of interesting results in terms of both activity and enantioselectivity have been reported by using various chiral sulfur-containing ligands. The best results have been reached with ligands in which the sulfur atom is part of a sulfoxide moiety.

The Diels–Alder reaction has been performed with a range of Lewis acids. Copper complexes are the most successfully used, but other metals such as iron, magnesium, palladium, nickel or ytterbium have proved to be efficient to catalyse this reaction.

The most active chiral catalysts used to date employing a chiral sulfur atom as a unique source of chirality were reported very recently by Ellman's group.[6] Hence, a novel bis(sulfinyl)imidoamidine (siam) ligand, readily available, was proved to catalyse, in the presence of $Cu(SbF_6)_2$, the Diels–Alder reaction of a

RSC Catalysis Series No. 2
Chiral Sulfur Ligands: Asymmetric Catalysis
By Hélène Pellissier
© Hélène Pellissier 2009
Published by the Royal Society of Chemistry, www.rsc.org

variety of dienophiles, such as crotyl-, cinnamoyl-, or activated β-carboxy-substituted dienophiles, with either cyclopentadiene or cyclohexadiene with exceptional levels of enantio- and diastereoselectivity (Scheme 5.1). Furthermore, the Cu(II)-siam complex was shown by X-ray analysis to exhibit a unique mode of binding, self-assembling to form a rarely observed M_2L_4 quadruple-stranded helicate. Additionally, while the siam ligand could coordinate to the metal through the N-, S-, and O-atoms, it was shown that, both in the crystalline state and in CH_2Cl_2 solution, the ligand was O-coordinated to the Cu. In 2003, the scope of this reaction was extended to the use of relatively unreactive acyclic dienes, such as 2-methylbutadiene or 2,3-dimethylbutadiene, which gave enantioselectivities of 93 and 92% ee, respectively, by reaction with *N*-acryloyloxazolidinone.[7] The selectivity of the catalyst system was, however, sensitive to the size of the substituents on the 2-position of the diene. As an example, while the reactivity (87% yield) was maintained for 2-phenylbutadiene, the selectivity remained low (45% ee).

In the same study, these authors have examined the importance of the substitution on the sulfinyl imidoamidine framework.[7] It was shown that the substitution of the internal nitrogen atom in the siam ligand by different groups with various electronic properties did not lead to a noticeable modification in the catalyst efficiency. In order to investigate the importance of the substituent on the sulfur atom, these authors have prepared another siam ligand depicted in Scheme 5.2, which led to a major decrease in the enantioselectivity (32% ee) when applied to the cycloaddition between cyclopentadiene and *N*-acryloyloxazolidinone. In addition, other novel sulfinyl imines depicted in Scheme 5.2 were also investigated in the same reaction, providing the corresponding cycloadduct with good yields but moderate enantioselectivities ($\leq 72\%$ ee). These authors have demonstrated the importance played by the copper counterion on the activity by using noncoordinating hexafluoroantimonate for accelerating the reaction.

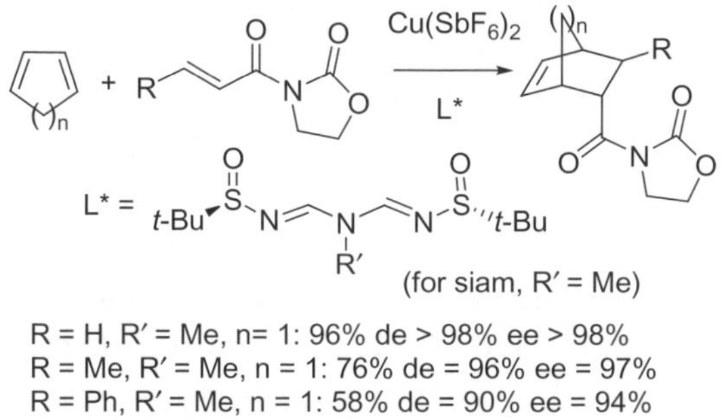

R = H, R' = Me, n= 1: 96% de > 98% ee > 98%
R = Me, R' = Me, n = 1: 76% de = 96% ee = 97%
R = Ph, R' = Me, n = 1: 58% de = 90% ee = 94%
R = CO$_2$Et, R' = Me, n = 1: 85% de = 94% ee = 96%
R = H, R' = Me, n = 2: 50% de = 96% ee = 90%
R = H, R' = *i*-Bu, n= 1: de = 98% ee = 97%
R = H, R' = Ph, n= 1: de = 98% ee = 96%
R = H, R' = CH$_2$CF$_3$, n= 1: de = 98% ee = 98%

Scheme 5.1 Cu-catalysed Diels–Alder reactions with siam ligands.

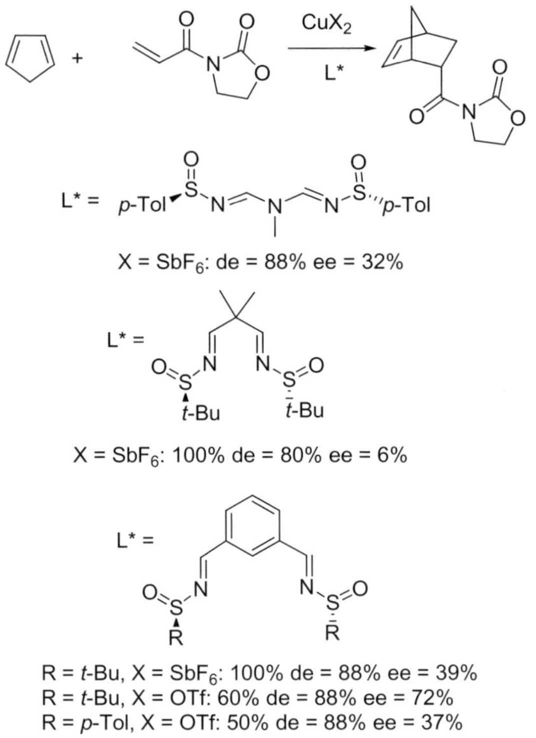

Scheme 5.2   Cu-catalysed Diels–Alder reaction with siam ligand and sulfinyl imine ligands.

Relatively few chiral sulfoxide ligands have been useful for catalytic asymmetric Diels–Alder reactions. The chirality in these structures is very often solely introduced via the sulfoxide functionality, demonstrating the important role of the sulfur atom for the enantioselectivity of the catalytic process. As an example, Hiroi *et al.* have developed new ligands bearing a chiral sulfinyl function and a 1,3-oxazoline ring with an asymmetric carbon centre, in which the chiral sulfinyl group has been revealed to play a crucial role in achieving a high enantioselectivity in asymmetric Diels–Alder reactions (Scheme 5.3).[8] The best enantioselectivities of up to 92% ee were obtained using MgI$_2$ as the Lewis-acid catalyst. A study of the mechanistic pathway showed that, for the first time, seven-membered chelates of sulfoxide-magnesium complexes (normally six-membered chelates were formed in previous oxazoline ligands) were involved, indicating particularly, the potential advantage of the chirality of the sulfoxide functionality for achieving a high enantioselectivity. These authors proposed a preferential attack from the *Re* face side opposite to the bulky substituents. The methoxy group in the sulfoxide naphthyl derivative was of the utmost importance for orienting the substituent, due to dipole–dipole repulsion from the sulfinyl group.

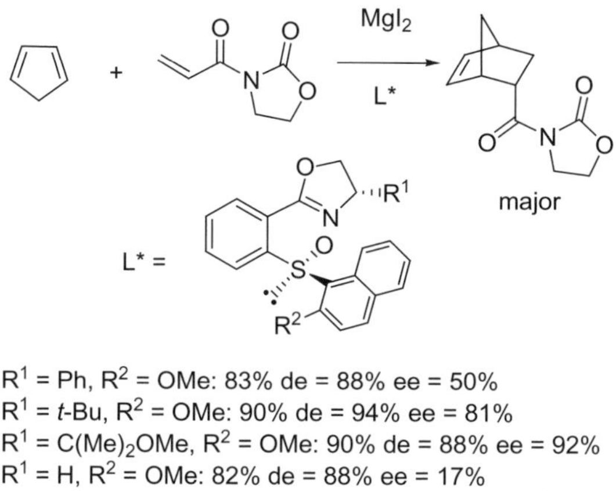

R^1 = Ph, R^2 = OMe: 83% de = 88% ee = 50%
R^1 = *t*-Bu, R^2 = OMe: 90% de = 94% ee = 81%
R^1 = C(Me)$_2$OMe, R^2 = OMe: 90% de = 88% ee = 92%
R^1 = H, R^2 = OMe: 82% de = 88% ee = 17%

Scheme 5.3　Mg-catalysed Diels–Alder reaction with sulfoxide-oxazoline ligands.

In the same study, these authors have studied a series of other ligands of the same type bearing various substituents on the oxazoline ring or on the sulfinyl function, both being bridged by a phenyl group (Scheme 5.4).[8] These experiments proved that 2-methoxy-1-naphthyl sulfoxides provided higher enantioselectivities than other aryl sulfoxides. Moreover, the steric hindrance generated by bulky substituents on the oxazoline ring was shown to be of major importance for the selectivity of the reaction (Scheme 5.4). Thus, both chiral centres seemed to be necessary to obtain a good enantioselectivity, since the loss of one of these two centres led to a major decrease in the enantioselectivity, as shown in Scheme 5.4. The essential role played by the sulfoxide moiety in the chirality was demonstrated by the low enantioselectivity (6% ee) obtained by using the corresponding sulfone ligand to the sulfoxide that gave 81% ee. In addition, the scope of the reaction was extended to the use of other similar ligands such as chiral 2-(arylsulfinylmethyl)-1,3-oxazoline derivatives, which gave, in the same conditions, a relatively low enantioselectivity ($\leq$32% ee). Nevertheless, when copper complexes were used to catalyse the reaction, a better, but still moderate, enantioselectivity was observed ($\leq$66% ee).[9] Moreover, the introduction of a counterion (triflate or hexafluoroantimonate) into the copper catalysts provided a higher degree of asymmetric induction ($\leq$75% ee).[10]

Another example of enantioselective Diels–Alder reaction using sulfoxide ligands was reported by Khiar *et al.*, who prepared C_2-symmetric ligands containing two oxygen-donor atoms, in 1993.[11] Thus, bis(sulfoxides) ligands with a C_2-symmetry axis and depicted in Scheme 5.5 were employed as chelates for the iron(III)-catalysed Diels–Alder reaction between cyclopentadiene and *N*-acryloyloxazolidinone, affording the corresponding expected diastereomers in good yields, high diastereoselectivities but with low to moderate enantioselectivities ($\leq$56% ee). As shown in Scheme 5.5, the enantioselectivity of the

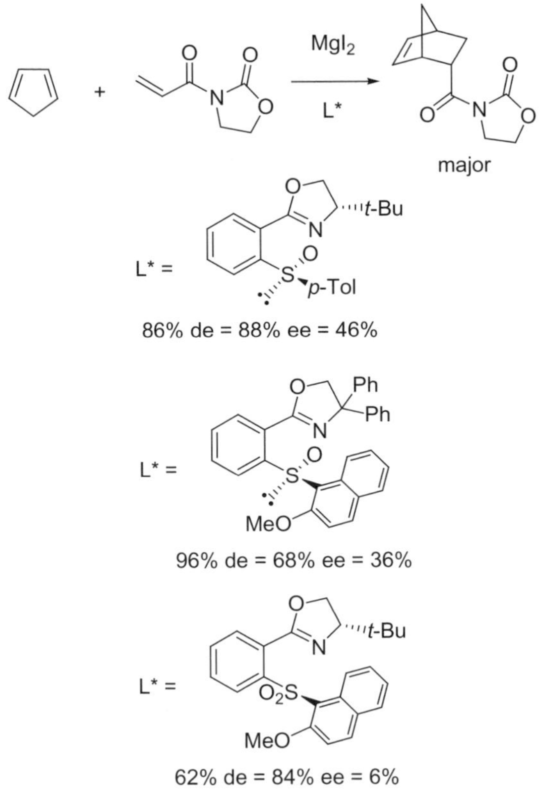

Scheme 5.4 Mg-catalysed Diels–Alder reaction with sulfoxide-oxazoline ligands.

reaction increased with the steric hindrance of the ligand. In order to explain the configuration of the major product, the authors have proposed a transition state in which the ligand and the dienophile were chelated to the metal with an octahedral geometry via the equatorial site. The less hindered face for the approach of the cyclopentadiene was thus located on the side opposite to the sterically hindered *p*-tolyl groups of the ligand.

With the aim of controlling the asymmetric Diels–Alder reaction by a double chelation of the Lewis acid by hydroxyl groups vicinal to the sulfoxide moiety, Llera *et al.* have prepared various chiral hydroxysulfoxides and evaluated their efficiency as ligands in the enantioselective Diels–Alder reaction of cyclopentadiene with *N*-acryloyloxazolidinone in the presence of MgI$_2$.[12] Only the ligand bearing two phenyl groups adjacent to the hydroxyl functionality allowed a high enantioselectivity of up to 88% ee to be obtained (Scheme 5.6). A donor–acceptor interaction between the aromatic rings and the electron-deficient double bond in the dienophile was proposed by these authors to be responsible of this enhanced enantioselectivity compared to the other ligands. When the hydroxysulfoxide ligands were substituted in the *para* position of the

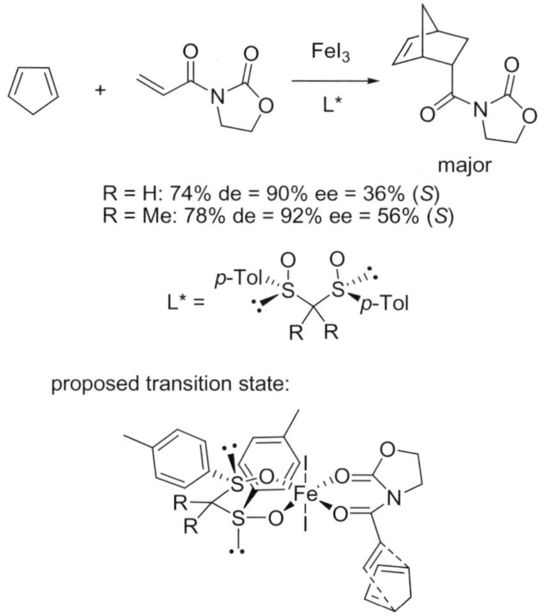
FeI$_3$
L*
major
R = H: 74% de = 90% ee = 36% (*S*)
R = Me: 78% de = 92% ee = 56% (*S*)
L* =
p-Tol
p-Tol
R R
proposed transition state:
Fe

Scheme 5.5 Fe-catalysed Diels–Alder reaction with bis(sulfoxides) ligands.

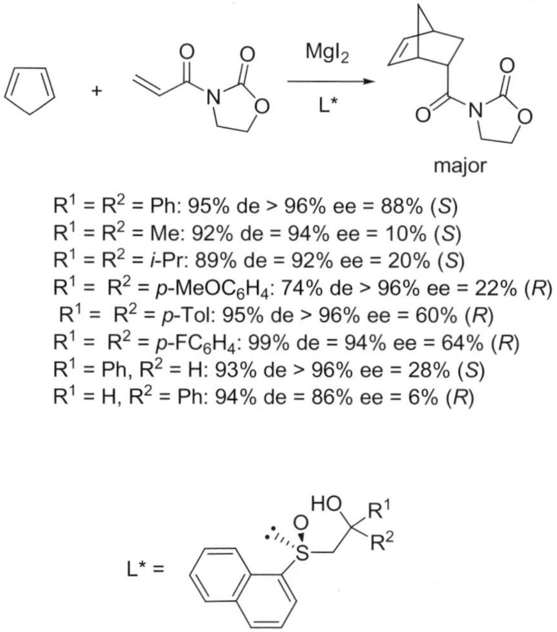
MgI$_2$
L*
major
R^1 = R^2 = Ph: 95% de > 96% ee = 88% (*S*)
R^1 = R^2 = Me: 92% de = 94% ee = 10% (*S*)
R^1 = R^2 = *i*-Pr: 89% de = 92% ee = 20% (*S*)
R^1 = R^2 = *p*-MeOC$_6$H$_4$: 74% de > 96% ee = 22% (*R*)
R^1 = R^2 = *p*-Tol: 95% de > 96% ee = 60% (*R*)
R^1 = R^2 = *p*-FC$_6$H$_4$: 99% de = 94% ee = 64% (*R*)
R^1 = Ph, R^2 = H: 93% de > 96% ee = 28% (*S*)
R^1 = H, R^2 = Ph: 94% de = 86% ee = 6% (*R*)
L* =
HO
R^1
R^2

Scheme 5.6 Mg-catalysed Diels–Alder reaction with hydroxysulfoxide ligands.

phenyl groups with methoxy substituents, the enantioselectivity of the reaction decreased dramatically (22% ee) probably due to an association between the methoxy groups and magnesium. Moreover, when only one phenyl group was present at the hydroxylic carbon, leading to the formation of an asymmetric carbon centre, a reversal in the sense of the stereoselectivity was observed when both diastereomeric ligands were used (with the configuration at the sulfur atom being maintained). The observed sense of the enantioselectivity was explained by the authors by proposing a 1:1:1 complex between the ligand, the metal and the dienophile with a tetrahedral arrangement of the oxygen atoms around the metal.

In 1996, Bolm *et al.* introduced another class of sulfur-containing ligands, salen-like bis(sulfoximines), the chirality of which was generally introduced via the use of chiral diamines in the salen series, whereas, in sulfoximines, the chirality was present via the sulfur atom.[13] The enantioselective (hetero) Diels–Alder reaction using chiral mono- or bis(sulfoximines) ligands has been extensively studied recently by these authors.[14,15] In particular, bis(sulfoximines) ligands have proved to be highly efficient in the Cu(II)-catalysed Diels–Alder cycloaddition between cyclopentadiene and *N*-acryloyloxazolidinone, providing the corresponding cycloadduct with up to 93% ee (Scheme 5.7).[16] It was shown that the presence of a methoxy group on the aryl substituent of the ligand allowed the enantioselectivity of the reaction to be increased. Moreover, the use of less coordinating counterions, such as perchlorate in chloroform as the solvent, also allowed the enantioselectivity to be increased. This reaction was also studied in the presence of more flexible ligands such as bis(sulfoximines) having an ethylene as the bridging unit, giving the cycloadduct in lower enantioselectivities than the first studied bis(sulfoximines) ligands ($\leq$83% ee).

In 2007, these authors reported a mechanistic study of this reaction using a C_2-symmetric bis(sulfoximines) ligand substituted at the sulfur atom by phenyl and methyl groups (R^1=Me, R^2=Ph, R^3=H in Scheme 5.7).[17] The reaction was followed by EXAFS (extended X-ray absorption fine structure), EPR (electron paramagnetic resonance), spectroscopy and UV-visible spectroscopy. The complexes formed between the parent CuX_2 (X=Cl, Br, OTf, SbF_6) salts, the chiral bis(sulfoximines) ligand and *N*-acryloyloxazolidinone as the substrate in CH_2Cl_2 were investigated in frozen and fluid solution. In all cases, penta- or hexacoordinated Cu(II) centres were established. The complexes with counterions indicating a high stereoselectivity (OTf, SbF_6) revealed one unique species in which acryloyloxazolidinone bound to pseudoequatorial positions (via O atoms), shifting the weakly coordinating counterions to axial locations. On the other hand, those lacking stereoselectivity (X=Cl, Br) formed two species in which the parent halogen anions remained at equatorial positions preventing the formation of geometries compatible with those found for X=OTf and SbF_6. In all cases, the bis(sulfoximines) ligand bound via nitrogen atoms. The results are collected in Scheme 5.8.

The scope of this methodology was extended to the enantioselective hetero-Diels–Alder reaction[18] between 1,3-cyclohexadiene and ethylglyoxylate,

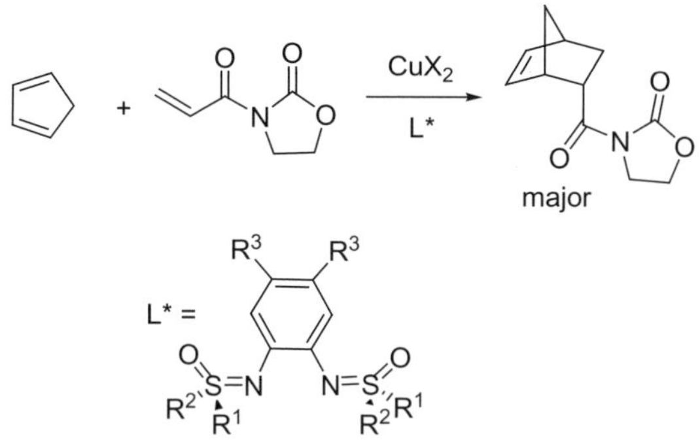

major

L* =

X = OTf, R^1 = R^3 = Me, R^2 = Ph:
98% de = 86% ee = 81%
X = OTf, R^1 = Me, R^2 = Ph, R^3 = H:
98% de = 86% ee = 75%
X = OTf, R^1 = H, R^2 = Me, R^3 = o-MeOC$_6$H$_4$:
98% de = 82% ee = 79%
X = ClO$_4$, R^1 = R^2 = Me, R^3 = o-MeOC$_6$H$_4$:
98% de = 78% ee = 93%

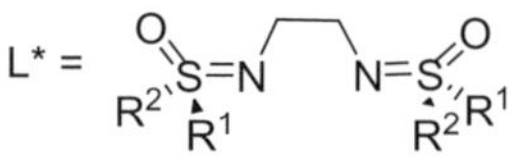

L* =

X = OTf, R^1 = Me, R^2 = Ph:
98% de = 86% ee = 66%
X = OTf, R^1 = i-Pr, R^2 = Ph:
98% de = 86% ee = 49%
X = OTf, R^1 = o-MeOC$_6$H$_4$, R^2 = Me:
98% de = 88% ee = 83%

Scheme 5.7 Cu-catalysed Diels–Alder reaction with C_2-symmetric bis(sulfoximines) ligands.

providing the corresponding cycloadduct in high yield, diastereo- and enantioselectivity, as shown in Scheme 5.9.[16a,19] Similarly, the same conditions could be applied to the cycloaddition between cyclohexadiene and diethyl mesoxalate that afforded the expected product in a high yield and an enantioselectivity of up to 98% ee. In order to understand how the reactants, solvent and concentration interacted together with the catalytically active complex, spectroscopic investigations were carried out using different techniques, including EXAFS, ESR and UV-vis spectroscopy.[20] It was concluded that the ligand was coordinated to the Cu(II) atom via the imine nitrogens and to the dienophile via the carbonyl oxygen atoms in a tetragonally distorted complex, affording a nonsymmetric square-pyramidal geometry.

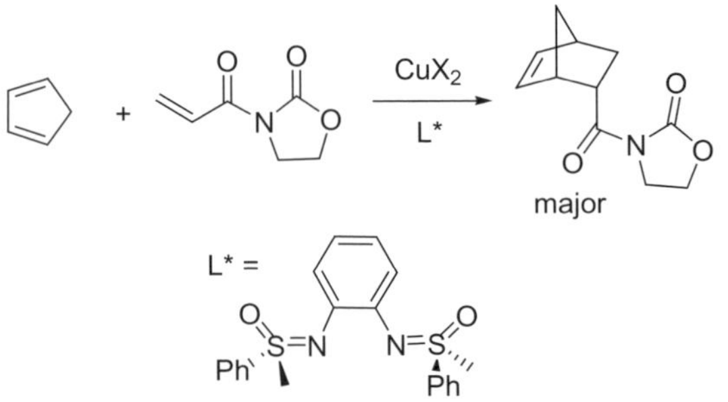

X = OTf: 98% de = 86% ee = 75%
X = SbF$_6$: 98% de = 86% ee = 73%
X = Cl: 98% de = 86% ee = 0%
X = Br: 98% de = 86% ee = 0%

Scheme 5.8 Cu-catalysed Diels–Alder reaction with C_2-symmetric bis(sulfoximines) ligand.

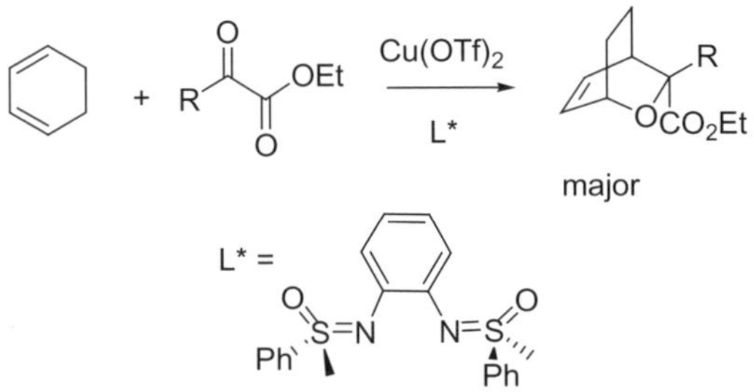

R = H: 81% de = 98% ee = 98%
R = CO$_2$Et: 92% de = 100% ee = 98%

Scheme 5.9 Cu-catalysed hetero-Diels–Alder reactions with C_2-symmetric bis(sulfoximines) ligand.

Since the two coordinating sulfoximine nitrogens were nonequivalent in the coordination mode described above, and in order to ascertain whether the C_2-symmetry of the efficient bis(sulfoximines) ligand was really essential, or if C_1-symmetric monosulfoximine derivatives could also be applied to asymmetric hetero-Diels–Alder reactions and lead to high enantioselectivities, these workers have developed the syntheses of various C_1-symmetric sulfoximines, in which the second donor atom was a quinolyl nitrogen.[21] The use of these novel ligands in the copper-catalysed hetero-Diels–Alder reaction between 1,3-cyclohexadiene and ethyl glyoxylate or diethyl mesoxalate led to the corresponding cycloadducts in good yields and with enantioselectivities of up to 96% ee by using as little as 1 mol% of the complex (Scheme 5.10). The steric

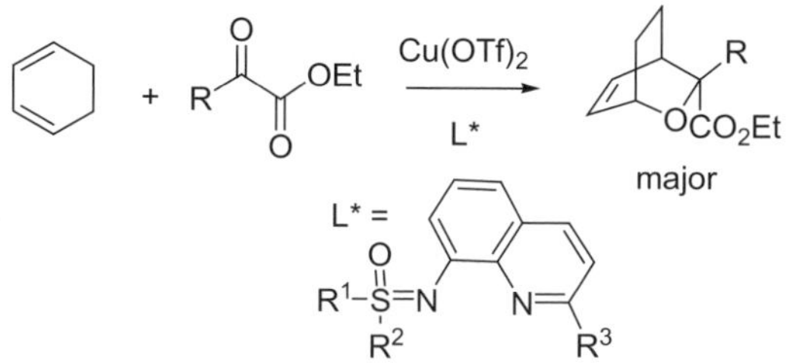

(*R*)-L*, R^1 = Me, R^2 = Ph, R = R^3 = H: 97% de = 94% ee = 75%
(*R*)-L*, R = CO_2Et, R^1 = Me, R^2 = Ph, R^3 = H: 80% de = 96% ee = 91%
(*R*)-L*, R^1 = Me, R^2 = *o*-$MeOC_6H_4$, R = R^3 = H: 98% de = 96% ee = 91%
(*R*)-L*, R = CO_2Et, R^1 = Me, R^2 = *o*-$MeOC_6H_4$, R^3 = H: 86% ee = 89%
(*R*)-L*, R^1 = *n*-Pent, R^2 = *o*-$MeOC_6H_4$, R = R^3 = H: 93% de = 94% ee = 86%
(*S*)-L*, R^1 = Me, R^2 = *o*-$MeOC_6H_4$, R = R^3 = H: 88% de = 96% ee = 91%
(*S*)-L*, R^1 = Ph, R^2 = *t*-Bu, R = R^3 = H: 18% de = 76% ee = 0%
(*S*)-L*, R^1 = *o*-$MeOC_6H_4$, R^2 = *t*-Bu, R = R^3 = H: 41% de = 84% ee = 0%

Scheme 5.10 Cu-catalysed hetero-Diels–Alder reactions with C_1-symmetric monosulfoximine ligands.

hindrance generated by the alkyl substituent on the sulfoximine moiety seemed to be of great importance for obtaining a high enantioselectivity, since the presence of a *t*-Bu group on the sulfoximine moiety led systematically to racemic cycloadducts isolated in low yields. According to an X-ray structure of one complex and $CuCl_2$, these authors have proposed a mechanistic model explaining the stereochemical outcome of the reaction. These ligands were supposed to act as *N,N*-bidentate ligands, leading to a copper complex with a distorted tetrahedral coordination geometry. However, based on these results, better values in terms of activity and enantioselectivity were obtained with C_2-symmetric bis(sulfoximines) compared to monosulfoximines. These studies clearly showed that C_2-symmetric bis(sulfoximines) associated to copper precatalysts were very efficient systems for affording either Diels–Alder or hetero-Diels–Alder cycloadducts in excellent yields and diastereo- and enantioselectivities.

In 2002, Sannicolo *et al.* developed enantioselective Diels–Alder reactions promoted by ligands in which the sulfur atom was part of an aromatic cycle, such as a thiophene moiety.[22] In this context, several C_2-symmetric biheteroaromatic diphosphines, including the thiophene moiety or not, were investigated as Pd(II) or Pt(II) ligands for the reaction between cyclopentadiene and *N*-2-alkenoyl-1,3-oxazolidin-2-ones. These authors compared their abilities to promote these reactions with the results reported by Gosh and Matsuda who have used BINAP as ligand associated to the same metals.[23] The authors assumed that the variation in the electronic properties at phosphorus resulting from the presence of the heteroaromatic backbone should allow a control of the stereoselection. According to the results collected in Scheme 5.11 concerning the Pd-catalysed process, the authors proposed that the more electron rich the diphosphine was, the more stereoselective and the faster the corresponding

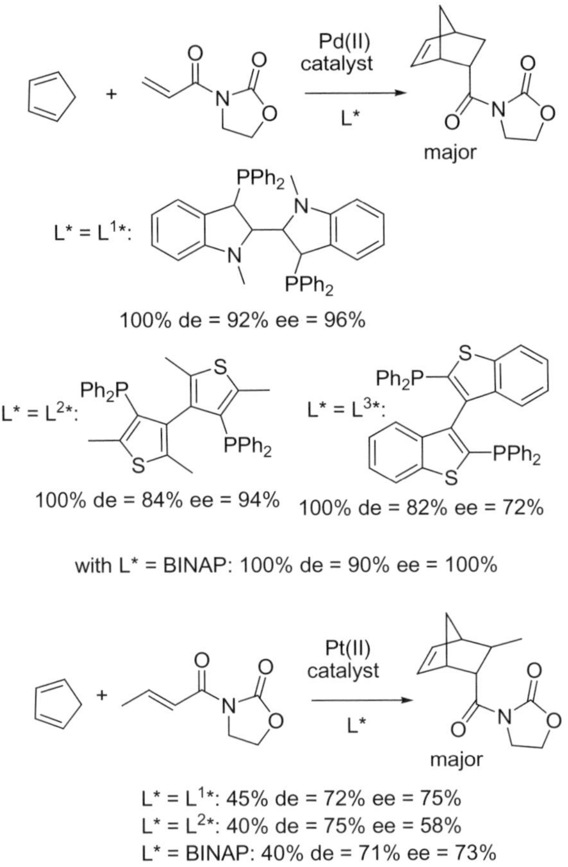

Scheme 5.11 Pd- and Pt-catalysed Diels–Alder reactions with C_2-symmetric biheteroaromatic diphosphines.

reaction. The electronic availability of each diphosphine was measured by electrochemical experiments to determine their oxidation potential (*vs.* Ag/Ag$^+$). The same observations were made by performing the Diels–Alder reaction between cyclopentadiene and N-(2-butenoyl)-2-oxazolidinone in the presence of Pt(II). The authors concluded that the ligand should possess a large electronic availability at phosphorus in order to efficiently carry out Diels–Alder reactions.

Another ligand including a thiophene moiety but lacking the C_2-symmetry and thus bearing electronically different phosphorus atoms was prepared by these authors, in 2001.[24] The electrochemical oxidative potential was obtained by cyclic voltammetry. The oxidation potential of the phosphine group located on the phenyl ring was found to be 0.74 V (*vs.* Ag/Ag$^+$) and the authors attributed a value of 0.91 V to the phosphine attached to the thiophene moiety. This second functionality is a rather electron-poor phosphine. As shown

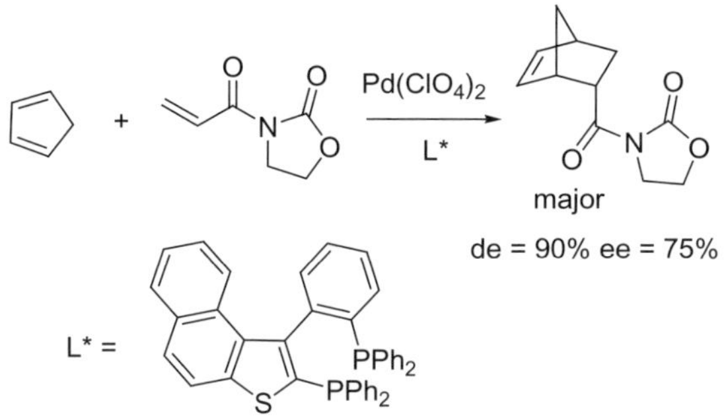

Scheme 5.12 Pd-catalysed Diels–Alder reaction with thiophene-containing ligand.

in Scheme 5.12, the application of this ligand to the Pd-catalysed Diels–Alder reaction of cyclopentadiene with *N*-acryloyloxazolidinone afforded the corresponding major *endo* cycloadduct in a high diastereoselectivity but a moderate enantioselectivity.

In 2001, Seebach *et al.* demonstrated that salen ligands could be immobilised on polystyrene by copolymerisation of suitable crosslinking styryl derivatives with styrene.[25] In the course of this study, it was shown that these chiral polymer-bound Cr-salens presented a high catalytic activity in the hetero-Diels–Alder reaction of Danishefsky's diene with aldehydes. More recently, Schulz *et al.* have reported the successful use of new electropolymerised chiral thiophene-salen-chromium complexes as heterogeneous catalysts for the hetero-Diels–Alder reaction of several aldehydes with Danishefsky's diene (Scheme 5.13).[26] This novel immobilisation procedure for the chiral salen-based-chromium catalysts occurred by electropolymerisation of the corresponding monomeric thiophene complexes. These insoluble catalysts were reused up to six times, affording the expected cycloadducts with an unchanged enantioselectivity of up to 88% ee along the recycling procedure.

In 2005, another class of chiral ligands, bis(thiazolines) derivatives, was prepared by Nishio *et al.* from chiral bis-(*N*-acylamino alcohols) by using Lawesson's reagent.[27] These new compounds have proved to be useful chiral ligands for the Zn-catalysed Diels–Alder reaction of 3-acryloyloxazolidine-2-one with cyclopentadiene, giving the corresponding cycloadducts as a 94:6 diastereomeric mixture, where the major diastereomer was formed with 92% ee (Scheme 5.14).

In 2005, Kunieda *et al.* prepared sterically congested "roofed" 2-thiazolines by thermal [4 + 2] cycloadditions of 2-thiazolone and cyclic dienes, subsequent hydrolytic ring cleavage with Ba(OH)$_2$, and final thiazoline ring formation following the general procedure for the preparation of corresponding oxazoline ligands.[28] Therefore, these authors synthesised a bis(thiazolines) derivative, a pyridylthiazoline, and a (2-diphenylphosphino)-phenylthiazoline, which were further investigated for the Cu-catalysed Diels–Alder reaction of 3-acryloyloxazolidine-2-one with cyclopentadiene, providing the corresponding *endo*

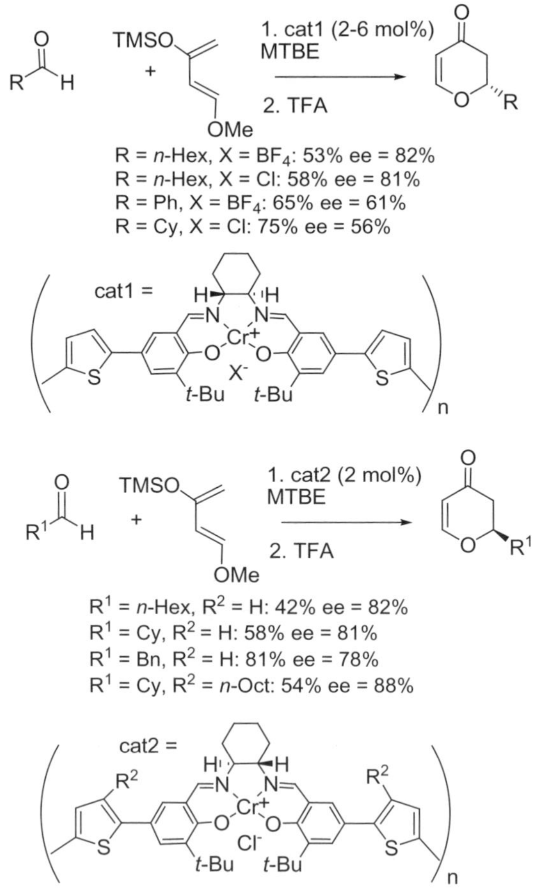
1. cat1 (2-6 mol%)
MTBE
2. TFA
TMSO
OMe
R = n-Hex, X = BF4: 53% ee = 82%
R = n-Hex, X = Cl: 58% ee = 81%
R = Ph, X = BF4: 65% ee = 61%
R = Cy, X = Cl: 75% ee = 56%
cat1 =
t-Bu
t-Bu
1. cat2 (2 mol%)
MTBE
2. TFA
R1 = n-Hex, R2 = H: 42% ee = 82%
R1 = Cy, R2 = H: 58% ee = 81%
R1 = Bn, R2 = H: 81% ee = 78%
R1 = Cy, R2 = n-Oct: 54% ee = 88%
cat2 =

Scheme 5.13 Electropolymerised Cr–salen-complexes for hetero-Diels–Alder reactions.

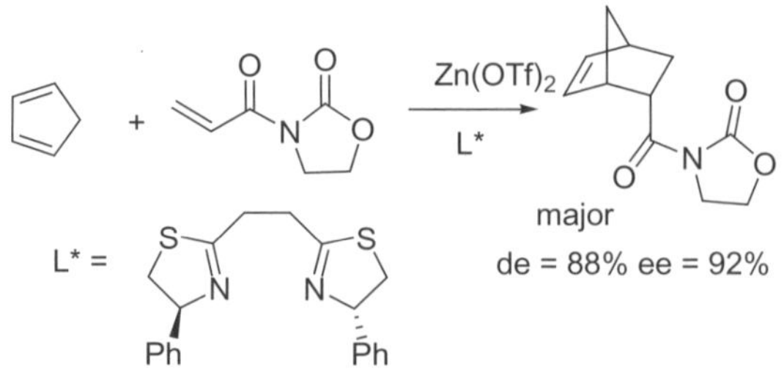
Zn(OTf)2
L*
major
de = 88% ee = 92%
L* =
Ph
Ph

Scheme 5.14 Zn-catalysed Diels–Alder reaction with bis(thiazolines) ligand.

cycloadduct as the major isomer (Scheme 5.15). The involvement of the bis (thiazolines) ligand and the pyridylthiazoline ligand in the presence of Cu(OTf)$_2$ gave low enantioselectivities ($\leq$ 16% ee), whereas the corresponding phosphino-thiazoline ligand led to the formation of the product with 76% ee by performing the reaction at 0 °C. When the reaction temperature was lowered to –60 °C, a significant increase in the enantioselectivity of the reaction was observed, since the cycloadduct was isolated with 92% ee. The corresponding oxazoline analogue of this latter ligand was also synthesised for comparison, giving under similar reaction conditions a lower enantioselectivity (Scheme 5.15). This result proved the beneficial effect of the presence of the sulfur atom in the heterocyclic ring on the *N,P*-chelating behaviour to the copper centre.

On the other hand, chiral cationic palladium-phosphinooxathiane complexes have been found to be efficient catalysts for the enantioselective Diels–Alder reaction of cyclopentadiene with acryloyl- and fumaroyl-1,3-oxazolidin-2-ones, giving the corresponding cycloadducts in good yields and high enantioselectivities of up to 93% ee (Scheme 5.16).[29] This result was the first example using an S/P-type phosphinooxathiane ligand for the enantioselective Diels–Alder reaction. The catalysts were very active, since the reaction could be performed at −78 °C and completed within a few hours. The structure of a palladium species coordinated with a phosphinooxathiane ligand was determined by X-ray analysis, proving phosphorus and sulfur coordination in a square-planar geometry.

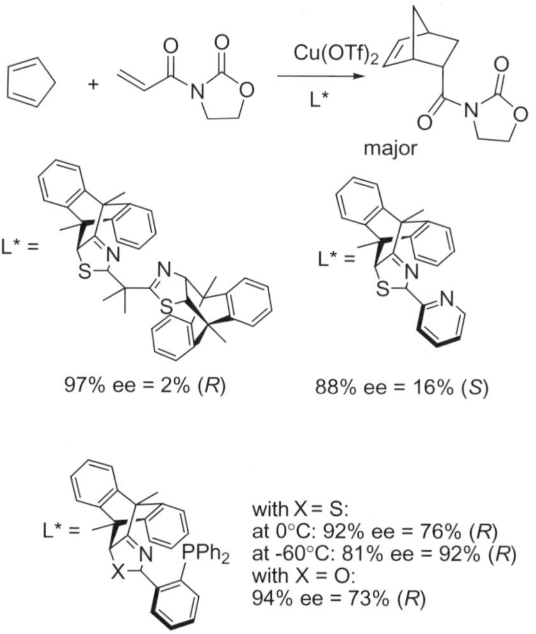

Scheme 5.15 Cu-catalysed Diels–Alder reaction with "roofed" 2-thiazoline ligands.

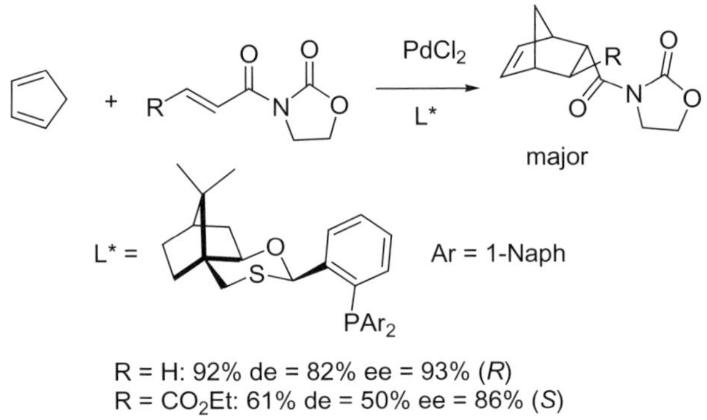
PdCl2
L*
major
L* =
Ar = 1-Naph
PAr2
R = H: 92% de = 82% ee = 93% (R)
R = CO2Et: 61% de = 50% ee = 86% (S)

Scheme 5.16 Pd-catalysed Diels–Alder reactions with phosphinooxathiane ligand.

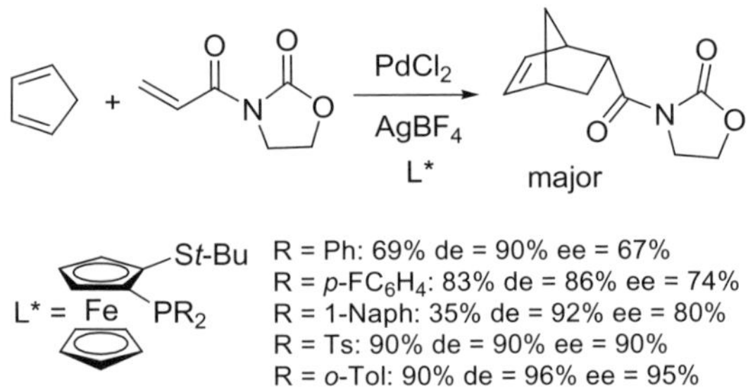
PdCl2
AgBF4
L*
major
St-Bu
L* = Fe
PR2
R = Ph: 69% de = 90% ee = 67%
R = p-FC6H4: 83% de = 86% ee = 74%
R = 1-Naph: 35% de = 92% ee = 80%
R = Ts: 90% de = 90% ee = 90%
R = o-Tol: 90% de = 96% ee = 95%

Scheme 5.17 Pd-catalysed Diels–Alder reaction with Fesulphos ligands.

In 2005, Carretero *et al.* reported a second example of chiral catalysts based on S/P-coordination employed in the catalysis of the enantioselective Diels–Alder reaction, namely palladium complexes of chiral planar 1-phosphino-2-sulfenylferrocenes (Fesulphos).[30] This new family of chiral ligands afforded, in the presence of $PdCl_2$, high enantioselectivities of up to 95% ee, in the asymmetric Diels–Alder reaction of cyclopentadiene with *N*-acryloyl-1,3-oxazolidin-2-one (Scheme 5.17). The S/P-bidentate character of the Fesulphos ligands has been proved by X-ray diffraction analysis of several metal complexes. When the reaction was performed in the presence of the corresponding copper-chelates, a lower and opposite enantioselectivity was obtained. This difference of results was explained by the geometry of the palladium (square-planar) and copper (tetrahedral) complexes.

In addition, the same group has used copper complexes of these ligands as efficient catalysts for enantioselective Cu-catalysed aza-Diels–Alder reactions[31] of *N*-sulfonyl imines with Danishefsky's dienes, providing the corresponding six-membered heterocycles with enantioselectivities of up to 80% ee.[32]

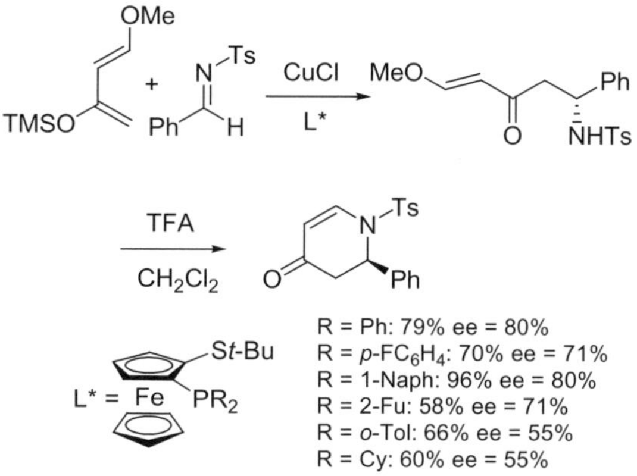

Scheme 5.18 Cu-catalysed aza-Diels–Alder reaction with Fesulphos ligands.

Actually, the reactions led to the formation of the corresponding acyclic Mannich-type addition products, which were further transformed into their corresponding Diels–Alder adducts by treatment with TFA (Scheme 5.18). The *S,P*-bidentate character of the ligands was proven by isolation of one complex and its X-ray analysis. Complexes prepared with CuCl afforded the expected product in up to 80% ee, whereas the use of CuBr as the catalytic precursor allowed an enantioselectivity of 97% ee to be obtained with the similar ligand.

In 2005, the scope of this process using the most efficient ligand was successfully extended to a wide range of *N*-sulfonyl imines derived from both aliphatic and aromatic aldehydes, providing the corresponding heterocyclic cycloadducts with high enantioselectivities of up to 97% ee, as shown in Scheme 5.19.[33] Interestingly, the enantioselectivity of the reaction was high in all cases of substrates, regardless of the substitution at the arylsulfonyl group.

Actually, the first studies that appeared in the literature dealing with the use of sulfur-containing ligands as chiral ligands for the enantioselective Diels–Alder reaction have concerned chiral bis(sulfonamides) in which the non-coordinating sulfur atom was only part of a functional group such as a triflate. Thus, Corey reported, in 1989, the first example of the highly enantioselective Diels–Alder reaction between cyclopentadiene and *N*-2-alkenoyl-1,3-oxazolidin-2-ones performed in the presence of a bis(trifluoromethanesulfonylamides) ligand, providing the corresponding cycloadducts with high yields and enantioselectivities of up to 95% ee (Scheme 5.20).[34] The reactions were performed in the presence of AlMe₃ and their scopes were extended to the use of substituted cyclopentadiene with similar excellent results.

In 1991, the practical value of this methodology was demonstrated by its application as the key step of a synthesis of a key intermediate for the syntheses of natural prostanoids (Scheme 5.21).[35]

In addition, the scope of this methodology was extended to the involvement of other bis(sulfonamides) ligands, acyclic dienes such as 2-methoxybutadiene

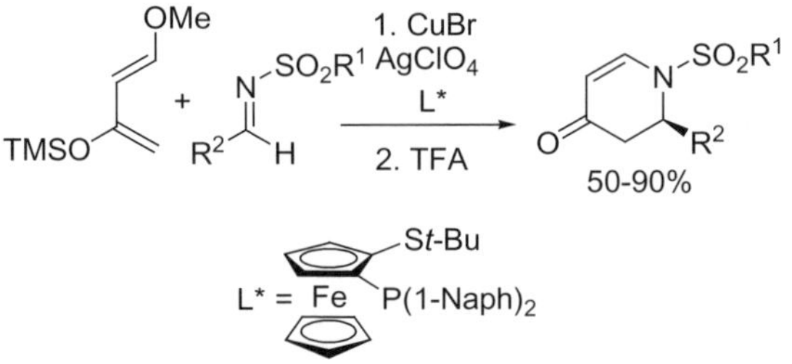

R^1 = 2-Thio, R^2 = Ph: ee = 93%
R^1 = *p*-$NO_2C_6H_4$, R^2 = Ph: ee = 90%
R^1 = *p*-$MeOC_6H_4$, R^2 = Ph: ee = 94%
R^1 = *p*-Tol, R^2 = Ph: ee = 97%
R^1 = *p*-Tol, R^2 = *p*-FC_6H_4: ee = 93%
R^1 = *p*-Tol, R^2 = *p*-$MeOC_6H_4$: ee = 91%
R^1 = *p*-Tol, R^2 = 2-Naph: ee = 93%
R^1 = *p*-Tol, R^2 = PhC≡CH: ee = 96%
R^1 = *p*-Tol, R^2 = *n*-Pr: ee = 82%

Scheme 5.19 Cu-catalysed aza-Diels–Alder reactions with Fesulphos ligand.

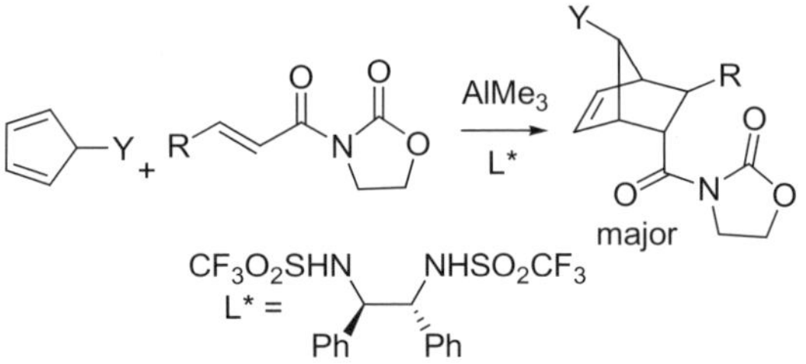

Y = R = H: 91% de > 96% ee = 95%
Y = H, R = Me: 88% de = 92% ee = 94%
Y = CH_2OBn, R = H: 93% ee > 95%

Scheme 5.20 Al-catalysed Diels–Alder reactions with bis(trifluoromethanesulfonylamides) ligand.

and C_2-symmetric dienophiles such as *N*-arylmaleimides.[36] The corresponding cycloadducts were isolated in moderate to high enantioselectivities according to the nature of the substituents borne by both the ligand and the dienophile, as shown in Scheme 5.22.

In 1995, these authors applied this methodology to the first total synthesis of the biosynthetically and unusual marine natural products, gracilins B and C.[37] Thus, the key step of this synthesis was the enantioselective Diels–Alder reaction of 2-((trimethylsilyl)methyl)-butadiene with *N*-(2-*tert*-butylphenyl)maleimide in

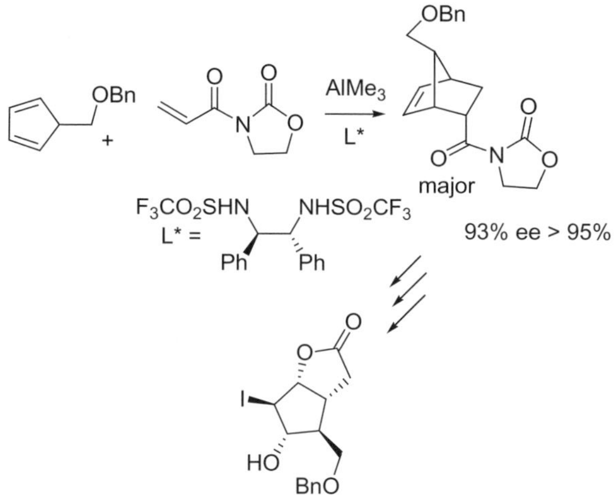

Scheme 5.21 Synthesis of key intermediate of prostanoids.

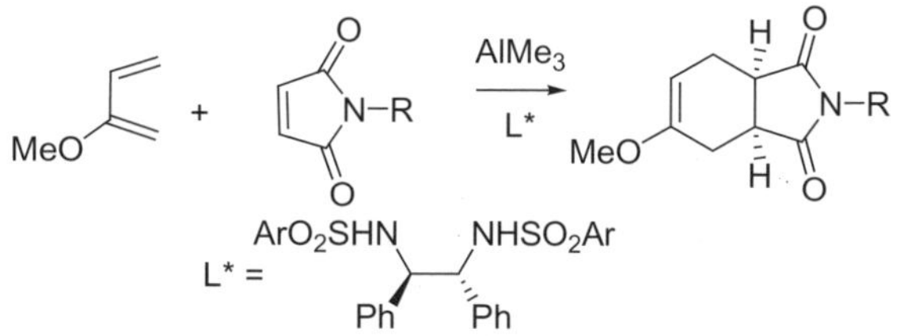

Ar = 3,5-(Me)$_2$C$_6$H$_3$, R = *o*-Tol: 98% ee = 93%
Ar = 3,5-(Me)$_2$C$_6$H$_3$, R = 4-Br-2-MeC$_6$H$_3$: 96% ee > 97%
Ar = 3,5-(Me)$_2$C$_6$H$_3$, R = Ph: ee = 62%
Ar = 3,5-(Me)$_2$C$_6$H$_3$, R = *o*-Tol: ee = 93%
Ar = 3,5-(Me)$_2$C$_6$H$_3$, R = *o*-(*t*-Bu)C$_6$H$_4$: ee = 95%
Ar = 3,5-(Me)$_2$C$_6$H$_3$, R = *o*-IC$_6$H$_4$: ee = 93%
Ar = R = Ph: ee = 30%
Ar = Ph, R = *o*-FC$_6$H$_4$: ee = 48%
Ar = Ph, R = *o*-Tol: ee = 58%
Ar = Ph, R = *o*-(*t*-Bu)C$_6$H$_4$: ee = 52%

Scheme 5.22 Al-catalysed Diels–Alder reactions between 2-methoxybutadiene and *N*-arylmaleimides with bis(sulfonamides) ligands.

the presence of a bis(sulfonamides) ligand combined with AlMe$_3$. As shown in Scheme 5.23, the corresponding key cycloadduct was obtained in high yield and enantioselectivity before being converted into the requested natural products.

Bis(trifluoromethanesulfonylamides) ligand was also used by Mikami *et al.*, in 1995, for the lanthanide-catalysed hetero-Diels–Alder reaction of Danishefsky's diene with butyl glyoxylate.[38] This reaction produced the corresponding

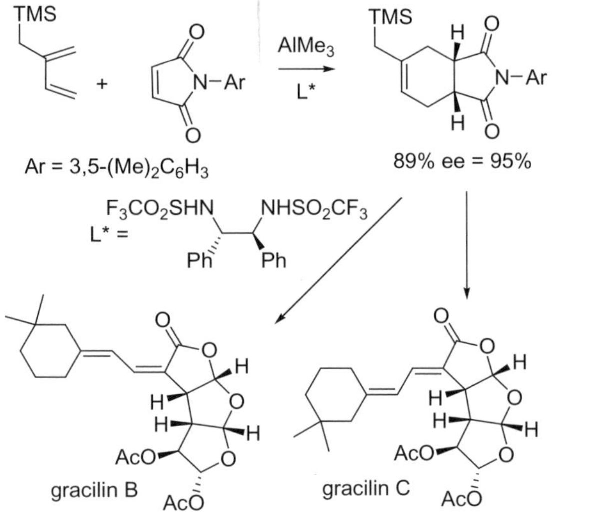

Scheme 5.23 Syntheses of gracilins B and C.

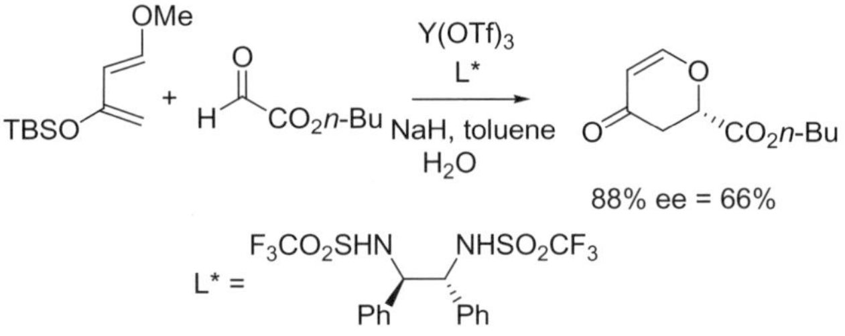

Scheme 5.24 Y-catalysed Diels–Alder reaction between 2-methoxybutadiene and *N*-arylmaleimides with bis(sulfonamides) ligand.

cycloadduct in a good yield and a moderate enantioselectivity in the presence of yttrium catalyst, which gave the best result (Scheme 5.24). A significant effect of water used as an additive was observed in increasing not only the enantio-selectivity but also the yield of the reaction.

On the other hand, several examples of chiral sulfonamides derived from chiral α-amino acids have been successfully employed as ligands for enantio-selective Diels–Alder reactions. Thus, Yamamoto and Takasu have easily prepared new chiral Lewis acids from borane and sulfonamides of various chiral α-amino acids, which were further studied for their abilities to promote the enantioselective Diels–Alder reaction between methacrolein and 2,3-dime-thyl-1,3-butadiene.[39] Since 2,4,6-triisopropylbenzenesulfonamide of α-amino-butyric acid gave the highest enantioselectivity, this catalyst was applied to the

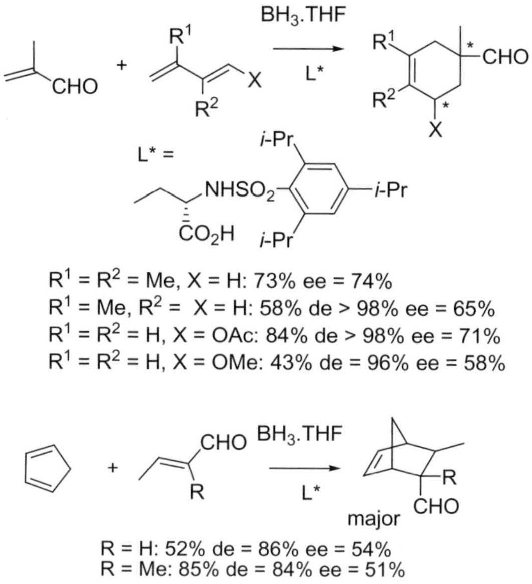
BH3.THF
R1
CHO
X
L*
R2
*
L* =
i-Pr
NHSO2
i-Pr
CO2H
R1 = R2 = Me, X = H: 73% ee = 74%
R1 = Me, R2 = X = H: 58% de > 98% ee = 65%
R1 = R2 = H, X = OAc: 84% de > 98% ee = 71%
R1 = R2 = H, X = OMe: 43% de = 96% ee = 58%
R
major
R = H: 52% de = 86% ee = 54%
R = Me: 85% de = 84% ee = 51%

Scheme 5.25 2,4,6-Triisopropylbenzenesulfonamide of α-aminobutyric acid ligand for B-catalysed Diels–Alder reactions.

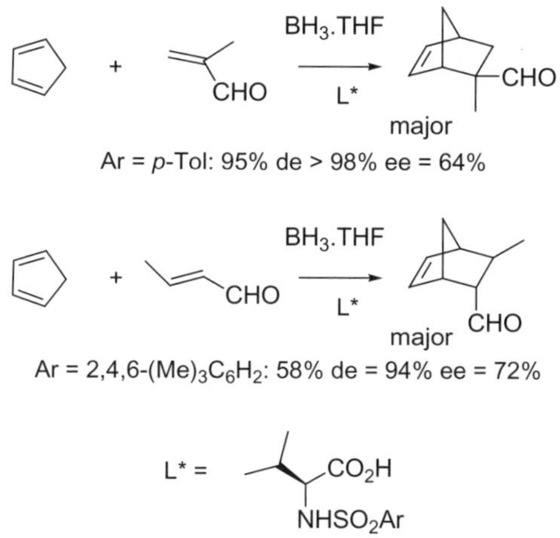
BH3.THF
CHO
L*
major
Ar = p-Tol: 95% de > 98% ee = 64%
Ar = 2,4,6-(Me)3C6H2: 58% de = 94% ee = 72%
L* =
CO2H
NHSO2Ar

Scheme 5.26 L-Valine-derived arylsulfonamide ligands for B-catalysed Diels–Alder reactions.

Diels–Alder reactions involving a variety of dienes and dienophiles, giving enantioselectivities of up to 71% ee, as shown in Scheme 5.25.

Similar reactions were also developed by Helmchen *et al.* using sulfonamides derived from L-valine.[40] Among Lewis acids such as borane, dialkylaluminum halides and titanium tetrachloride, borane allowed the best results to be obtained (Scheme 5.26).

In the same area, a (S)-tryptophan-derived oxazaborolidine including a *p*-tolylsulfonylamide function has been used by Corey *et al.* to catalyse the enantioselective Diels–Alder reaction between 2-bromoacrolein and cyclopentadiene to form the corresponding chiral product with an unprecedented high (> 99% ee) enantioselectivity (Scheme 5.27).[41] This highly efficient methodology was extended to various 2-substituted acroleins and dienes such as isoprene and furan.[42] In addition, it was applied to develop a highly efficient total synthesis of the potent antiulcer substance, cassiol, as depicted in Scheme 5.27.[43]

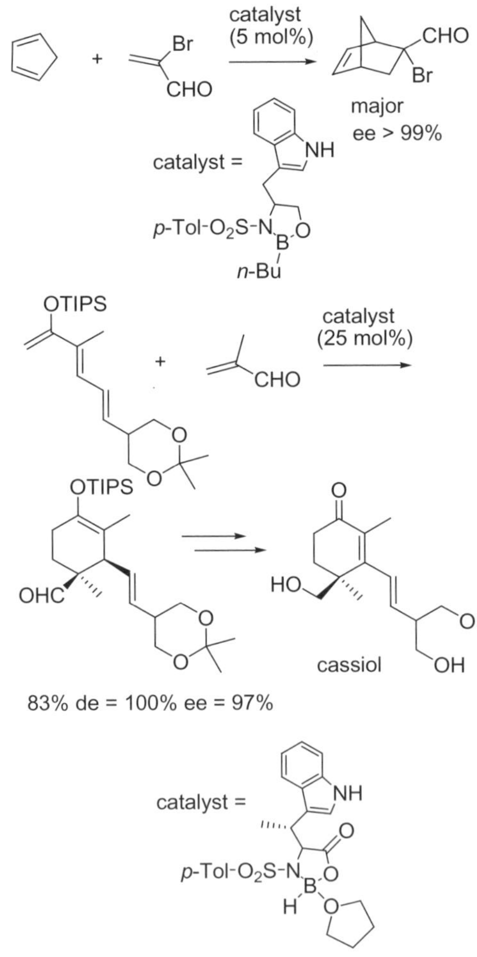

Scheme 5.27   L-Tryptophan-derived *p*-tolylsulfonylamide ligands for B-catalysed Diels–Alder reactions and synthesis of cassiol.

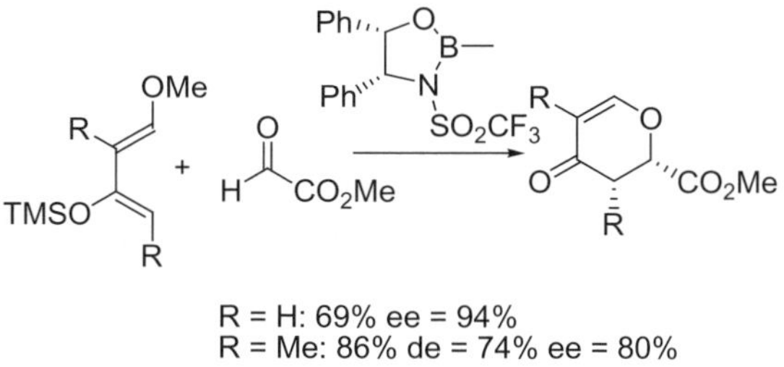

Scheme 5.28 β-Hydroxy sulfonamide ligand for B-catalysed Diels–Alder reactions.

Finally, Mikami and Motoyama have used a β-hydroxy sulfonamide ligand to catalyse the enantioselective B-catalysed Diels–Alder reaction of glyoxylate with Danishefsky's dienes.[44] A favourable transition-state assembly for a one-directional diene-approach from the site proximal to the sulfonylamino moiety was proposed to explain the observed high enantio- and *cis*(*endo*)-diastereoselectivity (Scheme 5.28).

In conclusion, a number of enantioselective Diels–Alder reactions have been successfully developed involving very different catalytic systems, in terms of metal as well as type of chiral sulfur-containing ligands. The oldest and excellent results are those based on the use of chiral sulfonamides and bis (sulfonamides) as ligands, which have been successfully used in combination with boron and aluminum. One of the most recent and efficient systems has been introduced by Ellman's group as a combination of chiral bis(sulfinyl)-imidoamidines with copper, which allowed either high yields and excellent enantioselectivities. On the other hand, copper has also been used by Bolm's group in association with highly efficient bis(sulfoximines) ligands. In addition, Carretero's group has successfully combined this metal with chiral sulfur-containing ferrocenyl ligands. Other metals such as zinc have proved their ability to catalyse the Diels–Alder reaction in the presence of bis(thiazolines) ligands for example. Moreover, bis(thiophenes)- and bis(benzothiophenes)-diphosphine BINAP analogues have allowed this type of reaction to be performed in the presence of palladium or platinum salts.

References

1. *Comprehensive Organic Synthesis*, B. M. Trost, I. Fleming, ed., Pergamon Press, Oxford, 1991, Vol. 5, W. Oppolzer in Chap. 4.1, and W. R. Roush in Chap. 4.4.
2. (a) E. J. Corey, *Angew. Chem., Int. Ed. Engl.*, 2002, **41**, 1650–1667; (b) D. A. Evans and J. S. Johnson, in *Comprehensive Asymmetric Catalysis*, ed. E. N. Jacobsen, A. Pfaltz, H. Yamamoto, Springer: Berlin, 1999, **Vol. 3**, p. 1177.
3. H. B. Kagan and O. Riant, *Chem. Rev.*, 1992, **92**, 1007–1019.
4. F. Fache, E. Schulz, M. L. Tommasino and M. Lemaire, *Chem. Rev.*, 2000, **100**, 2159–2231.

5. (a) D. A. Evans, S. J. Miller, T. Lectka and P. von Matt, *J. Am. Chem. Soc.*, 1993, **115**, 6460–6461; (b) D. A. Evans, S. J. Miller, T. Lectka and P. von Matt, *J. Org. Chem.*, 1999, **121**, 7559–7567.

6. T. D. Owens, F. J. Hollander, A. G. Olivier and J. A. Ellman, *J. Am. Chem. Soc.*, 2001, **123**, 1539–1540.

7. T. D. Owens, A. J. Souers and J. Ellman, *J. Org. Chem.*, 2003, **68**, 3–10.

8. K. Hiroi, K. Watanabe, I. Abe and M. Koseki, *Tetrahedron Lett.*, 2001, **42**, 7617–7619.

9. K. Watanabe, T. Hirasawa and K. Hiroi, *Chem. Pharm. Bull.*, 2002, **50**, 372–379.

10. K. Watanabe, T. Hirasawa and K. Hiroi, *Heterocycles*, 2002, **58**, 93–97.

11. N. Khiar, I. Fernandez and F. Alcudia, *Tetrahedron Lett.*, 1993, **34**, 123–126.

12. M. Ordonez, V. Guerrero-de la Rosa, V. Labastida and J. M. Llera, *Tetrahedron: Asymmetry*, 1996, **7**, 2675–2686.

13. C. Bolm, F. Bienewald and K. Harms, *Synlett*, 1996, 775–776.

14. H. Okamura and C. Bolm, *Chem. Lett.*, 2004, **33**, 482–487.

15. C. Bolm, M. Martin, O. Simic and M. Verrucci, *Chemtracts*, 2003, **16**, 660–666.

16. (a) C. Bolm and O. Simic, *J. Am. Chem. Soc.*, 2001, **123**, 3830–3831; (b) C. Bolm, M. Martin, O. Simic and M. Verrucci, *Org. Lett.*, 2003, **5**, 427–429.

17. C. Bolm, M. Martin, G. Gescheidt, C. Palivan, T. Stanoeva, H. Bertagnolli, M. Feth, A. Schweiger, G. Mitrikas and J. Harmer, *Chem. Eur., J.*, 2007, **13**, 1842–1850.

18. K. A. Jorgensen, *Eur. J. Org. Chem.*, 2004, **10**, 2093–2102.

19. C. Bolm, M. Verruci, O. Simic and C. P. R. Hackenberger, *Adv. Synth. Catal.*, 2005, **347**, 1696–1700.

20. C. Bolm, M. Martin, G. Gescheidt, C. Palivan, D. Neshchadin, H. Bertagnolli, M. Feth, A. Schweiger, V. Mitrikas and J. Harmer, *J. Am. Chem. Soc.*, 2003, **125**, 6222–6227.

21. C. Bolm, M. Verrucci, O. Simic, P. G. Cozzi, G. Raabe and H. Okamura, *Chem. Commun.*, 2003, 2826–2827.

22. G. Celentano, T. Benincori, S. Radaelli, M. Sada and F. Sannicolo, *J. Organomet. Chem.*, 2002, **643–644**, 424–430.

23. A. K. Ghosh and H. Matsuda, *Org. Lett.*, 1999, **1**, 2157–2159.

24. T. Benincori, S. Gladiali, S. Rizzo and F. Sannicolo, *J. Org. Chem.*, 2001, **66**, 5940–5942.

25. H. Sellner, J. K. Karjalainen and D. Seebach, *Chem. Eur. J.*, 2001, **7**, 2873–2887.

26. (a) M. Mellah, B. Ansel, F. Patureau, A. Voituriez and E. Schulz, *J. Mol. Catal.*, 2007, **272**, 20–25; (b) A. Zulauf, M. Mellah, R. Guillot and E. Schulz, *Eur. J. Org. Chem.*, 2008, 2118–2129.

27. T. Nishio, Y. Kodama and Y. Tsurumi, *Phosphorus, Sulfur, and Silicon*, 2005, **180**, 1449–1450.

28. M. Yamakuchi, H. Matsunaga, R. Tokuda, T. Ishizuka, M. Nakajima and T. Kunieda, *Tetrahedron Lett.*, 2005, **46**, 4019–4022.

29. H. Nakano, Y. Suzuki, C. Kabuto, R. Fujita and H. Hongo, *J. Org. Chem.*, 2002, **67**, 5011–5014.
30. O. G. Mancheno, R. G. Arrayas and J. C. Carretero, *Organometallics*, 2005, **24**, 557–561.
31. K. A. Jorgensen, *Angew. Chem., Int. Ed. Engl.*, 2000, **39**, 3558–3588.
32. O. G. Mancheno, R. G. Arrayas and J. C. Carretero, *J. Am. Chem. Soc.*, 2003, **126**, 456–457.
33. O. G. Mancheno, R. G. Arrayas and J. C. Carretero, *Phosphorus, Sulfur, and Silicon*, 2005, **180**, 1515–1516.
34. E. J. Corey, R. Imwinkelried, S. Pikul and Y. B. Xiang, *J. Am. Chem. Soc.*, 1989, **111**, 5493–5495.
35. E. J. Corey, N. Imai and S. Pikul, *Tetrahedron Lett.*, 1991, **32**, 7517–7520.
36. E. J. Corey, S. Sarshar and D.-H. Lee, *J. Am. Chem. Soc.*, 1994, **116**, 12089–12090.
37. E. J. Corey and M. A. Letavic, *J. Am. Chem. Soc.*, 1995, **117**, 9616–9617.
38. K. Mikami, O. Kotera, Y. Motoyama and H. Sakaguchi, *Synlett*, 1995, 975–977.
39. M. Takasu and H. Yamamoto, *Synlett*, 1990, 194–196.
40. D. Sartor, J. Saffrich and G. Helmchen, *Synlett*, 1990, 197–198.
41. (a) E. J. Corey, T.-P. Loh, T. D. Roper, M. D. Azimioara and M. C. Noe, *J. Am. Chem. Soc.*, 1992, **114**, 8290–8292; (b) E. J. Corey and T.-P. Loh, *J. Am. Chem. Soc.*, 1991, **113**, 8966–8967.
42. E. J. Corey and T.-P. Loh, *Tetrahedron Lett.*, 1993, **34**, 3979–3982.
43. E. J. Corey, A. Guzman-Perez and T.-P. Loh, *J. Am. Chem. Soc.*, 1994, **116**, 3611–3612.
44. Y. Motoyama and K. Mikami, *J. Chem. Soc., Chem. Commun.*, 1994, 1563–1564.

CHAPTER 6

Cyclopropanation

6.1 Introduction

The synthesis of chiral cyclopropanes remains a considerable challenge, especially due to the fact that cyclopropane rings are often found in a variety of natural products and biologically active compounds. Organic chemists have always been fascinated by the cyclopropane subunit,[1] which has played and continues to play a prominent role in organic chemistry.[2] Its strained structure, interesting bonding characteristics and value as an internal mechanistic probe have attracted the attention of the physical organic community. While the cyclopropane ring is a highly strained entity, it is nonetheless found in a wide variety of naturally occurring compounds including terpenes, pheromones, fatty acid metabolites and unusual amino acids.[3] Cyclopropane subunits also occur in many natural products of primary and secondary metabolism. Indeed, the prevalence of cyclopropane-containing compounds with biological activity, whether isolated from natural sources or rationally designed pharmaceutical agents, has inspired chemists to find novel and diverse approaches to their synthesis.[4] Naturally occurring and synthetic cyclopropanes bearing simple or complex functionalities are endowed with a large spectrum of biological properties, including enzyme inhibition, and insecticidal, antifungal, herbicidal, antimicrobial, antibiotic, antibacterial, antitumor and antiviral activities.[5] For example, they constitute a common structural motif in pyrethroids,[6] the anti-depressant, tranylcypromine,[7] papain and cysteine protease inhibitors,[8] potential antipsychotic substances,[9] anti-HIV agents,[10] and marine lactones.[11] So far, thousands of natural compounds and their derivatives carrying a cyclopropane group have been synthesised and described in the literature. Indeed, with representation in more than 4000 natural isolates and 100 therapeutic agents, the cyclopropane motif has long been established as a valuable platform for the development of new asymmetric technologies. In addition, the

RSC Catalysis Series No. 2
Chiral Sulfur Ligands: Asymmetric Catalysis
By Hélène Pellissier
Published by the Royal Society of Chemistry, www.rsc.org

rigidity of the three-membered ring renders this group an appealing structural unit for the preparation of molecules with a defined orientation of pendant functional groups.[12] Indeed, the strain associated with the three-membered ring allows cyclopropanes to undergo a variety of synthetically useful ring-opening reactions.[13] In recent years, most of the synthetic efforts involving cyclopropanes have focused on the enantioselective synthesis of these compounds. This has remained a challenge since it was demonstrated that members of the pyrethroid class of compounds were efficient insecticides.[14]

6.2 Transition-Metal-Catalysed Decomposition of Diazoalkanes

The cyclopropanation of olefins using the transition-metal-catalysed decomposition of diazoalkanes is one of the most extensively studied reactions in the organic chemist's arsenal.[15] Indeed, the synthesis of cyclopropanes by transition-metal-mediated carbene transfer from aliphatic diazo compounds to carbon–carbon double bonds is not only a major method for the preparation of cyclopropanes, but is also among the best developed and most general methods available to the synthetic organic chemist.[15e,16,17] Highly effective and stereocontrolled syntheses of functionalised cyclopropanes have been achieved, in particular, with catalysts based on copper, rhodium and, more recently, ruthenium. Outstanding levels of enantioselectivity have been achieved with some chiral catalysts such as copper(I) complexes with C_2-symmetric bis(oxazolines) ligands, and dinuclear rhodium(II) complexes of the type [$Rh_2L^*_4$], in which L* is a chiral bidentate carboxylate, amidate or phosphate ligand.[15,18] On the other hand, the diastereocontrol of the intermolecular cyclopropanation reaction is more difficult to handle, because the *cis/trans* or *syn/anti* selective formation of cyclopropanes is most often controlled by the particular olefin/diazo compound combination. Nevertheless, catalysts with cleverly designed ligands have been developed, which did allow highly selective *trans*- or *cis*-cyclopropanation in particular cases. The catalytic cycle of the carbenoid cyclopropanation reaction is outlined in Scheme 6.1, involving interaction of the catalyst with the diazo precursor to afford a metallocarbene complex as the central intermediate with concomitant release of nitrogen and subsequent transfer of the carbene to an appropriate substrate

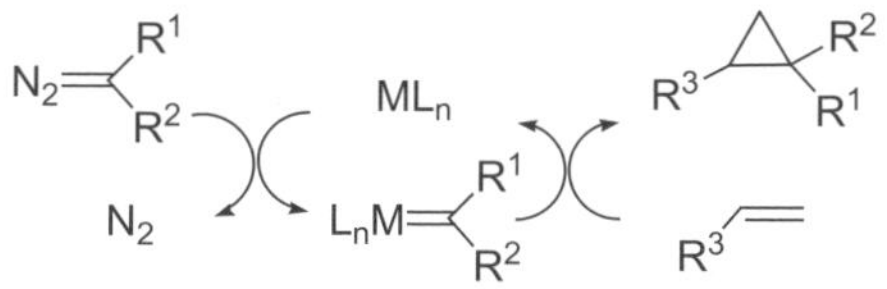

Scheme 6.1 Catalytic cycle of metal-catalysed carbenoid cyclopropanation reactions with diazo compounds.

such as an olefin. Enantiocontrol in the carbene-transfer step may be achieved by chiral ligands surrounding the metal centre of the catalyst.

Transition-metal-catalysed reactions of diazo compounds with alkenes have been widely used to prepare cyclopropanes, involving as the most common catalysts, copper,[19] rhodium or ruthenium complexes.[20] Most of the time (Cu, Rh, Ru or Os), the mechanism of the transition-metal-catalysed decomposition of α-diazocarbonyl compounds is believed to initially proceed via the formation of a metal-carbene complex.[21] In particular, the Cu-catalysed enantioselective cyclopropanation of alkenes is now well established, and chiral C_2-symmetric bidentate bis(oxazolines) ligands[22] are the most widely used ligands.[23] Many investigations have shown that the ligand structure has a strong influence on the stereoselectivity of the cyclopropanation. Even very small structural changes often have drastic and sometimes unpredictable effects on the enantioselectivity. In the last fifteen years, several types of chiral sulfur-containing ligands have been successfully applied to the asymmetric cyclopropanation of alkenes in the presence of metals such as rhodium, ruthenium or copper. Therefore, sulfide, thiophene, sulfoxide and sulfonamide ligands have allowed excellent enantioselectivities to be obtained for various cyclopropanes.

6.2.1   Sulfide Ligands

A few chiral sulfides have been used as chiral ligands in asymmetric rhodium-catalysed reactions of diazo compounds with alkenes. As an example, the chiral 1,3-oxathiane derived from camphorsulfonyl chloride was able to mediate the rhodium-catalysed cyclopropanation of electron-deficient alkenes with phenyldiazomethane, furnishing the corresponding cyclopropanes in good yields and very high enantioselectivities of up to 98% ee (Scheme 6.2).[24] The application of these conditions to ethyl diazoacetate (EDA) was, however, unsuccessful. Moreover, the process could not be easily scaled up, and the sulfide ligand could not be fully recovered. In order to solve these problems, these authors have developed a second process, involving the *in situ* generation of the diazo compound, together with the use of another class of chiral [2.2.2] bicyclic sulfides, depicted in Scheme 6.2.[25] In these conditions, the new process could be generalised to a broad range of electron-deficient alkenes, providing high enantioselectivities with all alkenes, allowing the sulfide ligand to be recovered in quantitative yield. The absolute configuration of the cyclopropane product, arising from the reaction between (*E*)-chalcone and the *p*-bromo derivative, was determined by X-ray analysis. This led the authors to propose that the sulfur ylide preferentially adopted a conformation in which the filled orbital on carbon was orthogonal to the lone pair on sulfur. This preferred conformation could explain the origin of the chiral induction.

6.2.2   Thiophene Ligands

A number of nonsulfur-containing bis(oxazolines) ligands with numerous structural diversities have been designed by several groups and successfully

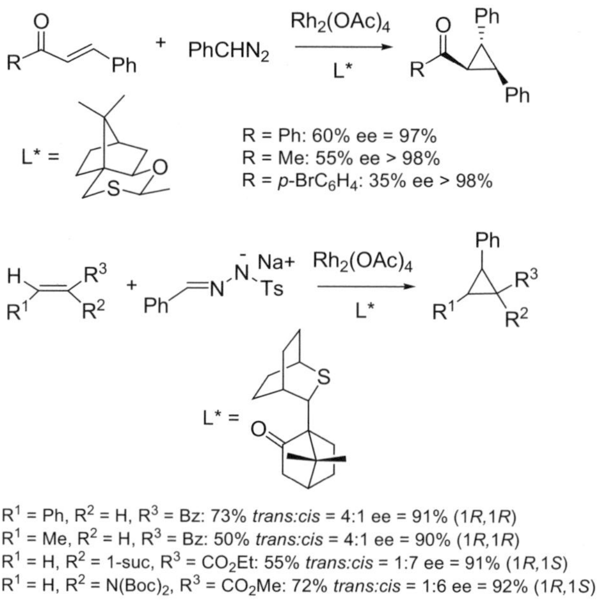

R = Ph: 60% ee = 97%
R = Me: 55% ee > 98%
R = p-BrC$_6$H$_4$: 35% ee > 98%

R^1 = Ph, R^2 = H, R^3 = Bz: 73% *trans:cis* = 4:1 ee = 91% (1*R*,1*R*)
R^1 = Me, R^2 = H, R^3 = Bz: 50% *trans:cis* = 4:1 ee = 90% (1*R*,1*R*)
R^1 = H, R^2 = 1-suc, R^3 = CO$_2$Et: 55% *trans:cis* = 1:7 ee = 91% (1*R*,1*S*)
R^1 = H, R^2 = N(Boc)$_2$, R^3 = CO$_2$Me: 72% *trans:cis* = 1:6 ee = 92% (1*R*,1*S*)

Scheme 6.2 Rh-catalysed cyclopropanations with sulfide ligands.

involved in the cyclopropanation of styrene with EDA.[26] In this context, Gao *et al.* have reported novel structure-defined chiral bis(oxazolines)thiophene ligands for the ruthenium-catalysed asymmetric cyclopropanation of alkenes with EDA.[27] A high enantioselectivity of up to 99% ee was observed in the case of the cyclopropanation of diphenylethene (Scheme 6.3). Compared to the results obtained with the corresponding analogous catalyst, *N,N,N*-Ru-pybox, in the cyclopropanation of 1,1-diphenylethene, better results were reported when *N,S,N*-Ru-thiobox was used. According to these authors, the differences could arise from different catalyst structures with, on the one hand, a N–S–N bond angle much larger than the N–N–N bond angle and, on the other hand, a stronger Ru–S coordination resulting from soft–soft interactions.

As an extension of this methodology, the efficiency of these ligands was also evaluated by these authors for the Cu-catalysed cyclopropanation of styrene derivatives with EDA, providing the corresponding cyclopropanes with similar enantioselectivities of up to 97% ee (Scheme 6.4).[28]

In 2005, Sannicolo *et al.* developed other novel bis(oxazolines) ligands displaying flexible and atropisomeric 3,3′-bithiophene backbones (Scheme 6.5).[29] The electronic and steric effects of these ligands on the Cu-catalysed cyclopropanation of styrene with EDA were examined. The electrochemical experiments with the determination of the oxidative potential were used as a reliable index of the electronic density on the nitrogen atom of the chelating groups of new and, for comparative purposes, of already known bis(oxazolines) ligands. All the chiral ligands depicted in Scheme 6.5 were found to be active in

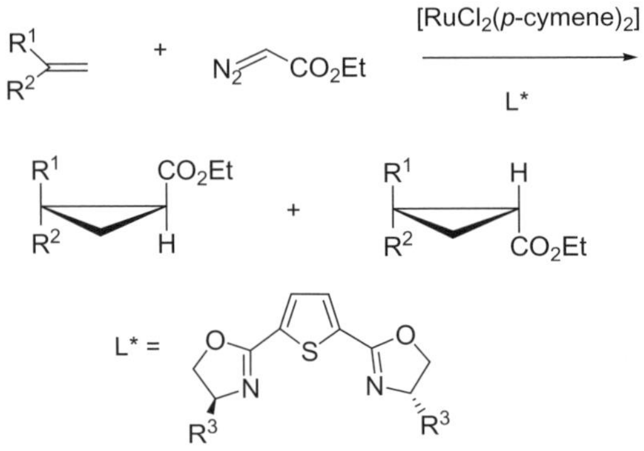

R^1 = R^2 = Ph, R^3 = Et: 70% ee > 99%
R^1 = R^2 = Ph, R^3 = *i*-Pr: 77% ee = 98%
R^1 = R^2 = Ph, R^3 = Bn: 72% ee = 96%
R^1 = R^2 = Ph, R^3 = *t*-Bu: 63% ee = 98%
R^1 = H, R^2 = Ph, R^3 = *i*-Pr: 83% *trans:cis* = 76:24
ee (*trans*) = 81% ee (*cis*) = 84%
R^1 = H, R^2 = Ph, R^3 = *t*-Bu: 79% *trans:cis* = 79:21
ee (*trans*) = 89% ee (*cis*) = 82%
R^1 = H, R^2 = *p*-MeOC$_6$H$_4$, R^3 = *t*-Bu: 78% *trans:cis* = 82:18
ee (*trans*) = 91% ee (*cis*) = 85%
R^1 = H, R^2 = *p*-ClC$_6$H$_4$, R^3 = *t*-Bu: 82% *trans:cis* = 80:20
ee (*trans*) = 87% ee (*cis*) = 83%

Scheme 6.3 Ru-catalysed cyclopropanations of alkenes with bis(oxazolines) thiophene ligands.

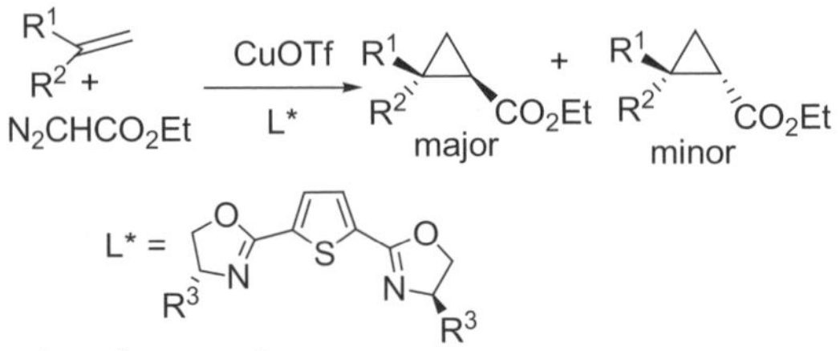

R^1 = R^2 = Ph, R^3 = Et: 84% ee = 91%
R^1 = R^2 = Ph, R^3 = *i*-Pr: 76% ee = 89%
R^1 = R^2 = R^3 = Ph: 56% ee = 86%
R^1 = R^2 = Ph, R^3 = *t*-Bu: 71% ee = 97%
R^1 = H, R^2 = Ph, R^3 = *t*-Bu: 78% de = 78%
ee (*trans*) = 88% ee (*cis*) = 84%
R^1 = H, R^2 = *p*-ClC$_6$H$_4$, R^3 = *t*-Bu: 80% de = 76%
ee (*trans*) = 85% ee (*cis*) = 82%
R^1 = H, R^2 = *p*-MeOC$_6$H$_4$, R^3 = *t*-Bu: 76% de = 82%
ee (*trans*) = 90% ee (*cis*) = 87%

Scheme 6.4 Bis(oxazolines)thiophene ligands for Cu-catalysed cyclopropanations of alkenes with EDA.

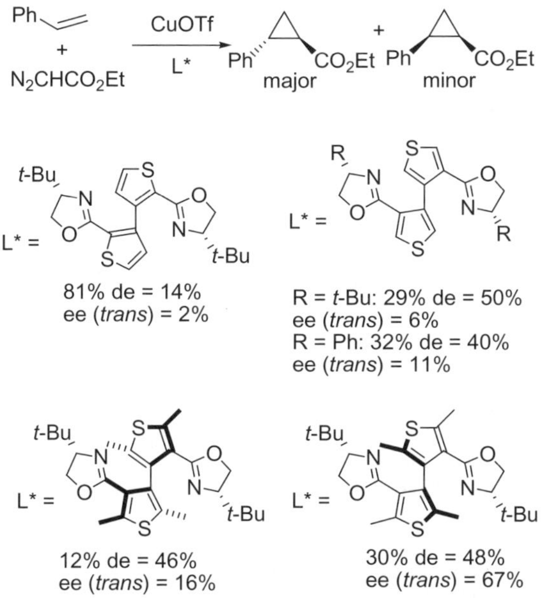

Scheme 6.5 Bis(oxazolines)bithiophene ligands for Cu-catalysed cyclopropanation of styrene with EDA.

the cyclopropanation, but the desired cyclopropane was generally isolated in modest yields. Moreover, these ligands provided low to moderate enantio-selectivities, since the best ligand gave an enantioselectivity of 67% ee. A similar result was obtained by using its corresponding biphenyl-based bis(oxazolines) analogue ligand, demonstrating that the difference in the electronic effect did not influence the styrene cyclopropanation. On the other hand, the steric properties of the chiral ligands seemed to play a crucial role in controlling both the stereoselection levels and the reaction rate, which was confirmed by the fact that the axially free chiral bithiophene ligands afforded nearly racemic cyclopropanes.

In 2004, ruthenium-catalysed asymmetric cyclopropanations of styrene derivatives with diazoesters were also performed by Masson *et al.*, using chiral 2,6-bis(thiazolines)pyridines.[30] These ligands were prepared from dithioesters and commercially available enantiopure 2-aminoalcohols. When the cyclo-propanation of styrene with diazoethylacetate was performed with these ligands in the presence of ruthenium, enantioselectivities of up to 85% ee were obtained (Scheme 6.6). The scope of this methodology was extended to various styrene derivatives and to isopropyl diazomethylphosphonate with good yields and enantioselectivities. The comparative evaluation of enantiocontrol for cyclopropanation of styrene with chiral ruthenium-bis(oxazolines), Ru-Pybox, and chiral ruthenium-bis(thiazolines), Ru-thia-Pybox, have shown many similarities with, in some cases, good enantiomeric excesses. The modification

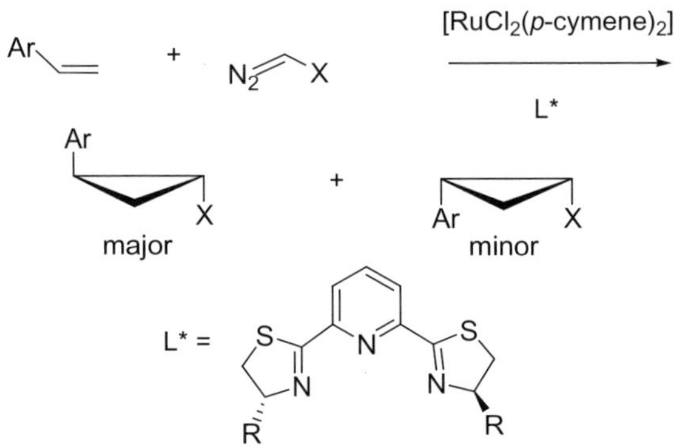

Ar = Ph, X = CO₂Et, R = Et: 87% *trans/cis* = 14
ee (*trans*) = 84% ee (*cis*) = 59%
Ar = Ph, X = CO₂Et, R = *i*-Pr: 89% *trans/cis* = 12
ee (*trans*) = 85% ee (*cis*) = 65%
Ar = p-CF₃C₆H₄, X = CO₂Et, R = Et: 83% *trans/cis* = 19
ee (*trans*) = 83% ee (*cis*) = 55%
Ar = p-Tol, X = CO₂Et, R = Et: 84% *trans/cis* = 10
ee (*trans*) = 81% ee (*cis*) = 46%
Ar = p-Tol, X = CO₂Et, R = *i*-Pr: 72% *trans/cis* = 10
ee (*trans*) = 81% ee (*cis*) = 54%
Ar = p-MeOC₆H₄, X = CO₂Et, R = Et: 96% *trans/cis* = 10
ee (*trans*) = 85% ee (*cis*) = 48%
Ar = Ph, X = PO(O*i*-Pr)₂, R = Et: 62% ee (*trans*) = 84%

Scheme 6.6 Ru-catalysed cyclopropanations of styrene derivatives with 2,6-bis (thiazolines)pyridine ligands.

of the five-membered ring did not greatly influence the enantioselectivity of the reaction.

6.2.3 Sulfoxide Ligands

In 2005, Nguyen *et al.* reported the first example of asymmetric cyclopropanation of olefins with EDA mediated by a combination of a (salen) ruthenium(II) catalyst and a catalytic amount of a chiral sulfoxide (Scheme 6.7).[31] These authors proposed that the mechanism explaining the asymmetric induction involved the axial coordination of the chiral sulfoxide to the ruthenium centre as the key induction step in the reaction stereoselectivity.

6.2.4 Sulfonamide Ligands

Since their first introduction by Brunner[32] and McKervey[33] as chiral catalysts for the asymmetric cyclopropanation of alkenes with diazo compounds, chiral di-rhodium tetra(*N*-arylsulfonylprolinates) complexes have been widely used by Davies,[34] in particular, in the context of these reactions. Therefore, the use of

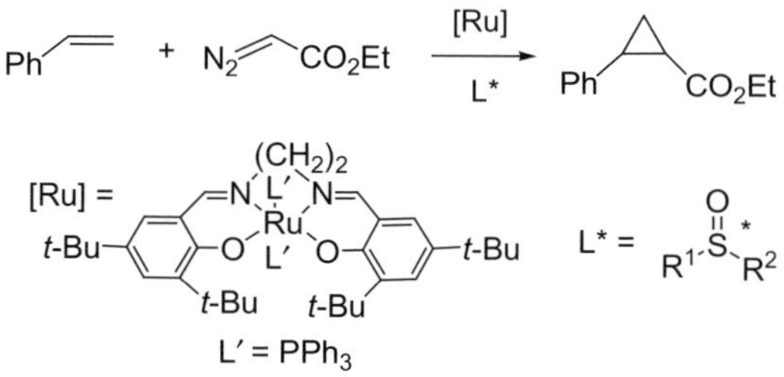

(R)-L*, R^1 = Me, R^2 = Ts: 92% *cis:trans* = 1:6.7
ee (*cis*) = 93% ee (*trans*) = 87%
(S)-L*, R^1 = Me, R^2 = Ts: 84% *cis:trans* = 1:7.5
ee (*cis*) = 50% ee (*trans*) = 42%
(R)-L*, R^1 = Bn, R^2 = Ts: 90% *cis:trans* = 1:7.3
ee (*cis*) = 56% ee (*trans*) = 45%
(S)-L*, R^1 = Me, R^2 = 2-Naph: 90% *cis:trans* = 1:6.7
ee (*cis*) = 41% ee (*trans*) = 45%

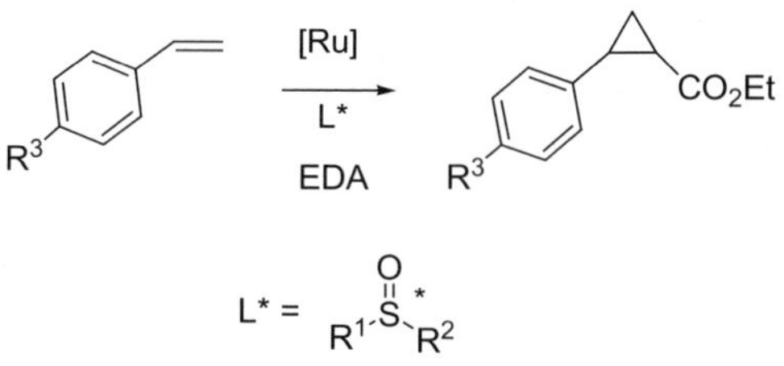

(R)-L*, R^1 = Me, R^2 = Ts, R^3 = OMe: 97%
cis:trans = 1:5.9 ee (*cis*) = 89% ee (*trans*) = 87%
(R)-L*, R^1 = Me, R^2 = Ts, R^3 = O*t*-Bu: 97%
cis:trans = 1:4.8 ee (*cis*) = 93% ee (*trans*) = 87%
(R)-L*, R^1 = Me, R^2 = Ts, R^3 = F: 86%
cis:trans = 1:5.8 ee (*cis*) = 80% ee (*trans*) = 83%
(R)-L*, R^1 = Me, R^2 = Ts, R^3 = CF$_3$: 98%
cis:trans = 1:7.8 ee (*cis*) = 79% ee (*trans*) = 86%

Scheme 6.7   Ru-catalysed cyclopropanations of alkenes with sulfoxide ligands.

these complexes has allowed the decomposition of various vinyldiazomethanes in the presence of alkenes to be achieved, providing the corresponding cyclopropanes in good yields and high diastereo- and enantioselectivities (Scheme 6.8).[35]

In addition, this methodology was extended to the cyclopropanation of a series of alkenes with phenyldiazomethane, giving rise to the corresponding cyclopropanes with high yields, diastereo- and enantioselectivities, as shown in Scheme 6.9.[36] It was shown that the diastereoselectivity of these reactions was not greatly altered by the type of rhodium carboxylate catalyst that was used.

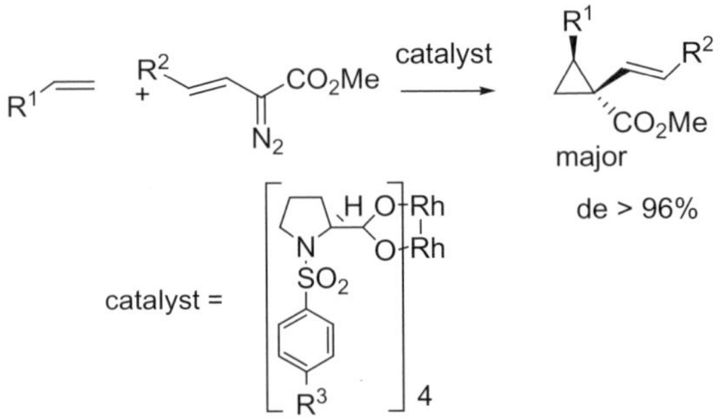

R^1 = R^2 = Ph, R^3 = *t*-Bu: 79% ee = 90%
R^1 = R^2 = Ph, R^3 = *n*-C$_{12}$H$_{25}$: 68% ee = 98%
R^1 = *p*-ClC$_6$H$_4$, R^2 = Ph, R^3 = *t*-Bu: 91% ee = 89%
R^1 = *p*-ClC$_6$H$_4$, R^2 = Ph, R^3 = *n*-C$_{12}$H$_{25}$: 70% ee > 97%
R^1 = *p*-MeOC$_6$H$_4$, R^2 = Ph, R^3 = *t*-Bu: 87% ee = 83%
R^1 = *p*-MeOC$_6$H$_4$, R^2 = Ph, R^3 = *n*-C$_{12}$H$_{25}$: 41% ee = 90%
R^1 = OAc, R^2 = Ph, R^3 = *t*-Bu: 40% ee = 76%
R^1 = OAc, R^2 = Ph, R^3 = *n*-C$_{12}$H$_{25}$: 26% ee = 95%
R^1 = OEt, R^2 = R^3 = Ph: 83% ee = 59%
R^1 = *n*-Bu, R^2 = R^3 = Ph: 63% ee > 90%
R^1 = Et, R^2 = R^3 = Ph: 65% ee > 95%
R^1 = *i*-Pr, R^2 = R^3 = Ph: 58% ee = 95%

Scheme 6.8   Rh-catalysed cyclopropanations of alkenes with vinyldiazomethanes in the presence of sulfonamide ligands.

Moreover, dirhodium tetrakis[(*S*)-(*N*-dodecylbenzenesulfonyl)prolinate] complex, (Rh$_2$[(*S*)-DOSP]$_4$), was found to catalyse the decomposition of methyl aryldiazoacetates in the presence of 1,1-diarylethylenes, resulting in the formation of the corresponding cyclopropanes with high enantioselectivities of up to 99% ee and moderate diastereoselectivities ($\leq$80% de).[37] The diastereoselectivity seemed to be influenced by the electronic properties of the substituents borne by the 1,1-diarylethylene. Thus, electron-donating groups, such as methoxy group, gave the best diastereoselectivities (Scheme 6.10). The synthetic utility of this cyclopropanation was demonstrated by the synthesis of both enantiomers of the cyclopropyl analogue of tamoxifen.

In 2000, Davies and Boebel demonstrated that alkynyldiazoacetates had a similar reactivity profile to aryl- and vinyldiazoacetates.[38] Therefore, in similar conditions to those employed above, the resulting carbenoids showed a high chemoselectivity and underwent highly stereoselective cyclopropanations (Scheme 6.11). These authors considered that the high stereoselectivity was due to a very demanding trajectory of the approach of the alkene to the carbenoid, with the approach occurring over the acceptor group. As the trajectory was based on an electronic argument, the same sense of enantioselectivity was observed from the alkynyldiazoacetates.

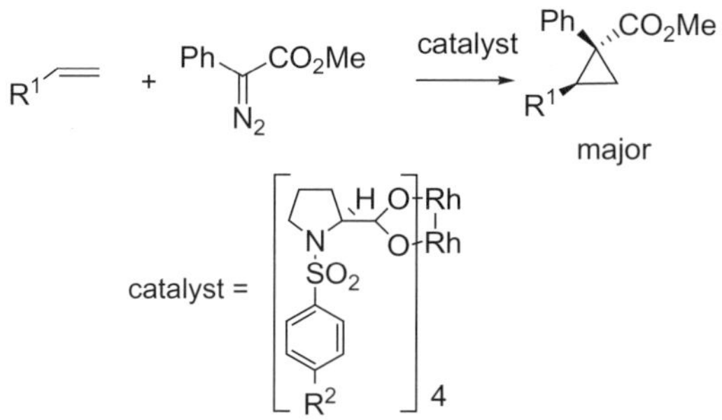

$R^1 = Ph$, $R^2 = t$-Bu: 90% de = 96% ee = 87%
$R^1 = Ph$, $R^2 = n$-$C_{12}H_{25}$: 79% de = 96% ee = 91%
$R^1 = p$-$MeOC_6H_4$, $R^2 = t$-Bu: 82% de = 96% ee = 88%
$R^1 = p$-ClC_6H_4, $R^2 = t$-Bu: 84% de = 96% ee = 85%
$R^1 = OEt$, $R^2 = n$-$C_{12}H_{25}$: 88% de = 94% ee = 66%
$R^1 = On$-Bu, $R^2 = n$-$C_{12}H_{25}$: 84% de = 94% ee = 64%
$R^1 = Et$, $R^2 = n$-$C_{12}H_{25}$: 86% de = 86% ee = 80%
$R^1 = n$-Bu, $R^2 = n$-$C_{12}H_{25}$: 86% de = 86% ee = 77%
$R^1 = i$-Pr, $R^2 = n$-$C_{12}H_{25}$: 63% de = 74% ee = 74%
$R^1 = CH_2OAc$, $R^2 = n$-$C_{12}H_{25}$: 85% de = 94% ee = 80%

Scheme 6.9 Rh-catalysed cyclopropanations of alkenes with phenyldiazoacetate in the presence of sulfonamide ligands.

As an application of the precedent methodologies, Davies *et al.* have developed asymmetric domino cyclopropanation Cope rearrangement reactions.[39] Therefore, when the decomposition of vinyl diazoacetates was performed in the presence of dienes and a similar catalyst ($Rh_2[(S)$-DOSP]$_4$) to that described above, it resulted in a direct and highly enantioselective method for the formation of *cis*-divinylcyclopropanes.[40] A combination of this process with a subsequent Cope rearrangement resulted in a highly enantioselective synthesis of a variety of cycloheptadienes containing multiple stereogenic centres (Scheme 6.12). The same methodology was extended to the enantioselective synthesis of various fused cycloheptadienes,[41] allowing the total synthesis of 5-*epi*-tremulenolide,[42] by using the intramolecular version of the process.[43]

In 1997, Davies *et al.* developed another type of sulfonamide ligands in which two of the prolinates were bridged together.[44] For such ligands, binding the dirhodium core forced the sulfonyl groups to adopt an α,β-arrangement leading to a D_2-symmetric complex. Therefore, a bridged dirhodium tetraprolinate, $Rh_2[(S)$-biTISP]$_2$, was able to catalyse the cyclopropanation of styrene with methyl phenyldiazoacetate with both high turnover number (92 000) and turnover frequency (4000 per h).[45] With a substrate/catalyst ratio of 100 000, for example, a 92% yield combined with a 85% ee were obtained for the cyclopropanation of styrene with methyl phenyldiazoacetate on a large scale (46 g). The same catalyst was applied to the stereoselective synthesis of cyclopropylphosphonates containing quaternary stereocentres by the reaction

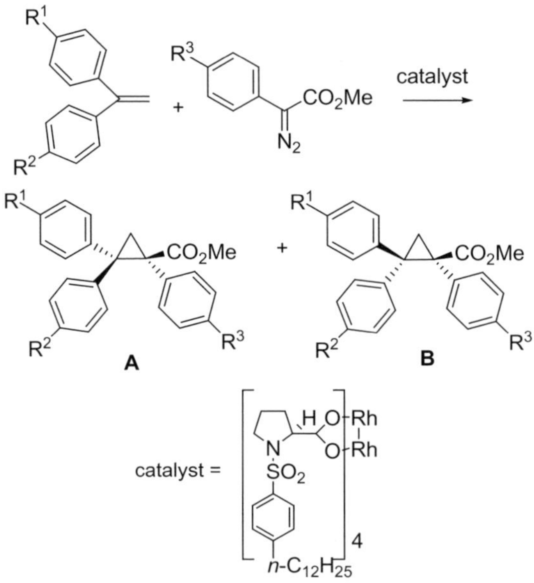

$R^1 = R^2 = R^3 = H$: 87% ee = 97%
$R^1 = NO_2$, $R^2 = R^3 = H$: 90% **A/B** = 55/45
ee (**A**) = 95% ee (**B**) = 91%
$R^1 = R^3 = H$, $R^2 = OMe$: 84% **A/B** = 87/13
ee (**A**) = 99% ee (**B**) = 96%
$R^1 = R^3 = H$, $R^2 = NHAc$: 75% **A/B** = 76/24
ee (**A**) = 92% ee (**B**) = 88%
$R^1 = NO_2$, $R^2 = OMe$, $R^3 = H$: 80% **A/B** = 88/12
ee (**A**) = 93% ee (**B**) = 74%
$R^1 = H$, $R^2 = OMe$, $R^3 = Cl$: 94% **A/B** = 89/11
ee (**A**) = 99% ee (**B**) = 95%
$R^1 = H$, $R^2 = R^3 = OMe$: 84% **A/B** = 90/10
ee (**A**) = 98% ee (**B**) = 94%

Scheme 6.10 Rh-catalysed cyclopropanations of 1,1-diarylethylenes with methyl aryldiazoacetates in the presence of sulfonamide ligand.

of dimethyl aryldiazomethylphosphonates, as shown in Scheme 6.13.[46] In addition, this catalyst could be immobilised on agitation in the presence of highly crosslinked polystyrene resins with a pyridine attachment.[47] The resulting heterogeneous complex was shown to be an effective catalyst for the cyclopropanation of styrene with methyl phenyldiazoacetate, providing the corresponding cyclopropane with an enantioselectivity of up to 88% ee.

In the same context, Müller *et al.* have reported the reaction of ethyl 3,3,3-trifluoro-2-diazopropionate with various olefins catalysed by dirhodium tetrakis[(*R*)-(*N*-dodecylbenzenesulfonyl)prolinate], $Rh_2[(R)$-DOSP]$_4$.[48] Yields and

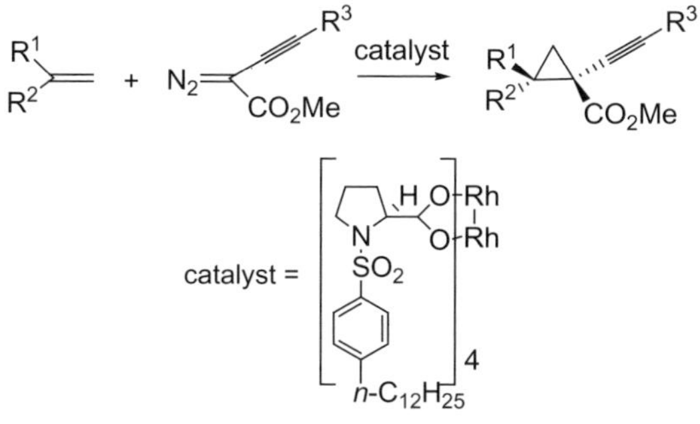

Scheme with reaction and catalyst structure.

$R^1 = H$, $R^2 = R^3 = Ph$: 68% de = 84% ee = 89%
$R^1 = H$, $R^2 = Ph$, $R^3 = Et$: 91% de = 98% ee = 56%
$R^1 = H$, $R^2 = Ph$, $R^3 = TMS$: 84% de = 88% ee = 65%
$R^1 = H$, $R^2 = On\text{-}Bu$, $R^3 = Ph$: 66% de > 94% ee = 87%
$R^1 = H$, $R^2 = OAc$, $R^3 = Ph$: 61% de > 94% ee = 95%
$R^1 = R^2 = R^3 = Ph$: 68% ee = 93%
$R^1 = R^2 = Ph$, $R^3 = Et$: 42% ee = 91%
$R^1 = R^2 = Ph$, $R^3 = TMS$: 68% ee = 93%

Scheme 6.11 Rh-catalysed cyclopropanations of alkenes with methyl alkynyldiazoacetates in the presence of sulfonamide ligand.

enantioselectivities of up to 72% and 40% ee, respectively, were obtained for the reaction between this diazo compound and 1,1-diphenylethylene, whereas the cyclopropanation of monosubstituted olefins with this diazo compound led to *cis/trans* mixtures of the corresponding cyclopropanes with a maximum enantioselectivity of 75% ee for 4-methoxystyrene. This catalyst was also used by Davies *et al.* to induce the decomposition of aryldiazoacetates in the presence of pyrroles or furans, resulting in the formation of mono- or bis(cyclopropanes) of the heterocycle, but with opposite enantioinduction (Scheme 6.14).[49] Indeed, an interesting effect in these reactions was that the enantioinduction was markedly influenced by the structure of the heterocyclic substrate. The cyclopropanation was considered to proceed in a concerted non-synchronous manner and depending upon which bond of the heterocycle initially interacted with the carbenoid, either face of the heterocycle could be attacked under the influence of the same chiral catalyst. The control was governed by a delicate interplay of steric and electronic influences. This methodology was applied to the total synthesis of a natural product, (+)-erogorgiaene, on the basis of the cyclopropanation of a dihydronaphthalene catalysed by $Rh_2[(S)\text{-}DOSP]_4$.[50]

On the other hand, Doyle *et al.* have developed methyl 2-oxoimidazolidine-4(*S*)-carboxylate ligands, containing 2-phenylcyclopropane attached at the 1-*N*-acyl site, such as the (4(*S*),2′(*R*),3′(*R*)-HMCPIM) ligand.[51] The resulting dirhodium complex led, for the cyclopropanation of styrene with EDA, to the corresponding cyclopropane with 68% ee and 59% yield, but with almost

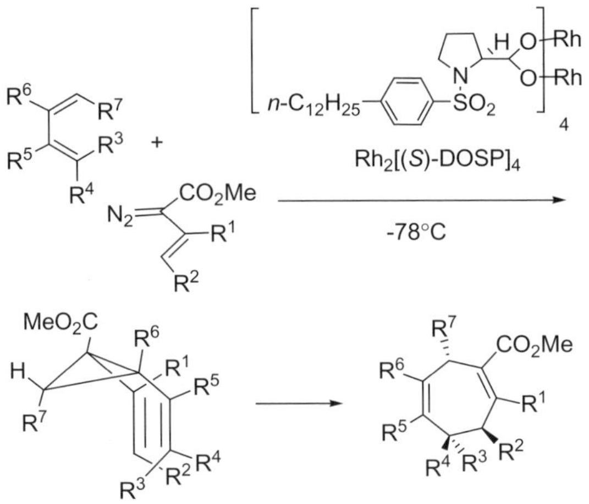

$R^1 = R^3 = R^5 = R^6 = R^7 = H$
$R^2 = Me, R^4 = Ph: 87\% \ ee = 98\%$
$R^1 = R^3 = R^5 = R^6 = R^7 = H$
$R^2 = R^4 = Ph: 83\% \ ee = 98\%$
$R^1 = R^3 = R^5 = R^6 = R^7 = H$
$R^2 = Me, R^4 = CH=CHPh: 84\% \ ee = 93\%$
$R^1 = R^3 = R^5 = R^6 = R^7 = H$
$R^2 = Me, R^4 = CH=CH_2: 56\% \ ee = 96\%$
$R^1 = R^4 = R^5 = R^6 = H, R^2 = CH=CHPh$
$R^3 = R^7 = CH_2: 90\% \ ee = 80\%$
$R^1 = R^4 = R^5 = R^6 = H, R^2 = CH=CH_2$
$R^3 = R^7 = CH_2: 92\% \ ee = 58\%$
$R^1 = R^4 = R^5 = R^6 = H, R^2 = Ph$
$R^3 = R^7 = CH_2: 93\% \ ee = 77\%$
$R^1 = OTBS, R^2 = R^4 = R^5 = R^6 = H$
$R^3 = R^7 = CH_2: 74\% \ ee = 97\%$

Scheme 6.12   Domino cyclopropanation Cope rearrangement reactions.

no diastereoselectivity. In 2004, Corey *et al.* reported the same reaction carried out in the presence of another dirhodium catalyst, $Rh_2(OAc)(DPTI)_3$, prepared from (R,R)-1,2-diphenylethylenediamine, affording the corresponding product with a high enantioselectivity, as shown in Scheme 6.15.[52] Surprisingly, the major diastereomer was the *cis* cyclopropane. In addition, a chiral fluorous complex, tetrakis-dirhodium(II)-(*S*)-*N*-(*n*-perfluorooctylsulfonyl) prolinate, has been prepared by Biffis *et al.* and subsequently used as catalyst in a homogeneous or fluorous biphasic fashion.[53] This catalyst displayed good chemo- and enantioselectivities in the cyclopropanation of styrene with ethyl phenyldiazoacetate, since 74% ee combined to 81% yield were obtained, when using perfluoro(methylcyclohexane) as the solvent.

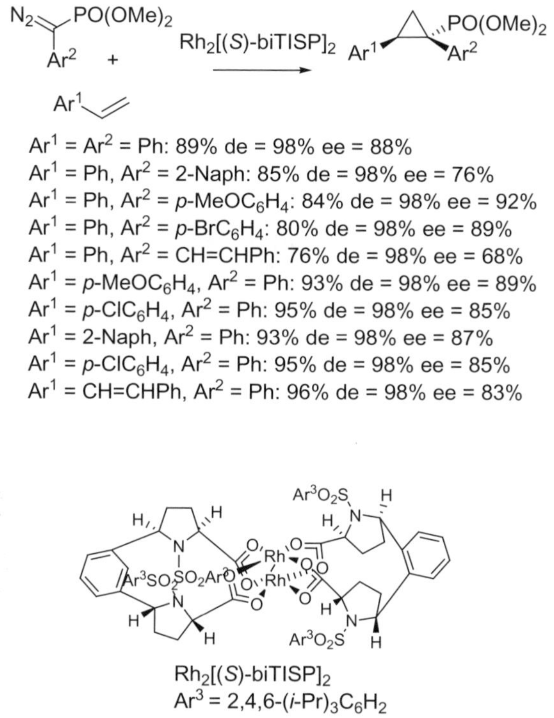

N2 PO(OMe)2
Ar2 +
Ar1
Rh2[(S)-biTISP]2
Ar1 PO(OMe)2 Ar2
Ar1 = Ar2 = Ph: 89% de = 98% ee = 88%
Ar1 = Ph, Ar2 = 2-Naph: 85% de = 98% ee = 76%
Ar1 = Ph, Ar2 = p-MeOC6H4: 84% de = 98% ee = 92%
Ar1 = Ph, Ar2 = p-BrC6H4: 80% de = 98% ee = 89%
Ar1 = Ph, Ar2 = CH=CHPh: 76% de = 98% ee = 68%
Ar1 = p-MeOC6H4, Ar2 = Ph: 93% de = 98% ee = 89%
Ar1 = p-ClC6H4, Ar2 = Ph: 95% de = 98% ee = 85%
Ar1 = 2-Naph, Ar2 = Ph: 93% de = 98% ee = 87%
Ar1 = p-ClC6H4, Ar2 = Ph: 95% de = 98% ee = 85%
Ar1 = CH=CHPh, Ar2 = Ph: 96% de = 98% ee = 83%
Ar3O2S
Ar3SO2SO2Ar3
Rh
Rh2[(S)-biTISP]2
Ar3 = 2,4,6-(i-Pr)3C6H2

Scheme 6.13 Cyclopropanations of alkenes catalysed by Rh_2[(*S*)-biTISP]$_2$.

In 2005, Doyle *et al.* reported an original sequence of two successive intramolecular cyclopropanations involving a bis(diazoacetates), using a sterically encumbered oxaimidazolidine carboxylate dirhodium(II) catalyst, Rh_2[(4*S*,*S*)-BSPIM]$_4$.[54] An excellent result, depicted in Scheme 6.16, was obtained resulting from a double diastereoselection.

In recent years, the variety of useful diazo substrates for asymmetric intramolecular cyclopropanation processes has really expanded. As another example, Charette and Wurz have reported the first example of an intramolecular cyclopropanation involving α-nitro-α-diazo carbonyl compounds.[55] This reaction, catalysed by Rh_2[(*S*)-DOSP]$_4$, led to the formation of nine-membered nitrocyclopropyl lactones in good yields and enantioselectivities with extremely high diastereoselectivities (Scheme 6.17). This novel methodology constituted an efficient entry into chiral functionalised macrocyclic-fused cyclopropane α-amino acids.

On the other hand, other chiral dirhodium(II) tetracarboxylate catalysts based on azetidine- and aziridine-2-carboxylic acids have been prepared by Zwanenburg *et al.* and submitted to the cyclopropanation of styrene with methyl phenyldiazoacetate.[56] However, the enantioselectivities obtained with these catalysts ($\leq$57% ee) did not match those obtained with the prolinate catalysts, as shown in Scheme 6.18.

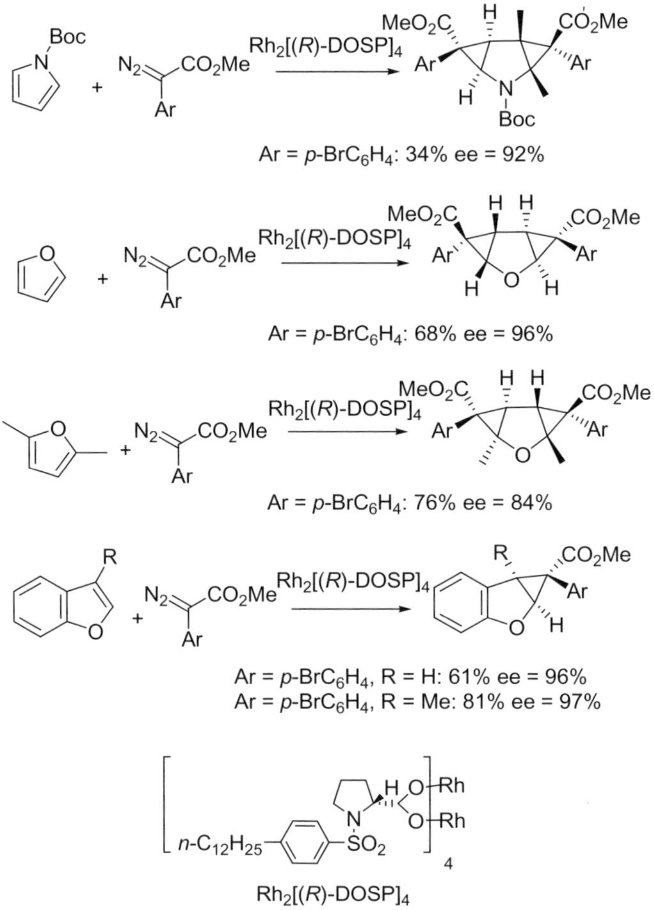

Ar = *p*-BrC$_6$H$_4$: 34% ee = 92%

Ar = *p*-BrC$_6$H$_4$: 68% ee = 96%

Ar = *p*-BrC$_6$H$_4$: 76% ee = 84%

Ar = *p*-BrC$_6$H$_4$, R = H: 61% ee = 96%
Ar = *p*-BrC$_6$H$_4$, R = Me: 81% ee = 97%

Rh$_2$[(*R*)-DOSP]$_4$

Scheme 6.14   Cyclopropanations of heterocyclic alkenes catalysed by Rh$_2$[(*R*)-DOSP]$_4$.

6.3   Simmons–Smith Cyclopropanation

More than 49 years ago, Simmons and Smith discovered that the reaction of alkenes with diiodomethane in the presence of activated zinc afforded cyclopropanes in high yields.[57] The reactive intermediate is an RZnCH$_2$I species. Several alternative methods for the preparation of this species have been developed such as those using ZnEt$_2$ or IZnCH$_2$I.[58] The features that have contributed to the popularity of the Simmons–Smith reagents include broad substrate generality, tolerance of a variety of functional groups, stereospecificity with respect to the alkene geometry, and the *syn*-directing/rate-enhancing effect observed with proximal oxygen atoms. These reactions have been shown to proceed through a "butterfly-type" transition structure.[59,60] In particular, the asymmetric cyclopropanation of various acyclic allylic

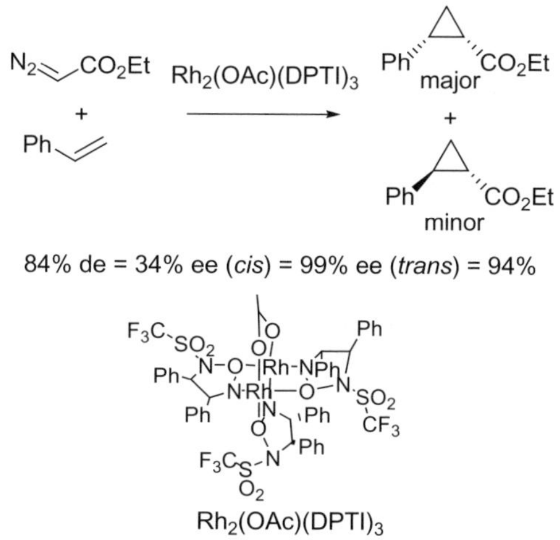

84% de = 34% ee (*cis*) = 99% ee (*trans*) = 94%

Rh₂(OAc)(DPTI)₃

Scheme 6.15 Cyclopropanation of styrene catalysed by Rh₂(OAc)(DPTI)₃.

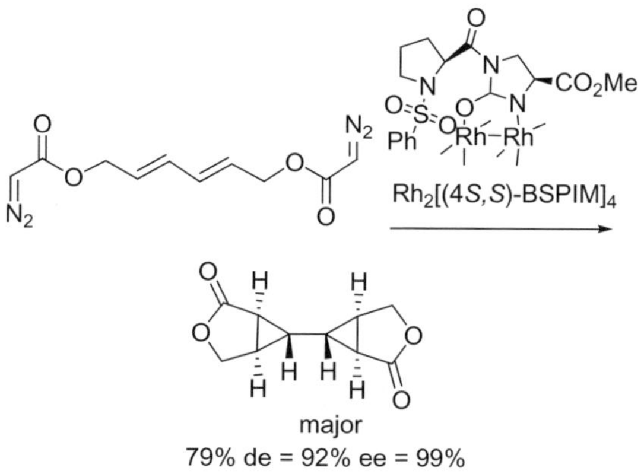

major
79% de = 92% ee = 99%

Scheme 6.16 Double intramolecular cyclopropanation of bis(diazoacetates) catalysed by Rh₂[(4S,S)-BSPIM]₄.

alcohols has been widely developed, using a heteroatom as a directing group.[61] The Simmons–Smith reaction with an allylic alcohol has distinct advantages over the reaction with a simple olefin in relation to the reaction rate and stereocontrol. Indeed, these reactions have been shown to be much faster than those of simple olefins (> 1000-fold) and, moreover, the reaction with a cyclic allylic alcohol took place in such a manner that the cyclopropane ring was formed on the same side as the hydroxyl group.[62] Numerous variants of the Simmons–Smith reagents have been explored, but little is known about their reaction pathways. An earlier mechanistic controversy focused on the

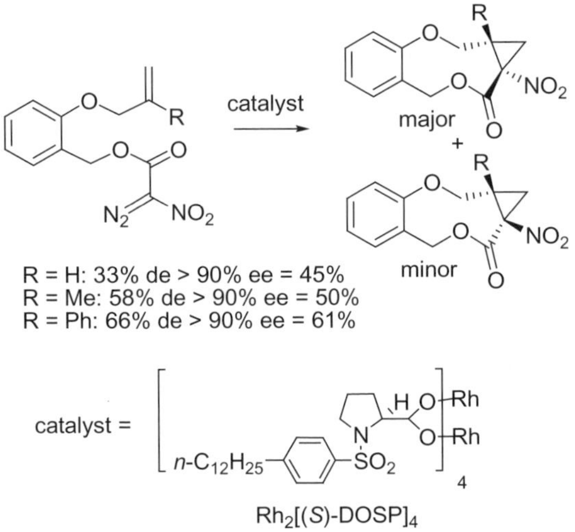

R = H: 33% de > 90% ee = 45%
R = Me: 58% de > 90% ee = 50%
R = Ph: 66% de > 90% ee = 61%

catalyst =

Rh₂[(S)-DOSP]₄

Scheme 6.17 Intramolecular cyclopropanations of α-nitro-α-diazo carbonyl compounds.

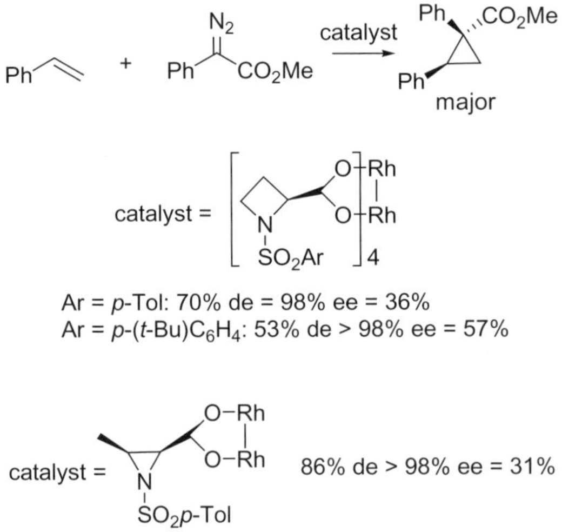

Ar = *p*-Tol: 70% de = 98% ee = 36%
Ar = *p*-(*t*-Bu)C₆H₄: 53% de > 98% ee = 57%

catalyst =   86% de > 98% ee = 31%

Scheme 6.18 Rh-catalysed cyclopropanation of styrene with sulfonamide ligands derived from aziridine- and azetidine-2-carboxylic acids.

dichotomy between a methylene-transfer mechanism and a carbometalation mechanism, in which the pseudotrigonal methylene group of iodomethylzinc iodide adds to an olefin π-bond and forms two new C–C bonds simultaneously, accompanying a 1,2-migration of the halide anion from the carbon atom to the zinc atom (path A of Scheme 6.19). There is experimental evidence for the

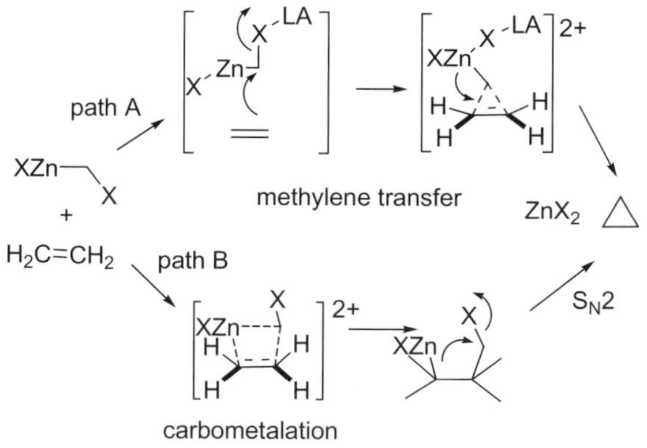

Scheme 6.19 Reaction pathways of the Simmons–Smith reaction.

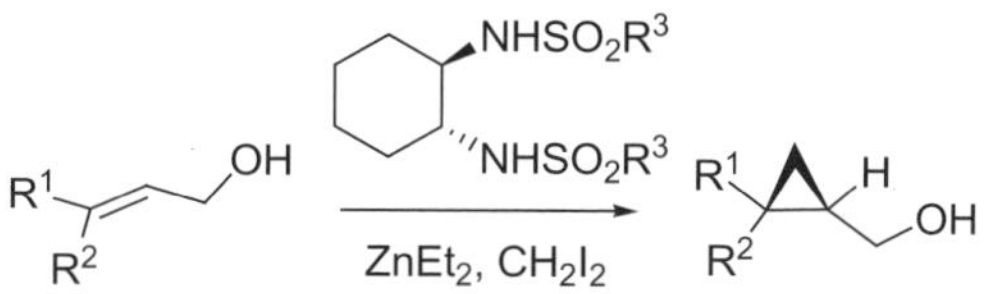

R^1 = Ph, R^2 = H, R^3 = p-NO$_2$C$_6$H$_4$: 82% ee = 76%
R^1 = H, R^2 = Ph, R^3 = o-NO$_2$C$_6$H$_4$: 82% ee = 51%
R^1 = H, R^2 = Ph, R^3 = p-NO$_2$C$_6$H$_4$: 71% ee = 75%
R^1 = BnCH$_2$, R^2 = H, R^3 = o-NO$_2$C$_6$H$_4$: 82% ee = 80%
R^1 = BnCH$_2$, R^2 = H, R^3 = p-NO$_2$C$_6$H$_4$: 100% ee = 82%
R^1 = BzOCH$_2$, R^2 = H, R^3 = p-NO$_2$C$_6$H$_4$: 70% ee = 36%
R^1 = H, R^2 = Sn(n-Bu)$_3$, R^3 = p-NO$_2$C$_6$H$_4$: 75% ee = 66%
R^1 = H, R^2 = SiMe$_2$Ph, R^3 = p-NO$_2$C$_6$H$_4$: 67% ee = 59%
R^1 = SiMe$_2$Ph, R^2 = H, R^3 = p-NO$_2$C$_6$H$_4$: 83% ee = 81%
R^1 = Sn(n-Bu)$_3$, R^2 = H, R^3 = p-NO$_2$C$_6$H$_4$: 94% ee = 86%
R^1 = Ph, R^2 = H, R^3 = Me: 91% ee = 80%
R^1 = H, R^2 = Ph, R^3 = Me: 81% ee = 81%
R^1 = BnCH$_2$, R^2 = H, R^3 = Me: 89% ee = 81%
R^1 = H, R^2 = BnCH$_2$, R^3 = Me: 93% ee = 72%
R^1 = Ph, R^2 = R^3 = Me: 91% ee = 73%
R^1 = R^3 = Me, R^2 = Ph: 88% ee = 81%
R^1 = BnCH$_2$, R^2 = R^3 = Me: 94% ee = 79%
R^1 = R^3 = Me, R^2 = BnCH$_2$: 98% ee = 66%
R^1 = Ph, R^2 = H, R^3 = n-Bu: 98% ee = 84%

Scheme 6.20 Simmons–Smith cyclopropanations of allylic alcohols with cyclohexanediamine-derived bis(sulfonamides) ligands.

Simmons–Smith reagent that contradicts path B, and path A has therefore been widely believed to represent the experimental reality. For lithium carbenoids, on the other hand, the alternative carbometalation/cyclisation pathway has received experimental support.[63] Actually, the factors that determine the reaction pathways of metal carbenoid addition to olefins are, therefore, still open to question.[64] In 2003, Nakamura *et al.* studied the reaction pathways of cyclopropanation using the Simmons–Smith reagent by means of the B3LYP hybrid density functional method, confirming that the methylene-transfer pathway was the favoured reaction course.[65] It took place through two stages, a S_N2-like displacement of the leaving group by the olefin, followed by a cleavage of the C–Zn bond to give the cyclopropane ring (Scheme 6.19).

Several catalytic systems have been reported for the enantioselective Simmons–Smith cyclopropanation reaction and, among these, only a few could be used in catalytic amounts. Chiral bis(sulfonamides) derived from cyclo-hexanediamine have been successfully employed as promoters of the enantio-selective Simmons–Smith cyclopropanation of a series of allylic alcohols.[66] Excellent results in terms of both yield and stereoselectivity were obtained even with disubstituted allylic alcohols, as shown in Scheme 6.20. Moreover, this methodology could be applied to the cyclopropanation of stannyl and silyl-substituted allylic alcohols, providing an entry to the enantioselective route to stannyl- and silyl-substituted cyclopropanes of potential synthetic inter-mediates. On the other hand, it must be noted that the presence of a methyl substituent at the 2-position of the allylic alcohol was not well tolerated and led to slow reactions and poor enantioselectivities (ee ≤ 50% ee).[66f]

In 1998, Kurt and Halm reported the preparation of resin-based bis(sulfo-namides) ligands in order to extend the precedent methodology to the solid phase.[67] Therefore, the solid-phase catalyst depicted in Scheme 6.21 was found to be able to mediate the Simmons–Smith cyclopropanation of cinnamyl alcohol with an enantioselectivity of 65% ee.

Other bis(sulfonamides) ligands based on more flexible diamines have been investigated by Denmark *et al.* as promoters for the enantioselective Simmons–Smith cyclopropanation of cinnamyl alcohol.[68] This study has revealed a

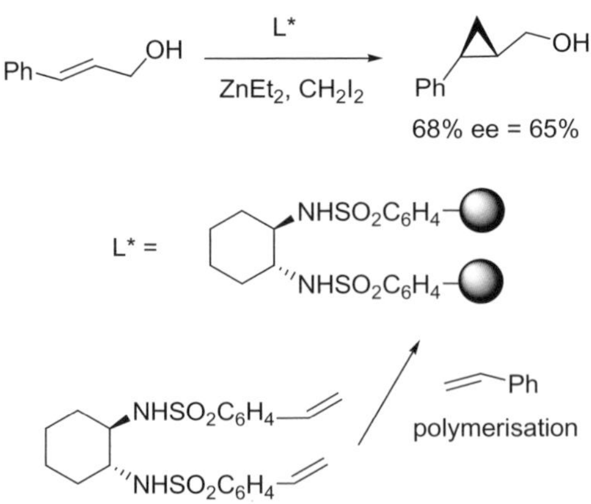

Scheme 6.21   Simmons–Smith cyclopropanation of cinnamyl alcohol with resin-based bis(sulfonamides) ligand.

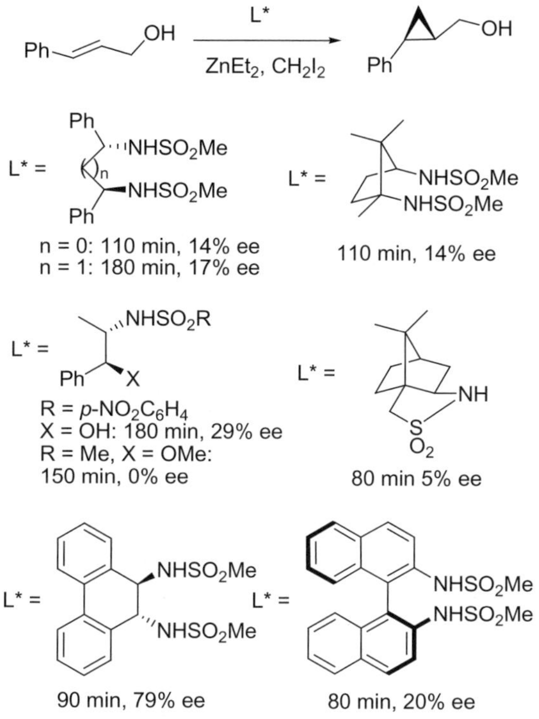

Scheme 6.22 Simmons–Smith cyclopropanation of cinnamyl alcohol with various bis(sulfonamides) ligands.

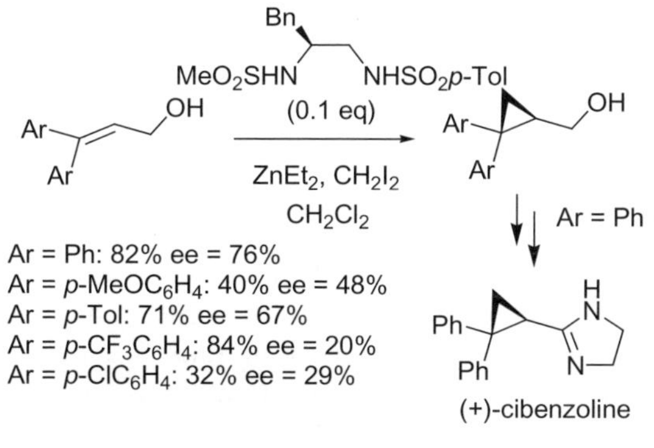

Scheme 6.23 Simmons–Smith cyclopropanations of allylic alcohols with (*S*)-phenylalanine-derived bis(sulfonamides) ligand.

marked sensitivity to the spatial relationship of the amine groups. Indeed, neither of the flexible bis(sulfonamides) ligands that were tested led to respectable levels of enantioselectivity (Scheme 6.22), demonstrating the harmful role of the skeletal flexibility.

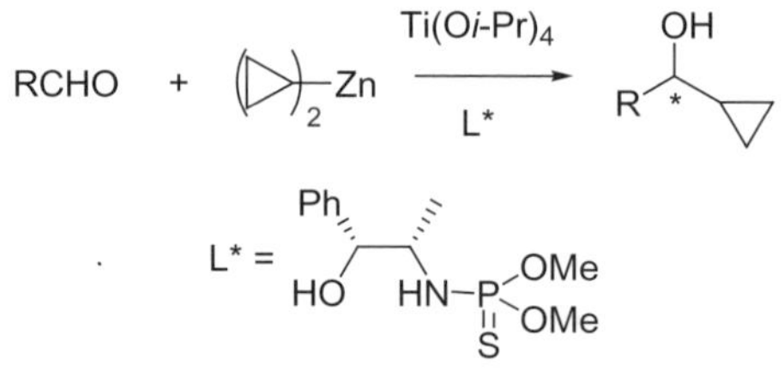

R = Ph: 93% ee = 96%
R = *p*-MeOC_6H_4: 99% ee = 97%
R = *p*-ClC_6H_4: 85% ee = 91%
R = 1-Naph: 89% ee = 96%
R = (*E*)-Ph-CH=CH: 94% ee = 64%

Scheme 6.24 Cyclopropanations of aldehydes with Cy_2Zn in the presence of thiophosphoramidate ligand.

In 2006, Imai *et al.* reported the syntheses of (+)-cibenzoline, an antiarrhythmic agent, and its analogues via enantioselective Simmons–Smith cyclopropanations using another chiral bis(sulfonamides) ligand derived from an α-amino acid such as (*S*)-phenylalanine.[69] A catalytic amount of this catalyst was, therefore, successfully employed to cyclopropanate a range of 3,3-diaryl-2-propen-1-ols in the presence of Et_2Zn and CH_2I_2, providing the corresponding cyclopropylmethanols with moderate to good enantioselectivities (Scheme 6.23). The chiral cyclopropane derived from 3,3-diphenyl-2-propen-1-ol was further converted into the expected (+)-cibenzoline.

In another context, Soai *et al.* have described the enantioselective cyclopropanation of various aldehydes using dicyclopropylzinc in the presence of a catalytic amount of a chiral thiophosphoramidate ligand derived from norephedrine and Ti(O*i*-Pr)$_4$, providing the corresponding cyclopropyl alkanols with high yields and enantioselectivities of up to 97% ee (Scheme 6.24).[70]

6.4 Conclusions

The most efficient method to achieve the asymmetric catalytic cyclopropanation of alkenes in the presence of chiral sulfur-containing ligands concerns the transition-metal-catalysed decomposition of diazoalkanes performed in the presence of alkenes. Since the first excellent results reported independently by Kobayashi and Davies using chiral bis(sulfonamides) ligands derived from chiral cyclohexanediamine for the enantioselective cyclopropanation of alkenes with diazo compounds, the best recent results have been obtained by Gao's group, who involved chiral bis(oxazolines)thiophene ligands in the presence of ruthenium. The second methodology based on the asymmetric Simmons–Smith cyclopropanation of allylic alcohols has been much less studied in the presence of chiral sulfur-containing ligands, apart from cyclohexanediamine-derived bis(sulfonamides) ligands.

References

1. (a) S. Patai and Z. Rappoport, in *The Chemistry of the Cyclopropyl Group*, Wiley and Sons, New York, 1987; (b) *Small Ring Compounds in Organic Synthesis VI*, A. de Meijere, ed., Springer: Berlin, Germany, 2000, Vol. 207.
2. M. Rubin, M. Rubina and V. Gevorgyan, *Chem. Rev.*, 2007, **107**, 3117–3179.
3. (a) J. Salaün, *Top. Curr. Chem.*, 2000, **207**, 1–67; (b) R. Faust, *Angew. Chem., Int. Ed. Engl.*, 2001, **40**, 2251–2253; (c) F. Gnad and O. Reiser, *Chem. Rev.*, 2003, **103**, 1603–1623.
4. (a) H. Lebel, J.-F. Marcoux, C. Molinaro and A. B. Charette, *Chem. Rev.*, 2003, **103**, 977–1050; (b) W. A. Donaldson, *Tetrahedron*, 2001, **57**, 8589–8627; (c) L. A. Wessjohan, W. Brandt and T. Thiemann, *Chem. Rev.*, 2003, **103**, 1625–1647; (d) H.-U. Reissig, *Angew. Chem., Int. Ed. Engl.*, 1996, **35**, 971–973; (e) J. Salaün, *Chem. Rev.*, 1989, **89**, 1247–1270; (f) R. C. Hartley and S. T. Caldwell, *J. Chem. Soc., Perkin Trans. I*, 2000, 477–502; (g) M. Lautens, W. Klute and W. Tam, *Chem. Rev.*, 1996, **96**, 49–92; (h) A. H. Hoveyda, D. A. Evans and G. C. Fu, *Chem. Rev.*, 1993, **93**, 1307–1370; (i) A. de Meijere, S. I. Kozhushkov and A. F. Khlebnivow, *Top. Curr. Chem.*, 2000, **207**, 89–147; (j) A. de Meijere, S. I. Kozhushkov and L. P. Hadjiarapoglou, *Top. Curr. Chem.*, 2000, **207**, 149–227; (k) I. L. Lysenko, H. G. Lee and J. K. Cha, *Org. Lett.*, 2006, **8**, 2671–2673; (l) H. Pellissier, *Tetrahedron*, 2008, **64**, 7041–7095; (m) P. Garcia, D. D. Martin, A. B. Anton, N. M. Garrido, I. S. Marcos, P. Basabe and J. G. Urones, *Mini-Rev. Org. Chem.*, 2006, **3**, 291–314.
5. H. W. Liu and C. T. Walsh, in *The Chemistry of The Cyclopropyl Group*, ed. Z. Rappoport, John Wiley, New York, 1997, p. 959.
6. Y. Nishii, N. Maruyama, K. Wakasugi and Y. Tanabe, *Bioorg. Med. Chem.*, 2001, **9**, 33–39.
7. R. Csuk, M. J. Schabel and Y. von Scholz, *Tetrahedron: Asymmetry*, 1996, **7**, 3505–3512.
8. J. S. Kumar, S. Roy and A. Datta, *Bioorg. Med. Chem. Lett.*, 1999, **9**, 513–514.
9. X. Zhang, K. Hodgetts, S. Rachwal, H. Zhao, J. W. F. Wasley, K. Craven, R. Brodbeck, A. Kieltyka, D. Hoffman, M. D. Bacolod, B. Girard, J. Tran and A. Thurkauf, *J. Med. Chem.*, 2000, **43**, 3923–3932.
10. (a) R. Csuk, A. Kern and K. Z. Mohr, *Naturforsch*, 1999, 1463–1468; (b) M. Högberg, P. Engelhardt, L. Vrang and H. Zhang, *Bioorg. Med. Chem. Lett.*, 2000, **10**, 265–268.
11. D. K. Mohapatra and A. Datta, *J. Org. Chem.*, 1998, **63**, 642–646.
12. (a) H. -U. Reissig and R. Zimmer, *Chem. Rev.*, 2003, **103**, 1151–1196; (b) J. Pietruszka, *Chem. Rev.*, 2003, **103**, 1051–1070; (c) M. Yu and B. L. Pagenkopf, *Tetrahedron*, 2005, **61**, 321–347.
13. H. N. C. Wong, M. -Y. Hon, C. -W. Tse, Y. -C. Yip, J. Tanko and T. Hudlicky, *Chem. Rev.*, 1989, **89**, 165–198.
14. D. Arlt, M. Jautelat and R. Lantzsch, *Angew. Chem., Int. Ed. Engl.*, 1981, **20**, 703–722.

15. (a) H. M. L. Davies and E. Antoulinakis, *Org. React.*, 2001, **57**, 1–326; (b) T. Rovis and D. A. Evans, *Prog. Inorg. Chem.*, 2001, **50**, 1–150; (c) H. Nishiyama, *Enantiomer*, 1999, **4**, 569–574; (d) V. K. Singh, A. Datta Gupta and G. Sekar, *Synthesis*, 1997, 137–149; (e) M. P. Doyle and D. C. Forbes, *Chem. Rev.*, 1998, **98**, 911–935.

16. G. Boche and J. C. W. Lohrenz, *Chem. Rev.*, 2001, **101**, 697–756.

17. M. P. Doyle, M. A. McKervey and T. Ye, in *Modern Catalytic Methods for Organic Synthesis with Diazo Compounds*, John Wiley and Sons, New York, 1998.

18. M. P. Doyle and M. Protopopova, *Tetrahedron*, 1998, **54**, 7919–7946.

19. W. Kirmse, *Angew. Chem., Int. Ed. Engl.*, 2003, **42**, 1088–1093.

20. C. A. Merlic and A. L. Zechman, *Synthesis*, 2003, **8**, 1137–1156.

21. (a) A. Pfaltz, in *Comprehensive Asymmetric Catalysis*, ed. E. N. Jacobsen, A. Pfaltz, H. Yamamoto, Springer-Verlag, Berlin, 1999, **Vol. II**, p. 513–538; (b) A. B. Charrette and H. Lebel, in *Comprehensive Asymmetric Catalysis*, ed. E. N. Jacobsen, A. Pfaltz, H. Yamamoto, Springer-Verlag, Berlin, 1999, **Vol. II**, p. 581–603.

22. G. Desimoni, G. Faita and K. A. Jorgensen, *Chem. Rev.*, 2006, **106**, 3561–3651.

23. H. A. McManus and P. J. Guiry, *Chem. Rev.*, 2004, **104**, 4151–4202.

24. V. K. Aggarwal, H. W. Smith, G. Hynd, R. V. H. Jones, R. Fieldhouse and S. E. Spey, *J. Chem. Soc., Perkin Trans. I*, 2000, 3267–3276.

25. V. K. Aggarwal, E. Alonso, G. Fang, M. Ferrara, G. Hynd and M. Porcelloni, *Angew. Chem., Int. Ed. Engl.*, 2001, **40**, 1433–1436.

26. (a) R. Annunziata, M. Benaglia, M. Cinquini, F. Cozzi and G. Pozzi, *Eur. J. Org. Chem.*, 2003, 1191–1197; (b) I. Atodiresei, I. Schiffers and C. Bolm, *Tetrahedron: Asymmetry*, 2006, **17**, 620–633; (c) K. Khanbabaee, S. Basceken and U. Flörke, *Tetrahedron: Asymmetry*, 2006, **17**, 2804–2812; (d) K. Khanbabaee, S. Basceken and U. Flörke, *Eur. J. Org. Chem.*, 2007, 831–837; (e) Z.-Y. Wang, D.-M. Du and J.-X. Xu, *Synth. Commun.*, 2005, **35**, 299–313; (f) D.-M. Du, B. Fu and W.-T. Hua, *Tetrahedron*, 2003, **59**, 1933–1938.

27. M. Z. Gao, D. Kong, A. Clearfield and R. A. Zingaro, *Tetrahedron Lett.*, 2004, **45**, 5649–5652.

28. M. Z. Gao, B. Wang, D. Kong, R. A. Zingaro, A. Clearfield and Z. L. Xu, *Synth. Commun.*, 2005, **35**, 2665–2673.

29. M. Benaglia, T. Benincori, P. Mussini, T. Pilati, S. Rizzo and F. Sannicolo, *J. Org. Chem.*, 2005, **70**, 7488–7495.

30. P. Le Maux, I. Abrunhosa, M. Berchel, G. Simmoneaux, M. Gulea and S. Masson, *Tetrahedron: Asymmetry*, 2004, **15**, 2569–2573.

31. J. A. Miller, B. A. Gross, M. A. Zhuravel, W. Jin and S. T. Nguyen, *Angew. Chem., Int. Ed. Engl.*, 2005, **44**, 3885–3889.

32. H. Brunner, H. Kluschanzoff and K. Wutz, *Bull. Soc. Chim. Belg.*, 1989, **98**, 63–72.

33. (a) M. Kennedy, M. A. McKervey, A. R. Maguire and G. H. P. Ross, *J. Chem. Soc., Chem. Commun.*, 1990, 361–362; (b) T. Ye, M. A.

McKervey, B. D. Brandes and M. P. Doyle, *Tetrahedron Lett.*, 1994, **35**, 7269–7272; (c) T. Ye, F. C. Garcia and M. A. McKervey, *J. Chem. Soc., Perkin Trans I*, 1995, 1373–1379; (d) F. C. Garcia, M. A. McKervey and T. Ye, *J. Chem. Soc., Chem. Commun.*, 1996, 1465–1466.

34. H. M. L. Davies, *Eur. J. Org. Chem.*, 1999, 2459–2469.
35. (a) H. M. L. Davies and D. K. Hutcheson, *Tetrahedron Lett.*, 1993, **34**, 7243–7246; (b) H. M. L. Davies, P. R. Bruzinski, D. H. Lake, N. Kong and M. J. Fall, *J. Am. Chem. Soc.*, 1996, **118**, 6897–6907; (c) H. M. L. Davies and D. K. Hutcheson, *Tetrahedron Lett.*, 1993, **34**, 7243–7246; (d) K. Yoshikawa and K. Achiwa, *Chem. Pharm. Bull.*, 1995, **43**, 2048–2053; (e) H. M. L. Davies, P. R. Bruzinski and M. J. Fall, *Tetrahedron Lett.*, 1996, **37**, 4133–4136.
36. (a) H. M. L. Davies and L. Rusiniak, *Tetrahedron Lett.*, 1998, **39**, 8811–8812; (b) M. P. Doyle, Q. -L. Zhou, C. Charnsangavej and M. A. Longoria, *Tetrahedron Lett.*, 1996, **37**, 4129–4132; (c) H. M. L. Davies, P. R. Bruzinski and M. J. Fall, *Tetrahedron Lett.*, 1996, **37**, 4133–4136.
37. H. M. L. Davies, T. Nagashima and J. L. Klino III, *Org. Lett.*, 2000, **2**, 823–826.
38. H. M. L. Davies and T. A. Boebel, *Tetrahedron Lett.*, 2000, **41**, 8189–8192.
39. H. M. L. Davies, Z. -Q. Peng and J. H. Houser, *Tetrahedron Lett.*, 1994, **35**, 8939–8942.
40. H. M. L. Davies, D. G. Stafford, B. D. Doan and J. H. Houser, *J. Am. Chem. Soc.*, 1998, **120**, 3326–3331.
41. H. M. L. Davies and M. L. Hodges, *J. Org. Chem.*, 2002, **67**, 5683–5689.
42. H. M. L. Davies and B. D. Doan, *J. Org. Chem.*, 1999, **64**, 8501–8508.
43. H. M. L. Davies and B. D. Doan, *Tetrahedron Lett.*, 1996, **37**, 3967–3970.
44. (a) H. M. L. Davies and N. Kong, *Tetrahedron Lett.*, 1997, **38**, 4203–4206; (b) H. M. L. Davies and S. A. Panaro, *Tetrahedron Lett.*, 1999, **40**, 5287–5290.
45. H. M. L. Davies and C. Venkataramani, *Org. Lett.*, 2003, **5**, 1403–1406.
46. H. M. L. Davies and G. H. Lee, *Org. Lett.*, 2004, **6**, 2117–2120.
47. H. M. L. Davies, A. M. Walji and T. Nagashima, *J. Am. Chem. Soc.*, 2004, **126**, 4271–4280.
48. P. Müller, S. Grass, S. P. Shahi and G. Bernardinelli, *Tetrahedron*, 2004, **60**, 4755–4763.
49. S. J. Hedley, D. L. Ventura, P. M. Dominiak, C. L. Nygren and H. M. L. Davies, *J. Org. Chem.*, 2006, **71**, 5349–5356.
50. H. M. L. Davies, X. Dai and M. S. Long, *J. Am. Chem. Soc.*, 2006, **128**, 2485–2490.
51. M. P. Doyle, J. P. Morgan, J. C. Fettinger, P. Y. Zavalij, J. T. Colyer, D. J. Timmons and M. D. Carducci, *J. Org. Chem.*, 2005, **70**, 5291–5301.
52. Y. Lou, M. Horikawa, R. A. Kloster, N. A. Hawryluk and E. J. Corey, *J. Am. Chem. Soc.*, 2004, **126**, 8916–8918.
53. A. Biffis, M. Braga, S. Cadamuro, C. Tubaro and M. Basato, *Org. Lett.*, 2005, **7**, 1841–1844.

54. M. P. Doyle, Y. Wang, P. Ghorbani and E. Bappert, *Org. Lett.*, 2005, **7**, 5035–5038.
55. A. B. Charette and R. Wurz, *J. Mol. Catal.*, 2003, **196**, 83–91.
56. W. A. J. Starmans, L. Thijs and B. Zwanenburg, *Tetrahedron*, 1998, **54**, 629–636.
57. (a) H. E. Simmons and R. D. Smith, *J. Am. Chem. Soc.*, 1958, **80**, 5323–5324; (b) H. E. Simmons and R. D. Smith, *J. Am. Chem. Soc.*, 1959, **81**, 4256–4264.
58. (a) E. Erdik, *Tetrahedron*, 1987, **43**, 2203–2212; (b) K. Takai, T. Kakiuchi and K. Utimoto, *J. Org. Chem.*, 1994, **59**, 2671–2673.
59. A. H. Hoveyda, D. A. Evans and G. C. Fu, *Chem. Rev.*, 1993, **93**, 1307–1370.
60. (a) W. -H. Fang, D. L. Phillips, D. -Q. Wang and Y. -L. Li, *J. Org. Chem.*, 2002, **67**, 154–160; (b) H. Hermann, J. C. W. Lohrenz, A. Kühn and G. Boche, *Tetrahedron*, 2000, **56**, 4109–4115.
61. A. B. Charette and J. -F. Marcoux, *Synlett*, 1995, 1197–1207.
62. (a) J. H. -H. Chan and B. Rickborn, *J. Am. Chem. Soc.*, 1968, **90**, 6406–6411; (b) J. A. Staroscik and B. Rickborn, *J. Org. Chem.*, 1972, **37**, 738–740.
63. H. C. Stiasny and R. W. Hoffmann, *Chem. Eur. J.*, 1995, **1**, 619–624.
64. G. Boche and J. C. W. Lohrenz, *Chem. Rev.*, 2001, **101**, 697–756.
65. M. Nakamura, A. Hirai and E. Nakamura, *J. Am. Chem. Soc.*, 2003, **125**, 2341–2350.
66. (a) H. Takahashi, M. Yoshioka, M. Ohno and S. Kobayashi, *Tetrahedron Lett.*, 1992, **33**, 2575–2578; (b) N. Imai, K. Sakamoto, H. Takahashi and S. Kobayashi, *Tetrahedron Lett.*, 1994, **35**, 7045–7048; (c) N. Imai, H. Takahashi and S. Kobayashi, *Chem. Lett.*, 1994, **23**, 177–180; (d) H. Takahashi, M. Yoshioka, M. Shibasaki, M. Ohno, N. Imai and S. Kobayashi, *Tetrahedron*, 1995, **51**, 12013–12026; (e) S. E. Denmark, B. L. Christenson, D. M. Coe and S. P. O'Connor, *Tetrahedron Lett.*, 1995, **36**, 2215–2218; (f) S. E. Denmark and S. P. O'Connor, *J. Org. Chem.*, 1997, **62**, 584–594; (g) S. E. Denmark and S. P. O'Connor, *J. Org. Chem.*, 1997, **62**, 3390–3401; (h) S. E. Denmark, S. P. O'Connor and S. R. Wilson, *Angew. Chem., Int. Ed. Engl.*, 1998, **37**, 1149–1151.
67. C. Halm and M. J. Kurth, *Angew. Chem., Int. Ed. Engl.*, 1998, **37**, 510–512.
68. S. E. Denmark, B. L. Christenson and S. P. O'Connor, *Tetrahedron Lett.*, 1995, **36**, 2219–2222.
69. T. Miura, Y. Murakami and N. Imai, *Tetrahedron: Asymmetry*, 2006, **17**, 3067–3069.
70. T. Shibata, H. Tabira and K. Soai, *J. Chem. Soc., Perkin Trans. 1*, 1998, 177–178.

CHAPTER 7
Heck-Type Reactions

The Heck reaction[1] consists in the Pd(0)-catalysed coupling of alkenes with an aryl or alkenyl halide or triflate in the presence of a base to form a substituted alkene (Scheme 7.1). The reaction is performed in the presence of an organopalladium catalyst. The halide or triflate is an aryl or a vinyl compound and the alkene contains at least one proton.

This reaction is an efficient method of carbon–carbon bond formation that tolerates various functional groups, such as ketones, esters, amides, ethers or heterocyclic rings. The particular ease of scale-up and the tolerance to water constitute other advantages of this reaction. There are many examples of the use of the Heck reaction in organic syntheses.[2] Another great advantage of the Heck reaction is that the substrate is not limited to activated alkenes but can be a simple olefin. Compared to other transition-metal-catalysed reactions, however, the effect of the ligands has been investigated less for this process.[3] The interest in the Heck reaction has recently increased dramatically. Perhaps the most significant progress to date is the development of numerous enantioselective variants of this reaction,[3b,4] which have been reported only quite recently. Indeed, the first reports of successful examples of the asymmetric Heck reaction were independently published in 1989 by Shibasaki *et al.* and Overman *et al.*, using BINAP and DIOP ligands, respectively.[5] The Heck reaction is, however, often limited by a low activity with certain classes of substrates and also by the formation of byproducts. For example, difficulties are encountered in controlling the regioselectivity of the reaction involving unsymmetrical alkenes. Indeed, there is a need for the design of new ligands to widen the scope and applications of enantioselective Heck reactions. Only a few examples in the recent literature deal with the enantioselective Heck reaction using chiral sulfur-containing ligands, such as thiophene-containing ligands, sulfur-containing ferrocenyl ligands, phosphorylated sulfides and sulfonamides. The most promising results have been obtained by Sannicolo *et al.* with diphosphines based on thiophene or benzothiophene backbones.[6] In this study,

RSC Catalysis Series No. 2
Chiral Sulfur Ligands: Asymmetric Catalysis
By Hélène Pellissier
© Hélène Pellissier 2009
Published by the Royal Society of Chemistry, www.rsc.org

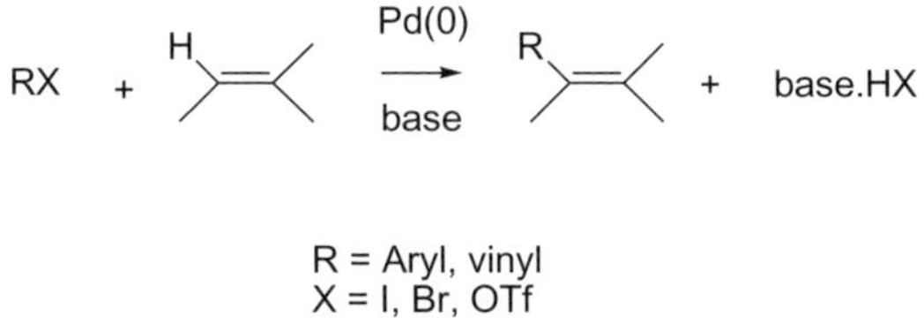

Scheme 7.1   The Heck reaction.

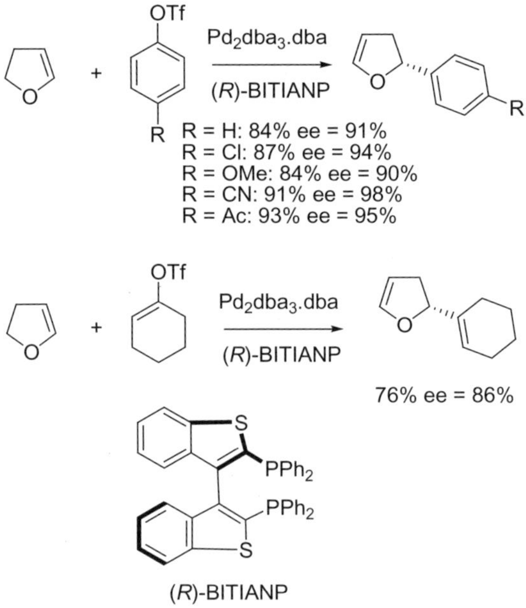

Scheme 7.2   Heck reactions of dihydrofuran with aryl or alkenyl triflates with (*R*)-BITIANP.

a new chiral ligand, (*R*)-(+)-2,2′-bis(diphenylphosphino)-3,3′-bi(ben-zo[b]thiophene), (*R*)-BITIANP, was successfully used in the enantioselective Heck reaction of dihydrofuran with aryl triflates or an alkenyl triflate, pro-viding the corresponding 2-substituted-2,3-dihydrofurans, with complete regioselectivities, high enantioselectivities (86–98% ee) and good yields (76–93%) (Scheme 7.2). In order to compare the efficiency of BITIANP with that of non-sulfur-containing BINAP, the reaction between dihydrofuran and phenyltriflate was also performed in the optimised conditions but in the pre-sence of BINAP, which led to worse results in terms of activity (42%) and both regio- and enantioselectivities (50% de, 33% ee). Steric and electronic reasons were mentioned by these authors to explain the difference observed between BINAP and BITIANP in terms of both activity and selectivity. As a more electron-rich ligand, BITIANP might cause a higher electron density at the phosphorus atom, enhancing the reaction rate by favouring the oxidative addition of the aryl triflate. Furthermore, the binding between the phosphorus and the palladium atoms is much stronger, allowing reactions to be run at higher temperatures without partial or complete dissociation of the ligand.

In 2000, the scope of this methodology was extended to the intermolecular Heck reaction of *N*-substituted 2-pyrrolines.[7] Thus, the reaction of *N*-methoxy-carbonyl-2-pyrroline with various aryl triflates in the presence of (*R*)-BITIANP as the ligand allowed the formation of the corresponding 5-substituted-pyrrolines as the major isomers with good yields, excellent regioselectivities (31/1) and enantioselectivities of up to 94% ee (Scheme 7.3). In addition, alkenyl triflates were also successfully used in the presence of this ligand, affording the corresponding 5-cyclohexenyl-2-pyrroline derivatives, albeit with a moderate regioselectivity (4/1) but with a high enantioselectivity (Scheme 7.3).

As an application of this methodology to the syntheses of natural products, Sannicolo *et al.* have reported the preparation of benzazepine and tetrahydroisoquinoline derivatives by silane-terminated intramolecular Heck reactions.[8] These workers argued that one main disadvantage of the Heck reaction

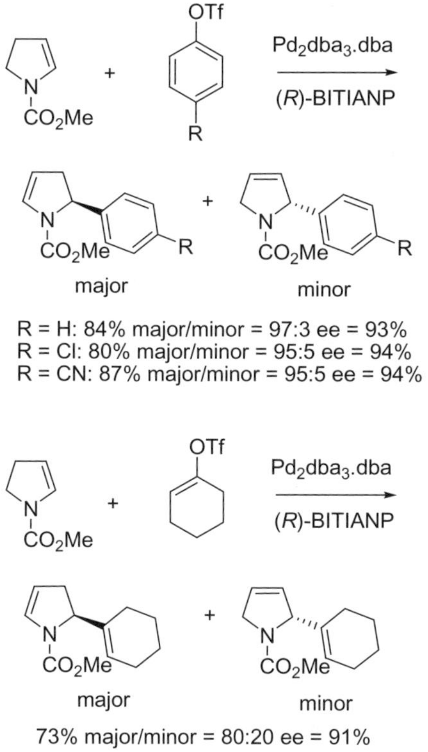
OTf
Pd_2dba_3.dba
(*R*)-BITIANP
N
CO_2Me
R
major
minor
R = H: 84% major/minor = 97:3 ee = 93%
R = Cl: 80% major/minor = 95:5 ee = 94%
R = CN: 87% major/minor = 95:5 ee = 94%
73% major/minor = 80:20 ee = 91%

Scheme 7.3 Heck reactions of *N*-methoxy-carbonyl-2-pyrroline with aryl or alkenyl triflates with (*R*)-BITIANP.

remained the low selectivity observed for the formation of the double bond by the last elimination step of the L_nPd–H species in the catalytic cycle. In this context, (E)- and (Z)-allylsilanes were prepared as the substrates for the Heck reaction in order to obtain a better control in the formation of the different side chains in the products. The Heck reactions of these allylsilanes were performed in the presence of (R)-BITIANP or another very efficient new chiral S/P ligand, (+)-4,4'-bis(diphenylphosphino)-2,2',5,5'-tetramethyl-3,3'-bithiophene, (+)-TMBTP, affording the corresponding benzazepine derivatives in good yields and excellent enantioselectivities, as shown in Scheme 7.4. Interestingly, the (E)-allylsilane was transformed in the presence of (R)-BITIANP into a mixture of two products, with the benzazepine containing the trimethylsilylvinyl side chain as the major compound, whereas the reaction of the analogous (Z)-allylsilane performed in the presence of (+)-TMBTP led to the unique formation of the expected benzazepine in a good yield and a high enantioselectivity of 92% ee. This highly efficient ligand was also applied to another (Z)-allylsilane, depicted in Scheme 7.4, allowing vinyl-substituted tetrahydroisoquinoline derivatives to be formed in good yields and enantioselectivities of up to 86% ee. The activity, regioselectivity and enantioselectivity obtained by using these two ligands were compared to those obtained with their analogous chiral diphosphine derivatives giving worse results. The superiority of the chiral sulfur-containing ligands was supposed to be due to electronic factors (high electronic density) arising from the sulfur transferred to the phosphorus atom for a stronger coordination.

On the other hand, Pfaltz *et al.* have demonstrated that chiral P/N-chelating ligands were also efficient for performing enantioselective intermolecular Heck reactions (Scheme 7.5).[9] In this context, Guiry *et al.* have reported the synthesis of corresponding chiral thiophene-containing phosphinooxazoline ligands called HETPHOX, in which the aromatic ring bearing the chelating phosphine group was replaced by a thiophene or a benzothiophene moiety for studying the influence of the modified electronic density of the phosphorus atom on the course of the catalytic transformation.[10] The best results were obtained by using thiophene-containing phosphinooxazoline ligands, which gave rise to the synthesis of the kinetic product as the major product in good to high yields and excellent enantioselectivities of up to 97% ee (Scheme 7.5). These values were comparable in both activity and selectivity to the results obtained by Pfaltz with the analogous non-sulfur-containing diphenylphosphinooxazoline ligands. The thiophene-containing phosphinooxazoline ligands were further examined in the Heck reaction of 2,2-dialkyl-2,3-dihydrofurans with phenyltriflate.[11] The best result (91% ee) was obtained for the phenylation of 2,2-dimethyl-2,3-dihydrofuran performed in the presence of the thiophene ligand bearing a *t*-Bu-substituted oxazoline ring. On the other hand, the use of benzothiophene-containing phosphinooxazoline ligands resulted in lower yield, regioselectivity and enantioselectivity, as shown in Scheme 7.5. In addition, the most efficient thiophene-containing phosphinooxazoline ligand bearing a *t*-Bu-substituted oxazoline ring was also applied to the cyclohexenylation of dihydrofuran, producing the corresponding product with high yield and enantioselectivity of

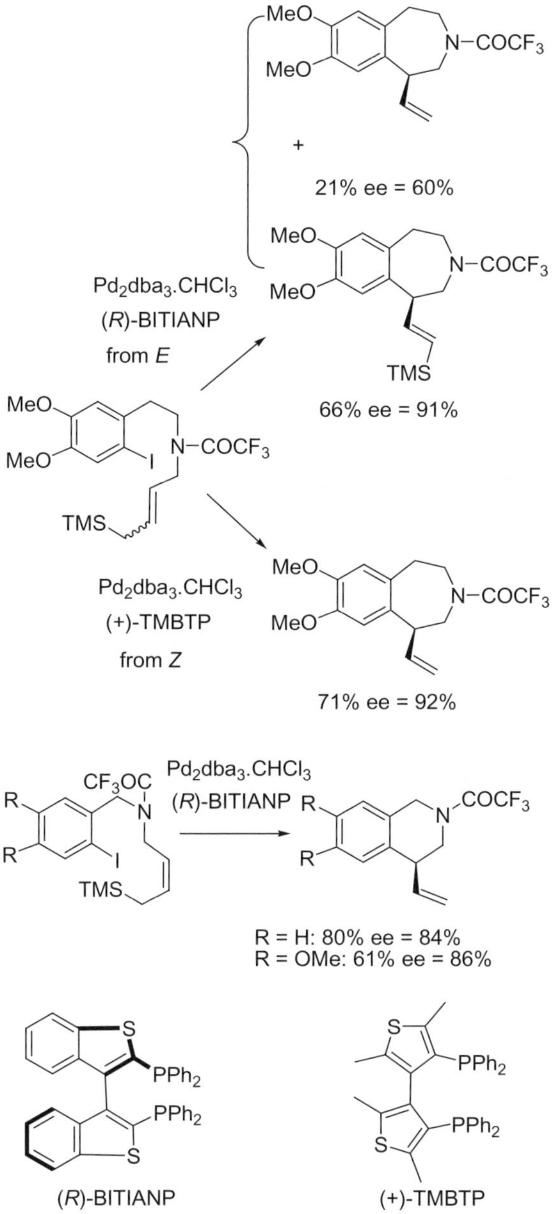

Scheme 7.4 Syntheses of benzazepines and tetrahydroisoquinolines by intramolecular silane-terminated Heck reactions with (*R*)-BITIANP and (+)-TMBTP ligands.

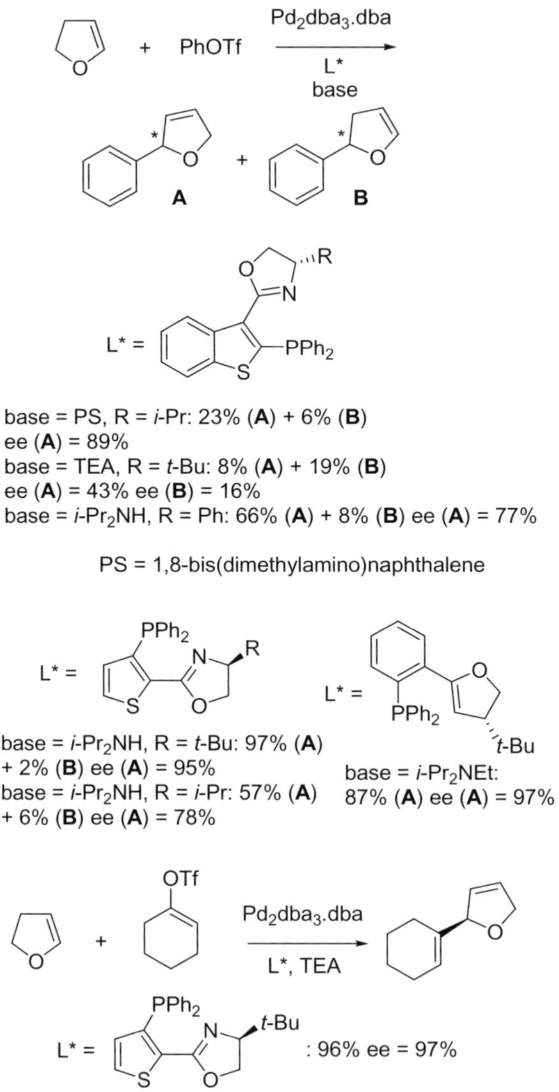

Pd2dba3.dba
PhOTf
L*
base
A
B
L* =
R
O
N
S
PPh2
base = PS, R = i-Pr: 23% (A) + 6% (B)
ee (A) = 89%
base = TEA, R = t-Bu: 8% (A) + 19% (B)
ee (A) = 43% ee (B) = 16%
base = i-Pr2NH, R = Ph: 66% (A) + 8% (B) ee (A) = 77%
PS = 1,8-bis(dimethylamino)naphthalene
base = i-Pr2NH, R = t-Bu: 97% (A)
+ 2% (B) ee (A) = 95%
base = i-Pr2NH, R = i-Pr: 57% (A)
+ 6% (B) ee (A) = 78%
t-Bu
base = i-Pr2NEt:
87% (A) ee (A) = 97%
OTf
Pd2dba3.dba
L*, TEA
: 96% ee = 97%

Scheme 7.5 Heck reactions of dihydrofuran with phenyl- or cyclohexenyl-triflates with (benzo)thiophene-containing phosphinooxazoline ligands.

up to 97% ee, but it was noted that the reaction needed a long reaction time of seven days (Scheme 7.5).

As an extension of this methodology, thiophene-containing phosphinooxazoline ligands were applied to the intramolecular version of the Heck reaction.[12] As depicted in Scheme 7.6, the intramolecular Heck cyclisation between an enamide function and an aryl triflate borne by the same substrate afforded a

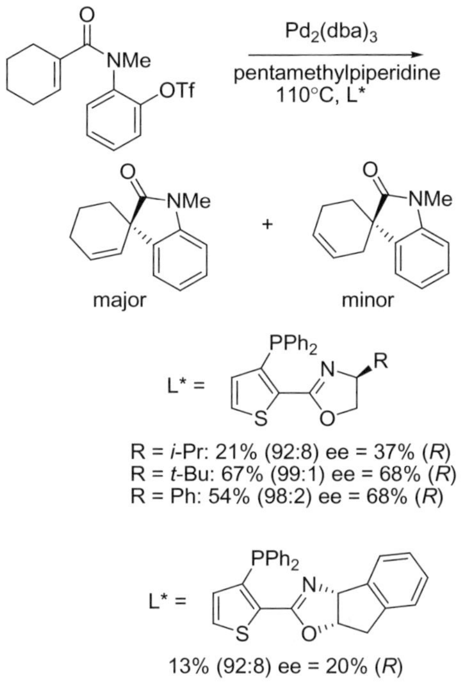

Scheme 7.6 Intramolecular Heck reaction with thiophene-containing phosphinooxazoline ligands.

mixture of two regioisomers with both moderate yield and enantioselectivity ($\leq 68\%$ ee), but with a high regioselectivity in all cases of ligands.

In 2004, Molander *et al.* developed another type of chiral sulfur-containing ligands for the intermolecular Heck reaction.[13] Thus, their corresponding novel cyclopropane-based phosphorus/sulfur palladium complexes proved to be active as catalysts for the reaction between phenyltriflate and dihydrofuran, providing at high temperature a mixture of the expected product and its isomerised analogue (Scheme 7.7). The major isomer **C** was obtained with a maximum enantioselectivity of 63% ee.

The third class of chiral sulfur-containing ligands that have been investigated for the Heck reaction is constituted by pseudo-C_2-symmetric S/P-hybrid ferrocenyl ligands reported by Kang *et al.*, in 2003.[14] As shown in Scheme 7.8, the reaction of dihydrofuran with aryltriflate favoured widely the 2,5-dihydrofuran derivative over its regioisomeric 2,3-dihydrofuran but, however, the enantioselectivity did not exceed 36% ee. Interestingly, Hayashi *et al.* have previously shown that the use of BINAP as the ligand provided an inversion in the regioselectivity of this reaction.[15]

Finally, Achiwa *et al.* have developed chiral (β-*N*-sulfonylaminoalkyl)phosphine ligands that proved to be efficient ligands for the Heck reaction of norbornene with phenyltriflate.[16] Indeed, the corresponding Heck product was

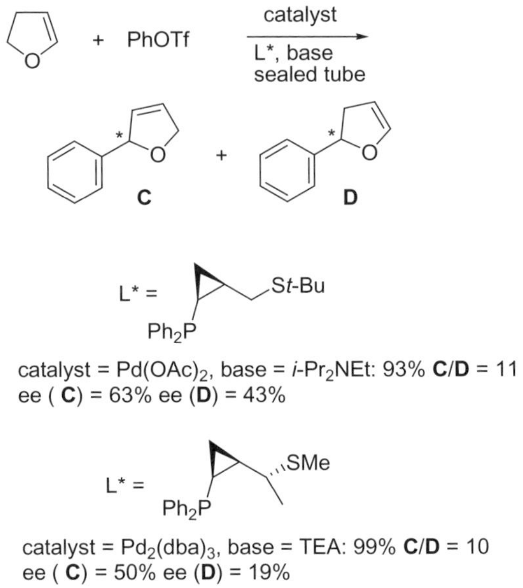

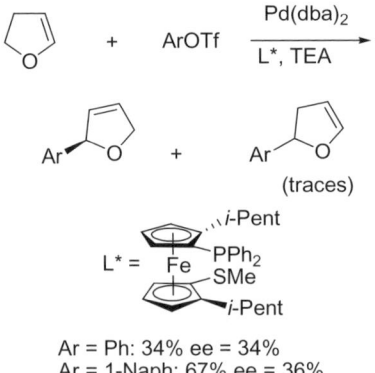

Scheme 7.7   Heck reaction of dihydrofuran with phenyltriflate with cyclopropane-based phosphorus/sulfur ligands.

Scheme 7.8   Heck reactions of dihydrofuran with aryltriflate with S/P-ferrocenyl ligand.

isolated in a good yield and an enantioselectivity of up to 71% ee, as shown in Scheme 7.9. Interestingly, the authors have performed the same reaction in the presence of (*S*)-BINAP as the ligand for comparison, and found that this catalyst was far less effective for this reaction (70%, 11% ee), while this ligand has been previously successfully used for other asymmetric Heck reactions.

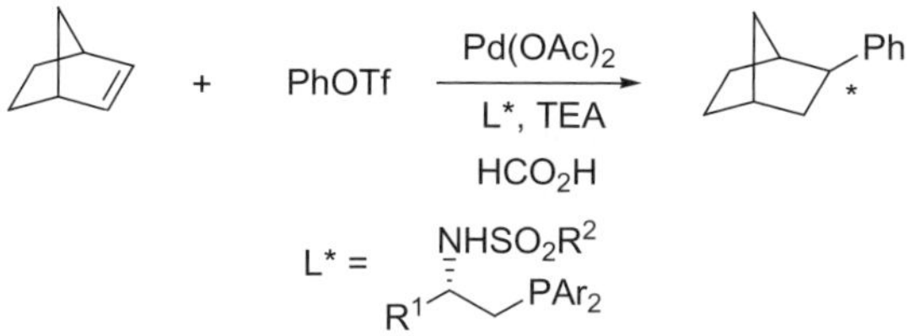

R^1 = *i*-Pr, R^2 = Me, Ar = Ph: 74% ee = 71%
R^1 = Bn, R^2 = Me, Ar = Ph: 89% ee = 60%
R^1 = *t*-Bu, R^2 = Me, Ar = Ph: 75% ee = 62%
R^1 = *i*-Pr, R^2 = Me, Ar = *p*-MeOC$_6$H$_4$: 85% ee = 63%
R^1 = *i*-Pr, R^2 = Me, Ar = *p*-ClC$_6$H$_4$: 81% ee = 68%
R^1 = *i*-Pr, R^2 = Ar = Ph: 66% ee = 61%

Scheme 7.9   Heck reaction of norbornene with phenyltriflate with (β-*N*-sulfonylaminoalkyl)phosphine ligands.

In conclusion, from its modest beginnings in the late 1980s, the asymmetric Heck reaction has become a powerful method for the synthesis of both tertiary and quaternary chiral carbon centres, with high enantioselectivities in some cases. So far, there are relatively few examples of asymmetric Heck reactions in which chiral sulfur-containing ligands have been successfully employed. Even if they lead to active complexes, these ligands generally provide modest results in terms of enantioselectivity. However, a spectacular result has been recently obtained by Sannicolo's group by using the benzothiophene analogous of BINAP, BITIANP, which allowed high yields, total regioselectivities and excellent enantioselectivities to be obtained.

References

1. (a) R. F. Heck, *Org. React.*, 1982, **27**, 345–390; (b) R. F. Heck, *Acc. Chem. Res.*, 1979, **12**, 146–151; (c) R. F. Heck, in *Palladium Reagents in Organic Synthesis*, Academic Press, New York, 1985; (d) R. F. Heck and J. P. Nolley, *J. Org. Chem.*, 1972, **37**, 2320–2322.
2. (a) L. S. Hegedus, in *Transition-metals in the Synthesis of Complex Organic Molecules*, University Science Books, Mill Valley, USA, 1994; (b) A. B. Dounay and L. E. Overman, *Chem. Rev.*, 2003, **103**, 2945–2963; (c) P. J. Guiry and D. Kiely, *Curr. Org. Chem.*, 2004, **8**, 781–794; (d) G. D. Daves and A. Hallberg, *Chem. Rev.*, 1989, **89**, 1433–1445; (e) C. Bolm, J. P. Hildebrand, K. Muniz and N. Hermanns, *Angew. Chem., Int. Ed. Engl.*, 2001, **40**, 3285–3308.
3. (a) W. Cabri and I. Candiani, *Acc. Chem. Res.*, 1995, **28**, 2–7; (b) A. de Meijere and F. E. Meyer, *Angew. Chem., Int. Ed. Engl.*, 1994, **33**, 2379–2411.
4. (a) H. G. Schmalz, *Nachr. Chem. Technol. Lab.*, 1994, **42**, 270–276; (b) M. Shibasaki, C. D. J. Boden and A. Kojima, *Tetrahedron*, 1997, **53**, 7371–7395; (c) M. Shibasaki, E. M. Vogl and T. Ohshima, *Adv. Synth.*

Catal., 2004, **346**, 1533–1552; (d) L. F. Tietze and F. Lotz, *Asymmetric Synthesis*, 2007, **1**, 147–152.

5. (a) Y. Sato, M. Sodeoka and M. Shibasaki, *J. Org. Chem.*, 1989, **54**, 4738–4739; (b) N. E. Carpenter, D. J. Kucera and L. E. Overman, *J. Org. Chem.*, 1989, **54**, 5846–5848.

6. L. F. Tietze, K. Thede and F. Sannicolo, *Chem. Commun.*, 1999, 1811–1812.

7. L. F. Tietze and K. Thede, *Synlett*, 2000, 1470–1472.

8. (a) L. F. Tietze, K. Thede, R. Schimpf and F. Sannicolo, *Chem. Commun.*, 2000, 583–584; (b) T. Bennincori, S. Rizzo and F. Sannicolo, *J. Heterocyclic. Chem.*, 2002, **39**, 471–485.

9. O. Loiseleur, P. Meier and A. Pfaltz, *Angew. Chem., Int. Ed. Engl.*, 1996, **35**, 200–202.

10. T. G. Kilroy, P. G. Cozzi, N. End and P. J. Guiry, *Synlett*, 2004, 106–110.

11. T. G. Kilroy, P. G. Cozzi, N. End and P. J. Guiry, *Synthesis*, 2004, 1879–1888.

12. M. O. Fitzpatrick, A. G. Coyne and P. J. Guiry, *Synlett*, 2006, **1**, 3150–3154.

13. G. A. Molander, J. P. Burke and P. J. Carroll, *J. Org. Chem.*, 2004, **69**, 8062–8069.

14. J. Kang, J. H. Lee and K. S. Im, *J. Mol. Catal.*, 2003, **196**, 55–63.

15. F. Ozawa, A. Kubo and T. Hayashi, *J. Am. Chem. Soc.*, 1991, **113**, 1417–1419.

16. (a) S. Sakuraba, K. Awano and K. Achiwa, *Synlett*, 1994, 291–292; (b) S. Sakuraba, T. Okada, T. Morimoto and K. Achiwa, *Chem. Pharm. Bull.*, 1995, **43**, 927–934.

CHAPTER 8
Hydrogenation

8.1 Introduction

The asymmetric hydrogenation of prochiral compounds catalysed by chiral transition-metal complexes has been in widespread use in stereoselective organic synthesis and some processes have found industrial applications.[1] Over the years, the scope of this process has been gradually extended in terms of both the reactant structure and the catalyst efficiency. Although phosphorus ligands such as diphosphines and diphosphinites are still among the most widely used chiral ligands in this process,[2] mixed donor ligands have only recently demonstrated their potential utility. In particular, the hydrogenation of unsaturated substrates, such as prochiral olefins or ketones, using complexes of chiral sulfur-donor ligands has recently been reported, providing excellent results in some cases.

8.2 Hydrogenation of Olefins

Various types of chiral sulfur-containing ligands have been investigated for the asymmetric hydrogenation of C–C double bonds, such as S/P ligands which have been the most promising. In this context, Hauptman *et al.* reported, in 1999, the rhodium-catalysed asymmetric hydrogenation of α-enamide esters in the presence of S/P ligands, in which the sulfur group was a sulfide.[3] All the rhodium complexes were proven to be active catalysts for the enantioselective hydrogenation reaction depicted in Scheme 8.1, but the enantioselectivities never exceeded 50% ee. No correlation could be established by the authors between the observed enantioselectivity and the S/P ligand structure or its electronic properties. On the other hand, the bidentate nature of the ligands was established by preparing various nickel, palladium, platinum or rhodium

RSC Catalysis Series No. 2
Chiral Sulfur Ligands: Asymmetric Catalysis
By Hélène Pellissier
© Hélène Pellissier 2009
Published by the Royal Society of Chemistry, www.rsc.org

square-planar metal complexes, which were structurally characterized by NMR spectroscopic data and X-ray analyses, particularly the palladium and rhodium complexes.

In 2004, a series of other chiral thioether-phosphine ligands based on a cyclopropane backbone were evaluated in the rhodium-catalysed hydrogenation of a dehydroamino acid by Molander *et al.*[4] As shown in Scheme 8.2, even if these ligands were generally active, only moderate enantioselectivities of up to 47% ee were obtained.

In general, of the mixed phosphorus-thioether ligands that have been used in the asymmetric hydrogenation of prochiral olefins, the thioether-phosphinite ligands have provided some of the best results. As an example, a new class of thioether-phosphinite ligands developed by Evans *et al.* has recently proved to be very efficient for the rhodium-catalysed asymmetric hydrogenation of a

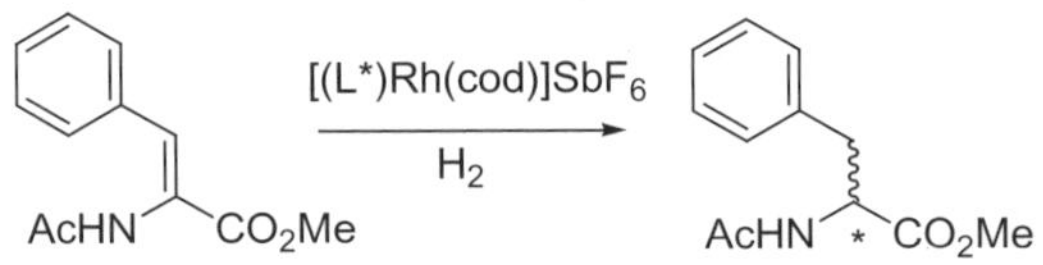

$R^1 = Cy, R^2 = R^3 = Me$: 100% ee = 27%
$R^1 = Cy, R^2 = Me, R^3 = Bn$: 100% ee = 50%
$R^1 = Cy, R^2 = Me, R^3 = CH_2(C_6(Me)_5)$: 100% ee = 18%
$R^1 = R^2 = Cy, R^3 = CH_2(C_6(Me)_5)$: 100% ee = 39%
$R^1 = Ph, R^2 = Me, R^3 = CH_2(C_6(Me)_5)$: 74% ee = 25%

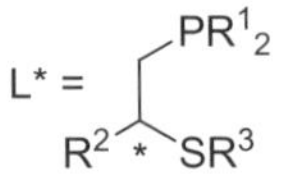

Scheme 8.1 Hydrogenation of α-enamide ester with thioether-phosphine ligands.

L* = [structure] SMe / PPh₂   L* = [structure] PPh₂ / ‴SMe   L* = [structure] PPh₂ / St-Bu
27% ee = 5%   93% ee = 46%   41% ee = 47%

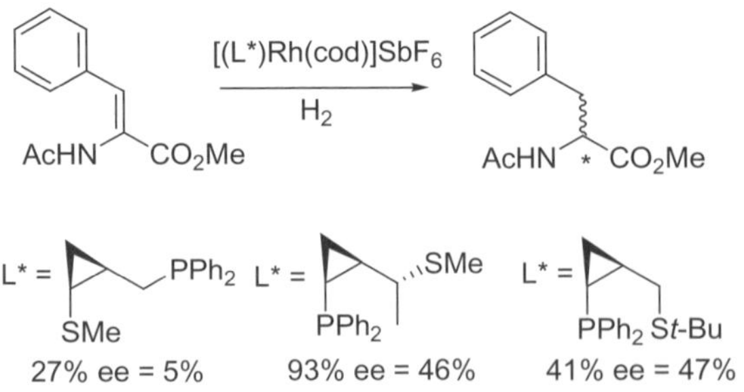

Scheme 8.2 Hydrogenation of dehydroamino acid with thioether-phosphine cyclopropanated ligands.

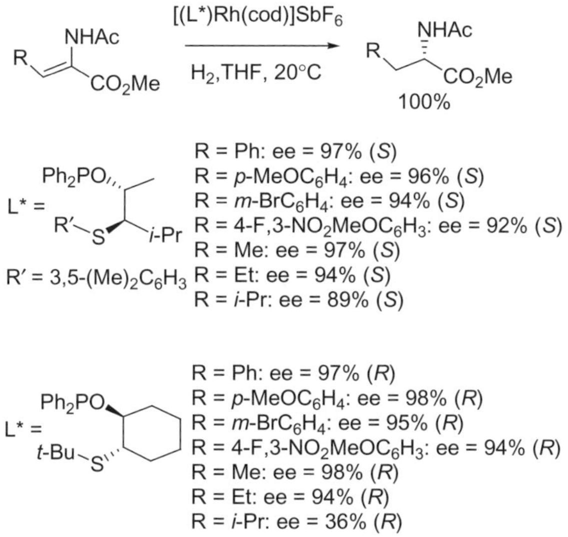

Scheme 8.3 Hydrogenations of substituted acetamidoacrylates with thioether-phosphinite ligands.

variety of substituted acetamidoacrylates, providing the expected products with enantioselectivities of up to 98% ee (Scheme 8.3).[5] The best enantioselectivities were obtained with the ligand bearing a bulky 3,5-dimethylphenyl substituent at sulfur, which forced this sulfur substituent into an *anti* orientation to minimize the steric hindrance. Moreover, these new catalysts were also interestingly described as tolerant to a wide range of *N*-protecting groups. Since the nature of this group had little effect on the selectivity of the reaction, these catalysts allowed the challenging hydrogenation of tetrasubstituted enamides in an enantioselective way to be achieved. A mechanistic approach for this process was proposed by these authors on the basis of NMR analyses and X-ray crystallographic determination of a catalyst–substrate complex. The proposed model involved a regioselective binding of the substrate due to the different *trans* influences between the phosphorus and sulfur atoms and the enantiofacial discrimination induced by the new sulfur stereocentre. It was demonstrated that the ligand induced a very high level of selectivity along the reaction by selecting only one out of the four possible diastereoisomers.

In order to extend the success encountered with thioether-phosphinite ligands, Diéguez *et al.* have recently developed the synthesis of carbohydrate-based thioether-phosphinite ligands, since carbohydrate ligands are known to be highly effective ligands for asymmetric hydrogenation.[6] Hence, thioether-phosphinite ligands containing a furanoside as a simple but effective backbone were tested in the rhodium- and iridium-catalysed asymmetric hydrogenations of α-acylaminoacrylates and itaconic acid derivatives, providing good to high enantioselectivities (Scheme 8.4). These ligands were substituted by different

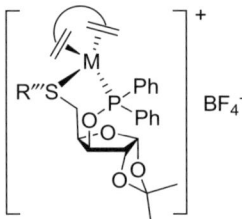

M	R'''	R	R'	R''	yield (%)	ee (%)
Rh	Ph	NHAc	H	CO$_2$Me	86	68 (R)
Ir	Ph	NHAc	H	CO$_2$Me	100	50 (R)
Rh	i-Pr	NHAc	H	CO$_2$Me	83	96 (R)
Ir	i-Pr	NHAc	H	CO$_2$Me	95	63 (R)
Rh	Me	NHAc	H	CO$_2$Me	75	54 (R)
Ir	Me	NHAc	H	CO$_2$Me	60	30 (R)
Rh	Ph	NHAc	Ph	CO$_2$Me	100	68 (R)
Ir	Ph	NHAc	Ph	CO$_2$Me	100	50 (R)
Rh	i-Pr	NHAc	Ph	CO$_2$Me	100	93 (R)
Rh	Me	NHAc	Ph	CO$_2$Me	75	45 (R)
Rh	Ph	CH$_2$CO$_2$H	H	CO$_2$H	83	47 (S)
Ir	Ph	CH$_2$CO$_2$H	H	CO$_2$H	85	32 (S)
Rh	i-Pr	CH$_2$CO$_2$H	H	CO$_2$H	83	55 (S)

active complex:

Scheme 8.4   Hydrogenations of olefins with furanoside thioether-phosphinite ligands.

aryl or alkyl groups at the sulfur atom. It was shown that the enantioselectivity depended on the steric bulk of the substituent at the sulfur atom. Hence, a bulky group on the thioether moiety, such as an *i*-Pr group, combined with the rhodium metal had a positive effect on the enantioselectivity, since enantio-selectivities of up to 96% ee could be observed in these conditions. The complexes were fully characterised by NMR analyses, which revealed that the reaction of these ligands with complex [M(cod)$_2$]BF$_4$ yielded the corresponding complexes [M(cod)(P–SR)]BF$_4$ present in only one diastereomeric form, having the sulfur substituent in a pseudoaxial disposition (Scheme 8.4). In addition, the same group has investigated the corresponding thioether-phosphinite

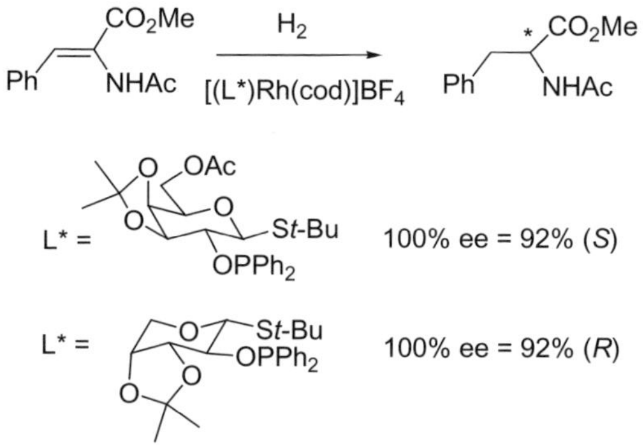

Scheme 8.5 Hydrogenation of methyl α-acetamidocinnamate with S/P ligands derived from carbohydrates.

ligands for the asymmetric iridium-catalysed hydrogenation of itaconic acid, giving good conversions and enantioselectivities of up to 51% ee.[7]

Other S/P ligands derived from carbohydrates and depicted in Scheme 8.5 were found by Khiar *et al.* to be efficient catalysts for the rhodium-catalysed methyl α-acetamidocinnamate hydrogenation, leading to protected (*S*)-phenylalanine in a quantitative yield and with an enantioselectivity of 92% ee.[8] On the other hand, the use of 2-phosphinite *tert*-butyl-thioarabinoside as the ligand afforded the corresponding (*R*)-isomer in 92% ee.

In 1998, Hauptman *et al.* reported the synthesis of novel C_2-symmetric bis(phosphinites) ligands having a tetrahydrothiophene backbone, which were further tested in the asymmetric rhodium-catalysed hydrogenation of methyl α-acetamidocinnamate, providing the corresponding product with an enantioselectivity of up to 55% ee (Scheme 8.6).[9] As demonstrated by conventional spectroscopic methods and by single-crystal X-ray analysis, the bis(diphenylphosphinites) ligand bound in an unusual tridentate mode, forming a thioether-bridged bimetallic rhodium species.

In 2000, Sannicolo *et al.* reported the synthesis of a new C_2-symmetric chelating ligand, 2,2′,5,5′-tetramethyl-4,4′-bis(diphenylphosphino)-3,3′-bithiophene (tetraMe-BITIOP) combining a high electronic density at phosphorus and a low bite-angle value for an efficient use in catalytic hydrogenation reactions.[10] The complexes of this electron-rich diphosphine ligand with ruthenium and rhodium were used as catalysts in the homogeneous hydrogenation reaction of prostereogenic double bonds of substituted acrylic acids and *N*-acetylenamino acids. The enantioselectivities were found to be excellent (up to 94% ee) in general, and comparable with the best results reported for the same reactions. Hence, this new ligand constituted one of the most efficient diphosphine chiral ligands available today for asymmetric hydrogenation. An important advantage offered by this ligand over its competitors was its very good synthetic accessibility (Scheme 8.7). The success of this ligand was

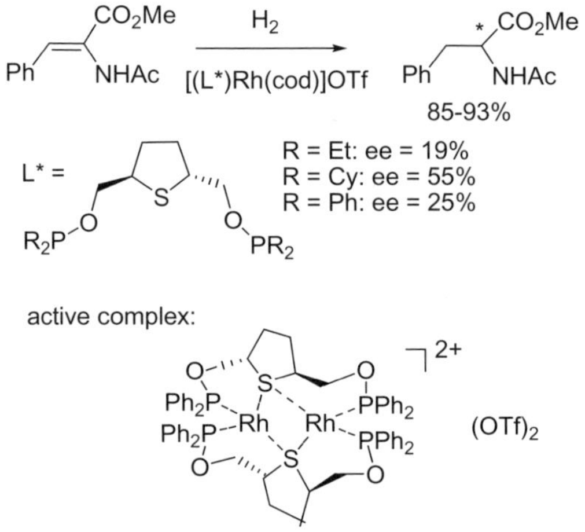
CO2Me
H2
[(L*)Rh(cod)]OTf
Ph
NHAc
*
CO2Me
Ph
NHAc
85-93%
L* =
S
R2P
O
O
PR2
R = Et: ee = 19%
R = Cy: ee = 55%
R = Ph: ee = 25%
active complex:
2+
(OTf)2

Scheme 8.6 Hydrogenation of methyl α-acetamidocinnamate with bis(phosphinites) ligands.

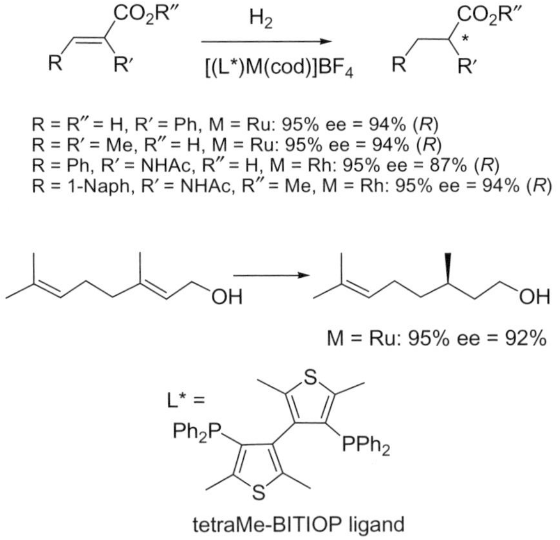
CO2R″
R
R′
H2
[(L*)M(cod)]BF4
CO2R″
*
R
R′
R = R″ = H, R′ = Ph, M = Ru: 95% ee = 94% (R)
R = R′ = Me, R″ = H, M = Ru: 95% ee = 94% (R)
R = Ph, R′ = NHAc, R″ = H, M = Rh: 95% ee = 87% (R)
R = 1-Naph, R′ = NHAc, R″ = Me, M = Rh: 95% ee = 94% (R)
OH
OH
M = Ru: 95% ee = 92%
L* =
Ph2P
PPh2
S
S
tetraMe-BITIOP ligand

Scheme 8.7 Hydrogenations of olefins with tetraMe-BITIOP.

extended to the asymmetric hydrogenation of α- and β-keto esters, providing enantioselectivities of up to 99% ee.

Another class of chiral diphosphine ligands bearing two interconnected thiophene rings, has been successfully developed by Sannicolo *et al.* These bis(diphenylphosphino)[*b*]thiophene ligands, called tetraMe-BITIANP and

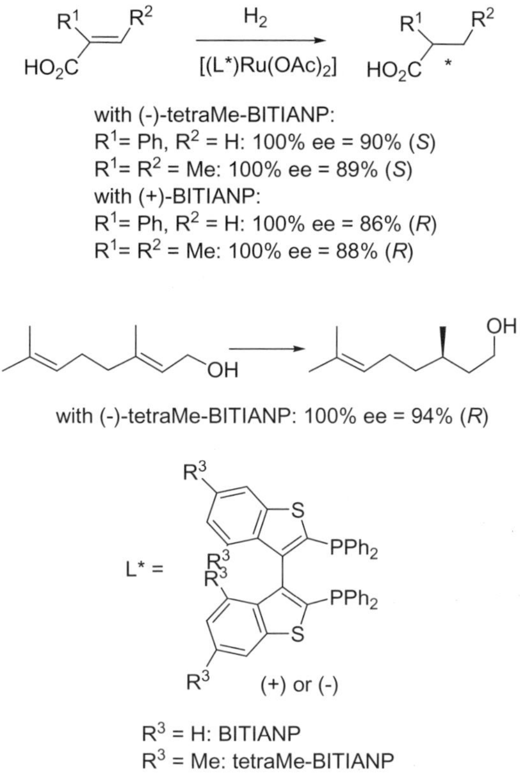

with (-)-tetraMe-BITIANP:
R^1 = Ph, R^2 = H: 100% ee = 90% (*S*)
R^1 = R^2 = Me: 100% ee = 89% (*S*)
with (+)-BITIANP:
R^1 = Ph, R^2 = H: 100% ee = 86% (*R*)
R^1 = R^2 = Me: 100% ee = 88% (*R*)

with (-)-tetraMe-BITIANP: 100% ee = 94% (*R*)

R^3 = H: BITIANP
R^3 = Me: tetraMe-BITIANP

Scheme 8.8   Hydrogenations of olefins with tetraMe-BITIANP and BITIANP.

BITIANP, were tested as ruthenium ligands for the asymmetric hydrogenation of various olefinic substrates. The results collected in Scheme 8.8 show that these novel ligands were able to induce high enantioselectivities of up to 94% ee.[11]

More recently, these authors have reported the synthesis of a new thiophene-based analogue of (*R*,*R*)-Me-DuPHOS called UlluPHOS.[12] The synthesis of this novel C_2-symmetric diphospholane ligand, containing a thiophene ring as the linker between the two phospholane units, was simpler than that available for the DuPHOS ligands family. This new ligand displayed geometric properties very similar to those found for Me-DuPHOS but quite a high electronic availability at phosphorus relative to the latter ligand. The facial recognition and enantioselection associated with rhodium and ruthenium complexes of UlluPHOS and Me-DuPHOS were shown to be similarly high in various hydrogenations of olefins (Scheme 8.9). The most important difference between these two ligands was found by comparing the reaction rates. Indeed, the authors have observed that the use of UlluPHOS considerably increased the activity of the complexes.

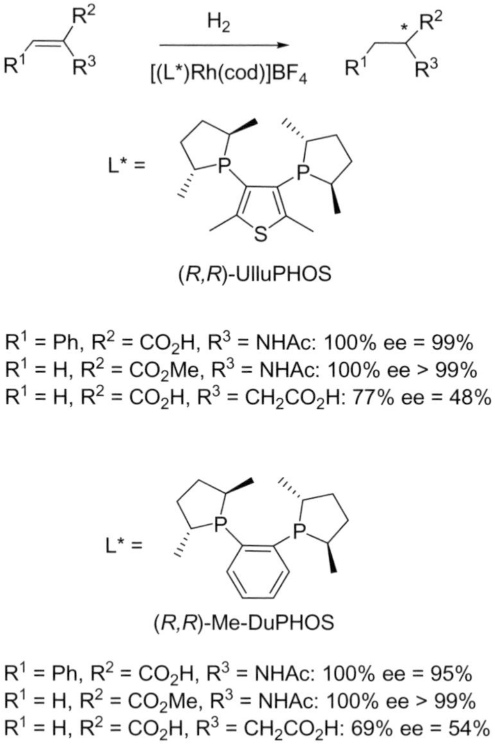

$L* =$

(*R,R*)-UlluPHOS

R^1 = Ph, R^2 = CO$_2$H, R^3 = NHAc: 100% ee = 99%
R^1 = H, R^2 = CO$_2$Me, R^3 = NHAc: 100% ee > 99%
R^1 = H, R^2 = CO$_2$H, R^3 = CH$_2$CO$_2$H: 77% ee = 48%

$L* =$

(*R,R*)-Me-DuPHOS

R^1 = Ph, R^2 = CO$_2$H, R^3 = NHAc: 100% ee = 95%
R^1 = H, R^2 = CO$_2$Me, R^3 = NHAc: 100% ee > 99%
R^1 = H, R^2 = CO$_2$H, R^3 = CH$_2$CO$_2$H: 69% ee = 54%

Scheme 8.9   Hydrogenations of olefins with UlluPHOS and Me-DuPHOS.

In 2006, Berens *et al.* reported the synthesis of novel benzothiophene-based DuPHOS analogues, which gave excellent levels of enantioselectivity when applied as the ligands to the asymmetric rhodium-catalysed hydrogenation of various olefins, such as dehydroamino acid derivatives, enamides and itaconates (Scheme 8.10).[13]

Other sulfur-containing ligands based on thiophene derivatives were involved by Cozzi *et al.* in the iridium-catalysed asymmetric hydrogenation of olefins.[14] The corresponding iridium complexes have proved to be highly active and the enantioselectivity was generally excellent, up to 99% ee, as shown in Scheme 8.11. These authors have also investigated the corresponding benzothiophene-based ligands, but the enantioselectivities were generally lower than when using the thiophene core. In addition, the hydrogenation of allylic alcohols was also investigated using both types of complexes, providing high yields and enantioselectivities of up to 94% ee.

On the other hand, Bolm *et al.* have reported, more recently, the use of BINOL-derived *N*-phosphino sulfoximines as ligands in the rhodium-catalysed hydrogenation of dimethyl itaconate and α-acetamidoacrylates, achieving excellent enantioselectivities of up to 99% ee (Scheme 8.12).[15] In the main

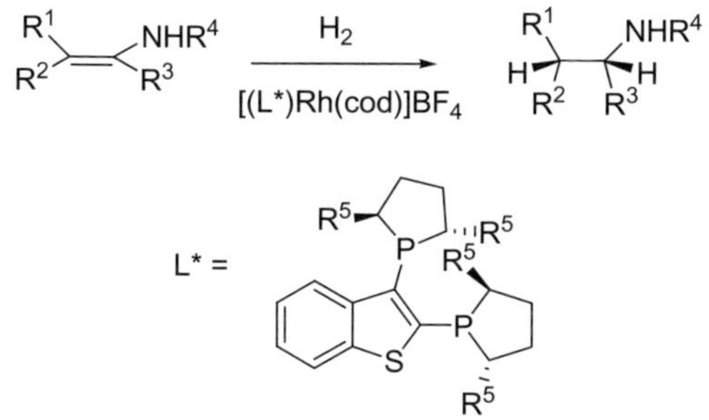

R^1 = Ph, R^2 = H, R^3 = CO$_2$Me, R^4 = Ac
R^5 = Me: ee = 95%
R^1 = R^2 = H, R^3 = CO$_2$H, R^4 = Ac
R^5 = Me: ee = 99%
R^1 = p-NO$_2$C$_6$H$_4$, R^2 = H, R^3 = CO$_2$Me, R^4 = Cbz
R^5 = Et: ee = 99%
R^1 = R^2 = H, R^3 = m-BnOPh, R^4 = Ac
R^5 = Et: ee = 90%
R^1 = H, R^2 = CO$_2$Me, R^3 = Et, R^4 = Ac
R^5 = Me: ee = 98%
R^1 = H, R^2 = CO$_2$Me, R^3 = i-Bu, R^4 = Ac
R^5 = Et: ee = 87%
R^1 = CO$_2$Me, R^2 = H, R^3 = R^5 = Me
R^4 = Ac: ee = 90%

Scheme 8.10 Hydrogenations of olefins with benzothiophene-based DuPHOS analogue ligands.

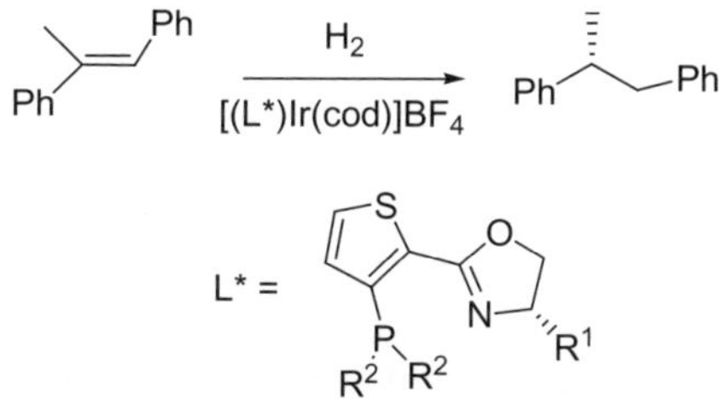

R^1 = i-Pr, R^2 = Ph: > 99% ee = 77%
R^1 = t-Bu, R^2 = Ph: > 99% ee = 95%
R^1 = t-Bu, R^2 = o-Tol: > 99% ee = 97%
R^1 = t-Bu, R^2 = Cy: > 99% ee = 99%

Scheme 8.11 Hydrogenation of olefin with thiophene-based ligands.

cases, it seemed that the configuration at the sulfur atom only had a small influence on the enantioselectivities of the reaction. The question of whether these ligands behaved in a monodentate manner, or whether a P/O-bidentate chelation was involved needed to be addressed.

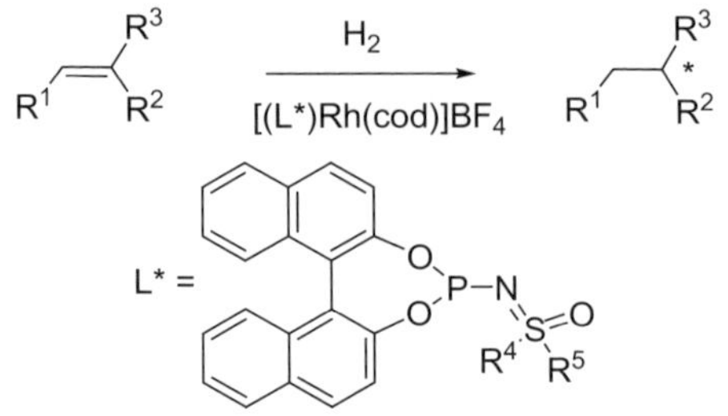

R^1 = H, R^2 = CH$_2$CO$_2$Me, R^3 = CO$_2$Me, (R)-L*, R^4 = Ph, R^5 = Me: ee > 99% (R)
R^1 = H, R^2 = CH$_2$CO$_2$Me, R^3 = CO$_2$Me, (S)-L*, R^4 = Ph, R^5 = Me: ee = 99% (S)
R^1 = H, R^2 = CH$_2$CO$_2$Me, R^3 = CO$_2$Me, (R)-L*, R^4 = R^5 = Me: ee > 99% (R)
R^1 = H, R^2 = CH$_2$CO$_2$Me, R^3 = CO$_2$Me, (R)-L*, R^4 = R^5 = Ph: ee = 98% (R)
R^1 = H, R^2 = CH$_2$CO$_2$Me, R^3 = CO$_2$Me, (R)-L*, R^4 = Ph, R^5 = Me: ee = 99% (S)
R^1 = H, R^2 = CH$_2$CO$_2$Me, R^3 = CO$_2$Me, (S)-L*, R^4 = Ph, R^5 = Me: ee = 98% (R)
R^1 = H, R^2 = CH$_2$CO$_2$Me, R^3 = CO$_2$Me, (R)-L*, R^4 = R^5 = Me: ee = 98% (R)
R^1 = H, R^2 = CH$_2$CO$_2$Me, R^3 = CO$_2$Me, (R)-L*, R^4 = R^5 = Ph: ee = 98% (S)
R^1 = Ph, R^2 = CO$_2$Me, R^3 = NHAc, (R)-L*, R^4 = Ph, R^5 = Me: ee = 97% (S)
R^1 = Ph, R^2 = CO$_2$Me, R^3 = NHAc, (S)-L*, R^4 = Ph, R^5 = Me: ee = 94% (R)
R^1 = Ph, R^2 = CO$_2$Me, R^3 = NHAc, (R)-L*, R^4 = R^5 = Me: ee = 95% (S)
R^1 = Ph, R^2 = CO$_2$Me, R^3 = NHAc, (R)-L*, R^4 = R^5 = Ph: ee = 94% (S)
R^1 = H, R^2 = Ph, R^3 = NHAc, (R)-L*, R^4 = Ph, R^5 = Me: ee = 87% (R)
R^1 = H, R^2 = Ph, R^3 = NHAc, (S)-L*, R^4 = Ph, R^5 = Me: ee = 76% (R)
R^1 = H, R^2 = Ph, R^3 = NHAc, (R)-L*, R^4 = R^5 = Me: ee = 85% (S)
R^1 = H, R^2 = Ph, R^3 = NHAc, (R)-L*, R^4 = R^5 = Ph: ee = 64% (S)

Scheme 8.12 Hydrogenations of olefins with BINOL-derived *N*-phosphino sulfoximine ligands.

In 2006, chiral bidentate phosphine thiazoles were prepared and successfully applied as ligands in the iridium-catalysed asymmetric hydrogenation of tri-substituted aryl olefins by Hedberg *et al.*[16] In order to evaluate the role of the cyclic backbone of the ligand on the efficiency of the hydrogenation, these authors prepared ligands with five-, six- and seven-membered cyclic backbones, which all gave excellent conversions and enantioselectivities, with a preference for the six-membered backbone (Scheme 8.13). Unsurprisingly, it was found that the steric bulk of the heteroaromatic substituent and the size of the phosphine substituents were important for achieving a high enantioselectivity.

In 2004, Schenkel and Ellman reported the application of P/N-sulfinyl imine ligands to iridium-catalysed asymmetric hydrogenation of olefins.[17] The asymmetry induced by this new class of ligands was only due to the chirality of the sulfur centre, arising from the sulfoxide group as a non-sulfur-coordinating moiety. In order to study the influence of the sulfinamide structure on the enantioselectivity of the hydrogenation of olefins, these authors have synthe-sised a collection of sulfinamides with a wide range of steric and electronic properties. It was shown that the increase of the steric bulk of the sulfinamide by the introduction of a 1-adamantane or a 3-ethylheptane group instead of a *tert*-butyl group did not allow the enantioselectivity to be improved (Scheme 8.14). These results were attributed to an unfavourable geometry around the metallic centre. Moreover, the use of arylsulfinyl imines clearly showed a negative impact on the enantioselectivity. The scope of this methodology was

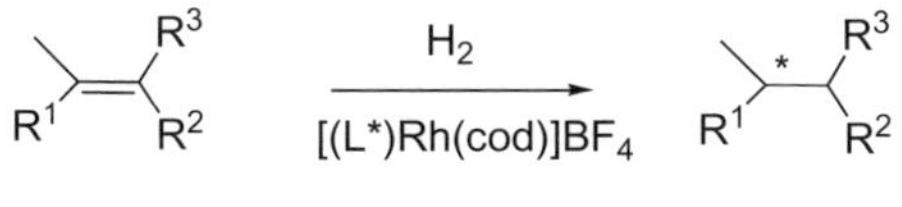

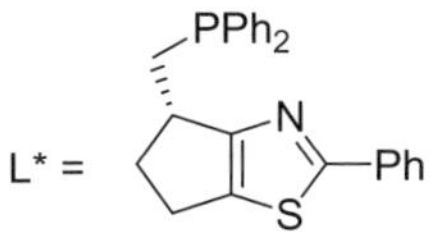

Scheme 8.13 Hydrogenations of olefins with phosphine thiazole ligands.

R^1 = R^3 = Ph, R^2 = H: 99% ee = 94%
R^1 = Ph, R^2 = H, R^3 = CO$_2$Et: 99% ee = 86%
R^1 = Ph, R^2 = CO$_2$Et, R^3 = H: 99% ee = 3%

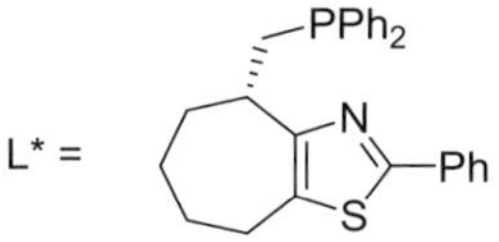

R^1 = R^3 = Ph, R^2 = H: 99% ee = 98%
R^1 = Ph, R^2 = H, R^3 = CO$_2$Et: 99% ee = 95%
R^1 = Ph, R^2 = CO$_2$Et, R^3 = H: 97% ee = 3%

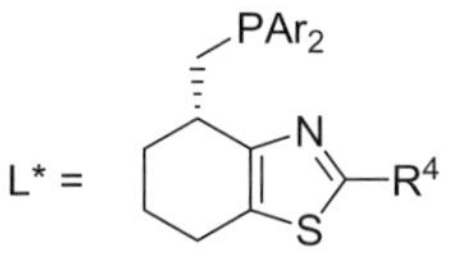

R^1 = R^3 = R^4 = Ar = Ph, R^2 = H: 99% ee > 99%
R^1 = R^4 = Ar = Ph, R^2 = H, R^3 = CO$_2$Et: 99% ee = 98%
R^1 = R^4 = Ar = Ph, R^2 = CO$_2$Et, R^3 = H: 85% ee = 0%
R^1 = R^3 = R^4 = Ph, R^2 = H, Ar = o-Tol: 99% ee > 99%
R^1 = R^4 = Ph, R^2 = H, R^3 = CO$_2$Et, Ar = o-Tol: 99% ee = 98%
R^1 = R^4 = Ph, R^2 = CO$_2$Et, R^3 = H, Ar = o-Tol: 99% ee = 72%
R^1 = R^3 = Ar = Ph, R^2 = R^4 = H: 99% ee = 77%
R^1 = Ar = Ph, R^2 = R^4 = H, R^3 = CO$_2$Et: 91% ee = 87%
R^1 = R^4 = Ar = Ph, R^2 = CO$_2$Et, R^3 = H: 99% ee = 95%
R^1 = R^3 = Ph, R^2 = R^4 = H, Ar = o-Tol: 99% ee = 48%
R^1 = Ph, R^2 = R^4 = H, R^3 = CO$_2$Et, Ar = o-Tol: 99% ee = 69%
R^1 = Ph, R^2 = CO$_2$Et, R^3 = R^4 = H, Ar = o-Tol: 99% ee = 95%

extended to the iridium-catalysed hydrogenation of α,β-unsaturated esters and allylic alcohols that gave, in the presence of the most performative *tert*-butanesulfinamide ligand, good enantioselectivities of 65 and 70% ee, respectively.

Several S/N ligands have also been investigated for the asymmetric hydrogenation of prochiral olefins. Thus, asymmetric enamide hydrogenations have been performed in the presence of S/N ligands and rhodium or ruthenium catalysts by Lemaire *et al.*, giving enantioselectivities of up to 70% ee.[18] Two

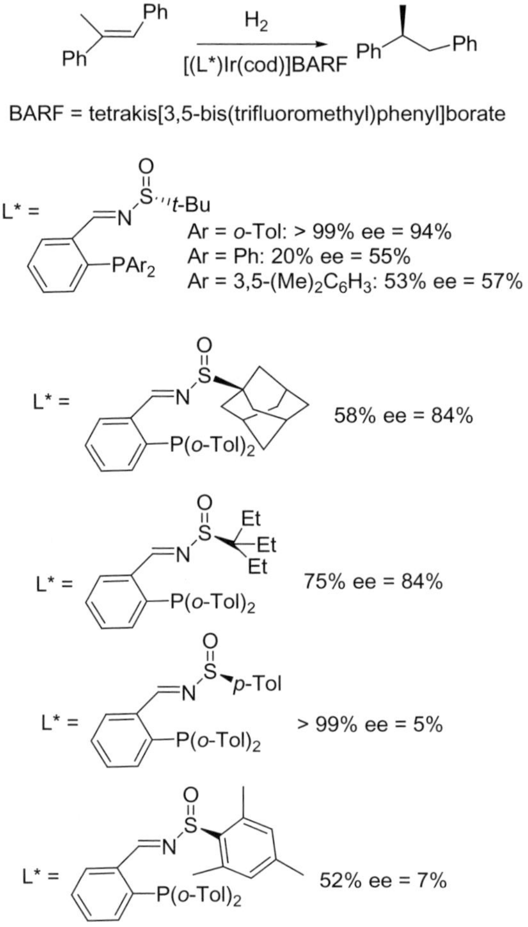

BARF = tetrakis[3,5-bis(trifluoromethyl)phenyl]borate

L* =

Ar = o-Tol: > 99% ee = 94%
Ar = Ph: 20% ee = 55%
Ar = 3,5-(Me)₂C₆H₃: 53% ee = 57%

L* =   58% ee = 84%

L* =   75% ee = 84%

L* =   > 99% ee = 5%

L* =   52% ee = 7%

Scheme 8.14 Hydrogenation of olefin with P/N-sulfinyl imine ligands.

new types of stable and readily available chiral ligands, mono- and dithioureas, have been tested, but only the dithiourea ligands provided a significant enantioselection. This result suggested that, for enamide hydrogenation, the C_2-symmetry of the dithiourea ligands was a key factor in the degree of control observed (Scheme 8.15). The use of these ligands was extended to the rhodium- and iridium-catalysed hydrogenations of phenylglyoxylate methyl ester, giving a good enantioselectivity with ee values of up to 72%.[19]

In addition, several S/S ligands were also investigated for the asymmetric hydrogenation of olefins. In 1977, James and McMillan reported the synthesis of various disulfoxide ligands, which were applied to the asymmetric ruthenium-catalysed hydrogenation of prochiral olefinic acid derivatives, such as itaconic acid.[20] These ligands, depicted in Scheme 8.16, were active to provide

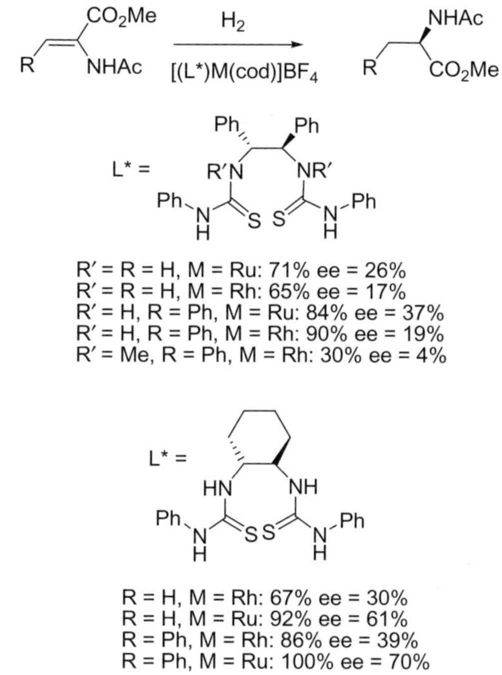
CO2Me
H2
NHAc
R
NHAc
[(L*)M(cod)]BF4
R
CO2Me
Ph
Ph
L* =
R′N
NR′
Ph
N
H
S
S
N
H
Ph
R′ = R = H, M = Ru: 71% ee = 26%
R′ = R = H, M = Rh: 65% ee = 17%
R′ = H, R = Ph, M = Ru: 84% ee = 37%
R′ = H, R = Ph, M = Rh: 90% ee = 19%
R′ = Me, R = Ph, M = Rh: 30% ee = 4%
L* =
HN
NH
Ph
N
H
S
S
N
H
Ph
R = H, M = Rh: 67% ee = 30%
R = H, M = Ru: 92% ee = 61%
R = Ph, M = Rh: 86% ee = 39%
R = Ph, M = Ru: 100% ee = 70%

Scheme 8.15 Hydrogenations of enamides with dithiourea ligands.

the corresponding products, but with low enantioselectivities (≤25% ee). Infrared and NMR studies indicated that the sulfoxide moieties were all S-bonded.

Chiral C_2-symmetric dithioether ligands possessing a 1,4-dioxane backbone and derived from tartrates have been employed as S/S ligands of rhodium(I) complexes.[21] The involvement of these novel catalysts in the asymmetric hydrogenation of *N*-acetyl-1-phenylethyleneamine provided good yields but low enantioselectivities (≤21% ee) as shown in Scheme 8.17. For comparison, the same reaction was performed in the presence of the corresponding diphenylphosphine analogous ligands, providing much better enantioselectivities of up to 70% ee.

In 1998, Ruiz *et al.* reported the synthesis of new chiral dithioether ligands based on a pyrrolidine backbone from (+)-L-tartaric acid.[22] Their corresponding cationic iridium complexes were further evaluated as catalysts for the asymmetric hydrogenation of prochiral dehydroamino acid derivatives and itaconic acid, providing enantioselectivities of up to 68% ee, as shown in Scheme 8.18.

In 1999, these authors reported the synthesis of the first family of sugar-derivative dithioethers containing sulfur as a unique donor atom.[23] These chiral C_1-symmetric dithioether ligands were tested in the iridium-catalysed

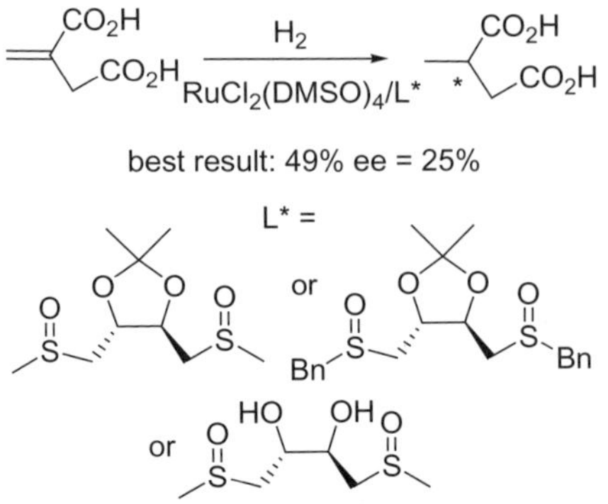

best result: 49% ee = 25%

L* =

Scheme 8.16   Hydrogenation of itaconic acid with disulfoxide ligands.

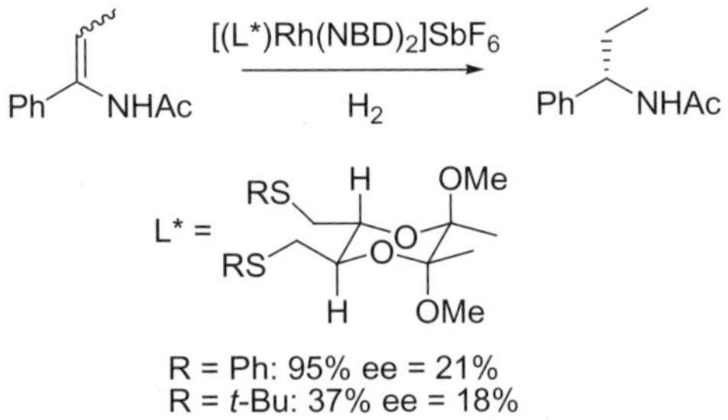

R = Ph: 95% ee = 21%
R = *t*-Bu: 37% ee = 18%

Scheme 8.17   Hydrogenation of *N*-acetyl-1-phenylethyleneamine with tartrate-derived dithioether ligands.

asymmetric hydrogenation of various acrylic acid derivatives, providing modest enantioselectivities (Scheme 8.19). The enantioselectivities were shown to be dependent on the nature of the substituent on the dithioethers. Indeed, the inductions were better in the case of using a bulky and electron-rich ligand, such as that bearing an isopropyl substituent, which allowed an enantioselectivity of up to 62% ee to be obtained for the hydrogenation of itaconic acid. In addition, the same authors have also investigated rhodium complexes of (*R*,*R*)-1-benzyl-3,4-dithioether pyrrolidine ligands for the asymmetric hydrogenation of acrylic acids, such as *Z*-α-acetamidocinnamic acid and itaconic acid but, however, these catalysts remained inactive.[24]

In 2004, Dieguez *et al.* reported the development of novel C_2-symmetric dithioether ligands derived from the corresponding binaphthyl or biphenanthryl diols.[25] Thus, various (*R*)-binaphthyl dithiols substituted by alkyl groups on the sulfur atom in order to increase the steric bulk were synthesised, and the corresponding mononuclear cationic Ir(I)–cyclooctadiene complexes were prepared and characterised (Scheme 8.20). NMR studies demonstrated that, in all cases, the coordination of the ligands proceeded with complete stereoselectivity at the

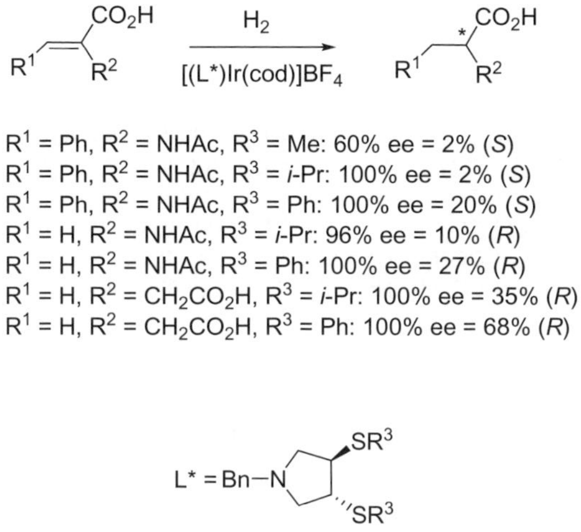
CO2H
H2
[(L*)Ir(cod)]BF4
R1
R2
R1 = Ph, R2 = NHAc, R3 = Me: 60% ee = 2% (S)
R1 = Ph, R2 = NHAc, R3 = i-Pr: 100% ee = 2% (S)
R1 = Ph, R2 = NHAc, R3 = Ph: 100% ee = 20% (S)
R1 = H, R2 = NHAc, R3 = i-Pr: 96% ee = 10% (R)
R1 = H, R2 = NHAc, R3 = Ph: 100% ee = 27% (R)
R1 = H, R2 = CH2CO2H, R3 = i-Pr: 100% ee = 35% (R)
R1 = H, R2 = CH2CO2H, R3 = Ph: 100% ee = 68% (R)
L* = Bn–N
SR3
SR3

Scheme 8.18 Hydrogenations of olefins with iridium complexes containing dithioether ligands with a pyrrolidine backbone.

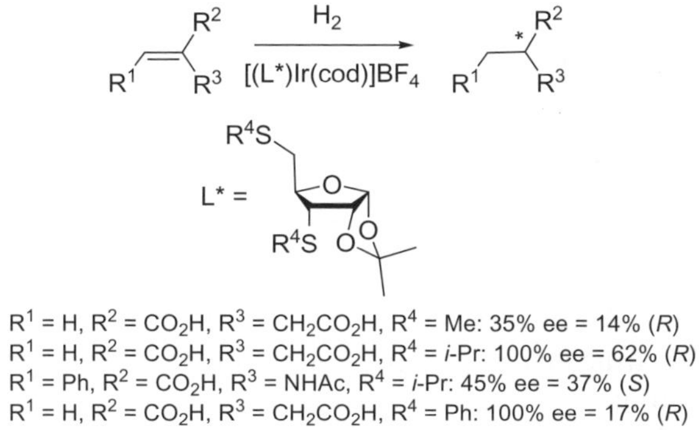
R2
H2
[(L*)Ir(cod)]BF4
R1
R3
R4S
L* =
R4S
R1 = H, R2 = CO2H, R3 = CH2CO2H, R4 = Me: 35% ee = 14% (R)
R1 = H, R2 = CO2H, R3 = CH2CO2H, R4 = i-Pr: 100% ee = 62% (R)
R1 = Ph, R2 = CO2H, R3 = NHAc, R4 = i-Pr: 45% ee = 37% (S)
R1 = H, R2 = CO2H, R3 = CH2CO2H, R4 = Ph: 100% ee = 17% (R)

Scheme 8.19 Hydrogenations of acrylic acid derivatives with sugar dithioether ligands.

newly generated *S*-stereocentres, affording only one stereoisomer, whatever the bulk of the alkyl substituent on the sulfur atom. The cationic complexes derived from these ligands were further investigated in the asymmetric hydrogenation of α,β-unsaturated acid derivatives. In all cases, the catalysts were active but poorly enantioselective, since the enantioselectivity never exceeded 5% ee.

Enantioselectivities of up to 47% ee were reported by Ruiz *et al.* in 1997 for the asymmetric hydrogenation of various prochiral dehydroamino acid derivatives and itaconic acid by using iridium cationic complexes of the novel chiral

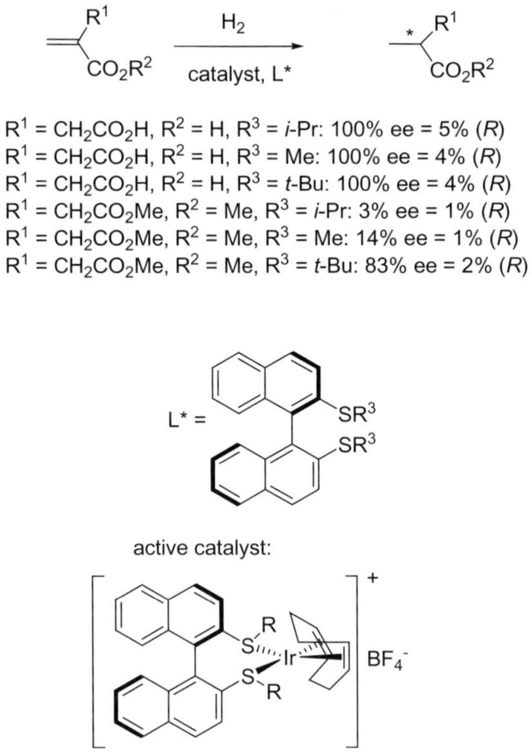
R1
CO2R2
H2
catalyst, L*
R1
CO2R2
R1 = CH2CO2H, R2 = H, R3 = i-Pr: 100% ee = 5% (R)
R1 = CH2CO2H, R2 = H, R3 = Me: 100% ee = 4% (R)
R1 = CH2CO2H, R2 = H, R3 = t-Bu: 100% ee = 4% (R)
R1 = CH2CO2Me, R2 = Me, R3 = i-Pr: 3% ee = 1% (R)
R1 = CH2CO2Me, R2 = Me, R3 = Me: 14% ee = 1% (R)
R1 = CH2CO2Me, R2 = Me, R3 = t-Bu: 83% ee = 2% (R)
L* =
SR3
SR3
active catalyst:
R
S
Ir
S
R
BF4-
+

Scheme 8.20 Hydrogenations of olefins with dithioether ligands derived from BINOL.

dithioether ligands depicted in Scheme 8.21.[26] The best results were obtained with the ligand bearing an *i*-Pr group, perhaps due to the greater hindrance and rigidity of this group.

A series of other cationic iridium complexes containing chiral dithioether ligands have more recently been prepared by Martin *et al.*, in order to study the influence of the sulfur substituents and the metallacycle size on the acetamidoacrylic acid hydrogenation reaction.[27] A mixture of diastereoisomers was obtained when the complexes were formed with BENOBS-R, DIOS-R and DMPS-R ligands (Scheme 8.22), probably due to the sulfur inversion process with decomposition of the complexes. This behaviour was not observed in the case of using a stable ligand containing a bis(metallacycles), such as DISOSEtS, which gave, however, a low enantioselectivity ($\leq$15% ee).

On the other hand, James reported, in 1976, the use of a chiral sulfoxide as a ligand of ruthenium for the asymmetric hydrogenation of itaconic acid, providing a low enantioselectivity of 12% ee (Scheme 8.23).[28]

Finally, Sinou *et al.* have shown that rhodium(I) catalysts formed with various chiral sulfonated diphosphines were efficient catalysts for the

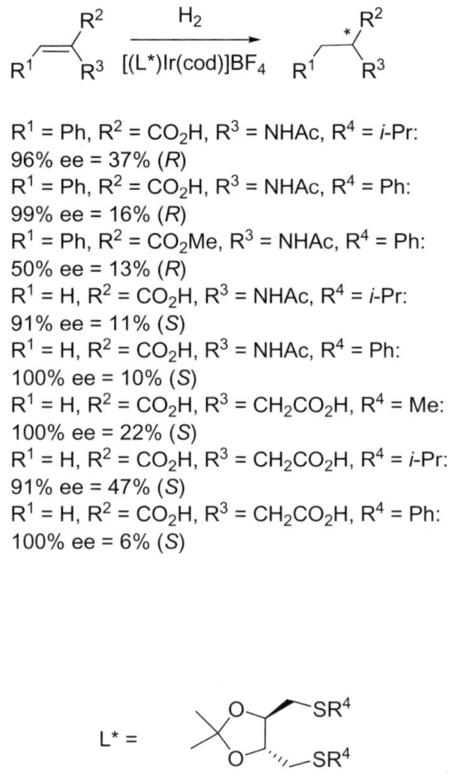

$$R^1R^2C{=}CR^3 \xrightarrow[\text{[(L*)Ir(cod)]BF}_4]{\text{H}_2} R^1 \overset{*}{C}R^2R^3$$

R^1 = Ph, R^2 = CO$_2$H, R^3 = NHAc, R^4 = *i*-Pr:
96% ee = 37% (*R*)
R^1 = Ph, R^2 = CO$_2$H, R^3 = NHAc, R^4 = Ph:
99% ee = 16% (*R*)
R^1 = Ph, R^2 = CO$_2$Me, R^3 = NHAc, R^4 = Ph:
50% ee = 13% (*R*)
R^1 = H, R^2 = CO$_2$H, R^3 = NHAc, R^4 = *i*-Pr:
91% ee = 11% (*S*)
R^1 = H, R^2 = CO$_2$H, R^3 = NHAc, R^4 = Ph:
100% ee = 10% (*S*)
R^1 = H, R^2 = CO$_2$H, R^3 = CH$_2$CO$_2$H, R^4 = Me:
100% ee = 22% (*S*)
R^1 = H, R^2 = CO$_2$H, R^3 = CH$_2$CO$_2$H, R^4 = *i*-Pr:
91% ee = 47% (*S*)
R^1 = H, R^2 = CO$_2$H, R^3 = CH$_2$CO$_2$H, R^4 = Ph:
100% ee = 6% (*S*)

L* =

Scheme 8.21 Hydrogenations of olefins with iridium complexes containing dithioether ligands.

asymmetric hydrogenation of a series of enamides using an aqueous-organic
two-phase solvent system.[29] As shown in Scheme 8.24, quantitative yields
combined with enantioselectivities of up to 87% ee were observed for these
reactions.

8.3 Hydrogenation of Ketones

A useful way to synthesise chiral alcohols is the reduction of prochiral carbonyl
compounds by using molecular hydrogen associated to a catalytic process
based on transition-metal complexes, generally of palladium, ruthenium, rhodium or iridium. In this context, several types of chiral sulfur-containing
ligands have been investigated for the asymmetric hydrogenation of ketones
with variable success. As an example, Lemaire *et al.* have developed the
asymmetric hydrogenation of acetophenone in the presence of palladium
complexes of chiral dithioether ligands, providing the corresponding alcohol in

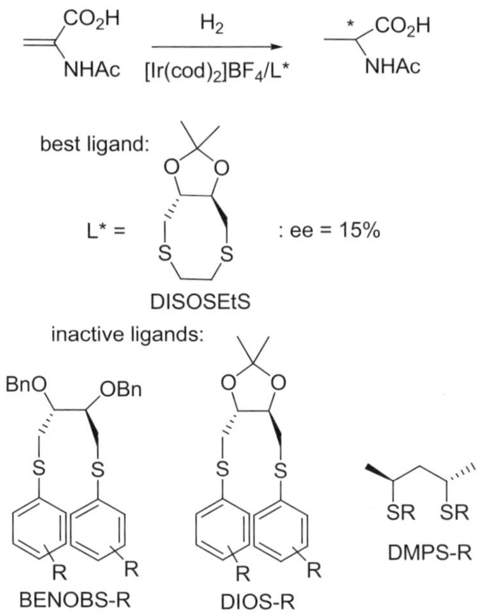

Scheme 8.22 Hydrogenation of acetamidoacrylic acid with iridium complexes containing dithioether ligands.

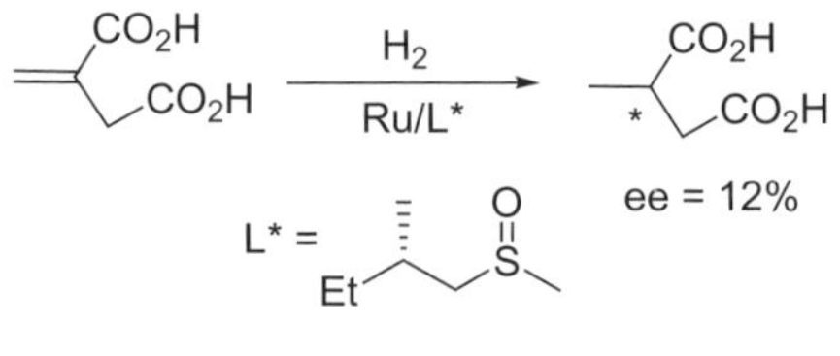

Scheme 8.23 Hydrogenation of itaconic acid with sulfoxide ligand.

only a low yield and no enantioselectivity, even when the ligand possessed a conformationally constrained structure, such as that derived from DIOP (Scheme 8.25).[30] The influence of the nature of the heteroatom was evaluated by changing one sulfur atom by a nitrogen atom, thus allowing the conversion to be improved up to 91% but, the asymmetric induction never exceeded 3% ee (Scheme 8.25).

In the same study, these authors also described the synthesis of other S/N ligands derived from various amino acids, such as proline, valine and cysteine.[30] These dithioether and azathioether ligands were further tested as potential palladium ligands for the reduction of acetophenone, but no significant induction was observed in each case of ligand (Scheme 8.26).

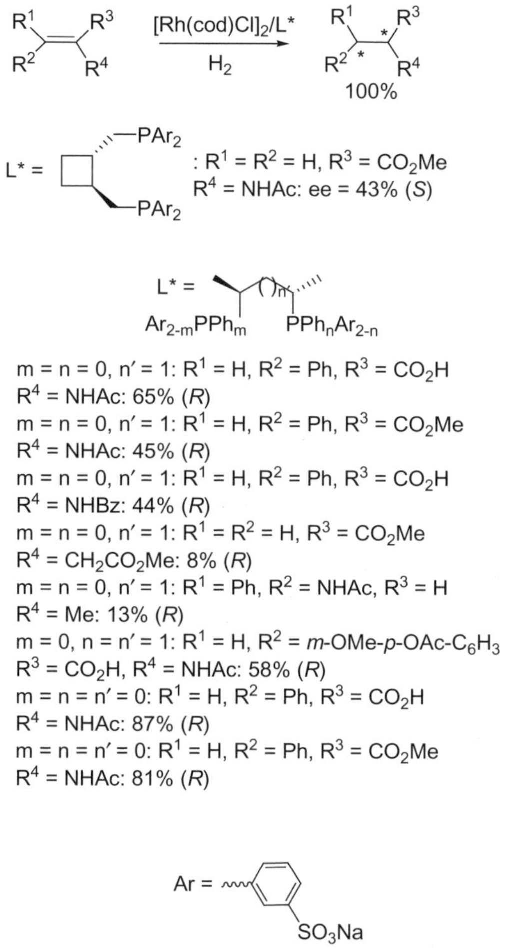

Scheme 8.24 Hydrogenations of enamides with sulfonated diphosphines.

In 1999, better results were reported by Lemaire *et al.* by using sulfur-containing ligands based on a dithiourea backbone in the asymmetric hydrogenation of phenylglyoxylate methyl ester.[31] In the presence of rhodium complexes of the two ligands depicted in Scheme 8.27, the reaction provided moderate enantioselectivities of up to 54% ee. It was noted that both the activity and the enantioselectivity slightly increased when using a diphenyldithiourea ligand compared with a cyclohexyldithiourea ligand. These ligands had the same isothiocyanate moiety but different chiral diamine parts, which suggested that the coordination did not occur exclusively via the sulfur atom. Moreover, the same reaction was performed in the presence of the corresponding iridium complexes, providing a dramatic loss of both the activity

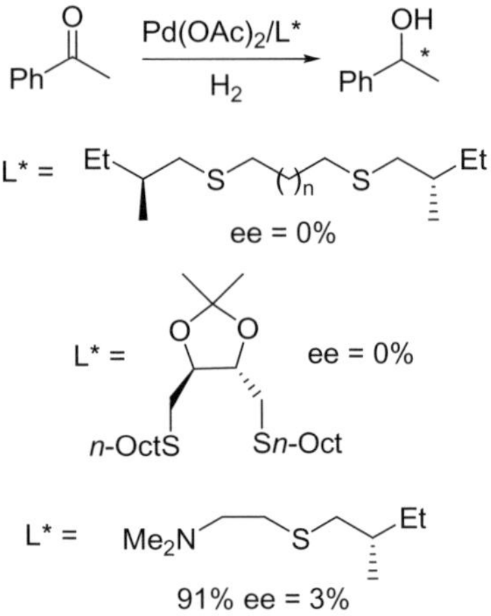

Scheme 8.25 Hydrogenation of acetophenone with S/S and S/N ligands.

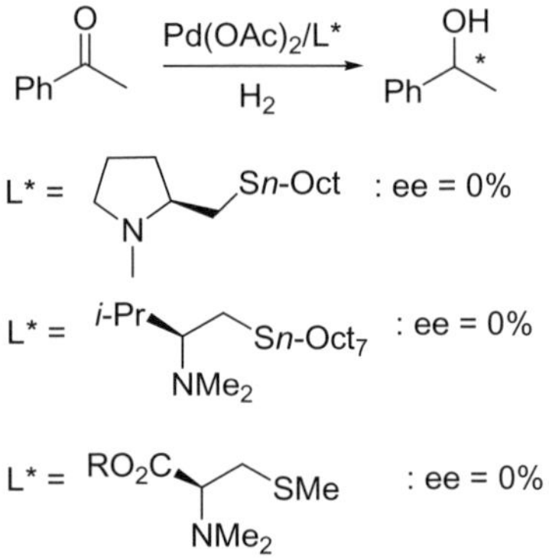

Scheme 8.26 Hydrogenation of acetophenone with S/N ligands.

and enantioselectivity when the dithioureas were used instead of the corresponding diamines. Thus, it seemed that the sulfur atom probably also contributed to form the metallic complexes.

Another class of chiral sulfur-containing ligands, diphosphine ligands based on a thiophene backbone, has been successfully developed by Sannicolo et al.[11,32] For the first time, bis(diphenylphosphines) bidentate ligands incorporating two benzothiophene moieties were tested as ruthenium ligands for the asymmetric hydrogenation of various α- and β-oxoesters. The results collected in Scheme 8.28 highlighted that these novel ligands induced excellent enantioselectivities of up to > 99% ee.

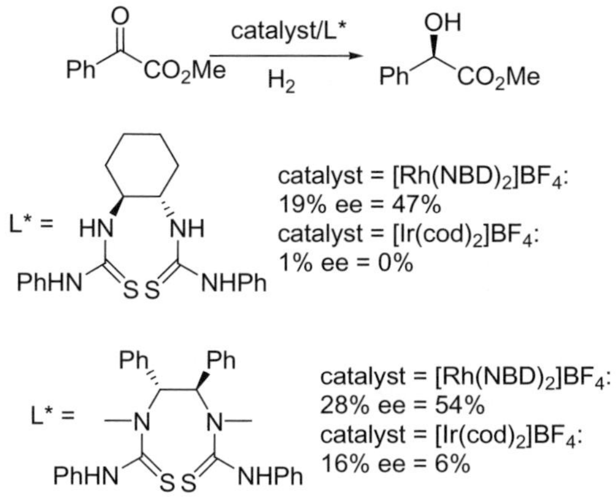
catalyst/L*
H2
Ph CO2Me
OH
L* =
HN NH
PhHN S S NHPh
catalyst = [Rh(NBD)2]BF4:
19% ee = 47%
catalyst = [Ir(cod)2]BF4:
1% ee = 0%
Ph Ph
N N
catalyst = [Rh(NBD)2]BF4:
28% ee = 54%
catalyst = [Ir(cod)2]BF4:
16% ee = 6%

Scheme 8.27 Hydrogenation of phenylglyoxylate methyl ester with dithiourea ligands.

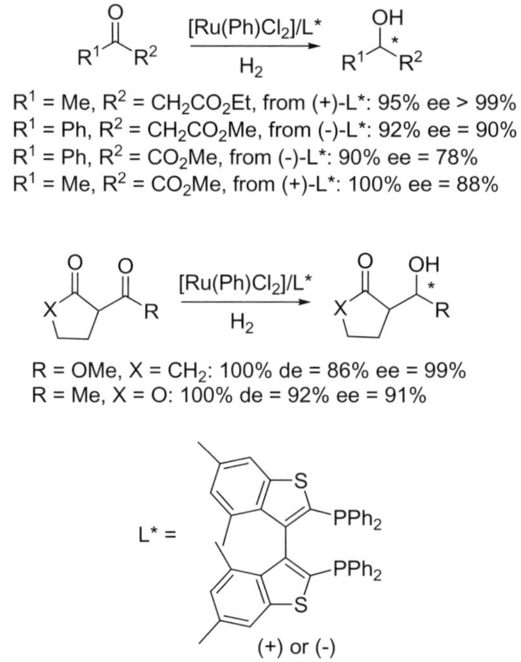
[Ru(Ph)Cl2]/L*
H2
R1 R2
OH
R1 = Me, R2 = CH2CO2Et, from (+)-L*: 95% ee > 99%
R1 = Ph, R2 = CH2CO2Me, from (-)-L*: 92% ee = 90%
R1 = Ph, R2 = CO2Me, from (-)-L*: 90% ee = 78%
R1 = Me, R2 = CO2Me, from (+)-L*: 100% ee = 88%
X R
R = OMe, X = CH2: 100% de = 86% ee = 99%
R = Me, X = O: 100% de = 92% ee = 91%
L* =
S
PPh2
(+) or (-)

Scheme 8.28 Hydrogenations of α- and β-oxoesters with diphosphine-dibenzothiophene ligand.

In 2000, these authors also developed a very efficient diphosphine-bithiophene ligand, tetraMe-BITIOP, which is depicted in Scheme 8.29.[10] The ruthenium complex of this electron-rich diphosphine was used as the catalyst in asymmetric hydrogenation reactions of prostereogenic carbonyl functions of α-

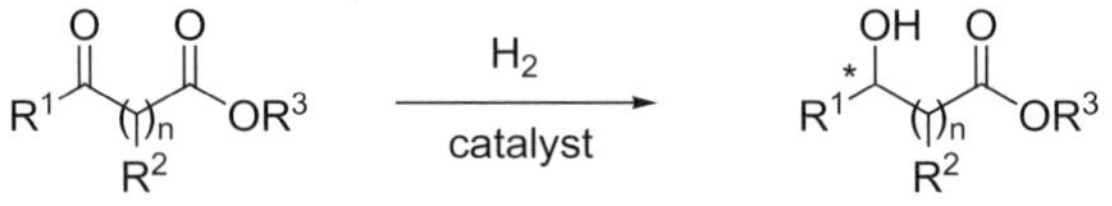

R^1 = Me, R^2 = H, R^3 = Et, n = 1, M = Ru, X = Y = Cl:
ee = 98% (S)
R^1 = Me, R^2 = H, R^3 = Et, n = 1, M = Ru, X = p-cymene
Y = Z = I: ee = 97% (S)
R^1 = Me, R^2 = Cl, R^3 = Et, n = 1, M = Ru, X = Ph
Y = Z = Cl: de = 82% ee = 94%
R^1 = CF$_3$, R^2 = H, R^3 = Et, n = 1, M = Ru, X = p-cymene
Y = Z = I: ee = 85% (R)
R^1 = Ph, R^2 = H, R^3 = Et, n = 1, M = Ru, X = Y = Cl:
ee = 93% (S)
R^1 = Me, R^2 = CH$_2$NHBz, R^3 = Et, n = 1, M = Ru
X = p-cymene, Y = Z = I: de = 94% ee = 99% (3S,2R)
R^1,R^2 = (CH$_2$)$_3$, R^3 = Me, n = 1, M = Ru, X = Y = Cl:
de = 84% ee = 99% (S,S)
R^1,R^2 = (CH$_2$)$_3$, R^3 = Me, n = 1, M = Ru, X = p-cymene
Y = Z = I: de = 88% ee = 99% (S,S)
R^1 = Ph, R^2 = H, R^3 = Me, n = 0, M = Ru, X = Ph
Y = Z = Cl: ee = 89% (R)
R^1 = Ph-(CH$_2$)$_2$, R^2 = H, R^3 = Et, n = 0, M = Ru
X = Ph, Y = Z = Cl: ee = 91% (S)

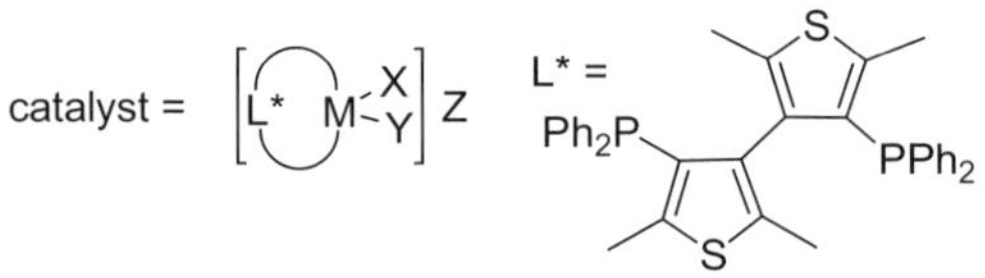

Scheme 8.29 Hydrogenations of α- and β-keto esters with diphosphine-bithiophene ligand.

and β-keto esters, generally affording the desired products with excellent enantioselectivities of up to 99% ee in all cases (Scheme 8.29).

In 2000, another type of diphosphine-thiophene ligand was developed by the same group for the asymmetric hydrogenation of α- and β-keto esters.[33] The enantioselectivities obtained by using this C_1-symmetric ligand were often similar to those obtained with the C_2-symmetric analogues, as shown in Scheme 8.30. Thus, the enantioselectivities observed for the hydrogenation of β-keto esters and methyl phenylglyoxylate were fully comparable to those obtained by using the most efficient DuPHOS and BINAP ligands.

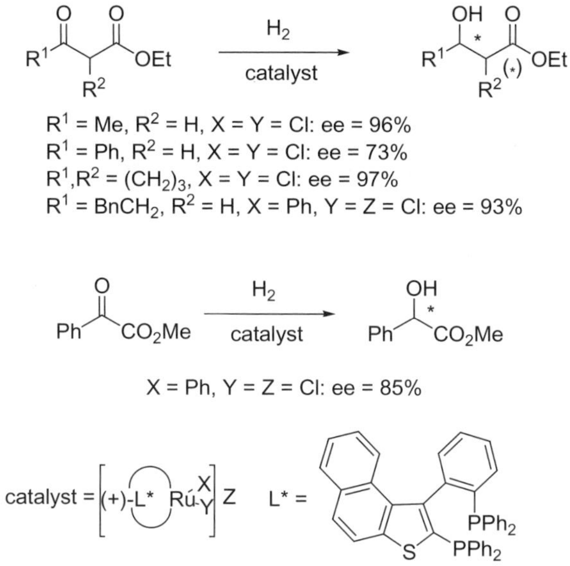

R^1 = Me, R^2 = H, X = Y = Cl: ee = 96%
R^1 = Ph, R^2 = H, X = Y = Cl: ee = 73%
R^1,R^2 = (CH$_2$)$_3$, X = Y = Cl: ee = 97%
R^1 = BnCH$_2$, R^2 = H, X = Ph, Y = Z = Cl: ee = 93%

X = Ph, Y = Z = Cl: ee = 85%

catalyst = [(+)-L* Ru^X_Y] Z   L* =

Scheme 8.30 Hydrogenations of α- and β-keto esters with C_1-symmetric diphosphine-thiophene ligand.

As an extension of this work, Sannicolo *et al.* have developed the syntheses of a wide range of novel C_1-symmetric chelating diphosphines with stereogenic axes displaying a mixed aryl–heteroaryl atropisomeric backbone.[34] Among these ligands, the ligand bearing a methoxy group on the phenyl moiety has proved to be the most efficient to catalyse, in combination with a ruthenium complex, the asymmetric hydrogenation of β-keto esters, since enantioselectivities of up to 99% ee were obtained (Scheme 8.31). It must be noted that the main advantageous features of this novel class of ligands were their easy synthetic accessibility and low cost. Indeed, these authors have estimated their preliminary cost at about one tenth of DuPHOS, one fifth of BINAP, and one third of tetraMe-BITIOP and, in some cases, even less.

More recently, these authors have reported the synthesis of a new thiophene-based analogue of (*R*,*R*)-Me-DuPHOS called UlluPHOS.[12] The facial recognition and enantioselection associated with ruthenium complexes of UlluPHOS and Me-DuPHOS were shown to be similarly high in various hydrogenations of β-keto esters (Scheme 8.32). The most important difference between these two ligands was found by comparing the reaction rates. Indeed, the authors have observed that the use of UlluPHOS considerably increased the activity of the complexes.

In addition, Peruzzini *et al.* developed, in 2007, iridium complexes of planar-chiral ferrocenyl phosphine-thioether ligands that were tested in the hydrogenation of simple alkyl aryl ketones.[35] These complexes were diastereoselectively generated in high yields (85–90%) by addition of the corresponding

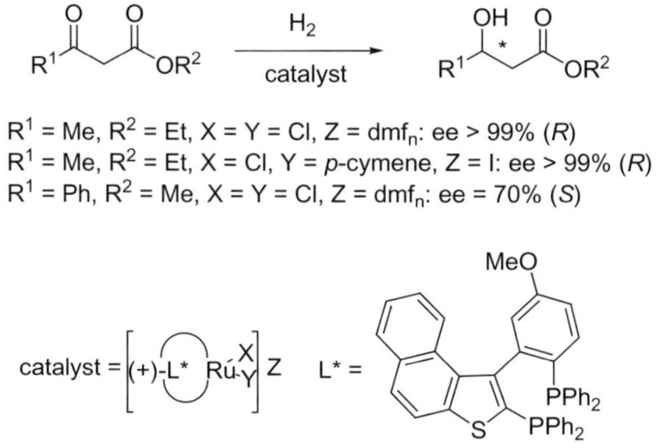

R^1 = Me, R^2 = Et, X = Y = Cl, Z = dmf$_n$: ee > 99% (R)
R^1 = Me, R^2 = Et, X = Cl, Y = p-cymene, Z = I: ee > 99% (R)
R^1 = Ph, R^2 = Me, X = Y = Cl, Z = dmf$_n$: ee = 70% (S)

catalyst = [(+)-L* Ru$_Y^X$] Z   L* =

Scheme 8.31 Hydrogenations of β-keto esters with diphosphine-thiophene ligand.

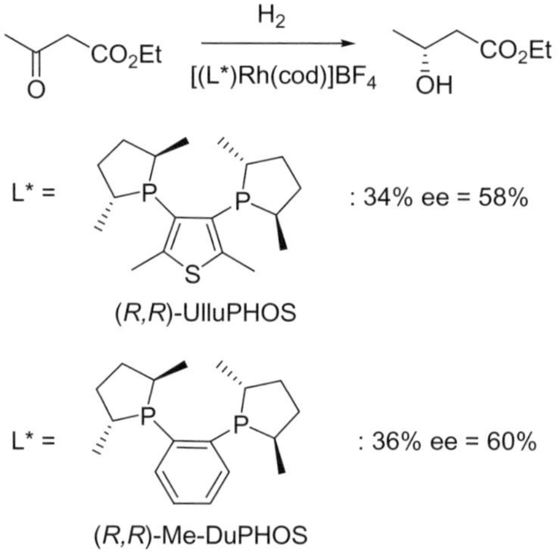

L* =   : 34% ee = 58%

(R,R)-UlluPHOS

L* =   : 36% ee = 60%

(R,R)-Me-DuPHOS

Scheme 8.32 Hydrogenation of β-keto ester with UlluPHOS and Me-DuPHOS.

bidentate ligands to [Ir(cod)Cl]$_2$. As shown in Scheme 8.33, excellent enantioselectivities of up to 99% ee combined with complete conversions were obtained by applying these highly active catalytic systems having turnover numbers of up to 915.

Finally, Sinou *et al.* have employed a chiral tetrasulfonated diphosphine as a ligand in the Rh-catalysed reduction of the Schiff base depicted in Scheme 8.34.[29] In the presence of an aqueous-organic two-phase solvent system, a quantitative yield combined with a moderate enantioselectivity of 34% ee were obtained for this reaction.

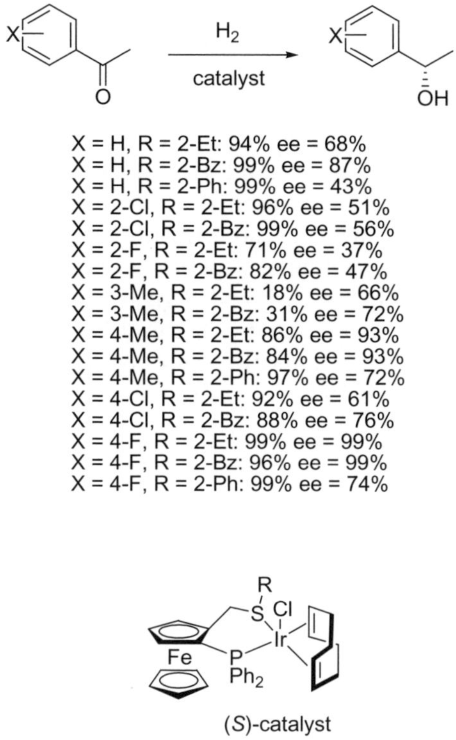
X
H2
catalyst
O
X
OH
X = H, R = 2-Et: 94% ee = 68%
X = H, R = 2-Bz: 99% ee = 87%
X = H, R = 2-Ph: 99% ee = 43%
X = 2-Cl, R = 2-Et: 96% ee = 51%
X = 2-Cl, R = 2-Bz: 99% ee = 56%
X = 2-F, R = 2-Et: 71% ee = 37%
X = 2-F, R = 2-Bz: 82% ee = 47%
X = 3-Me, R = 2-Et: 18% ee = 66%
X = 3-Me, R = 2-Bz: 31% ee = 72%
X = 4-Me, R = 2-Et: 86% ee = 93%
X = 4-Me, R = 2-Bz: 84% ee = 93%
X = 4-Me, R = 2-Ph: 97% ee = 72%
X = 4-Cl, R = 2-Et: 92% ee = 61%
X = 4-Cl, R = 2-Bz: 88% ee = 76%
X = 4-F, R = 2-Et: 99% ee = 99%
X = 4-F, R = 2-Bz: 96% ee = 99%
X = 4-F, R = 2-Ph: 99% ee = 74%
R
S
Cl
Ir
Fe
P
Ph2
(S)-catalyst

Scheme 8.33 Hydrogenations of alkyl aryl ketones with ferrocenyl phosphine-thioether ligands.

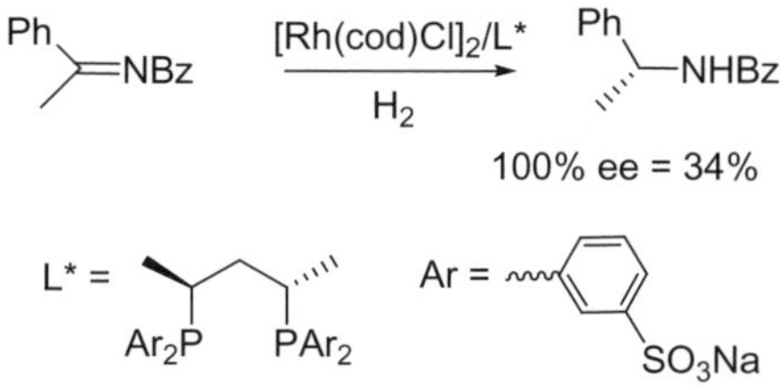
Ph
=NBz
[Rh(cod)Cl]2/L*
H2
Ph
NHBz
100% ee = 34%
L* =
Ar2P
PAr2
Ar =
SO3Na

Scheme 8.34 Hydrogenation of a Schiff base with tetrasulfonated diphosphine ligand.

8.4 Conclusions

In recent years, the asymmetric hydrogenation of prochiral olefins have been developed in the presence of various chiral sulfur-containing ligands combined with rhodium, iridium or more rarely ruthenium catalysts. The best results have been obtained by using S/P ligands, with enantioselectivities of up to 99% ee in some cases. Thus, various chiral thioether-phosphinite ligands developed independently by the groups of Evans, Dieguez and Khiar have allowed enantioselectivities of up to 98% ee to be obtained for the hydrogenation of various olefins. Moreover, Bolm's group. has obtained enantioselectivities of up to 99% ee for similar reactions by using BINOL-derived *N*-phosphino sulfoximine ligands. In addition, a wide range of thiophene-based diphosphine ligands, such as tetraMe-BITIOP ligand, UlluPHOS ligand or benzothiophene-based DUPHOS analogues have been developed by Sannicolo's group and proved to be the most promising chelates for the asymmetric hydrogenation of olefins, providing enantioselectivities superior to 99% ee.

Concerning the reduction of carbonyl derivatives, the best ligands of rhodium or ruthenium have proved to be C_2-chiral biphosphines based on thiophene and developed by Sannicolo's group, providing in some cases enantioselectivities of up to 99% ee. On the other hand, the catalytic system based on iridium complexes of chiral planar ferrocenyl phospino-thioether ligands, recently developed by Peruzzini's group, is so far the best one for the iridium-catalysed hydrogenation of ketones. Much less success was found when using other types of sulfur-containing ligands such as S/S, S/O or S/N ligands.

References

1. (a) R. Noyori, in *Asymmetric Catalysis in Organic Synthesis*, Wiley, New York, 1994; (b) I. Ojima, (ed.), in *Catalytic Asymmetric Synthesis*, Wiley, New York, 2000; (c) E. N. Jacobsen, A. Pfaltz and H. H. Yamamoto, (Ed.), in *Comprehensive Asymmetric Catalysis*, **Vol. 1**, Springer, Berlin, 1999, pp. 121–247.
2. R. Noyori, H. Takaya and T. Ohta, in *Catalytic Asymmetric Synthesis*, ed. I. Ojima, VCH Publishers, Weinheim, 1993, pp. 1–39.
3. E. Hauptman, P. J. Fagan and W. Marshall, *Organometallics*, 1999, **18**, 2061–2073.
4. G. A. Molander, J. P. Burke and P. J. Carroll, *J. Org. Chem.*, 2004, **69**, 8062–8069.
5. D. A. Evans, F. E. Michael, J. S. Tedrow and K. R. Campos, *J. Am. Chem. Soc.*, 2003, **125**, 3534–3543.
6. E. Guimet, M. Diéguez, A. Ruiz and C. Claver, *J. Chem. Soc., Dalton Trans.*, 2005, 2557–2562.
7. O. Pàmies, M. Diéguez, G. Net, A. Ruiz and C. Claver, *Organometallics*, 2000, **19**, 1488–1496.
8. N. Khiar, B. Suarez, M. Stiller, V. Valdivia and I. Fernandez, *Phosphorus, Sulfur, and Silicon*, 2005, **180**, 1253–1258.
9. E. Hauptman, R. Shapiro and W. Marshall, *Organometallics*, 1998, **17**, 4976–4982.
10. T. Benincori, E. Cesarotti, O. Piccolo and F. Sannicolo, *J. Org. Chem.*, 2000, **65**, 2043–2047.
11. T. Benincori, E. Brenna, F. Sannicolo, L. Trimarco, P. Antognazza, E. Cesarotti, F. Demartin and T. Pilati, *J. Org. Chem.*, 1996, **61**, 6244–6251.

12. T. Benincori, T. Pilati, S. Rizzo, F. Sannicolo, M. J. Burk, L. de Ferra, E. Ulluci and O. Piccolo, *J. Org. Chem.*, 2005, **70**, 5436–5441.
13. U. Berens, U. Englert, S. Geyser, J. Runsink and A. Salzer, *Eur. J. Org. Chem.*, 2006, 2100–2109.
14. P. G. Cozzi, F. Menges and S. Kaiser, *Synlett*, 2003, 833–836.
15. M. T. Reetz, O. G. Bondarev, H.-J. Gais and C. Bolm, *Tetrahedron Lett.*, 2005, **46**, 5643–5646.
16. C. Hedberg, K. Källström, P. Brandt, L. K. Hansen and P. G. Andersson, *J. Am. Chem. Soc.*, 2006, **128**, 2995–3001.
17. L. B. Schenkel and J. A. Ellman, *J. Org. Chem.*, 2004, **69**, 1800–1802.
18. M. L. Tommasino, M. Casalta, J. A. Breuzard and M. Lemaire, *Tetrahedron: Asymmetry*, 2000, **11**, 4835–4841.
19. M. L. Tommasino, C. Thomazeau, F. Touchard and M. Lemaire, *Tetrahedron: Asymmetry*, 1999, **10**, 1813–1819.
20. B. R. James and R. S. McMillan, *Can. J. Chem.*, 1977, **55**, 3927–3932.
21. W. Li, J. P. Waldkirch and X. Zhang, *J. Org. Chem.*, 2002, **67**, 7618–7623.
22. M. Diéguez, A. Ruiz, C. Claver, M. M. Pereira and A. M. d'A. Rocha Gonsalves, *J. Chem. Soc., Dalton Trans.*, 1998, 3517–3522.
23. O. Pàmies, M. Diéguez, G. Net, A. Ruiz and C. Claver, *J. Chem. Soc., Dalton Trans.*, 1999, 3439–3444.
24. M. Dieguez, A. Ruiz, C. Claver, M. M. Pereira, M. T. Flor, J. C. Bayon, M. E. S. Serra and A. M. d'A. Rocha Gonzalves, *Inorg. Chim. Acta*, 1999, **295**, 64–70.
25. M. Dieguez, A. Ruiz, C. Claver, F. Doro, M. G. Sanna and S. Gladiali, *Inorg. Chim. Acta*, 2004, **357**, 2957–2964.
26. M. Diéguez, A. Orejon, A. M. Masdeu-Bulto, R. Echarri, S. Castillon, C. Claver and A. Ruiz, *J. Chem. Soc., Dalton Trans.*, 1997, 4611–4618.
27. L. Flores-Santos, E. Martin, A. Aghmiz, M. Diéguez, C. Claver, A. M. Masdeu-Bulto and M. A. Munoz-Hernandez, *Eur. J. Inorg. Chem.*, 2005, **1**, 2315–2323.
28. B. R. James, R. S. McMillan and K. J. Reimer, *J. Mol. Catal.*, 1976, **1**, 439–441.
29. Y. Amrani, L. Lecomte and D. Sinou, *Organometallics*, 1989, **8**, 542–547.
30. F. Fache, P. Gamez, F. Nour and M. Lemaire, *J. Mol. Catal.*, 1993, **85**, 131–141.
31. M. L. Tommasino, C. Thomazeau, F. Touchard and M. Lemaire, *Tetrahedron: Asymmetry*, 1999, **10**, 1813–1819.
32. T. Benincori, E. Brenna, F. Sannicolo, L. Trimarco, P. Antognazza and E. Cesarotti, *J. Chem. Soc., Chem. Commun.*, 1995, 685–686.
33. T. Benincori, E. Cesarotti, O. Piccolo and F. Sannicolo, *J. Org. Chem.*, 2001, **66**, 5940–5942.
34. F. Sannicolo, T. Benincori, S. Rizzo, S. Gladiali, S. Pulacchini and G. Zotti, *Synthesis*, 2001, **15**, 2327–2336.
35. E. Le Roux, R. Malacea, E. Manoury, R. Poli, L. Gonsalvi and M. Peruzzini, *Adv. Synth. Catal.*, 2007, **349**, 309–313.

CHAPTER 9
Hydrogen Transfer

Asymmetric reduction of C=O and C=N bonds forming chiral alcohols and amines, respectively, is among the most fundamental molecular transformations[1] and has been a well-explored reaction in asymmetric catalysis over the past ten years.[2] In nature, oxidoreductases such as horse liver alcohol dehydrogenase catalyse the transfer hydrogenation of carbonyl compounds to alcohols using cofactors like NADH or NADPH.[3] Such biochemical reactions are normally very stereoselective. However, organic synthesis needs economically and technically more beneficial methods that are very general. A reaction using nonhazardous organic molecules provides a useful complement to catalytic reduction using molecular hydrogen, particularly for small- to medium-scale reactions. Transfer hydrogenation is operationally simple and the selectivities including functional-group differentiation may be different from those of hydrogenation. The first successful enantioselective transfer hydrogenations of ketones using *i*-PrOH as the source of hydrogen were reported in the 1990s by Pfaltz,[4] Genêt,[5] Lemaire,[6] and Evans,[7] among others.[8] Most studies were carried out by using the chirally modified transition metals such as ruthenium, rhodium, and iridium as catalyst precursors.[9] Hydride-transfer reactions generally use sources of hydrogen other than molecular hydrogen such as cyclohexene, cyclohexadiene, alcohols, acids, cyclic ethers and other molecules that are fairly readily dehydrogenated. This method avoids all the risks inherent to molecular hydrogen, and the chemoselectivities can be modulated by the proper choice of the hydride donor. The reaction occurs under relatively mild conditions, giving the desired alcohol generally in high yields and good enantioselectivities. Different substrates can be hydrogenated with the appropriate catalysts and hydrogen donors.[10] Considerable efforts have been devoted to the study of the asymmetric version of these processes,[11] the most widely studied of which is, perhaps, the asymmetric hydrogenation of ketones with alcohols, using normally *i*-PrOH, as the source of hydrogen. Chiral S/N ligands have been the most implicated

RSC Catalysis Series No. 2
Chiral Sulfur Ligands: Asymmetric Catalysis
By Hélène Pellissier
© Hélène Pellissier 2009
Published by the Royal Society of Chemistry, www.rsc.org

ligands in this type of reaction. The earliest enantioselective version of the hydrogen-transfer reduction using chiral sulfur-containing ligands was developed by James *et al.*, in 1986.[12] These authors studied the ability of chiral sulfoxide-containing rhodium complexes to promote the enantioselective transfer hydrogenation of ketones. Thus, the use of a chiral sulfoxide derived from methionine as ligand of rhodium in the presence of *i*-PrOH as the source of hydrogen allowed an enantioselectvity of up to 75% ee to be obtained for the reduction of 1-*p*-tolylacetone (Scheme 9.1).

A new class of efficient ligands, chiral aminosulf(ox)ides was developed, in 2000, by Petra *et al.* for the iridium-catalysed asymmetric transfer hydrogenation of acetophenone in the presence of formic acid as the hydrogen donor.[13] In this study, the authors assumed a bidentate S/N-coordination on the basis of the fact that ethanolamine, which is representative for N/O-coordination, did not catalyse the reduction of acetophenone under these conditions. The use of aminosulfide ligands based on (*R*)-cysteine in the iridium-catalysed asymmetric transfer hydrogenation of acetophenone provided the corresponding alcohol in low enantioselectivities ($\leq$18% ee) as shown in Scheme 9.2. On the other hand, the involvement in the same reaction of the corresponding sulfoxide-containing β-amino alcohols as the ligands permitted the fast formation of the expected alcohol with enantioselectivities of up to 65% ee (Scheme 9.2).

The scope of this methodology was extended by these authors to more sterically hindered ketones that provided the corresponding alcohols with enhanced enantioselectivities.[13] As shown in Scheme 9.3, the results demonstrated that the steric and electronic properties of the substrates influenced the reaction course.

In the same study, these authors have prepared another series of aminosulf(ox)ide ligands based on the (Nor)ephedrine and 2-aminodiphenylethanol skeletons, bearing two chiral centres in the carbon backbone.[13] Their application to the iridium-catalysed hydrogen-transfer reduction of acetophenone generally gave better yields, but the enantioselectivity never exceeded 65% ee (Scheme 9.4).

In 2001, a screening study for the enantioselective reduction of various aryl ketones was developed by Petra *et al.* in the presence of amino sulfide

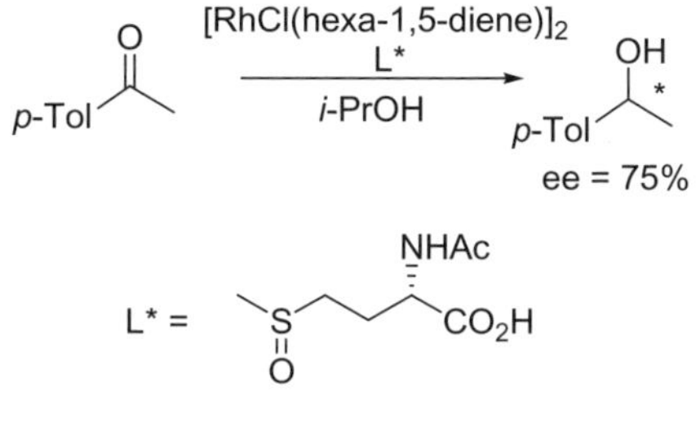

Scheme 9.1 Reduction of 1-*p*-tolylacetone with sulfoxide ligand.

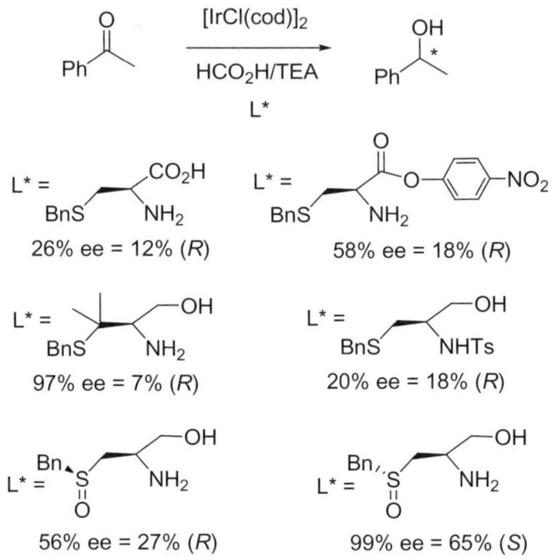

Scheme 9.2   Ir-catalysed reduction of acetophenone with aminosulf(ox)ide ligands.

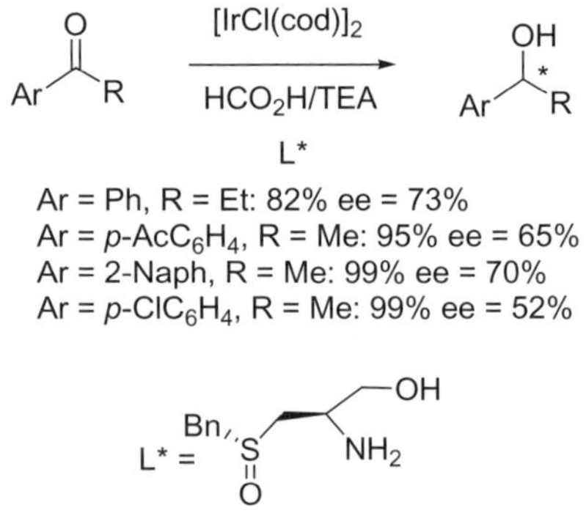

Scheme 9.3   Ir-catalysed reductions of ketones with aminosulfoxide ligand.

(1*R*,2*S*)-2-amino-1,2-diphenyl-1-benzylthioethane ligands and *i*-PrOH as the source of hydrogen.[14] The corresponding chiral alcohols could be isolated with moderate to high enantioselectivities of up to 97% ee, as shown in Scheme 9.5. The choice of the substrate did not markedly affect the outcome of the reactions. In an attempt to extend the scope of this methodology, the authors applied the same conditions to the reduction of various dialkyl ketones, but only low enantioselectivities ($\leq$33% ee) were obtained in these cases (Scheme 9.5). However, it must be noted that an enantioselectivity of 90% ee was

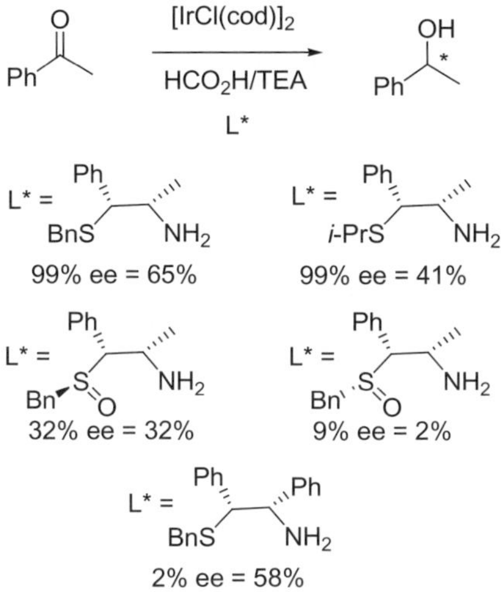

Scheme 9.4 Ir-catalysed reduction of acetophenone with aminosulf(ox)ide ligands.

obtained for the reduction of an acetylenic ketone, as depicted in Scheme 9.5. On the other hand, the application of the same conditions to functionalised ketones, such as 2-chloroacetophenone or 3-chloropropiophenone, did not result in the formation of the corresponding alcohols.

In 2003, two new classes of nitrogen- and sulfur-containing ligands, 2-aza-norbornyl amino sulfides and their corresponding sulfoxides, have been synthesised by Andersson *et al.* and further evaluated in the asymmetric iridium-catalysed transfer hydrogenation of acetophenone.[15] When formic acid was employed as the hydrogen donor, the use of 2-azanorbornyl amino sulfide ligand depicted in Scheme 9.6 led to a complete conversion into the corresponding (*R*)-alcohol in 15% ee when the reaction was carried out at 60 °C. Surprisingly, when the reaction was carried out at room temperature, the conversion was much lower (30%) but interestingly led to the formation of the (*S*)-product as the major enantiomer with an enantioselectivity of 28% ee. No explanation was proposed in order to rationalize these unexpected results. On the other hand, when *i*-PrOH was used as the source of hydrogen, the enantioselectivity dramatically increased up to 63% ee ((*R*)-enantiomer). Moreover, the use of the corresponding sulfoxide ligand in similar conditions allowed a better enantioselectivity of 80% ee to be obtained (Scheme 9.6), whereas the employment of formic acid as the source of hydrogen resulted in no reaction in this case of ligand.

In the same study, several ligands variously functional on both the nitrogen and the sulfur atoms have been developed, providing a new class of cyclo-hexylamino sulfide ligands derived from cyclohexene oxide.[15] All the ligands depicted in Scheme 9.7 were evaluated for the Ir-catalysed hydride-transfer reduction of acetophenone in the presence of *i*-PrOH as the hydrogen donor, providing enantioselectivities of up to 70% ee.

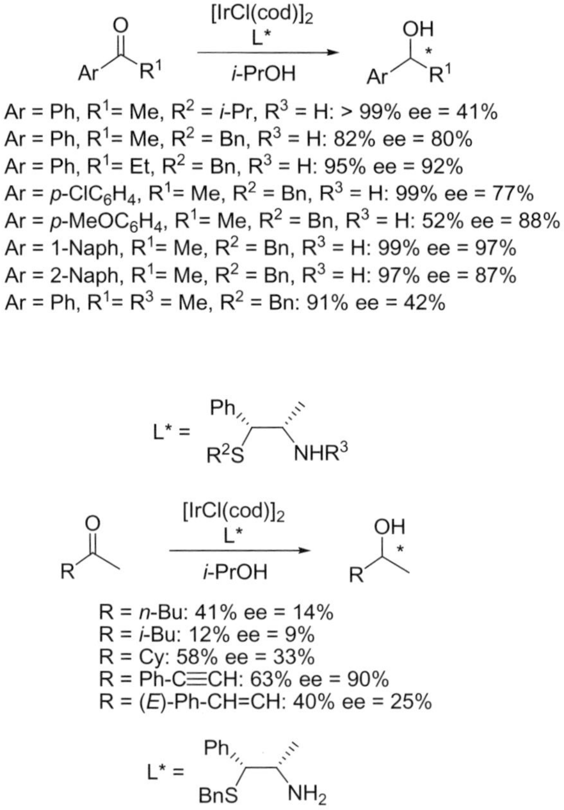
[IrCl(cod)]2
L*
i-PrOH
O
Ar R1
OH
*
Ar R1
Ar = Ph, R1= Me, R2 = i-Pr, R3 = H: > 99% ee = 41%
Ar = Ph, R1= Me, R2 = Bn, R3 = H: 82% ee = 80%
Ar = Ph, R1= Et, R2 = Bn, R3 = H: 95% ee = 92%
Ar = p-ClC6H4, R1= Me, R2 = Bn, R3 = H: 99% ee = 77%
Ar = p-MeOC6H4, R1= Me, R2 = Bn, R3 = H: 52% ee = 88%
Ar = 1-Naph, R1= Me, R2 = Bn, R3 = H: 99% ee = 97%
Ar = 2-Naph, R1= Me, R2 = Bn, R3 = H: 97% ee = 87%
Ar = Ph, R1= R3 = Me, R2 = Bn: 91% ee = 42%
L* =
Ph
R2S
NHR3
[IrCl(cod)]2
L*
i-PrOH
O
R
OH
*
R
R = n-Bu: 41% ee = 14%
R = i-Bu: 12% ee = 9%
R = Cy: 58% ee = 33%
R = Ph-C≡CH: 63% ee = 90%
R = (E)-Ph-CH=CH: 40% ee = 25%
L* =
Ph
BnS
NH2

Scheme 9.5 Ir-catalysed reductions of ketones with aminosulfide ligands.

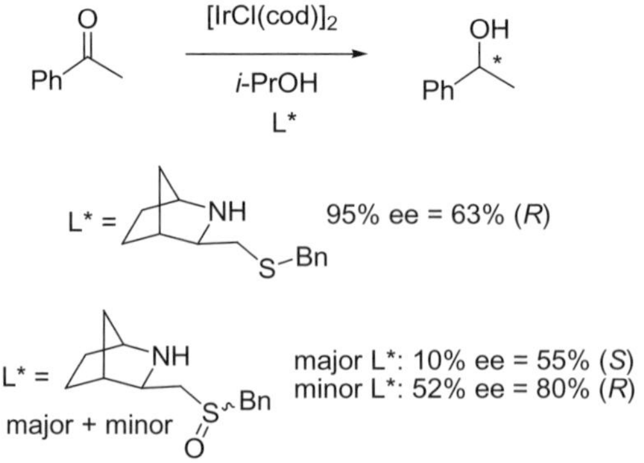
[IrCl(cod)]2
i-PrOH
L*
O
Ph
OH
*
Ph
L* =
NH
S
Bn
95% ee = 63% (R)
L* =
NH
S
Bn
O
major + minor
major L*: 10% ee = 55% (S)
minor L*: 52% ee = 80% (R)

Scheme 9.6 Ir-catalysed reduction of acetophenone with 2-azanorbornyl amino-sulf(ox)ide ligands.

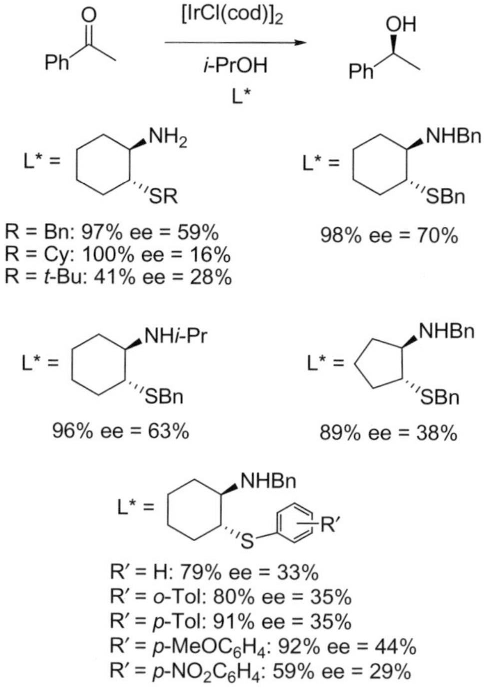

Scheme 9.7 Ir-catalysed reduction of acetophenone with cyclohexylamino sulfide ligands.

In 2004, the same group reported the synthesis of another new class of chiral S/N ligands based on the camphor scaffold on the basis of a simple five-step sequence starting from commercially available enantiopure camphor sulfonic acid.[16] Actually, two series of ligands involving six- and five-membered chelating rings were prepared and further evaluated in the asymmetric transfer hydrogenation of acetophenone in the presence of an iridium catalyst and *i*-PrOH as the hydrogen donor. The best enantioselectivities of up to 80% ee were obtained by using ligands having an exocyclic primary amine moiety, which generated a six-membered chelate with the metal, while ligands containing an intracyclic secondary amine, forming a five-membered chelate with iridium, led to lower enantioselectivities of up to 60% ee (Scheme 9.8). Indeed, both higher enantioselectivity and activity were observed in the case of using 1,3-aminothiol ligands despite the large freedom of this system giving a six-membered chelate with the metal compared to the 1,2-aminothiol ligand system offering a more rigid five-membered chelate. These surprising results suggested that the formation of the catalyst and/or its reactivity were very dependent on the bulkiness of the ligand substitution on the nitrogen and sulfur atoms.

More recently, Zaitsev and Adolfsson have reported the preparation and application of novel chiral S/N ligands for rhodium- and ruthenium-catalysed

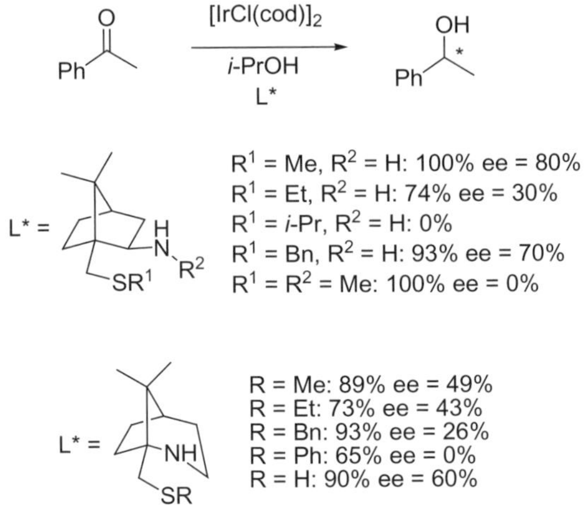

Scheme 9.8   Ir-catalysed reduction of acetophenone with camphor-derived amino sulfide ligands.

reductions of aryl ketones under transfer hydrogenation conditions.[17] It was noted that a subtle change in the ligand structure, replacing the carbonyl oxygen with sulfur in simple α-amino acid amides, resulted in a dramatic activity and selectivity improvement in the reduction of ketones. In addition, this novel class of simple, modular and highly efficient thioamide ligands had a remarkable feature based on the switch of the product enantioselectivity observed when the amide functionality was replaced by the corresponding thioamide when chiral induction occurred. According to the authors, this inversion might occur from a different coordination mode and suggested another hydrogen-transfer pathway. In the presence of *i*-PrOH as the hydrogen donor, the Rh- or Ru-catalysed reduction of various ketones allowed high yields and enantioselectivities of up to 97% ee to be obtained by using only 0.25 mol% catalyst loading (Scheme 9.9). These results are representative once again of the positive influence of sulfur ligands in asymmetric catalysis. Indeed, a little change in the ligand structure, such as changing the oxygen atom by a sulfur atom, had a dramatic influence on both the activity and selectivity of the catalysts.

Chiral thiourea ligands have the advantages, among others, of being easily accessible from their corresponding diamines and easy to use since they do not need to be handled or stored under an inert atmosphere. In this context, Lemaire *et al.* demonstrated, in 1997, that chiral mono- and dithiourea ligands could be efficient ligands for the asymmetric Ru-catalysed hydride transfer reduction of ketones in the presence of *i*-PrOH used as the hydrogen donor.[18] As shown in Scheme 9.10, the best enantioselectivities of up to 94% ee were obtained by using the dithiourea ligands. As the bulkiness of the alkyl group of the ketone was raised from methyl to ethyl to isopropyl, and the steric

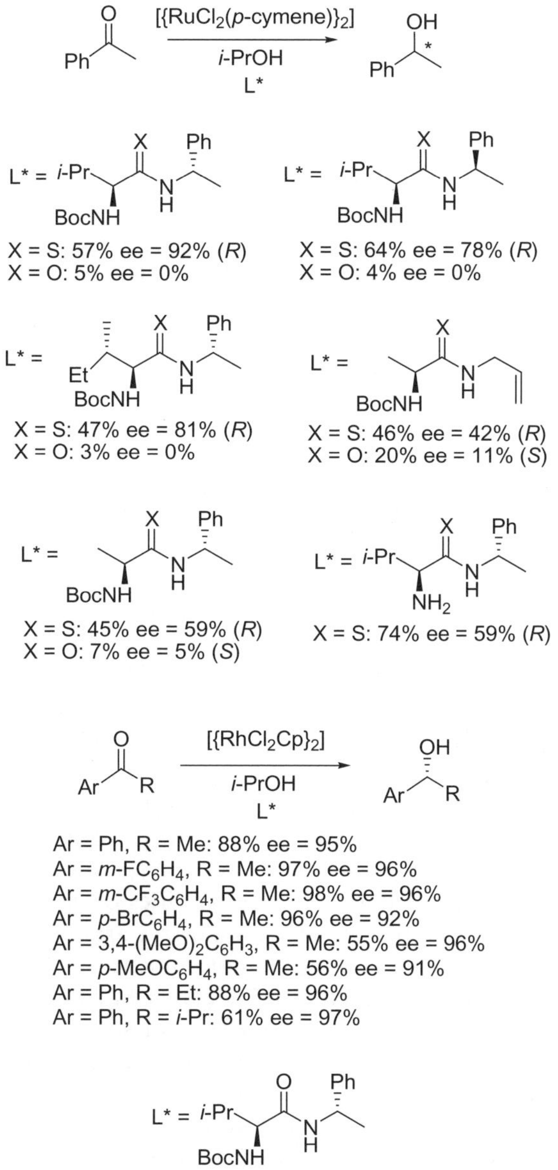

Scheme 9.9 Rh- and Ru-catalysed reductions of ketones with thioamide ligands.

differentiation between phenyl and alkyl groups decreased, it was shown that the enantioselectivity increased. On the other hand, with the *tert*-butyl analogue, the enantioselectivity was reversed since the bulkier *tert*-butyl group forced the substrate to approach by its *Si* face instead of the *Re* face for the other ketones.

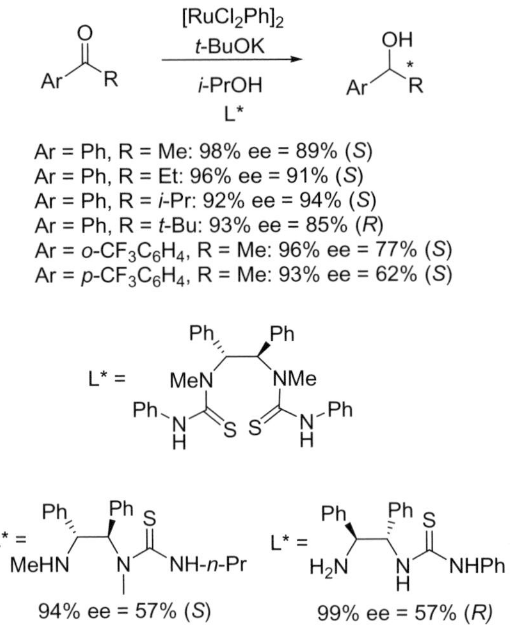

$$[RuCl_2Ph]_2$$
$$t\text{-BuOK}$$
$$\xrightarrow{}$$
$$i\text{-PrOH}$$
$$L^*$$

Ar = Ph, R = Me: 98% ee = 89% (S)
Ar = Ph, R = Et: 96% ee = 91% (S)
Ar = Ph, R = i-Pr: 92% ee = 94% (S)
Ar = Ph, R = t-Bu: 93% ee = 85% (R)
Ar = o-CF₃C₆H₄, R = Me: 96% ee = 77% (S)
Ar = p-CF₃C₆H₄, R = Me: 93% ee = 62% (S)

94% ee = 57% (S)   99% ee = 57% (R)

Scheme 9.10 Ru-catalysed reductions of ketones with (di)thiourea ligands.

In the same context, Lemaire *et al.* have reported the rhodium-catalysed hydride-transfer reduction of ketones using various dithiourea ligands bearing an aromatic ring on their terminal nitrogen atoms.[19] Various electron-donating and electron-withdrawing groups were evaluated for the Rh-catalysed reduction of acetophenone, providing enantioselectivities of up to 75% ee (Scheme 9.11). The results indicated that the steric hindrance around the nitrogen atoms bound to the aromatic rings had only a minor influence. A dramatic decrease of the enantioselectivity was observed ($\leq 15\%$ ee) when an electron-withdrawing group was introduced. In this work, the effects of structural modifications of both the ligand and the substrate on the enantioselectivity and conversion have been studied, demonstrating that the coordination of the dithioureas took place through the sulfur atom. As a consequence of this coordination, the thiourea should be considered as an S ligand rather than an N ligand.

In another context, chiral thioimidazolidine ligands have been successfully applied to the ruthenium-catalysed asymmetric hydrogen transfer of several aryl ketones by Kim *et al.*, furnishing the corresponding chiral alcohols with high yields and enantioselectivities of up to 77% ee (Scheme 9.12).[20]

Excellent enantioselectivities have also been reported by End *et al.* by using chiral phosphino-oxazoline ligands containing thiophene or benzothiophene backbones in the Ru-catalysed asymmetric hydrogen transfer of acetophenone.[21] These novel ligands, called HetPHOX, gave generally lower catalytic activity in this reaction than their corresponding phosphino-oxazoline

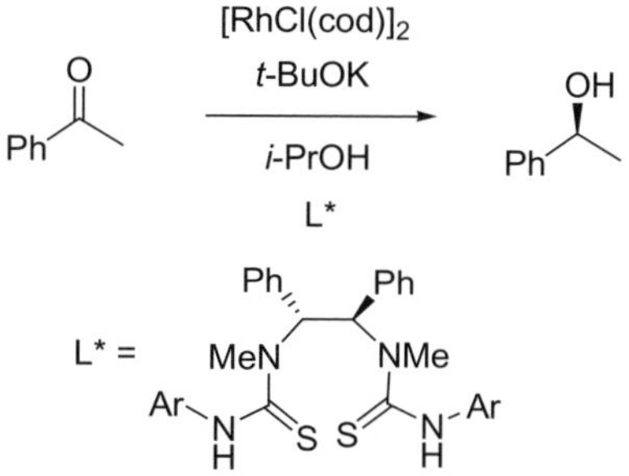

Ar = Ph: 97% ee = 63%
Ar = 2,6-(Me)$_2$C$_6$H$_3$: 92% ee = 66%
Ar = 1-Naph: 98% ee = 65%
Ar = *p*-MeOC$_6$H$_4$: 96% ee = 71%
Ar = *p*-(NMe$_2$)C$_6$H$_4$: 99% ee = 65%
Ar = 2,4-(MeO)$_2$C$_6$H$_3$: 97% ee = 75%
Ar = *p*-NO$_2$C$_6$H$_4$: 42% ee = 14%
Ar = *p*-CF$_3$C$_6$H$_4$: 60% ee = 15%
Ar = *p*-CNC$_6$H$_4$: 85% ee = 12%

Scheme 9.11   Rh-catalysed reduction of acetophenone with dithiourea ligands.

analogues (PHOX) but the expected alcohols were isolated with better enantioselectivities of up to 99% ee in some cases (Scheme 9.13).

In 2007, Gao *et al.* reported the development of a novel family of excellent chiral tetradentate S/N/N/S ligands including thiophene moieties.[22] Coupled with iridium complex, these ligands have proved to be efficient catalyst systems for the asymmetric transfer hydrogenation of a wide range of aromatic ketones in the presence of *i*-PrOH as the hydrogen donor. Indeed, the corresponding alcohols were obtained with high yields and enantioselectivities of up to 96% ee, as shown in Scheme 9.14. The reactions could be performed in air and the catalytic experiments were greatly simplified. In order to determine the role of the sulfur in the thiophene group of the ligand depicted in Scheme 9.14, these authors have synthesised the corresponding chiral ligand, (1*R*,2*R*)-*N*¹,*N*²-bis(furan-2-ylmethyl)cyclohexane-1,2-diamine, which bore oxygen instead of sulfur. In the same reaction conditions the use of this ligand in the asymmetric transfer hydrogenation of propiophenone afforded the corresponding alcohol with both lower yield and enantioselectivity of 25% and 75% ee respectively, instead of 94% yield and 87% ee in the case of the sulfur ligand. This result indicated that the sulfur in the thiophene group of the sulfur ligand played an important role in the asymmetric transfer hydrogenation process.

On the other hand, one of the first chiral sulfur-containing ligands employed in the asymmetric transfer hydrogenation of ketones was introduced by Noyori *et al.*[23] Thus, the use of *N*-tosyl-1,2-diphenylethylenediamine (TsDPEN) in combination with ruthenium for the reduction of various aromatic ketones in the presence of *i*-PrOH as the hydrogen donor, allowed the corresponding alcohols to be obtained in both excellent yields and enantioselectivities, as

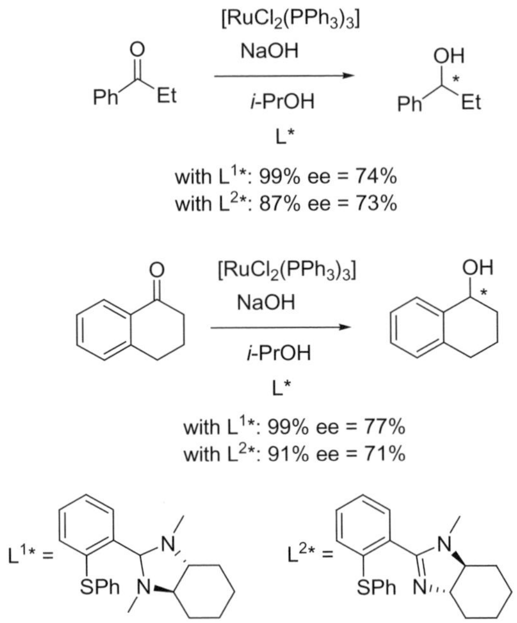

Scheme 9.12 Ru-catalysed reductions of ketones with thioimidazolidine ligands.

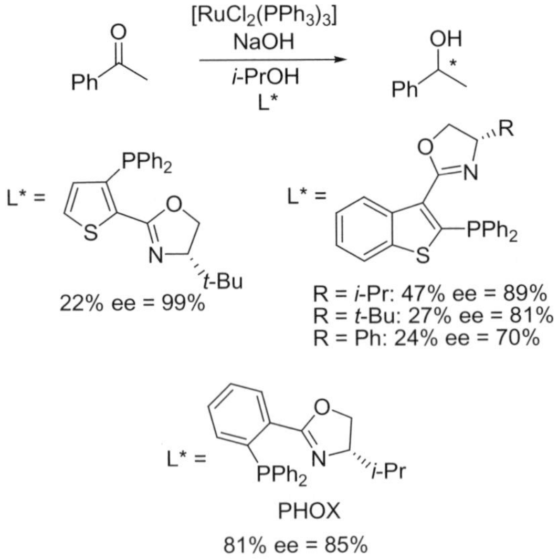

Scheme 9.13 Ru-catalysed reduction of acetophenone with phosphino-oxazoline ligands.

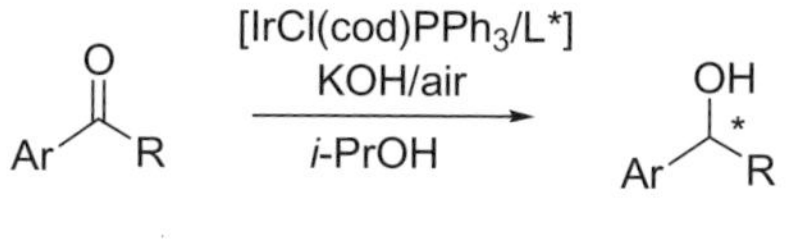

with (*R,R*)-L*:
Ar = Ph, R = Me: 96% ee = 83% (*S*)
Ar = Ph, R = Et: 94% ee = 87% (*S*)
Ar = Ph, R = *n*-Pr: 94% ee = 88% (*S*)
Ar = Ph, R = *n*-Bu: 97% ee = 90% (*S*)
Ar = Ph, R = *i*-Pr: 92% ee = 95% (*S*)
Ar = *o*-Tol, R = Me: 96% ee = 91% (*S*)
Ar = *m*-Tol, R = Me: 94% ee = 82% (*S*)
Ar = *m*-ClC$_6$H$_4$, R = Me: 99% ee = 71% (*S*)
with (*S,S*)-L*:
Ar = Ph, R = *i*-Pr: 92% ee = 94% (*R*)
Ar = Ph, R = Cy: 97% ee = 96% (*R*)
Ar = *o*-Tol, R = Me: 92% ee = 90% (*R*)
Ar = *m*-AcC$_6$H$_4$, R = Me: 94% ee = 89% (*R*)

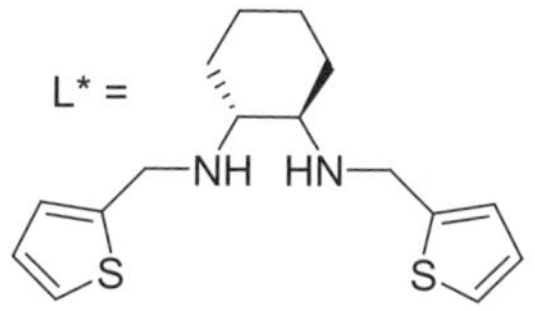

Scheme 9.14 Ir-catalysed reductions of ketones with S/N/N/S ligands.

shown in Scheme 9.15. Screening experiments showed that the reduction gave even better results when performed in the presence of a 5:2 HCO$_2$H–TEA mixture in acetonitrile instead of *i*-PrOH as the hydrogen donor (Scheme 9.15).

The scope of this methodology, based on the use of HCO$_2$H–TEA mixture as the hydrogen donor, has been successfully extended to the reduction of imines, thus providing the corresponding chiral amines in both high yields and enantioselectivities, as shown in Scheme 9.16.[2,24] In this case, the chiral active complexes were formed from different chiral *N*-sulfonylated 1,2-diphenylethylenediamine ligands and [RuCl$_2$(η^6-arene)]$_2$.

As another successful application of Noyori's TsDPEN ligand, Yan *et al.* reported the synthesis of antidepressant duloxetine, in 2008.[25] Thus, the key step of this synthesis was the asymmetric transfer hydrogenation of 3-(dimethylamino)-1-(thiophen-2-yl)propan-1-one performed in the presence of (*S,S*)-TsDPEN–Ru(II) complex and a HCO$_2$H–TEA mixture as the hydrogen donor. The reaction afforded the corresponding chiral alcohol in both high yield and enantioselectivity, which was further converted in two steps into expected (*S*)-duloxetine, as shown in Scheme 9.17.

In order to improve the performance of Noyori's catalytic system, Ru(II)-TsDPEN, which is very efficient but suffers from a long reaction time and a low activity in some cases, Mohar *et al.* have modified the diamine ligand by

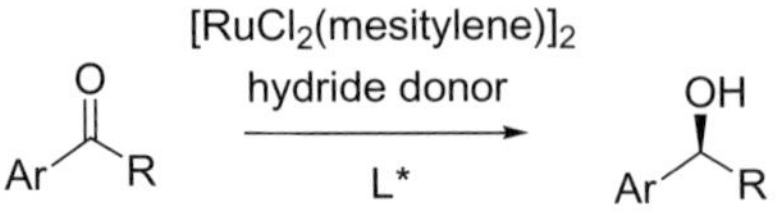

$$\text{Ar} \overset{\text{O}}{\underset{}{\|}} \text{R} \xrightarrow[\text{L*}]{\substack{[\text{RuCl}_2(\text{mesitylene})]_2 \\ \text{hydride donor}}} \text{Ar} \overset{\text{OH}}{\underset{}{\|}} \text{R}$$

with hydride donor = *i*-PrOH/KOH:
Ar = Ph, R = Me: 95% ee = 97%
Ar = Ph, R = Et: 94% ee = 97%
Ar = *o*-Tol, R = Me: 53% ee = 91%
Ar = *m*-ClC$_6$H$_4$, R = Me: 99% ee = 98%
Ar = *p*-ClC$_6$H$_4$, R = Me: 95% ee = 93%
Ar = *m*-MeOC$_6$H$_4$, R = Me: 96% ee = 96%
Ar = *p*-MeOC$_6$H$_4$, R = Me: 53% ee = 72%
Ar = 1-Naph, R = Me: 92% ee = 93%
Ar = 2-Naph, R = Me: 93% ee = 98%
Ar = Ph, R = CH$_2$: 45% ee = 91%
Ar = Ph, R = (CH$_2$)$_2$: 65% ee = 97%
with hydride donor = HCO$_2$H/TEA:
Ar = Ph, R = Me: 99% ee = 98%
Ar = *m*-ClC$_6$H$_4$, R = Me: 99% ee = 98%
Ar = *p*-ClC$_6$H$_4$, R = Me: 99% ee = 95%
Ar = *p*-CNC$_6$H$_4$, R = Me: 99% ee = 90%
Ar = *m*-MeOC$_6$H$_4$, R = Me: 99% ee = 98%
Ar = *p*-MeOC$_6$H$_4$, R = Me: 99% ee = 97%
Ar = Ph, R = Et: 96% ee = 97%
Ar = Ph, R = (CH$_2$)$_3$CO$_2$Et: 99% ee = 95%
Ar = 1-Naph, R = Me: 93% ee = 83%
Ar = 2-Naph, R = Me: 99% ee = 96%
Ar = *p*-MeOC$_6$H$_4$, R = *p*-CNC$_6$H$_4$: 54% ee = 66%
Ar = Ph, R = (CH$_2$)$_2$: 99% ee = 99%
Ar = Ph, R = (CH$_2$)$_3$: 99% ee = 99%

$$\text{L*} = \quad \overset{\text{Ph} \diagdown \ \ \text{''NHSO}_2\text{-}p\text{-Tol}}{\underset{\text{Ph} \diagup \ \ \text{''NH}_2}{\bigwedge}}$$

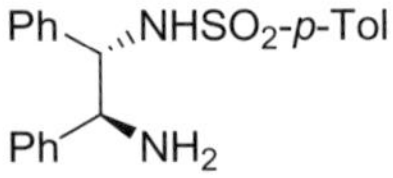

TsDPEN

Scheme 9.15   Ru-catalysed reductions of ketones with TsDPEN ligand.

introducing electron-withdrawing fluorosulfonyl groups.[26] Hence, (η6-arene)-
N-perfluorosulfonyl-1,2-diamines   and   *N*-(*N,N*-dialkylamino)sulfamoyl-1,2-
diamines were successfully used as ligands in the ruthenium-catalysed hydride
transfer reduction of various aromatic ketones, α-keto esters and α,α,α-tri-
fluoromethyl ketones in high yields and excellent enantioselectivities when the
reactions were carried out in the presence of a HCO$_2$H–TEA mixture as the
hydrogen donor. The results are collected in Scheme 9.18.

The corresponding 2-amino(sulfonamido)cyclohexane ligands were also
studied by Knochel *et al.*,[27] and later by Mohar *et al.*,[26a,b] as chiral ligands for
the ruthenium-catalysed hydride transfer reduction of various aromatic

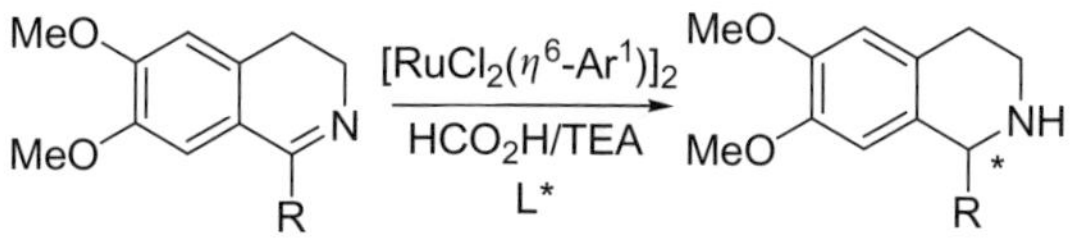

from (S,S)-L*:
R = Me, Ar¹ = p-cymene, Ar² = p-Tol: 99% ee = 95% (R)
R = Ph, Ar¹ = benzene, Ar² = 1-Naph: 99% ee = 84% (R)
from (R,R)-L*:
R = 3,4-(MeO)$_2$C$_6$H$_3$CH$_2$, Ar¹ = p-cymene
Ar² = 2,4,6-(Me)$_3$C$_6$H$_2$: 90% ee = 95% (S)
R = 3,4-(MeO)$_2$C$_6$H$_3$(CH$_2$)$_2$, Ar¹ = p-cymene
Ar² = 2,4,6-(Me)$_3$C$_6$H$_2$: 99% ee = 92% (S)
R = 3,4-(MeO)$_2$C$_6$H$_3$, Ar² = 1-Naph: 99% ee = 84% (S)

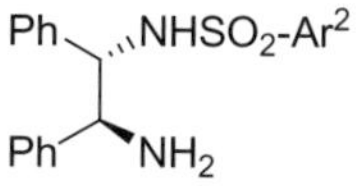

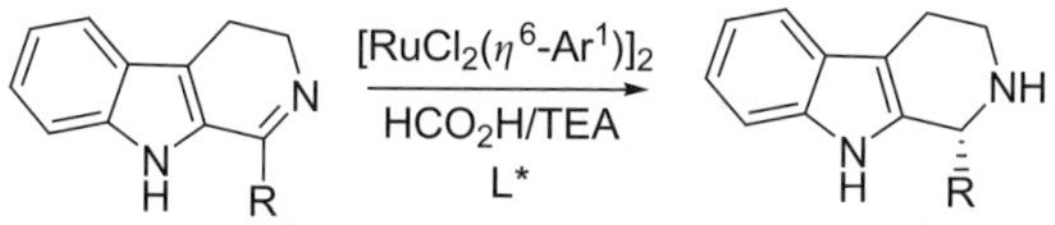

from (S,S)-L*:
R = Me, Ar¹ = p-cymene, Ar² = p-Tol: 86% ee = 97%
R = Ph, Ar¹ = p-cymene, Ar² = p-Tol: 83% ee = 96%

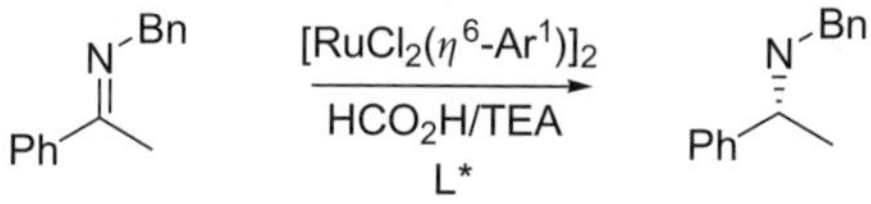

from (S,S)-L*:
Ar¹ = benzene, Ar² = 2,4,6-(Me)$_3$C$_6$H$_2$: 72% ee = 77%

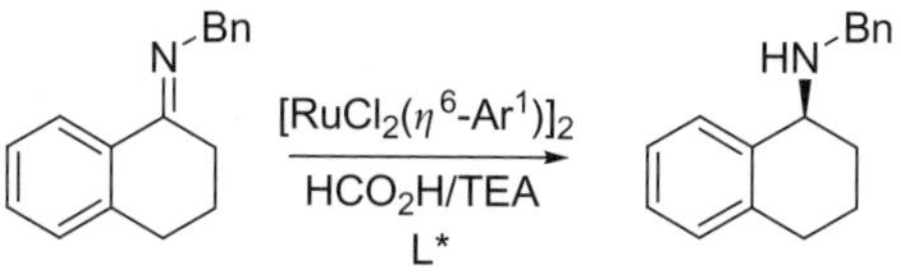

from (S,S)-L*:
R = Ph, Ar¹ = benzene, Ar² = 1-Naph: 90% ee = 89%

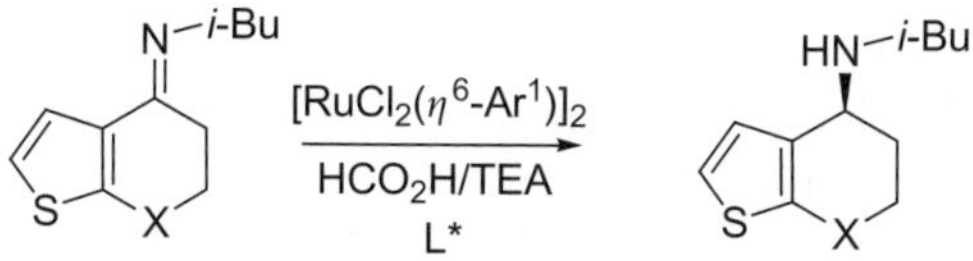

from (S,S)-L*:
X = S, Ar¹ = benzene, Ar² = 1-Naph: 82% ee = 85%
X = SO$_2$, Ar¹ = benzene, Ar² = 1-Naph: 84% ee = 88%

Scheme 9.16 Ru-catalysed reductions of imines with *N*-sulfonylated 1,2-dipheny-lethylenediamine ligands.

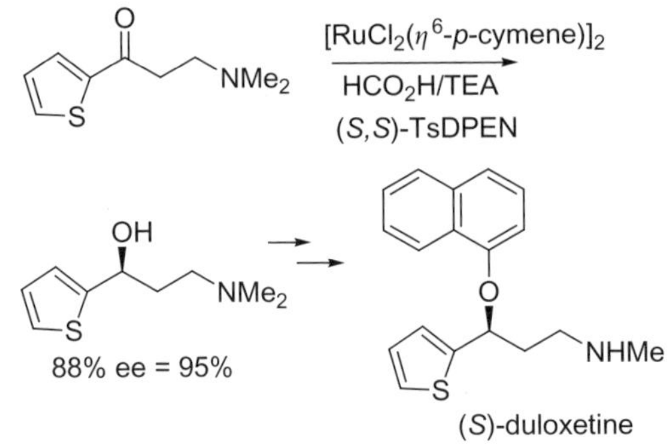

Scheme 9.17 Synthesis of (*S*)-duloxetine.

ketones, providing high activity and excellent enantioselectivities, as shown in Scheme 9.19.

There has been considerable interest in the development of water-soluble ligands, which allow metal-catalysed reactions to take place in water.[28] In this context, water-soluble analogues of Noyori's TsDPEN ligand and Knochel's 2-amino(sulfonamido)cyclohexane ligand, containing an additional sulfonic acid group, were examined by Williams *et al.* in the ruthenium-catalysed reduction of aromatic ketones, achieving a high enantioselectivity and a moderate activity by using *i*-PrOH as the hydrogen donor.[29] The results collected in Scheme 9.20 show that the water-soluble analogue of Noyori's ligand combined with ruthenium complex gave rise to a higher enantioselectivity, while the ruthenium water-soluble analogue of Knochel's ligand system demonstrated higher activity.

Finally, the use of S/P ligands derived from (*R*)-binaphthol has been considered by Gladiali *et al.* in the asymmetric rhodium-catalysed hydrogen-transfer reduction of acetophenone performed in the presence of *i*-PrOH as the hydrogen donor.[30] It was noted that racemisation occurred when the reaction time increased and consequently the corresponding alcohol was obtained in only low enantioselectivities ($\leq 5\%$ ee) as shown in Scheme 9.21. Similar results were more recently reported by these authors by using iridium combined with the same ligands.[31]

In recent years, many efforts have been devoted to obtain chiral compounds by dynamic kinetic resolution (DKR).[32] This process combines the resolution of kinetic resolution with an *in situ* equilibration or racemisation of a chirally labile substrate. In DKR, the enantiomers of a racemic substrate are induced to equilibrate at a rate that is faster than that of the slow-reacting enantiomer in reaction with a chiral reagent. If the enantioselectivity is sufficient, then isolation of a highly enriched nonracemic product is possible with a theoretical yield of 100% based on the racemic substrate. In this context, the ruthenium-catalysed transfer hydrogenation has been used in the DKR of a wide range of α-substituted ketones in the presence of *N*-tosyl-1,2-diphenylethylenediamine (TsDPEN) as sulfur-containing ligand and a HCO₂H–TEA

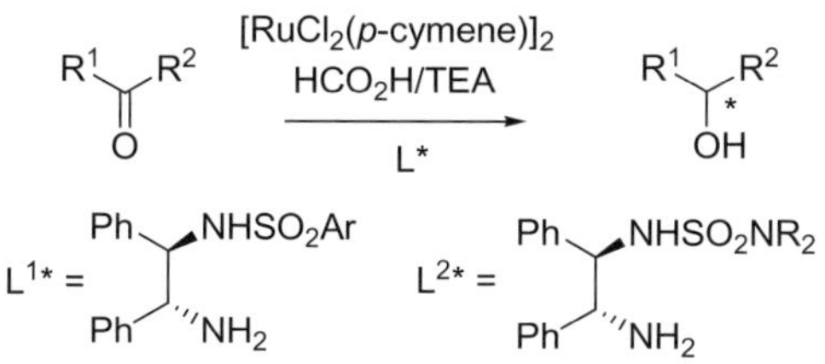

With L* = L^{1*}:

R^1 = Ph, R^2 = CO$_2$Me, Ar = p-NO$_2$C$_6$H$_4$:
100% ee = 93%

R^1 = Ph, R^2 = CO$_2$Me, Ar = 2,4,6-(MeO)$_3$C$_6$H$_2$:
100% ee = 96%

R^1 = p-FC$_6$H$_4$, R^2 = CO$_2$Et, Ar = p-NO$_2$C$_6$H$_4$:
100% ee = 77%

R^1 = p-FC$_6$H$_4$, R^2 = CO$_2$Et, Ar = 2,4,6-(MeO)$_3$C$_6$H$_2$:
100% ee = 84%

R^1 = p-MeOC$_6$H$_4$, R^2 = CO$_2$Et, Ar = p-NO$_2$C$_6$H$_4$:
58% ee = 95%

R^1 = Fu, R^2 = CO$_2$Et, Ar = p-NO$_2$C$_6$H$_4$:
100% ee = 94%

R^1 = Fu, R^2 = CO$_2$Et, Ar = 2,4,6-Cl$_3$C$_6$H$_2$:
100% ee = 87%

R^1 = Ph(CH$_2$)$_2$, R^2 = CO$_2$Et, Ar = o-NO$_2$C$_6$H$_4$:
100% ee = 76%

R^1 = Me, R^2 = CO$_2$Et, Ar = o-NO$_2$C$_6$H$_4$:
100% ee = 86%

R^1 = Me, R^2 = CO$_2$Et, Ar = 2,4,6-(i-Pr)$_3$C$_6$H$_2$:
100% ee = 83%

R^1 = i-Pr, R^2 = CO$_2$Et, Ar = 2,4,6-(i-Pr)$_3$C$_6$H$_2$:
100% ee = 44%

R^1 = MeO$_2$C(CH$_2$)$_2$, R^2 = CO$_2$Me, Ar = o-NO$_2$C$_6$H$_4$:
100% ee = 63%

R^1 = Ph, R^2 = CF$_3$, Ar = p-Tol:
100% ee = 42%

R^1 = Bn, R^2 = CF$_3$, Ar = 2,4,6-(i-Pr)$_3$C$_6$H$_2$:
100% ee = 98%

R^1 = 3,4-(MeO)$_2$C$_6$H$_3$, R^2 = CH(CO$_2$Me)NHMe, Ar =(CF$_2$)$_3$CF$_3$:
100% de = 90% ee = 99%

With L* = L^{2*}:

R^1 = Ph, R^2 = R = Me: 80% ee = 96%

R^1 = Ph, R^2 = R,R = (CH$_2$)$_5$: 51% ee = 95%

R^1 = Ph, R^2 = CO$_2$Me, R = Cy: 100% ee = 85%

R^1 = Ph, R^2 = CO$_2$Me, R = Me: 100% ee = 81%

R^1 = Ph, R^2 = CH$_2$CO$_2$Et, R = Me: 100% ee = 98%

Scheme 9.18   Ru-catalysed reductions of ketones with (η6-arene)-*N*-perfluorosulfonyl-1,2-diamine and *N*-(*N,N*-dialkylamino)sulfamoyl-1,2-diamine ligands.

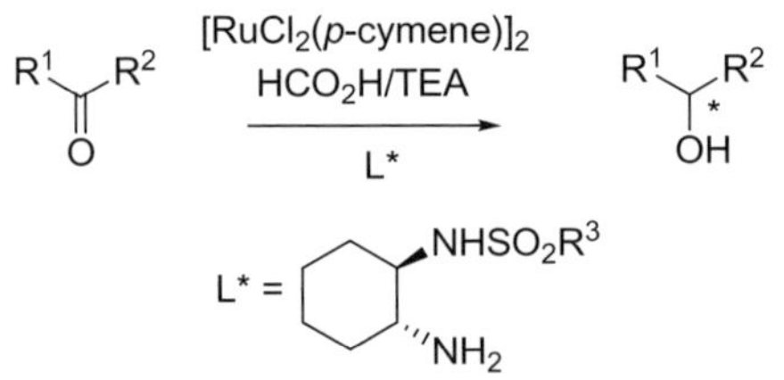

R^1 = Ph, R^2 = Me, R^3 = *p*-Tol: 99% ee = 94%
R^1 = Ph, R^2 = Me, R^3 = 2,4,6-(*i*-Pr)$_3$C$_6$H$_2$: 99% ee = 89%
R^1 = Ph, R^2 = Me, R^3 = 2-Naph: 99% ee = 95%
R^1 = Ph, R^2 = Me, R^3 = 1-Naph: 99% ee = 96%
R^1 = Me, R^2 = R^3 = 1-Naph: 99% ee = 96%
R^1 = Ph, R^2 = *i*-Pr, R^3 = 1-Naph: 54% ee = 67%
R^1 = 3,4-(MeO)$_2$C$_6$H$_3$, R^2 = (CO$_2$Me)CHNHMe:
R^3 = (CF$_2$)$_3$CF$_3$: 100% de = 90% ee = 99%

Scheme 9.19   Ru-catalysed reductions of ketones with 2-amino(sulfonamido)cyclo-hexane ligands.

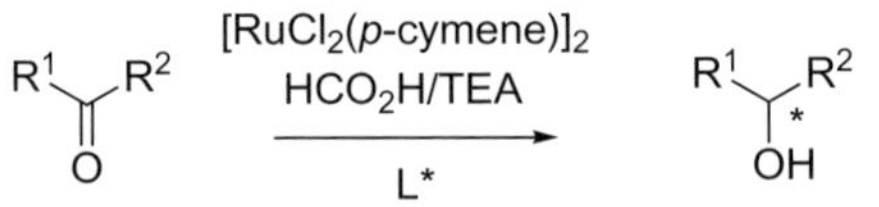

R^1 = Me, R^2 = Ph, L* = L^{1*}: 96% ee = 94% (*S*)
R^1 = Me, R^2 = Ph, L* = L^{2*}: 91% ee = 88% (*R*)
R^1 = Me, R^2 = 2-Naph, L* = L^{1*}: 94% ee = 95% (*S*)
R^1 = Me, R^2 = 2-Naph, L* = L^{2*}: 87% ee = 90% (*R*)
R^1 = Me, R^2 = *p*-CF$_3$C$_6$H$_4$, L* = L^{1*}: 100% ee = 81% (*S*)
R^1 = Me, R^2 = *p*-CF$_3$C$_6$H$_4$, L* = L^{2*}: 100% ee = 88% (*R*)
R^1 = Me, R^2 = *m*-CF$_3$C$_6$H$_4$, L* = L^{1*}: 90% ee = 87% (*S*)
R^1 = Me, R^2 = *m*-CF$_3$C$_6$H$_4$, L* = L^{2*}: 91% ee = 81% (*R*)
R^1 = Me, R^2 = *p*-MeOC$_6$H$_4$, L* = L^{1*}: 31% ee = 91% (*S*)
R^1 = Me, R^2 = *p*-MeOC$_6$H$_4$, L* = L^{2*}: 35% ee = 83% (*R*)

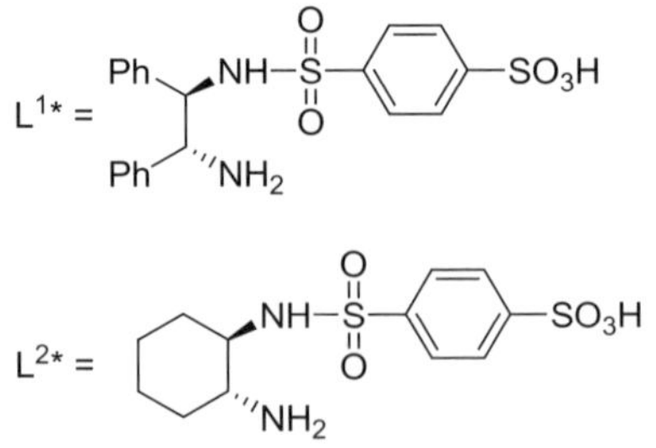

Scheme 9.20   Ru-catalysed reductions of ketones with water-soluble analogues of Noyori's and Knochel's ligands.

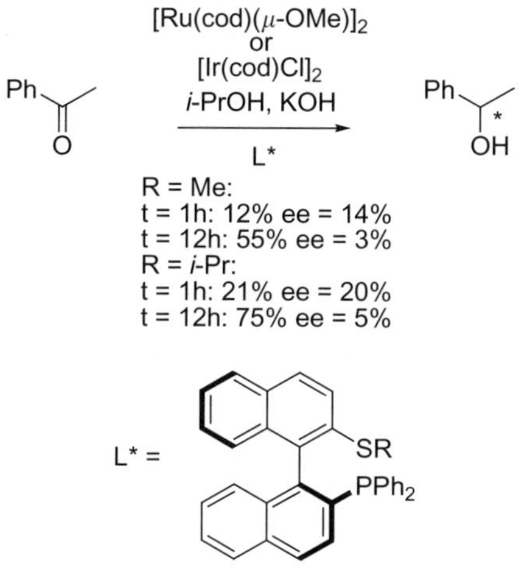

Scheme 9.21 Rh- and Ru-catalysed reductions of acetophenone with BINOL-derived S/P ligands.

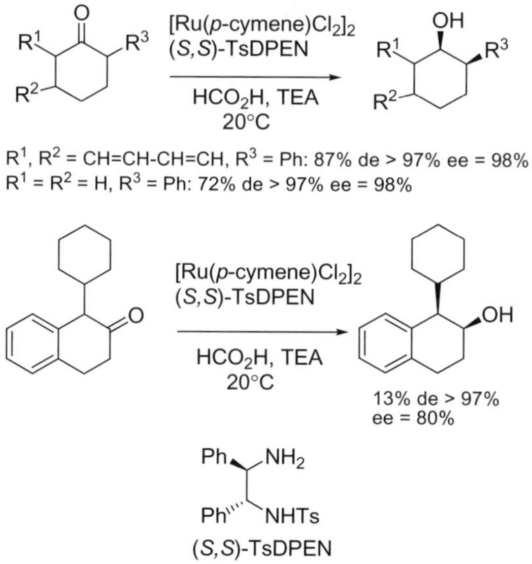

Scheme 9.22 Ru-catalysed DKR-reductions of cyclic ketones with TsDPEN ligand.

mixture as the hydrogen donor. Therefore, an (S,S)-TsDPEN-based-ruthenium complex has been involved in the DKR-transfer hydrogenation process of various cyclic ketones, providing the corresponding *syn* products with diastereoselectivities of up to 97% de and enantioselectivities of up to 98% ee, as shown in Scheme 9.22.[33]

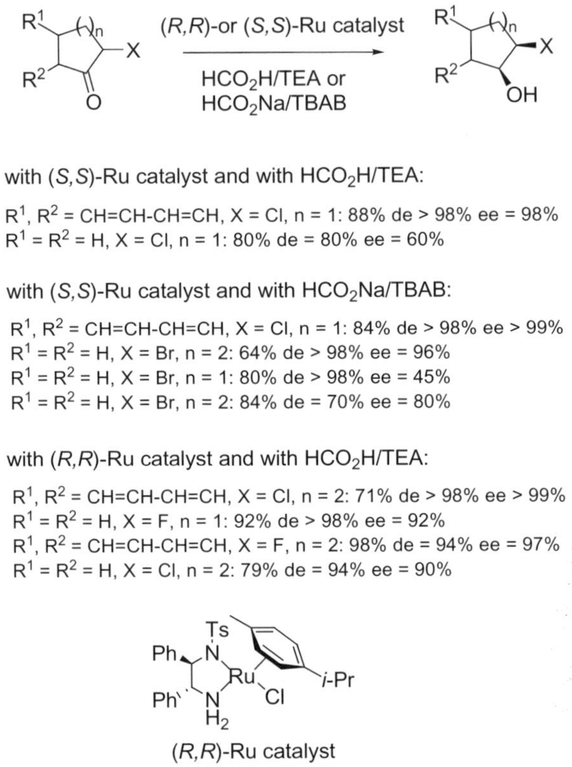

(R,R)-or (S,S)-Ru catalyst
HCO2H/TEA or
HCO2Na/TBAB
R1
R2
X
O
OH
n
with (S,S)-Ru catalyst and with HCO2H/TEA:
R1, R2 = CH=CH-CH=CH, X = Cl, n = 1: 88% de > 98% ee = 98%
R1 = R2 = H, X = Cl, n = 1: 80% de = 80% ee = 60%
with (S,S)-Ru catalyst and with HCO2Na/TBAB:
R1, R2 = CH=CH-CH=CH, X = Cl, n = 1: 84% de > 98% ee > 99%
R1 = R2 = H, X = Br, n = 2: 64% de > 98% ee = 96%
R1 = R2 = H, X = Br, n = 1: 80% de > 98% ee = 45%
R1 = R2 = H, X = Br, n = 2: 84% de = 70% ee = 80%
with (R,R)-Ru catalyst and with HCO2H/TEA:
R1, R2 = CH=CH-CH=CH, X = Cl, n = 2: 71% de > 98% ee > 99%
R1 = R2 = H, X = F, n = 1: 92% de > 98% ee = 92%
R1, R2 = CH=CH-CH=CH, X = F, n = 2: 98% de = 94% ee = 97%
R1 = R2 = H, X = Cl, n = 2: 79% de = 94% ee = 90%
Ts
Ph
N
Ru
i-Pr
Ph
N
Cl
H2
(R,R)-Ru catalyst

Scheme 9.23 Ru-catalysed DKR-reductions of cyclic α-haloketones with TsDPEN ligand.

In 2006, Lassaletta *et al.* extended the scope of this methodology to a variety of cyclic α-haloketones, offering an efficient tool for the synthesis of chiral halohydrins, including bromo-, chloro-, and even fluorohydrins in good-to-excellent yields and stereoselectivities, using either a HCO_2H/TEA or a HCO_2H/TBAB mixture as the hydrogen source (Scheme 9.23).[34]

In the course of their study on the hydrogenation of 2-alkyl-1,3-diketones, Cossy *et al.* have achieved the synthesis of the C14–C25 fragment of biologically active natural bafilomycin A_1 in eleven steps in a sequence involving two enantioselective transfer-hydrogenation steps induced by chiral ruthenium complexes including TsDPEN ligand.[35] These reactions set the C15, C16, C21 and C22 stereogenic centres via DKR, as depicted in Scheme 9.24.

The first example of an asymmetric reduction of C=N bonds proceeding via DKR was reported in 2005 by Lassaletta *et al.*[36] In this process, the transfer hydrogenation of 2-substituted bicyclic and monocyclic ketimines could be accomplished via DKR by using a HCO_2H/TEA mixture as the hydrogen source and a chiral ruthenium complex including TsDPEN ligand,

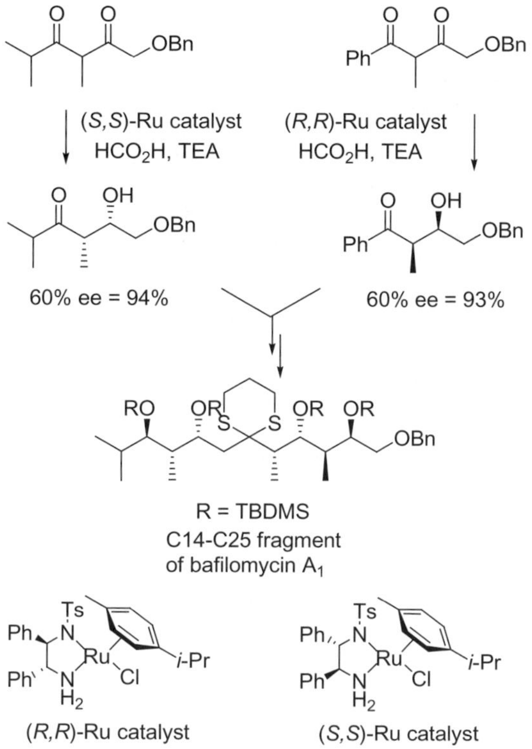

Scheme 9.24 Synthesis of the C14–C25 fragment of bafilomycin A$_1$.

which afforded the corresponding *cis*-cycloalkylamines with moderate-to-excellent levels of diastereo- and enantioselectivities (Scheme 9.25).

In conclusion, since the excellent earlier results obtained by using Noyori's TsDPEN catalyst and its derivatives, several chiral S/N ligands have been successfully applied to the asymmetric transfer hydrogenation of ketones and more rarely imines. The most relevant studies are those of Petra's group based on the use of simple aminosulfide ligands, those of Lemaire's group involving dithiourea ligands, those of Zaitsev/Adolfsson's group employing thioamide ligands, and more recently those of Gao's group introducing a new type of S/N/N/S ligands. All of these novel chiral S/N ligands have allowed enantioselectivities of up to 94–97% ee to be obtained for the reduction of ketones. Most research studies about asymmetric transfer hydrogenation of ketones and derivatives have been carried out by using Ru, Rh and Ir as catalyst precursors.

Even if chiral S/N ligands have proved their efficiency for promoting metal-catalysed asymmetric transfer hydrogenation, up to now they do not compete with the N/N ligands, at least in terms of enantioselectivity.[37]

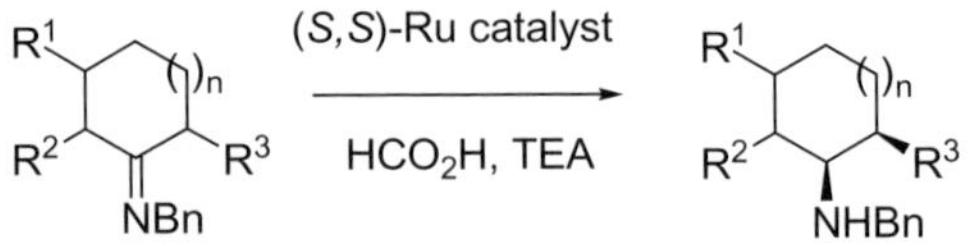

R^1, R^2 = CH=CH-CH=CH, R^3 = Me, n = 0:
70% de > 98% ee = 96%
R^1, R^2 = CH=CH-C(Me)=CH, R^3 = allyl, n = 0:
67% de > 98% ee = 92%
R^1, R^2 = CH=CH-C(Me)=CH, R^3 = Me, n = 0:
82% de > 98% ee = 97%
R^1 = R^2 = H, R^3 = allyl, n = 1:
75% de = 86% ee = 63%
R^1 = R^2 = H, R^3 = (CH$_2$)$_2$CN, n = 1:
60% de = 88% ee = 72%
R^1 = R^2 = H, R^3 = Ph, n = 1:
55% de > 98% ee = 50%

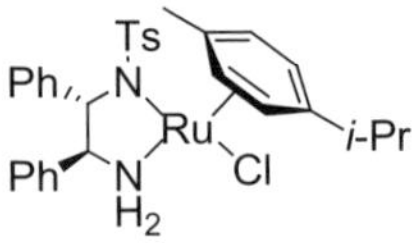

(S,S)-Ru catalyst

Scheme 9.25   Ru-catalysed DKR-reductions of cyclic ketimines with TsDPEN ligand.

References

1. R. Noyori, in *Asymmetric Catalysis in Organic Synthesis*, John Wiley and Sons, New York, 1994, Chap. 2.
2. (a) R. Noyori and S. Hashiguchi, *Acc. Chem. Res.*, 1997, **30**, 97–102; (b) M. J. Palmer and M. Wills, *Tetrahedron: Asymmetry*, 1999, **10**, 2045–2061.
3. (a) J. B. Jones, *Tetrahedron*, 1986, **42**, 3351–3403; (b) E. Schoffers, A. Golebiowski and C. R. Johnson, *Tetrahedron*, 1996, **52**, 3769–3826.
4. D. Müller, G. Umbricht, B. Weber and A. Pfaltz, *Helv. Chim. Acta*, 1991, **74**, 232–240.
5. J.-P. Genêt, V. Ratovelomanana-Vidal and C. Pinel, *Synlett*, 1993, 478–480.
6. P. Gamez, F. Fache and M. Lemaire, *Tetrahedron: Asymmetry*, 1995, **6**, 705–718.
7. D. A. Evans, S. G. Nelson, M. R. Gagné and A. R. Muci, *J. Am. Chem. Soc.*, 1993, **115**, 9800–9801.
8. (a) R. L. Chowdhury and J.-E. Bäckvall, *J. Chem. Soc., Chem. Commun.*, 1991, 1063–1064; (b) P. Krasik and H. Alper, *Tetrahedron*, 1994, **50**,

4347–4354; (c) Q. Jiang, D. Van Plew, S. Murtuza and X. Zhang, *Tetrahedron Lett.*, 1996, **37**, 797–800; (d) T. Langer and G. Helmchen, *Tetrahedron Lett.*, 1996, **37**, 1381–1384; (e) T. Ohkuma, H. Ooka, S. Hashigucchi, T. Ikariya and R. Noyori, *J. Am. Chem. Soc.*, 1995, **117**, 2675–2676; (f) R. Noyori and T. Okhuma, *Angew. Chem., Int. Ed. Engl.*, 2001, **40**, 40–73; (g) R. Noyori, *Angew. Chem., Int. Ed. Engl.*, 2002, **41**, 2008–2022; (h) T. Ohkuma, C. A. Sandoval, R. Srinivasan, Q. Lin, Y. Wei, K. Muniz and R. Noyori, *J. Am. Chem. Soc.*, 2005, **127**, 8288–8289.

9. (a) S. Gladiali and E. Alberico, *Chem. Soc. Rev.*, 2006, **106**, 226–236; (b) S. E. Clapham, A. Hadzovic and R. H. Morris, *Coord. Chem. Rev.*, 2004, **248**, 2201–2237; (c) Q.-H. Fan, Y.-M. Li and A. S. C. Chan, *Chem. Rev.*, 2002, **102**, 3385–3466.

10. R. A. W. Johnstone, A. H. Wilby and I. D. Entwistle, *Chem. Rev.*, 1985, **85**, 129–170.

11. G. Zassinovich, G. Mestroni and S. Gladiali, *Chem. Rev.*, 1992, **92**, 1051–1069.

12. P. Kvintovics, B. R. James and B. Heil, *J. Chem. Soc., Chem. Commun.*, 1986, 1810–1811.

13. D. G. I. Petra, P. C. J. Kamer, A. L. Spek, H. E. Schoemaker and P. W. N. M. van Leeuwen, *J. Org. Chem.*, 2000, **65**, 3010–3017.

14. A. Hage, D. G. I. Petra, J. A. Field, D. Schipper, J. B. P. A. Wijnberg, P. C. J. Kamer, J. N. H. Reek, P. W. N. M. van Leeuwen, R. Wever and H. E. Schoemaker, *Tetrahedron: Asymmetry*, 2001, **12**, 1025–1034.

15. J. K. Ekegren, P. Roth, K. Källström, T. Tarnai and P. G. Andersson, *Org. Biomol. Chem.*, 2003, **1**, 358–366.

16. A. Gayet, C. Bolea and P. G. Andersson, *Org. Biomol. Chem.*, 2004, **2**, 1887–1893.

17. A. B. Zaitsev and H. Adolfsson, *Org. Lett.*, 2006, **8**, 5129–5132.

18. (a) F. Touchard, F. Fache and M. Lemaire, *Tetrahedron: Asymmetry*, 1997, **8**, 3319–3326; (b) F. Touchard, P. Gamez, F. Fache and M. Lemaire, *Tetrahedron Lett.*, 1997, **38**, 2275–2278.

19. (a) F. Touchard, M. Bernard, F. Fache and M. Lemaire, *J. Mol. Catal.*, 1999, **140**, 1–11; (b) M. Bernard, F. Delbecq, F. Fache, P. Sautet and M. Lemaire, *Eur. J. Org. Chem.*, 2001, 1589–1596.

20. G. J. Kim, S.-H. Kim, P. H. Chong and M. A. Kwon, *Tetrahedron Lett.*, 2002, **43**, 8059–8062.

21. N. End, C. Stoessel, U. Berens, P. di Pietro and P. G. Cozzi, *Tetrahedron: Asymmetry*, 2004, **15**, 2235–2239.

22. X.-Q. Zhang, Y.-Y. Li, H. Zhang and J.-X. Gao, *Tetrahedron: Asymmetry*, 2007, **18**, 2049–2054.

23. S. Hashiguchi, A. Fujii, J. Takehara, T. Ikariya and R. Noyori, *J. Am. Chem. Soc.*, 1995, **117**, 7562–7563.

24. N. Uematsu, A. Fujii, S. Hashiguchi, T. Ikariya and R. Noyori, *J. Am. Chem. Soc.*, 1996, **118**, 4916–4917.

25. S. Z. He, X. M. Li, J. Dai and M. Yan, *Chinese Chem. Lett.*, 2008, **19**, 23–25.

26. (a) B. Mohar, A. Valleix, J.-R. Desmurs, M. Felemez, A. Wagner and C. Mioskowski, *Chem. Commun.*, 2001, 2572–2573; (b) D. Sterk, M. S. Stephan and B. Mohar, *Tetrahedron: Asymmetry*, 2002, **13**, 2605–2608; (c) D. Sterk, M. S. Stephan and B. Mohar, *Tetrahedron Lett.*, 2004, **45**, 535–537.
27. K. Püntener, L. Schwink and P. Knochel, *Tetrahedron Lett.*, 1996, **37**, 8165–8168.
28. F. Joo and A. Katho, *J. Mol. Catal.*, 1997, **116**, 3–26.
29. C. Bubert, J. Blacker, S. M. Brown, J. Crosby, S. Fitzjohn, J. P. Muxworthy, T. Thorpe and J. M. J. Williams, *Tetrahedron Lett.*, 2001, **42**, 4037–4039.
30. S. Gladiali, A. Dore and D. Fabbri, *Tetrahedron: Asymmetry*, 1994, **5**, 1143–1146.
31. S. Gladiali, S. Medici, G. Pirri, S. Pulacchini and D. Fabbri, *Can. J. Chem.*, 2001, **79**, 670–678.
32. (a) H. Pellissier, *Tetrahedron*, 2008, **64**, 1563–1601; (b) H. Pellissier, *Tetrahedron*, 2003, **59**, 8291–8327.
33. P. Peach, D. J. Cross, J. A. Kenny, I. Mann, I. Houson, L. Campbell, T. Walsgrove and M. Wills, *Tetrahedron*, 2006, **62**, 1864–1876.
34. A. Ros, A. Magriz, H. Dietrich, R. Fernandez, E. Alvarez and J. M. Lassaletta, *Org. Lett.*, 2006, **8**, 127–130.
35. (a) F. Eustache, P. I. Dalko and J. Cossy, *J. Org. Chem.*, 2003, **68**, 9994–10002; (b) F. Eustache, P. I. Dalko and J. Cossy, *Tetrahedron Lett.*, 2003, **44**, 8823–8826.
36. A. Ros, A. Magriz, H. Dietrich, M. Ford, R. Fernandez and J. M. Lassaletta, *Adv. Synth. Catal.*, 2005, **347**, 1917–1920.
37. D. Rechavi and M. Lemaire, *Chem. Rev.*, 2002, **102**, 3467–3494.

CHAPTER 10

Miscellaneous Reactions

10.1 Introduction

In addition to asymmetric allylic substitution, conjugate addition, addition of organometallic reagents to carbonyl compounds, Diels–Alder reaction, cyclopropanation, Heck-type reactions, hydrogenation and hydrogen transfer, a great number of asymmetric miscellaneous reactions have been successfully developed by using chiral sulfur-containing ligands. The impressive number of reports dealing with the use of such ligands in a broad variety of reaction types is representative of the extensive efforts that have been made to develop these ligands, especially in recent years.

10.2 Hydroformylation

Hydroformylation is one of the most studied reactions in homogeneous catalysis.[1] The regio- and enantioselective syntheses of optically pure branched aldehydes, in the case of involving functionalised olefins, are important challenges for this reaction. Cobalt was the first metal used in the asymmetric hydroformylation and rhodium was rapidly studied afterwards in the presence of chiral phosphines. In addition, ruthenium, iridium and palladium were also involved in this reaction, but Pt/SnCl$_2$ and rhodium catalysts[2] are currently the most promising asymmetric systems. The first report on the use of chiral sulfur-containing ligands in the rhodium-catalysed asymmetric hydroformylation appeared in 1993.[3] In this preliminary work, Claver et al. demonstrated the potential of a new chiral ligand, 1,1′-binapthalene-2,2′-dithiol (binas) and its analogous dimethyl sulfide, Me$_2$binas, to promote the enantioselective Rh-catalysed hydroformylation of styrene. Indeed, the application of the dinuclear neutral and cationic complexes of these ligands to the hydroformylation of

RSC Catalysis Series No. 2
Chiral Sulfur Ligands: Asymmetric Catalysis
By Hélène Pellissier
© Hélène Pellissier 2009
Published by the Royal Society of Chemistry, www.rsc.org

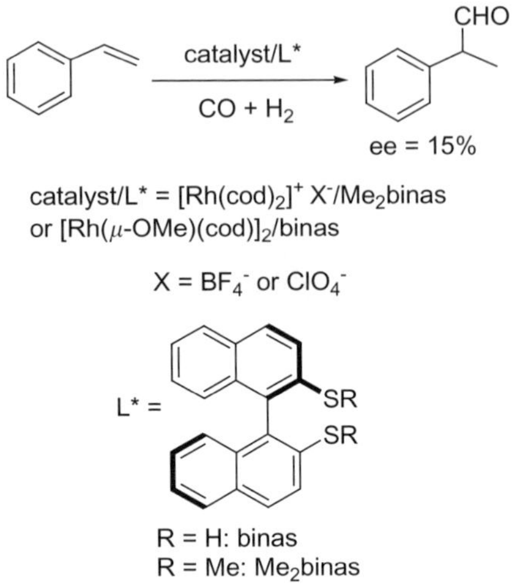

Scheme 10.1   Rh-catalysed hydroformylation of styrene with binas and Me$_2$binas ligands.

styrene led to a very good activity but a low enantioselectivity ($\leq 15\%$ ee), as shown in Scheme 10.1.

It has been well established that dinuclear bridged dithiolate rhodium complexes behaved as efficient catalyst precursors in the hydroformylation reaction. However, a phosphorus ligand, namely triphenylphosphine is required in most of the reported experiments to obtain high regioselectivities. In this context, these workers have developed the combination of highly efficient dinuclear dithiolato bridged rhodium complexes with diphenyl phosphine chiral auxiliary ligands, such as **BDPP** (2,4-bis(diphenylphosphine)pentane), to be used as catalytic precursor systems for the hydroformylation reaction of styrene.[4] In order to study the influence of the dinuclear framework of the thiolate precursor, various thiolate and dithiolate bridged rhodium complexes were combined with chiral diphosphines. As shown in Scheme 10.2, the combination of **BDPP** with rhodium dithiolate systems allowed enantioselectivities of up to 43% ee to be obtained along with regioselectivities of up to 94% in favour of 2-phenylpropanal. Moreover, it was shown that the combination of chiral (+)-or (−)-**DIOS** and (+)- or (−)-**BDPP** afforded no improvement in the enantioselectivity of the reaction (ee $\leq 34\%$ ee).

In 1999, Casado *et al.* developed heterotetranuclear complexes (TiRh$_3$) depicted in Scheme 10.3 with bridging sulfido ligands combined with P-donor ligands.[5] These complexes were further tested as catalysts for the asymmetric hydroformylation reaction of styrene. In this process, [CpTi((μ_3-S)3{Rh(tfbb}$_3$] was efficiently active under mild conditions (10 bar, CO/H$_2 = 1$ atm, 353 K). In order to explore the effect of the added phosphorus ligand and the possibilities of this system for the asymmetric hydroformylation of styrene, achiral diphosphines such as dppe (1,2-bis(diphenylphosphine)ethane) and

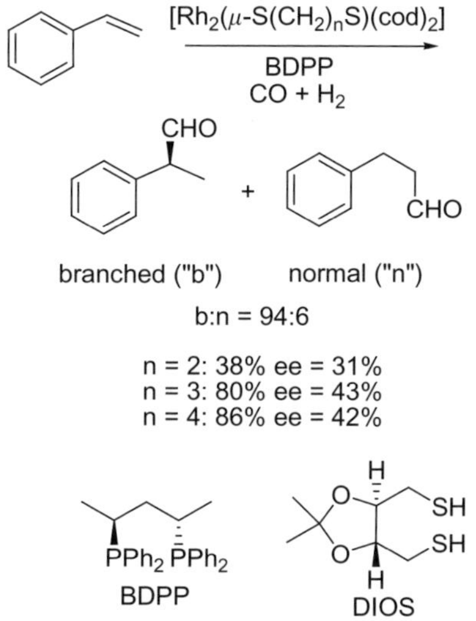

Scheme 10.2 Rh-catalysed hydroformylation of styrene with dithiolate rhodium complexes.

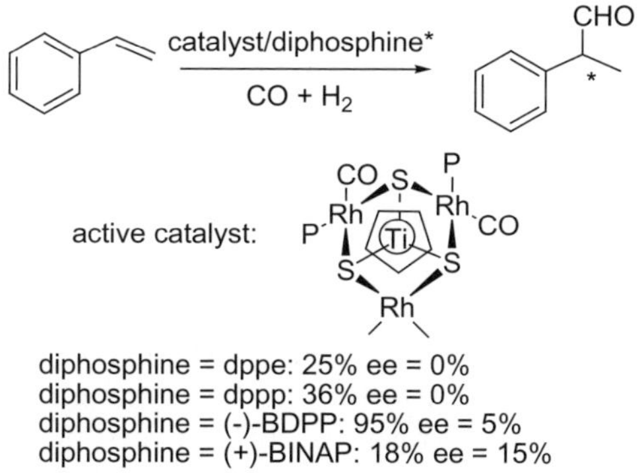

Scheme 10.3 TiRh₃-catalysed hydroformylation of styrene with bridging sulfido ligands.

dppp (1,3-bis(diphenylphosphine)propane) have been used without chiral inductions. Moreover, chiral diphosphines were studied, but only BINAP chiral diphosphines showed a modest enantioselectivity of 15% ee.

In 2000, Claver *et al.* reported the synthesis of novel chiral S/P ligands with a xylofuranose backbone.[6] These thioether-phosphite ligands derived from carbohydrates were investigated for the rhodium-catalysed hydroformylation of styrene but, in spite of good conversions (>99%) combined with excellent

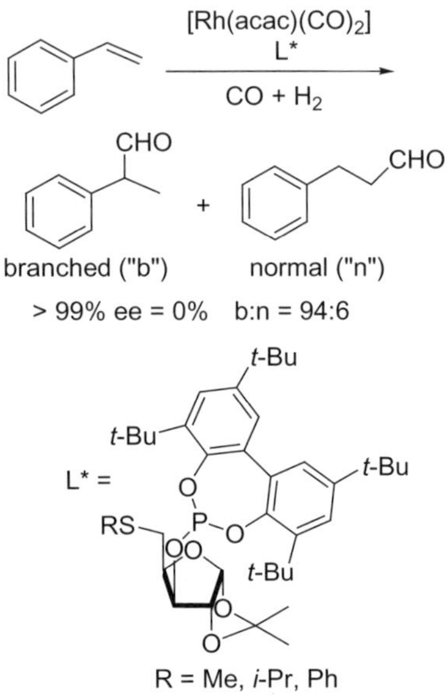

Scheme 10.4 Rh-catalysed hydroformylation of styrene with xylofuranose-derived thioether-phosphite ligands.

regioselectivities in 2-phenylpropanal (94%), enantioselectivities close to 0% ee were found for all the ligands (Scheme 10.4).

Rhodium complexes of (R,R)-1-benzyl-3,4-dithioether-pyrrolidines were also prepared by these authors, who further investigated them as ligands of rhodium complexes in the hydroformylation of styrene but, in all experiments, the enantioselectivity was lower than 3% ee, whereas the chemoselectivity was of 97% (Scheme 10.5).[7]

In 2000, better results were obtained by Bonnet *et al.* by using readily available chiral thioureas as new ligands in the asymmetric rhodium-catalysed hydroformylation of styrene.[8] In general, the conversion of styrene and enantioselectivities were modest, but when the reaction was carried out in heptane as the solvent, an enantioselectivity of 41% ee was obtained (Scheme 10.6).

10.3 1,3-Dipolar Cycloaddition

The 1,3-dipolar cycloaddition, also known as the Huisgen cycloaddition,[9] is a classic reaction in organic chemistry consisting in the reaction of a dipolarophile with a 1,3-dipolar compound that allows the production of various five-membered heterocycles.[10] This reaction represents one of the most productive fields of modern synthetic organic chemistry.[11] Most dipolarophiles are alkenes, alkynes, and molecules possessing related heteroatom functional

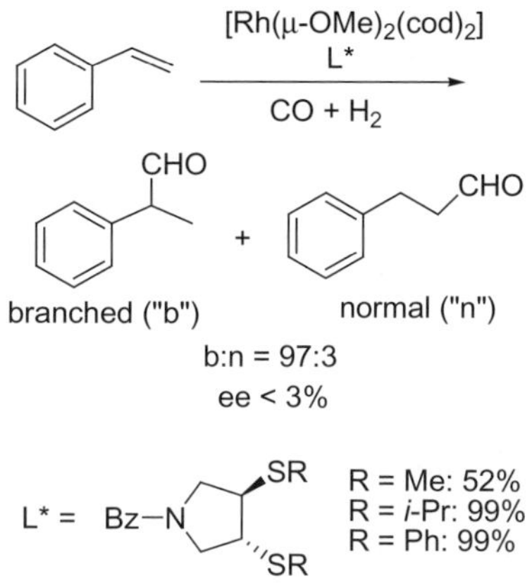

Scheme 10.5 Rh-catalysed hydroformylation of styrene with dithioether-pyrrolidine ligands.

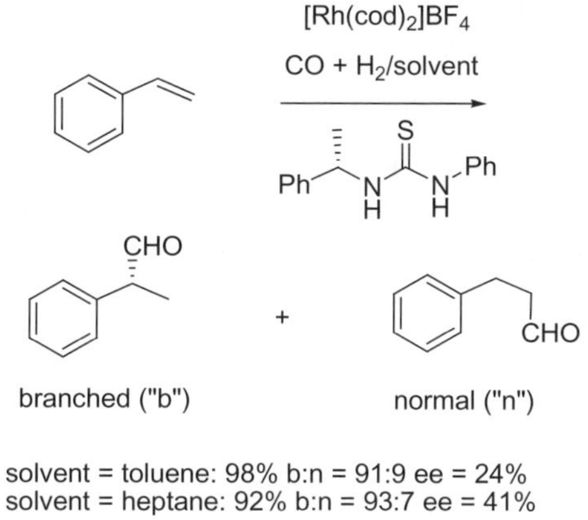

solvent = toluene: 98% b:n = 91:9 ee = 24%
solvent = heptane: 92% b:n = 93:7 ee = 41%

Scheme 10.6 Rh-catalysed hydroformylation of styrene with thiourea ligand.

groups (such as carbonyls and nitriles). The 1,3-dipoles can be basically divided into two different types, the allyl anion type such as nitrones, azomethine ylides, nitro compounds, bearing a nitrogen atom in the middle of the dipole, carbonyl ylides, carbonyl imines, bearing an oxygen atom in the middle of the dipole, and, the linear propargyl/allenyl anion type such as nitrile oxides, nitrilimines, nitrile ylides, diazoalkanes, or azides. Two π-electrons of the dipolarophile and four electrons of the dipolar compound participate in a concerted pericyclic shift. The addition is stereoconservative (suprafacial), and the reaction is therefore a $[2_S + 4_S]$ cycloaddition. In recent years, asymmetric 1,3-dipolar cycloadditions have become one of the most powerful tools for the construction of enantiomerically pure five-membered heterocycles.[12] Up to four stereo-centres can be introduced in a stereoselective manner in only one single step. In addition, a range of different substituents can be included in the dipole and the dipolarophile, resulting in a broad range of possible cycloadducts, which can serve as useful synthetic building blocks. The enantioselectivity can be controlled by either choosing a chiral 1,3-dipole, a chiral dipolarophile or a chiral catalyst of which the latter probably has greatest potential. Over the years a variety of chiral Lewis acids have been evaluated for dipolar cycloadditions. The most successful of these catalysts have employed copper, silver, nickel, aluminum, zinc and lanthanide Lewis acids, generally in combination with one of the privileged classes of chiral ligands such as bisoxazolines, BINAPs, and PyBoxs. In addition, several chiral sulfur-containing ligands have been implicated in asymmetric 1,3-dipolar cycloadditions. As an example, Scheeren et al. have reported the use of chiral oxazaborolidines derived from N-tosyl-L-α-amino acids to promote the 1,3-dipolar cycloaddition of nitrones with ketene acetals.[13] The chiral Lewis acid activated the nitrone by complexing the oxygen atom of the nitrone and lowering the LUMO (Lowest Unoccupied Molecular Orbital) energy. The electron-rich ketene, 1,1-diethoxypropene, led to a LUMO(nitrone)-HOMO(alkene) (HOMO: Highest Occupied Molecular Orbital) controlled 1,3-dipolar cycloaddition, providing the expected cycloadduct in high yields and enantioselectivities of up to 73% ee, as shown in Scheme 10.7.

Since catalysts immobilised on hydrophilic silica gel often give superior performances to their polymer-bound or polymer-incorporated analogues for multiple applications, Heckel and Seebach have immobilised TADDOL derivatives on hydrophobic controlled-pore glass (CPG) silica gel.[14] Indeed, CPG is

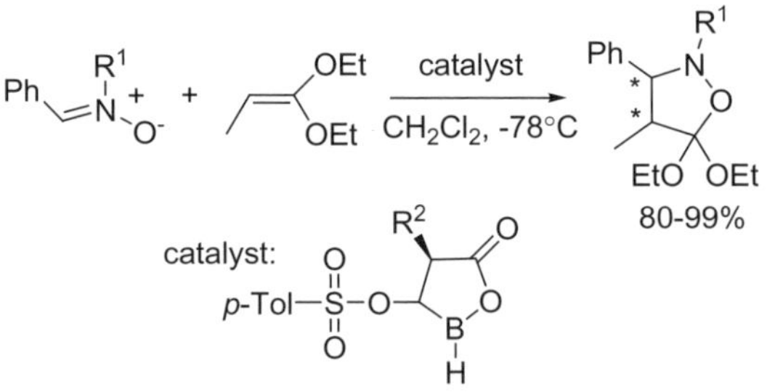

R¹ = Ph, R² = p-(BnO)C₆H₄: ee = 62% (S,S)
R¹ = Ph, R² = i-Pr: ee = 70% (R,R)
R¹ = Ph, R² = i-Bu: ee = 73% (R,R)
R¹ = Bn, R² = i-Bu: ee = 45% (R,R)
R¹ = Bn, R² = Ph: ee = 17% (R,R)
R¹ = R² = Bn: ee = 11% (R,R)
R¹ = Bn, R² = CH₂Bn: ee = 59% (R,R)
R¹ = Bn, R² = p-(BnO)C₆H₄-CH₂: ee = 0% (R,R)

Scheme 10.7   1,3-Dipolar cycloadditions of nitrones with 1,1-diethoxypropene catalysed by oxazaborolidines derived N-tosyl-L-α-amino acids.

a rigid support that offers an open and accessible pore structure in all possible solvents and over a wide temperature and pressure range. These novel CPG-immobilised catalysts were prepared, as summarised in Scheme 10.8, and then applied to the asymmetric 1,3-dipolar cycloaddition of diphenyl nitrone to 3-crotonoyl-1,3-oxazolidinone, providing the corresponding cycloadduct in a moderate enantioselectivity, but in a quantitative yield and with a preferential formation of the *exo* cycloadduct, as shown in Scheme 10.8. A seasoning of the catalyst material was shown in this reaction. Indeed, even if it were necessary to

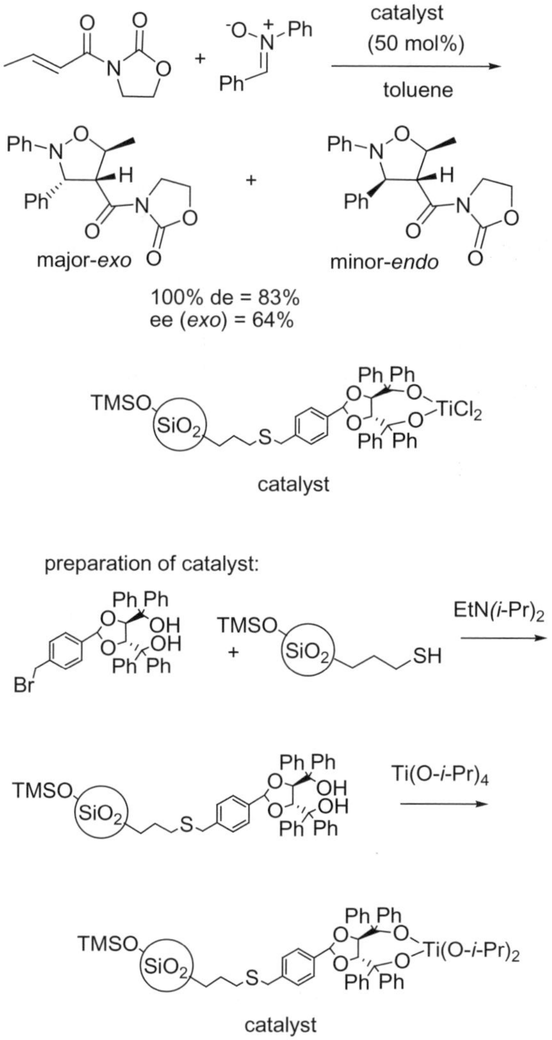

Scheme 10.8 1,3-Dipolar cycloaddition in presence of CPG-immobilised Ti-TADDOLate.

use 0.5 equivalent of the immobilised catalyst initially in order to match performance, after three runs, the reaction rate and the enantio- and diastereoselectivities dropped considerably. An acidic washing after each subsequent run completely restored the performance, so that, after a total of seven runs, the amount of the catalyst could be reduced to 0.4, 0.3, 0.2 and 0.1 equivalents in the following runs, with identical good results.

In another context, Davies *et al.* have developed the Rh$_2$(*S*-DOSP)$_4$-catalysed decomposition of vinyldiazocarbonyl derivatives in the presence of vinyl ethers.[15] The corresponding chiral cyclopentenecarboxylates were formed in high enantioselectivities of up to 99% ee with the full control of the relative stereochemistry at up to three contiguous stereogenic centres, as depicted in Scheme 10.9.

Of particular interest is the reaction occurring between azomethine ylides and activated alkenes to form functionalised pyrrolidines,[16] which represent key units in medicinal chemistry and are highly valuable synthetic building blocks.[17] In recent years, the development of various chiral catalysts for the enantioselective version of this reaction has been a great challenge. Methodologies using Cu-based reagents or Cu-Lewis acids are valuable synthetic tools with a wide application in the synthesis of complex molecules. It is not just the efficiency but also the diversity of these methods that make copper one of the most important transition metals in organic synthesis. In recent years, Cu-Lewis acids in combination with a variety of chiral ligands[18] have been evaluated in the enantioselective cycloadditions of azomethine ylides. In this context, Carretero *et al.* have described a novel family of readily available planar chiral S/P ligands, Fesulphos ligands,[19] whose copper(I) complexes have shown

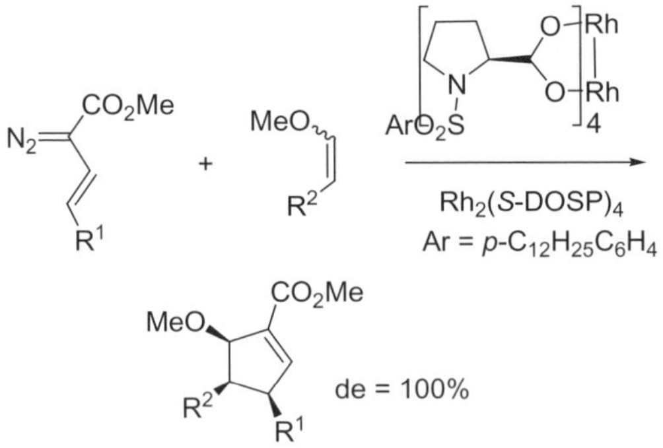

R^1 = Me, R^2 = Ph: 79% ee = 98%
R^1 = Me, R^2 = *p*-BrC$_6$H$_4$: 71% ee > 99%
R^1 = Me, R^2 = (*E*)-CH=CHEt: 49% ee = 99%
R^1 = *n*-Bu, R^2 = Ph: 55% ee > 94%
R^1 = H, R^2 = Ph: 61% ee = 82%
R^1 = H, R^2 = (*E*)-CH=CHEt: 74% ee = 79%

Scheme 10.9 Rh$_2$(*S*-DOSP)$_4$-catalysed 1,3-dipolar cycloadditions of vinyldiazocarbonyl derivatives with vinyl ethers.

excellent performance in asymmetric catalysis. Therefore, Carretero *et al.* demonstrated, in 2005, that the combination of copper(I) salts and a Fesulphos ligand resulted in a highly reactive catalyst system, displaying exceptional enantioselectivities and a broad scope in the asymmetric 1,3-dipolar cycload-dition of azomethine ylides generated from iminoesters and maleimide dipo-larophiles (Scheme 10.10).[20] In addition, it is important to note that these reactions were performed in the presence of only 0.5–3 mol% of catalyst loading.

In order to explore more deeply the scope and generality of this new meth-odology, other dipolarophiles of a varied nature, such as dimethyl maleate, dimethyl fumarate and fumarodinitrile, were also examined, proving to be excellent substrates for this reaction, since they provided high asymmetric inductions (76–99% ee), although the *endo/exo* selectivity was poorer than that obtained with the maleimide dipolarophiles (Scheme 10.11).[20] For comparison, Komatsu *et al.* have studied the same reactions in the presence of Cu(OTf)$_2$ and BINAP derivatives as the chiral ligands, providing, in this case, an *exo* selectivity (≤90% de) and a high but lower enantioselectivity (≤87% ee).[21]

The development of more efficient and environmentally friendly methodol-ogies in asymmetric catalysis is a very important area of research in chemistry.

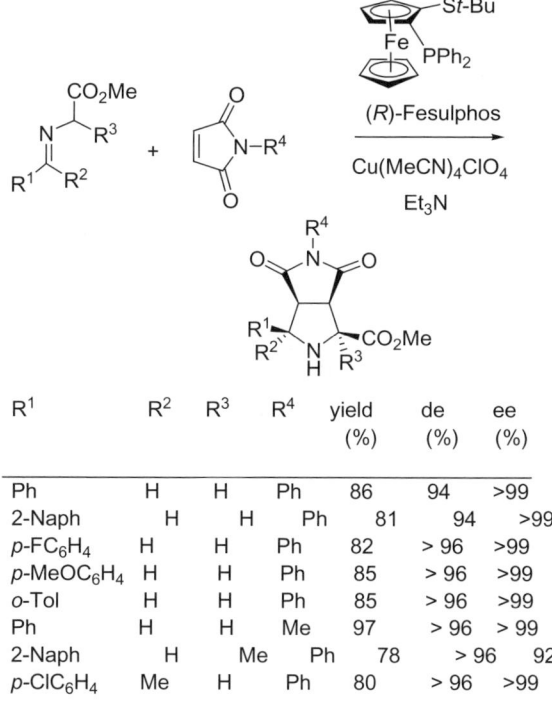

R^1	R^2	R^3	R^4	yield (%)	de (%)	ee (%)
Ph	H	H	Ph	86	94	>99
2-Naph	H	H	Ph	81	94	>99
p-FC$_6$H$_4$	H	H	Ph	82	> 96	>99
p-MeOC$_6$H$_4$	H	H	Ph	85	> 96	>99
o-Tol	H	H	Ph	85	> 96	>99
Ph	H	H	Me	97	> 96	> 99
2-Naph	H	Me	Ph	78	> 96	92
p-ClC$_6$H$_4$	Me	H	Ph	80	> 96	>99

Scheme 10.10 1,3-Dipolar cycloadditions of azomethine ylides with maleimides in the presence of Cu(I)-Fesulphos catalyst system.

An alternative that has received a great deal of attention in recent years is the immobilisation of a chiral catalyst on a nonsoluble support (polystyrene resins, silica gel, zeolites, *etc.*), thereby creating a chiral heterogeneous catalyst. Unlike homogeneous catalysts, these supported complexes can be recovered from the

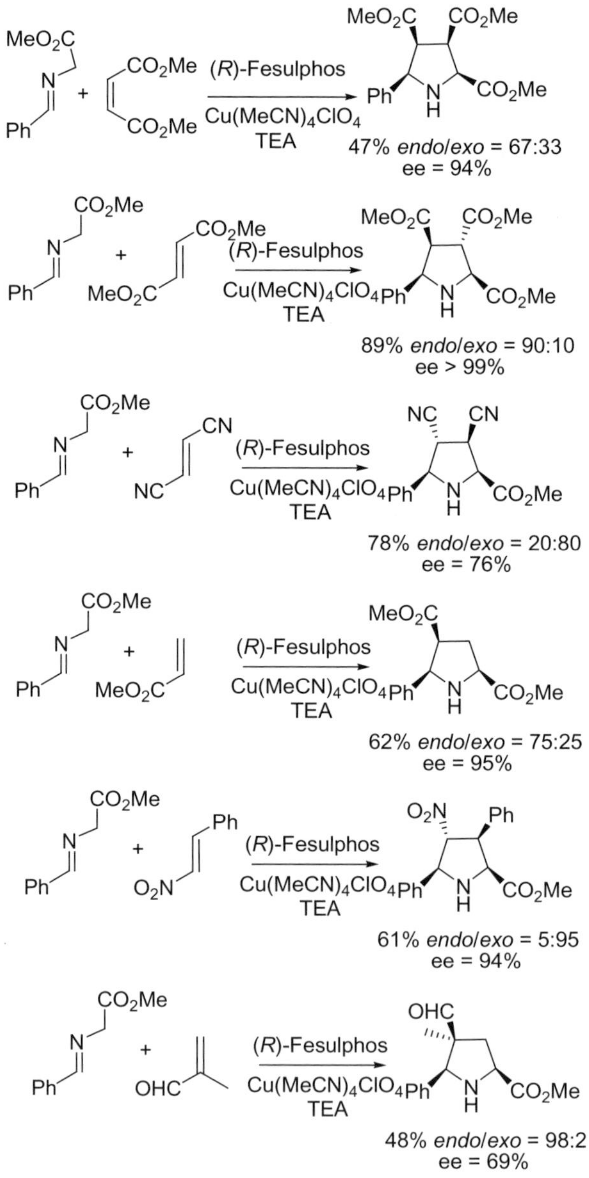

Scheme 10.11 1,3-Dipolar cycloadditions of azomethine ylides with various dipolarophiles in the presence of a Cu(I)-Fesulphos catalyst system.

reaction mixture by simple filtration and can be easily recycled.[22] In this context and as an application of Fesulphos ligands, Carretero *et al.* have recently developed polymer-supported Fesulphos ligands that have been successfully applied in asymmetric catalysis.[23] These polystyrene-supported Fesulphos ligands acted as very efficient catalysts in 1,3-dipolar cycloadditions of azomethine ylides with *N*-phenylmaleimide, providing the corresponding cycloadducts with excellent catalytic activities and enantioselectivities of up to 99% ee (Scheme 10.12). Furthermore, these supported ligands could be recovered and reused in successive catalytic 1,3-dipolar cycloadditions without further addition of Cu(I) maintaining excellent enantioselectivities in up to three runs.

In 2007, Zeng and Zhou developed other chiral ferrocene-derived S/P ligands, which were further efficiently used in AgOAc-catalysed asymmetric 1,3-dipolar cycloadditions of azomethine ylides.[24] The roles of the planar chirality and electronic properties of the phosphorous substituents were studied, allowing enantioselectivities of up to 93% ee along with high *endo* selectivities to be obtained. As shown in Scheme 10.13, a wide range of imines derived from aromatic aldehydes were reacted with various dipolarophiles, such as *N*-phenylmaleimide and dimethyl maleate, yielding the corresponding cycloadducts in both high yields and enantioselectivities.

In 2007, a novel C_2-symmetric diphenylthiophosphoramide ligand was found to be a fairly efficient chiral ligand for the Cu(I)-promoted 1,3-dipolar cycloaddition of imines and pyrrole-2,5-dione derivatives to give the corresponding adducts in moderate to good enantioselectivities and good yields (Scheme 10.14).[25]

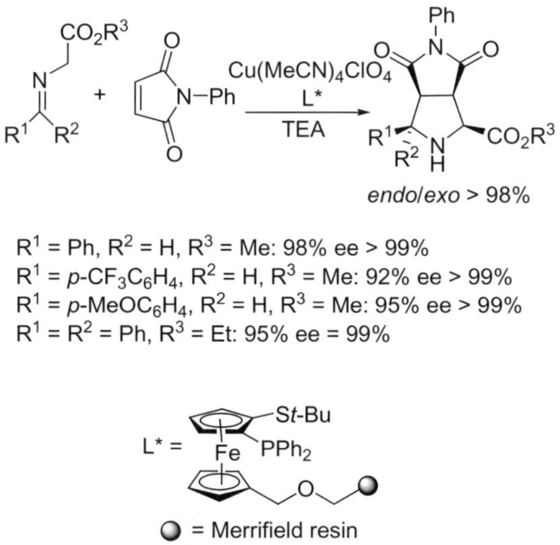
CO2R3
N
R1 R2
+
N–Ph
Cu(MeCN)4ClO4
L*
TEA
Ph
CO2R3
endo/exo > 98%
R1 = Ph, R2 = H, R3 = Me: 98% ee > 99%
R1 = p-CF3C6H4, R2 = H, R3 = Me: 92% ee > 99%
R1 = p-MeOC6H4, R2 = H, R3 = Me: 95% ee > 99%
R1 = R2 = Ph, R3 = Et: 95% ee = 99%
L* =
St-Bu
Fe
PPh2
= Merrifield resin

Scheme 10.12 1,3-Dipolar cycloadditions of azomethine ylides with polystyrene-supported Fesulphos ligand.

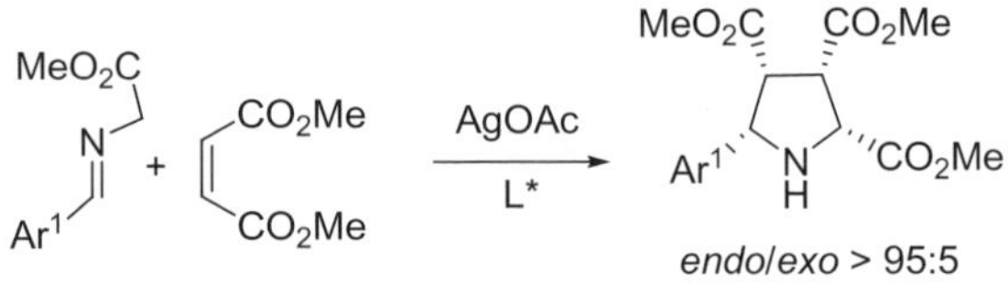

with Ar^2 = *p*-CF$_3$C$_6$H$_4$ and R = *p*-ClC$_6$H$_4$:
Ar^1 = Ph: 93% ee = 87%
Ar^1 = 2-Naph: 86% ee = 89%
Ar^1 = *p*-FC$_6$H$_4$: 96% ee = 89%
Ar^1 = *p*-MeOC$_6$H$_4$: 86% ee = 88%
Ar^1 = *o*-Tol: 96% ee = 84%
Ar^1 = *p*-ClC$_6$H$_4$: 96% ee = 87%

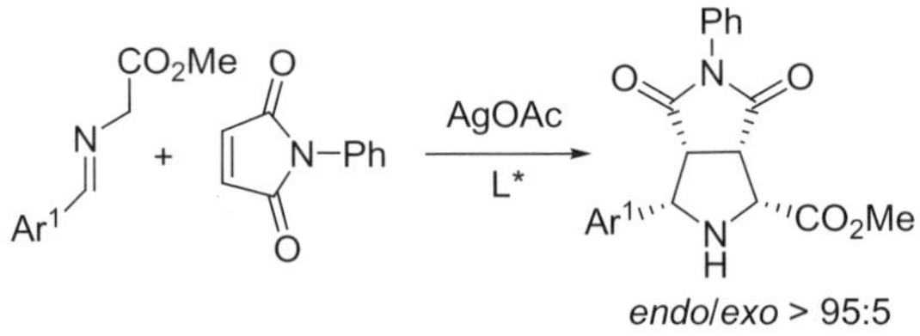

with Ar^2 = 3,5-(CF$_3$)$_2$C$_6$H$_3$ and R = *p*-MeOC$_6$H$_4$:
Ar^1 = Ph, Ar^2 = : 98% ee = 92%
Ar^1 = 2-Naph: 97% ee = 93%
Ar^1= *p*-FC$_6$H$_4$: 96% ee = 93%
Ar^1 = *p*-MeOC$_6$H$_4$: 98% ee = 86%
Ar^1 = *o*-Tol: 95% ee = 90%

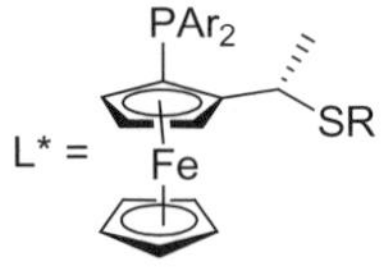

Scheme 10.13 1,3-Dipolar cycloadditions of azomethine ylides with ferrocene-derived S/P ligands and AgOAc.

10.4 [2 + 2] Cycloaddition

The existence of ketenes was established over a hundred years ago,[26] and, in recent years, asymmetric synthesis based on [2 + 2] cycloadditions of ketenes with carbonyl compounds to form chiral β-lactones has been achieved with high yields and high stereoselectivities.[27] In 1994, Miyano *et al.* reported the use of C_2-symmetric bis(sulfonamides) as ligands of trialkylaluminum complexes to promote the asymmetric [2 + 2] cycloaddition of ketenes with aldehydes.[28] The corresponding oxetanones were obtained in good yields and enantioselectivities

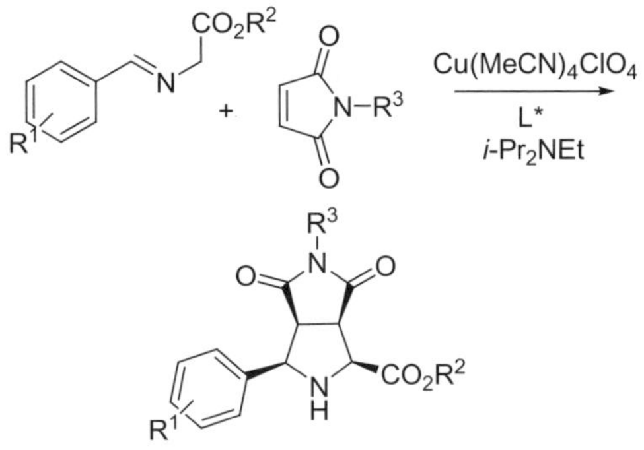

R^1 = *p*-Cl, R^2 = Me, R^3 = Ph: 82% ee = 74%
R^1 = *p*-Br, R^2 = Me, R^3 = Ph: 85% ee = 58%
R^1 = *o*-Cl, R^2 = Me, R^3 = Ph: 82% ee = 57%
R^1 = H, R^2 = Me, R^3 = Ph: 89% ee = 47%
R^1 = *p*-Me, R^2 = Me, R^3 = Ph: 76% ee = 26%
R^1 = *p*-Cl, R^2 = Et, R^3 = Ph: 72% ee = 62%
R^1 = *m*-Br, R^2 = Me, R^3 = Ph: 60% ee = 52%
R^1 = H, R^2 = R^3 = Me: 77% ee = 79%
R^1 = *p*-Cl, R^2 = R^3 = Me: 86% ee = 77%
R^1 = *p*-Cl, R^2 = Me, R^3 = Bn: 83% ee = 68%

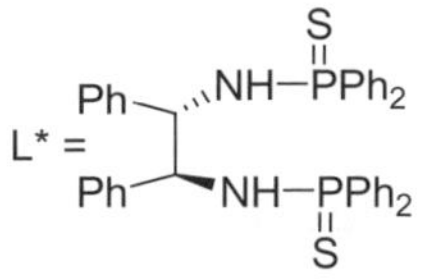

Scheme 10.14 1,3-Dipolar cycloadditions of azomethine ylides with C_2-symmetric diphenylthiophosphoramide ligand.

of up to 74% ee, as shown in Scheme 10.15. The bulky aldehydes gave the higher enantioselectivities compared with those obtained with small aldehydes. In all of these reactions, the acylation of the catalyst by the ketene did not occur. A steric model depicted in Scheme 10.15 was proposed to explain the stereoselectivity of the reaction. The aldehyde seemed to coordinate to the catalyst by using the lone-pair electrons of the carbonyl oxygen in a manner *anti* to the substituent R as suggested by Nevalainen[29] for a simple diazaaluminolidine–formaldehyde complex. The bulky aldehydes preferred such a configuration to avoid the steric repulsion between the bulky R group and the diazaaluminolidine moiety, resulting in the obtention of the higher selectivities compared with those obtained from the small aldehydes. Then, the nucleophilic attack of the ketene to the activated aldehydes proceeded preferentially from the *Re* face to avoid the bulky bis(trifluoromethyl)phenyl group that was located *anti* to the nearby phenyl group of the catalyst relative to the plane of the diazaaluminolidine ring.

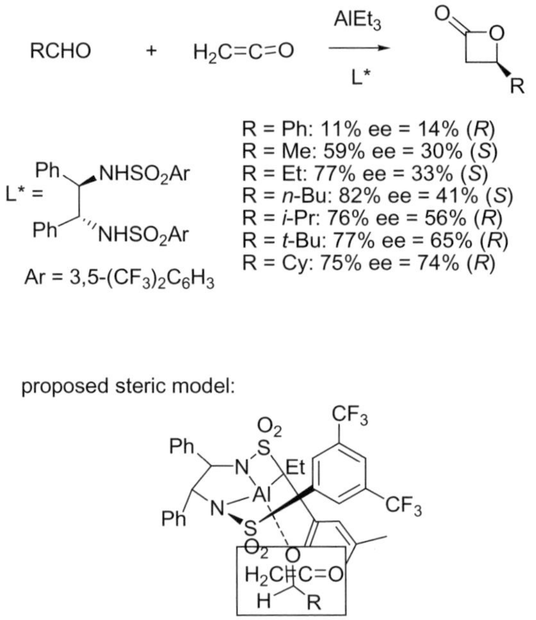

RCHO   +   H₂C=C=O $\xrightarrow[\text{L*}]{\text{AlEt}_3}$

R = Ph: 11% ee = 14% (*R*)
R = Me: 59% ee = 30% (*S*)
R = Et: 77% ee = 33% (*S*)
R = *n*-Bu: 82% ee = 41% (*S*)
R = *i*-Pr: 76% ee = 56% (*R*)
R = *t*-Bu: 77% ee = 65% (*R*)
R = Cy: 75% ee = 74% (*R*)

L* =   Ph–CH(NHSO₂Ar)–CH(NHSO₂Ar)–Ph

Ar = 3,5-(CF₃)₂C₆H₃

proposed steric model:

Scheme 10.15   [2 + 2] Cycloadditions of ketene with aldehydes catalysed by bis (sulfonamides)–aluminum complexes.

10.5   Carbonyl Ene Reaction

The carbonyl ene reaction is the reaction occurring between an alkene bearing an allylic hydrogen (the ene) and a carbonyl compound. This reaction is accompanied by the migration of the double bond and a 1,5-hydrogen shift.[30] The catalytic asymmetric version of the carbonyl ene reaction has been practised since 1980s, with the first genuine catalyst being derived from titanium(IV) complexes and enantiomerically pure diols such as BINOL.[31] Late transition-metal catalysts derived from chiral ligands have also been employed with success as copper(II) complexes of chiral nitrogen ligands worthy of special mention. These have extended the scope to some less-reactive enophiles and allowed fewer nucleophilic alkenes to be employed. Very few examples of this type of asymmetric reactions have been performed in the presence of chiral sulfur-containing ligands. As an example, Bolm *et al.* reported, in 2005, the use of chiral aminosulfoximines as ligands in Cu-catalysed carbonyl-ene reactions between activated carbonyl compounds as enophiles and 1,1-disubstituted olefins.[32] This important C–C bond-forming process yielded allylic α-hydroxy esters which are useful building blocks in organic synthesis.[33] It must be noted that, so far, the catalytic systems capable of mediating the intermolecular addition of olefins to activated ketones are rare.[34] Thus, the successful application of the optimised aminosulfoximine systems, depicted in Scheme 10.16, to the Cu-catalysed carbonyl-ene reactions of pyruvates with 1,1-disubstituted

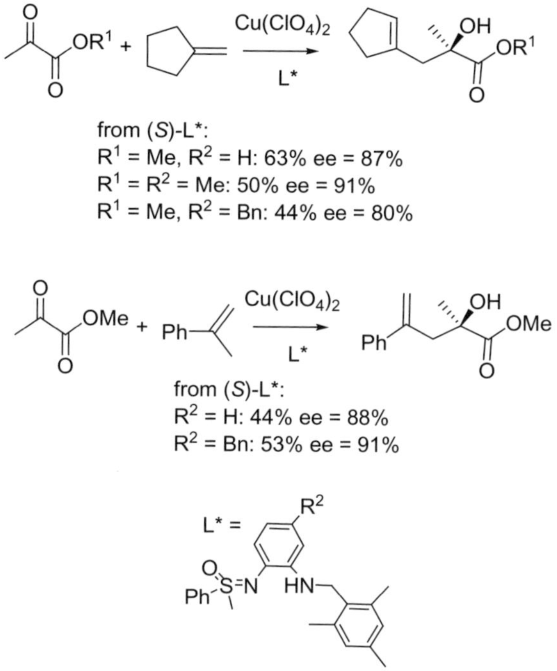

from (S)-L*:
R^1 = Me, R^2 = H: 63% ee = 87%
R^1 = R^2 = Me: 50% ee = 91%
R^1 = Me, R^2 = Bn: 44% ee = 80%

from (S)-L*:
R^2 = H: 44% ee = 88%
R^2 = Bn: 53% ee = 91%

Scheme 10.16 Carbonyl-ene reactions of pyruvates with 1,1-disubstituted olefins with aminosulfoximine ligand.

olefins was demonstrated, providing the corresponding hydroxy esters with high enantioselectivities of up to 91% ee and moderate yields. These enantioselectivities reached up to the values achieved with Evans's well-established Cu(II)-*t*-Bu-Box catalyst system.[34]

10.6   Addition of Diethylzinc to Imines

Efficient and asymmetric preparation of amines is one of the most promising methodologies in organic synthesis.[35] In addition to the asymmetric reduction of imines, the enantioselective addition of organometallic reagents to imines is a convenient route to obtain optically active amines.[36] Optically active α-branched amines are important chiral building blocks and are abundantly present in biologically active compounds. Very good enantioselectivities have been reported by using a wide range of chiral ligands, such as amines, amino alcohols, phosphines, amino phosphines, peptides, *etc.* for the addition to imines of various organometallic reagents, such as alkyllithium, organozinc and allylstannane reagents among others. In 2000, Shi and Sui were interested in the synthesis and applications of novel ligands based on the axially chiral

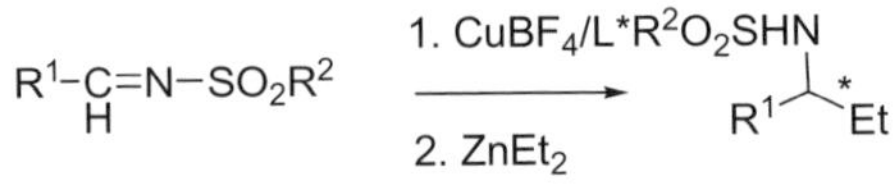

$$R^1-\underset{H}{C}=N-SO_2R^2 \quad \xrightarrow[\text{2. ZnEt}_2]{\text{1. CuBF}_4/\text{L}^*\text{R}^2\text{O}_2\text{SHN}} \quad R^1\overset{*}{\diagup}Et$$

R^1 = Ph, R^2 = p-Tol: 93% ee = 86%
R^1 = R^2 = p-Tol: 95% ee = 86%
R^1 = m-Tol, R^2 = p-Tol: 93% ee = 86%
R^1 = p-MeOC$_6$H$_4$, R^2 = p-Tol: 81% ee = 90%
R^1 = p-ClC$_6$H$_4$, R^2 = p-Tol: 96% ee = 91%
R^1 = m-ClC$_6$H$_4$, R^2 = p-Tol: 93% ee = 86%
R^1 = p-CF$_3$C$_6$H$_4$, R^2 = p-Tol: 95% ee = 93%
R^1 = p-BrC$_6$H$_4$, R^2 = p-Tol: 91% ee = 92%
R^1 = m-FC$_6$H$_4$, R^2 = p-Tol: 94% ee = 85%
R^1 = p-FC$_6$H$_4$, R^2 = p-Tol: 96% ee = 90%
R^1 = 2-Fu, R^2 = p-Tol: 95% ee = 82%
R^1 = 1-Naph, R^2 = p-Tol: 93% ee = 88%
R^1 = Ph, R^2 = Me: 96% ee = 87%
R^1 = p-ClC$_6$H$_4$, R^2 = (CH$_2$)$_2$TMS: 94% ee = 80%
R^1 = i-Pr, R^2 = p-Tol: 86% ee = 63%

L* =

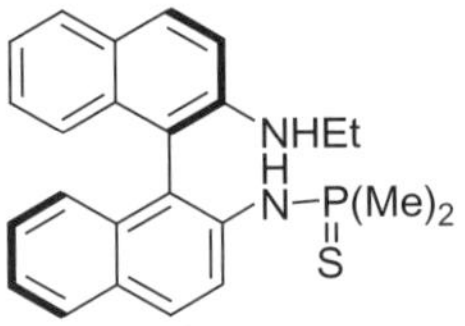

NHEt
N—P(Me)$_2$
‖
S

Scheme 10.17 Additions of ZnEt$_2$ to *N*-sulfonylimines with binaphthylthiopho-
sphoramide ligand.

binaphthalenediamine (BINAM).[37] In the field of catalytic asymmetric synthesis, BINAM has been much less popular than the widely used other axially
chiral binapthalene skeletons such as BINOL[38] and NOBIN,[39] or other chiral
diamines such as 1,2-cyclohexanediamine and 1,2-diphenylethenediamine.[40] In
2003, Shi and Wang developed a new family of axially chiral binaphthylthiophosphoramide ligands that were easily available, very stable and recoverable.[41] These novel ligands were successfully applied to the Cu-catalysed
asymmetric addition of ZnEt$_2$ to *N*-sulfonylimines, leading to the corresponding chiral amines with high yields and moderate to good enantioselectivities. The best results obtained with the most efficient ligand are collected in
Scheme 10.17. It was demonstrated that the presence of the sulfur on phosphorus was crucial for this catalytic reaction because the corresponding axially
chiral binaphthyldiphenylphosphoramide ligand showed a very low catalytic
activity for this reaction. Moreover, ^{13}C NMR and ^{31}P NMR studies proved

that Cu(I) could be potentially coordinated by both the sulfur and nitrogen atoms of the ligand.

The scope of this methodology was extended by these authors to the asymmetric addition of $ZnEt_2$ to various diphenylphosphinoyl imines in the presence of $Cu(OTf)_2$ combined with a chiral binaphthyldiphenylphosphoramide ligand.[42] The corresponding diphenylphosphinic amides were achieved in generally good yields and enantioselectivities, as shown in Scheme 10.18. In this study, the sulfur heteroatom on phosphorus was shown to be crucial for this process to be effective since the corresponding axially chiral *N*-(binaphthyl-2-yl)diphenylphosphoramide showed no enantioselectivity for this type of reactions. The addition product was formed in only 25% yield along with a large amount of unreacted starting materials under the same conditions.

In order to expand this novel type of chiral thiophosphoramide ligand to other systems, Shi and Zhang decided to use homochiral 1,2-cyclohexanediamine or 1,2-diphenylethylenediamine as a chiral source to prepare their corresponding thiophosphoramide ligands.[43] These novel ligands were subsequently investigated for the asymmetric addition of $ZnEt_2$ to *N*-sulfonylimines. When these reactions were performed in the presence of the chiral thiophosphoramide ligand derived from (1*R*,2*R*)-cyclohexanediamine depicted in Scheme 10.19, the corresponding amines were isolated in high yields and moderate to good enantioselectivities.

Ar = Ph: 80% ee = 85%
Ar = *p*-Tol: 85% ee = 82%
Ar = *m*-Tol: 89% ee = 82%
Ar = *p*-MeOC$_6$H$_4$: 61% ee = 66%
Ar = *p*-FC$_6$H$_4$: 83% ee = 85%
Ar = *p*-ClC$_6$H$_4$: 84% ee = 84%
Ar = *p*-BrC$_6$H$_4$: 80% ee = 84%
Ar = *o*-BrC$_6$H$_4$: 89% ee = 82%
Ar = 1-Naph: 82% ee = 83%

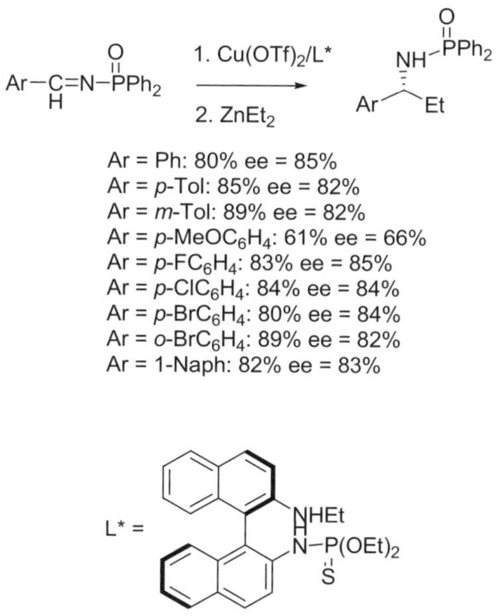

Scheme 10.18   Additions of $ZnEt_2$ to diphenylphosphinoyl imines with binaphthyl-thiophosphoramide ligand.

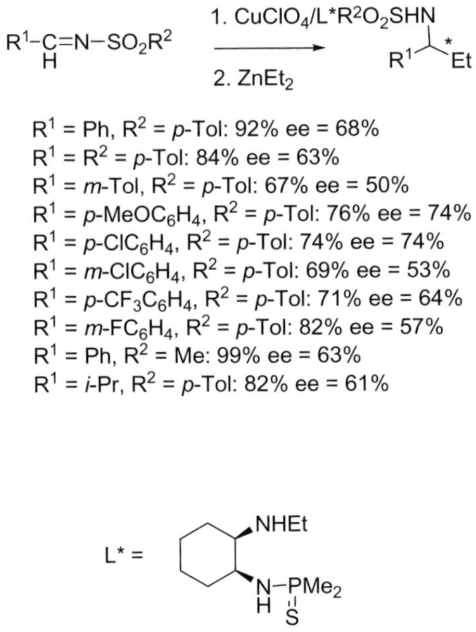

$$R^1-\underset{H}{C}=N-SO_2R^2 \quad \xrightarrow[\text{2. ZnEt}_2]{\text{1. CuClO}_4/\text{L*R}^2\text{O}_2\text{SHN}} \quad R^1 \overset{*}{\underset{}{\bigwedge}} Et$$

R^1 = Ph, R^2 = *p*-Tol: 92% ee = 68%
R^1 = R^2 = *p*-Tol: 84% ee = 63%
R^1 = *m*-Tol, R^2 = *p*-Tol: 67% ee = 50%
R^1 = *p*-MeOC$_6$H$_4$, R^2 = *p*-Tol: 76% ee = 74%
R^1 = *p*-ClC$_6$H$_4$, R^2 = *p*-Tol: 74% ee = 74%
R^1 = *m*-ClC$_6$H$_4$, R^2 = *p*-Tol: 69% ee = 53%
R^1 = *p*-CF$_3$C$_6$H$_4$, R^2 = *p*-Tol: 71% ee = 64%
R^1 = *m*-FC$_6$H$_4$, R^2 = *p*-Tol: 82% ee = 57%
R^1 = Ph, R^2 = Me: 99% ee = 63%
R^1 = *i*-Pr, R^2 = *p*-Tol: 82% ee = 61%

L* =

Scheme 10.19   Additions of ZnEt$_2$ to *N*-sulfonylimines with (1*R*,2*R*)-cyclohex-anediamine-derived thiophosphoramide ligand.

Only in recent years have limited organometallic systems been developed to catalyse the addition of alkynes to imines to produce propargylamines, which are important synthetic intermediates in the construction of many biologically active nitrogen-containing compounds and natural products. In contrast to the enantioselective alkynylation of aldehydes in which considerable progress has been made in recent years, the enantioselective alkynylzinc addition to imines still remains less developed. In this area, Chan *et al.* developed, in 2007, a new class of chiral tridentate *N*-tosylatedaminoimine ligands that were used in the Cu-catalysed enantioselective addition of phenylacetylene to imines.[44] The corresponding propargylic amines were isolated with both good enantioselectivities and yields, as shown in Scheme 10.20.

10.7   Allylstannation of Ketones

The majority of catalytic enantioselective allylation reactions involve the chiral Lewis-acid-catalysed additions of allylsilanes or allylstannanes to carbonyl compounds.[45] Monothiobinaphthol has been used by Woodward *et al.* as a chiral promoter in the enantioselective catalytic allylation of aryl ketones with impure Sn(allyl)$_4$, prepared from allyl chloride, air-oxidised magnesium and SnCl$_4$.[46] Therefore, the allylation of arylketones in these conditions was achieved very efficiently, since the corresponding allylic alcohols were formed in

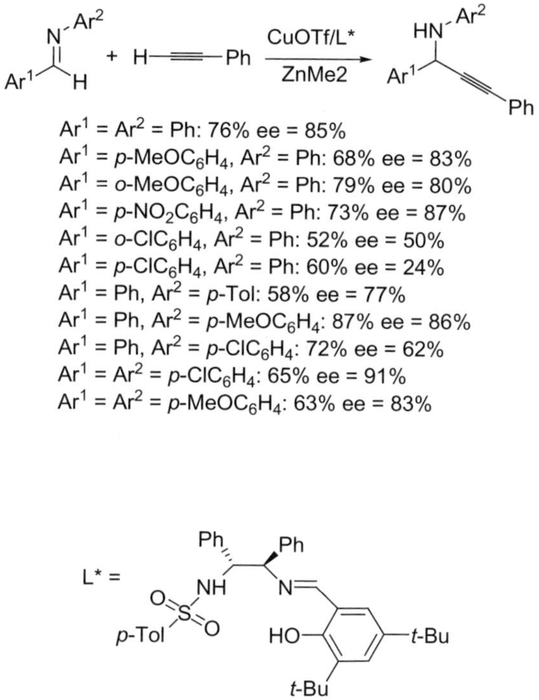

Ar^1 = Ar^2 = Ph: 76% ee = 85%
Ar^1 = p-MeOC$_6$H$_4$, Ar^2 = Ph: 68% ee = 83%
Ar^1 = o-MeOC$_6$H$_4$, Ar^2 = Ph: 79% ee = 80%
Ar^1 = p-NO$_2$C$_6$H$_4$, Ar^2 = Ph: 73% ee = 87%
Ar^1 = o-ClC$_6$H$_4$, Ar^2 = Ph: 52% ee = 50%
Ar^1 = p-ClC$_6$H$_4$, Ar^2 = Ph: 60% ee = 24%
Ar^1 = Ph, Ar^2 = p-Tol: 58% ee = 77%
Ar^1 = Ph, Ar^2 = p-MeOC$_6$H$_4$: 87% ee = 86%
Ar^1 = Ph, Ar^2 = p-ClC$_6$H$_4$: 72% ee = 62%
Ar^1 = Ar^2 = p-ClC$_6$H$_4$: 65% ee = 91%
Ar^1 = Ar^2 = p-MeOC$_6$H$_4$: 63% ee = 83%

Scheme 10.20 Cu-catalysed additions of phenylacetylene to imines with *N*-tosylate-daminoimine ligand.

good yields and enantioselectivities of up to 92% ee. However, it was noted that the enantioselectivity of these reactions decreased when the conversion increased, maybe due to achiral background reactions occurring in anhydrous media. Fortunately, the presence of traces of water was shown to inhibit these background reactions and increased drastically the catalyst activity, thus allowing both high yield and enantioselectivity to be obtained (Scheme 10.21). Indeed, in the same process, highly purified or commercial Sn(allyl)$_4$ yielded the corresponding products with enantioselectivities of only 35–50% ee. The origin of these effects was the presence of small amounts of the compounds, EtSn(allyl)$_3$, ClSn(allyl)$_3$, and ClSnEt(allyl)$_2$, in the tetraallyltin sample in addition to that of water that inhibited achiral background reactions.

10.8 Allylsilylation of Carbonyl Compounds

The first use of chiral sulfoxides as Lewis-base catalysts in the allylation of aldehydes with allyltrichlorosilane was reported in 2003.[47] The formation of the

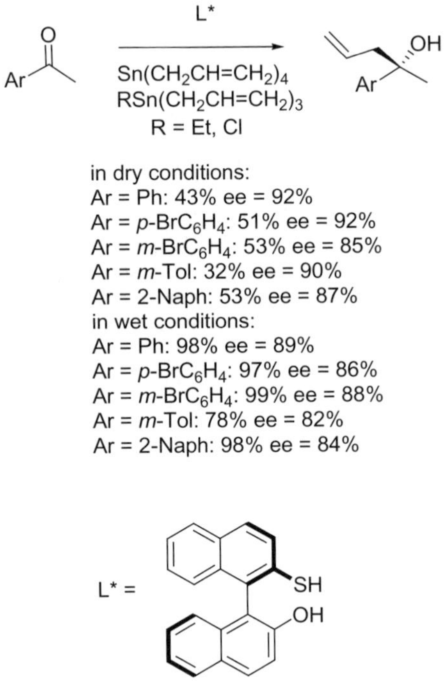
L*
Sn(CH2CH=CH2)4
RSn(CH2CH=CH2)3
R = Et, Cl
Ar
OH
in dry conditions:
Ar = Ph: 43% ee = 92%
Ar = p-BrC6H4: 51% ee = 92%
Ar = m-BrC6H4: 53% ee = 85%
Ar = m-Tol: 32% ee = 90%
Ar = 2-Naph: 53% ee = 87%
in wet conditions:
Ar = Ph: 98% ee = 89%
Ar = p-BrC6H4: 97% ee = 86%
Ar = m-BrC6H4: 99% ee = 88%
Ar = m-Tol: 78% ee = 82%
Ar = 2-Naph: 98% ee = 84%
L* =
SH
OH

Scheme 10.21 Allylstannations of aryl ketones with monothiobinaphthol ligand.

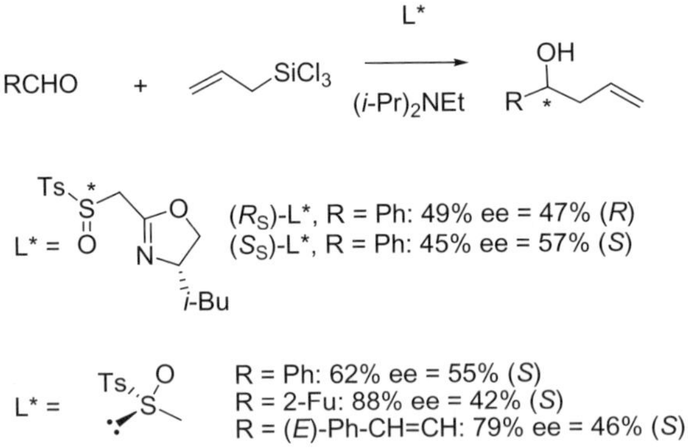
RCHO
+
SiCl3
L*
(i-Pr)2NEt
OH
R
L* =
Ts
i-Bu
(RS)-L*, R = Ph: 49% ee = 47% (R)
(SS)-L*, R = Ph: 45% ee = 57% (S)
R = Ph: 62% ee = 55% (S)
R = 2-Fu: 88% ee = 42% (S)
R = (E)-Ph-CH=CH: 79% ee = 46% (S)

Scheme 10.22 Allylsilylations of aldehydes with sulfoxide ligands.

corresponding homoallylic alcohols could be obtained in satisfactory yields and with moderate enantioselectivities, as shown in Scheme 10.22.

Moreover, Kobayashi *et al.* introduced chiral sulfoxides into the reactions of *N*-acylhydrazones with allyltrichlorosilanes as highly efficient neutral coordinating-organocatalysts (Scheme 10.23).[48] The corresponding chiral homoallylic

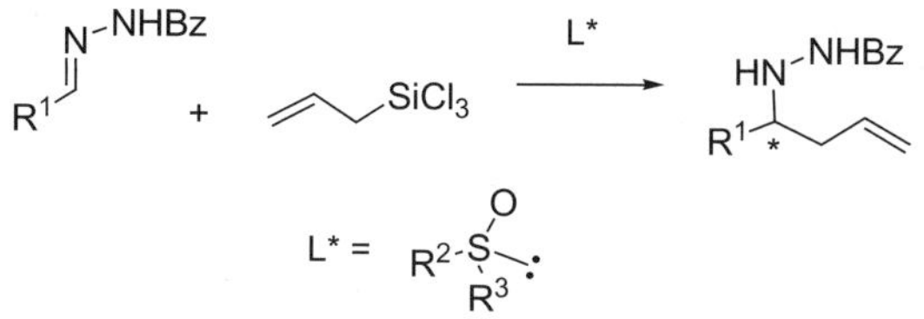

(R)-L*, R^1 = CH$_2$Bn, R^2 = Me, R^3 = Ts: 73% ee = 93% (R)
(R)-L*, R^1 = CH$_2$Bn, R^2 = Et, R^3 = Ts: 77% ee = 50% (R)
(S)-L*, R^1 = CH$_2$Bn, R^2 = o-MeOC$_6$H$_4$, R^3 = Me: 79% ee = 42% (S)
(S)-L*, R^1 = CH$_2$Bn, R^2 = p-MeOC$_6$H$_4$, R^3 = Me: 91% ee = 69% (S)
(R)-L*, R^1 = R^2 = Me, R^3 = Ts: 78% ee = 90% (R)
(R)-L*, R^1 = n-Hept, R^2 = Me, R^3 = Ts: 81% ee = 88% (R)
(R)-L*, R^1 = i-Pr, R^2 = Me, R^3 = Ts: 80% ee = 98% (R)
(R)-L*, R^1 = c-Hex, R^2 = Me, R^3 = Ts: 77% ee = 91% (R)
(R)-L*, R^1 = p-MeOC$_6$H$_4$, R^2 = Me, R^3 = Ts: 82% ee = 81% (S)
(R)-L*, R^1 = p-ClC$_6$H$_4$, R^2 = Me, R^3 = Ts: 69% ee = 89% (S)

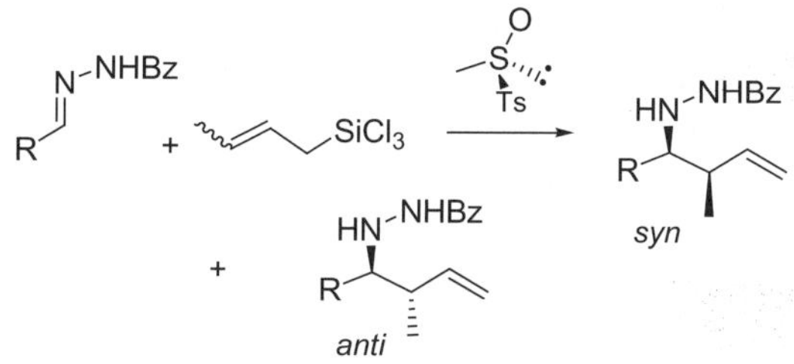

R = CH$_2$Bn, (Z)-silane: syn <1% + anti > 99% (ee = 91%)
R = CH$_2$Bn, (E)-silane: syn = 99% (ee = 89%) + anti = 1%
R = Me, (Z)-silane: syn <1% + anti > 99% (ee = 73%)
R = Me, (E)-silane: syn = 98% (ee = 82%) + anti = 2%
R = n-Hept, (Z)-silane: syn <1% + anti > 99% (ee = 86%)
R = n-Hept, (E)-silane: syn = 95% (ee = 91%) + anti = 5%

Scheme 10.23 Allylsilylations of *N*-acylhydrazones with sulfoxide ligands.

amine derivatives were prepared in these conditions with high enantioselectivities. The scope of the reaction was extended to asymmetric crotylations with (*Z*)- and (*E*)-crotyltrichlorosilanes, which proved to be highly stereospecific, since the (*E*)-crotylsilane afforded the corresponding *syn*-adducts, whereas the (*Z*)-crotylsilane led to the corresponding *anti*-adducts with excellent diastereoselectivities and good to high enantioselectivities.

10.9 Aldol-Type Reaction

Chiral sulfur-containing ligands have also been involved in other reactions such as metal-catalysed enantioselective Mukaiyama-type aldol reactions.[49]

Throughout the last two decades, the enantioselective metal-catalysed Mukaiyama-type aldol reaction has become a useful tool in organic synthesis.[50] This elegant approach allows the convenient preparation of valuable enantiomerically enriched alcohols by using chiral Lewis acids in catalytic amounts. Consequently, a multitude of catalysts derived from metals such as tin, boron, titanium, zirconium, copper,[51] silver and scandium have been reported. High enantioselectivities have been generally observed for the addition of enolsilanes to aldehydes, while the number of powerful catalytic systems applicable to the reaction involving activated ketones such as pyruvates has still remained rather limited. Thus, for the latter process, the development of new chiral Lewis acids still represents a major challenge because the resulting tertiary α-hydroxy ester derivatives are desirable building blocks for biologically active molecules.[52] However, the access to enantiopure products containing tertiary alcohol groups with stereogenic centres is commonly restricted.[53] In recent years, several types of chiral sulfur-containing ligands have been successfully applied to metal-catalysed enantioselective Mukaiyama-type aldol reactions. As an example, Bolm and Langner described, in 2004, the synthesis of new chiral C_1-symmetric benzene-bridged aminosulfoximines, which were capable of serving as efficient ligands in the Mukaiyama-type aldol reaction between 1-phenyl-1-(trimethylsilyloxy)ethane and a pyruvate derivative.[54] The corresponding aldol products bearing quaternary centres, which are commonly difficult to prepare in enantiomerically enriched form, have been obtained with enantioselectivities of up to 99% ee and high yields (Scheme 10.24). The scope of this methodology was extended to the use of other enolsilanes, yielding the corresponding products with a similar efficiency (Scheme 10.24).[55] The high yields (up to 90%) and the outstanding enantioselectivities (up to 99% ee) compared well with the well-established $[Cu(II)\text{-}t\text{-Bubox}]^{2+}$ and $[Sn(II)\text{-pybox}]^{2+}$ complexes.[56]

As an extension of this work, these authors have applied this catalyst system to vinylogous asymmetric Mukaiyama-type aldol reactions, involving silyl vinyl ketene acetals and pyruvate esters.[57] These reactions afforded the corresponding γ,δ-unsaturated α-hydroxy diesters with quaternary centres in high yields and enantioselectivities of up to 99% ee (Scheme 10.25). It was shown that the presence of CF_3CH_2OH as an additive facilitated the turnover of the catalyst.

In 1991, Kobayashi *et al.* prepared novel chiral S/N ligands for the tin-mediated aldol reaction of silyl enol ethers with aldehydes.[58] As an example, the reaction of benzaldehyde afforded the expected *syn* aldol product as the major product with a good yield and an enantioselectivity of up to 92% ee (Scheme 10.26). Moreover, other aldehydes such as substituted benzaldehydes or aliphatic unsaturated aldehydes were converted into their corresponding aldol products with enantioselectivities of more than 90% ee. It was checked that the corresponding diamine ligands provided less active complexes for the same reactions.

Asymmetric Mukaiyama aldol reactions have also been performed in the presence of Lewis-acid lanthanoid complexes combined with a chiral sulfonamide ligand.[59] Similar enantioselectivities of about 40% ee were obtained for all

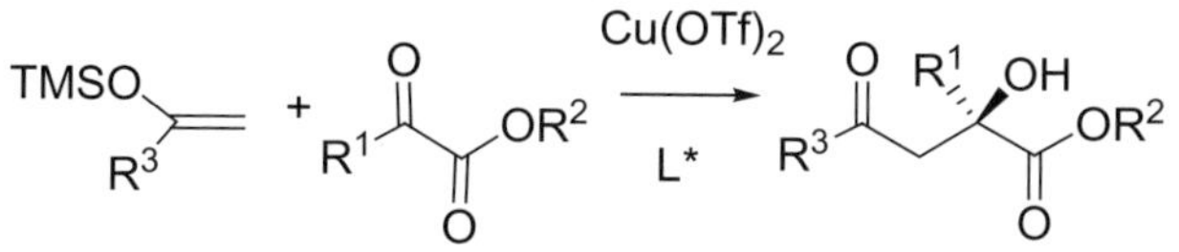

R = R′ = R^3 = Ph, R^1 = R^2 = R″ = Me: 64% ee = 70%
R = 1-Naph, R^1 = R^2 = R″ = Me, R^3 = R′ = Ph: 77% ee = 83%
R = 2-$MeOC_6H_4$, R^1 = R^2 = R″ = Me, R^3 = R′ = Ph: 72% ee = 86%
R = Mes, R^1 = R^2 = R″ = Me, R^3 = R′ = Ph: 88% ee = 93%
R = 2,4,6-(*i*-Pr)$_3C_6H_2$, R^1 = R^2 = R″ = Me, R^3 = R′ = Ph:
99% ee = 93%
R = 2,4,6-(*i*-Pr)$_3C_6H_2$, R^1 = R″ = Me, R^2 = Bn, R^3 = R′ = Ph:
86% ee = 98%
R = 2,4,6-(*i*-Pr)$_3C_6H_2$, R^1 = R″ = Me, R^2 = *i*-Pr, R^3 = R′ = Ph:
90% ee = 99%
R = 2,4,6-(*i*-Pr)$_3C_6H_2$, R^1 = Et, R^2 = R″ = Me, R^3 = R′ = Ph:
78% ee = 89%
R = 2,4,6-(*i*-Pr)$_3C_6H_2$, R^1 = CH_2Bn, R^2 = Et, R^3 = R′ = Ph
R″ = Me: 86% ee = 96%
R = 2,4,6-(*i*-Pr)$_3C_6H_2$, R^2 = Bn, R′ = Ph, R^1 = R^3 = R″ = Me:
71% ee = 91%
R = 2,4,6-(*i*-Pr)$_3C_6H_2$, R′ = Ph, R^3 = *t*-BuS, R^1 = R^2 = R″ = Me:
66% ee = 77%
R^1 = R^2 = R′ = Me, R^3 = *t*-BuS, R = Mes, R″ = *o*-$MeOC_6H_4$:
86% ee = 91%
R^1 = R′ = Me, R^2 = Bn, R^3 = *t*-BuS, R = Mes, R″ = *o*-$MeOC_6H_4$:
79% ee = 93%
R^1 = R′ = Me, R^2 = *i*-Pr, R^3 = *t*-BuS, R = Mes, R″ = *o*-$MeOC_6H_4$:
82% ee = 98%

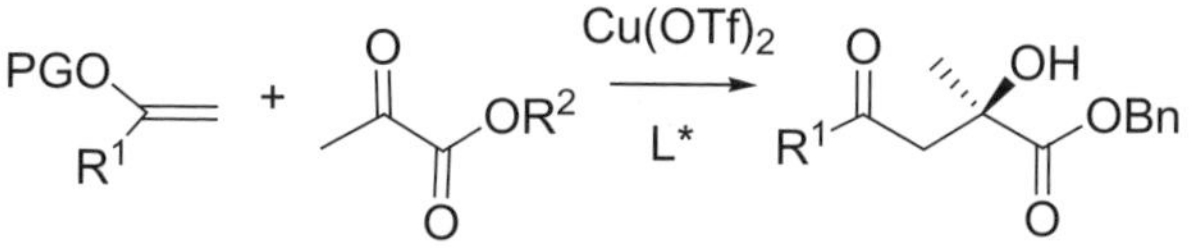

R = 2,4,6-(*i*-Pr)$_3C_6H_2$, R′ = Ph, R″ = R^1 = Me, R^2 = Bn
PG = TMS: 71% ee = 91%
R = CH_2Mes, R' = R^2 = Me, R" = *o*-$MeOC_6H_4$, R^1 = *t*-BuS
PG = Si*t*-$BuMe_2$: 76% ee = 91%
R = CH_2Mes, R' = Me, R" = *o*-$MeOC_6H_4$, R^1 = *t*-BuS
R^2 = Bn, PG = Si*t*-$BuMe_2$: 58% ee = 91%
R = CH_2Mes, R′ = Me, R″ = *o*-$MeOC_6H_4$, R^1 = *t*-BuS
R^2 = *i*-Pr, PG = Si*t*-$BuMe_2$: 76% ee = 93%

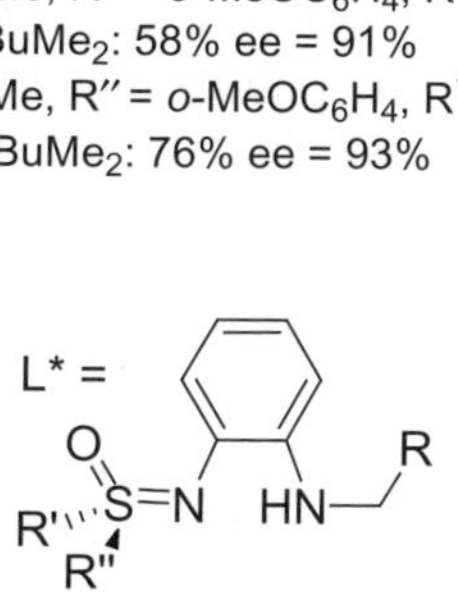

Scheme 10.24 Cu-catalysed Mukaiyama-type aldol reactions with C_1-symmetric benzene-bridged aminosulfoximine ligands.

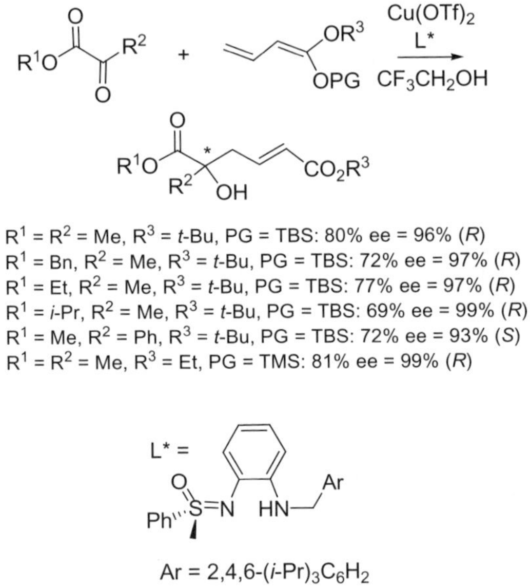

R¹ = R² = Me, R³ = *t*-Bu, PG = TBS: 80% ee = 96% (*R*)
R¹ = Bn, R² = Me, R³ = *t*-Bu, PG = TBS: 72% ee = 97% (*R*)
R¹ = Et, R² = Me, R³ = *t*-Bu, PG = TBS: 77% ee = 97% (*R*)
R¹ = *i*-Pr, R² = Me, R³ = *t*-Bu, PG = TBS: 69% ee = 99% (*R*)
R¹ = Me, R² = Ph, R³ = *t*-Bu, PG = TBS: 72% ee = 93% (*S*)
R¹ = R² = Me, R³ = Et, PG = TMS: 81% ee = 99% (*R*)

L* =

Ar = 2,4,6-(*i*-Pr)₃C₆H₂

Scheme 10.25   Cu-catalysed vinylogous Mukaiyama-type aldol reactions with C_1-symmetric benzene-bridged aminosulfoximine ligand.

the La, Eu and Yb catalysts but, however, the yields decreased from Yb, Eu to La. Consequently, the best results collected in Scheme 10.27 were obtained in the presence of Yb catalyst, with enantioselectivities of up to 49% ee combined with high yields. In all cases of aldehydes, the reaction led to a mixture of the expected silylated products and their corresponding OH-free products.

In 1990, Togni and Häusel reported the synthesis of chiral sulfur-containing ferrocenylphosphine ligands derived from *N*-methyl-ephedrine, which were further employed for the asymmetric Au(I)-catalysed aldol reaction of benzaldehyde with methyl isocyanoacetate, resulting in the formation of the corresponding chiral oxazolines.[60] The best ligand depicted in Scheme 10.28 allowed the formation of the expected *trans* oxazoline in a high enantioselectivity of up to 89% ee. These authors presumed that this ligand existed in a preferred conformation and that the methyl and phenyl substituents did not have a major influence in the stereoselective transition state. In addition, they showed that the introduction of the sulfur atom in the side chain had a positive effect upon the activity.

The nitroaldol reaction or Henry reaction[61] is a powerful and highly versatile carbon–carbon bond-forming reaction, allowing a plethora of key molecular frameworks, such as β-hydroxynitroalkanes, 1,2-amino alcohols or α-hydroxy carboxylic acids to be synthesised in a straightforward manner.[62] Therefore, the development of practical catalytic asymmetric versions of this reaction is still largely desirable.[63] The first catalytic asymmetric nitroaldol reaction was reported in 1992,[64] but despite its long history, relatively few chiral ligands have

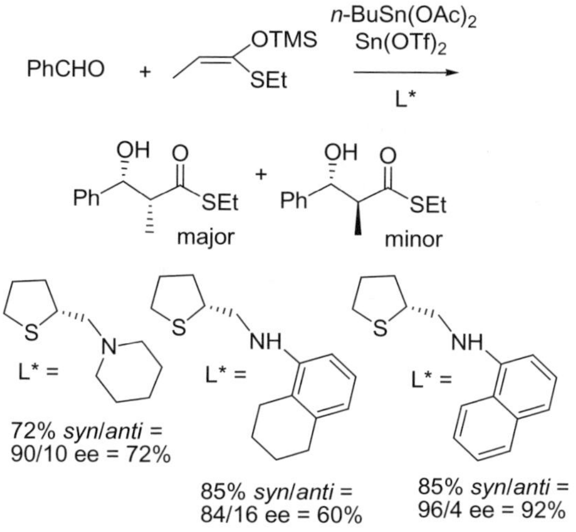

PhCHO + (structure: OTMS, SEt) → *n*-BuSn(OAc)$_2$, Sn(OTf)$_2$, L*

(structures)

Ph (OH, O, SEt) **major** + Ph (OH, O, SEt) **minor**

L* = (structure)
72% syn/anti = 90/10 ee = 72%

L* = (structure)
85% syn/anti = 84/16 ee = 60%

L* = (structure)
85% syn/anti = 96/4 ee = 92%

Scheme 10.26 Sn-catalysed Mukaiyama aldol reaction with S/N ligands.

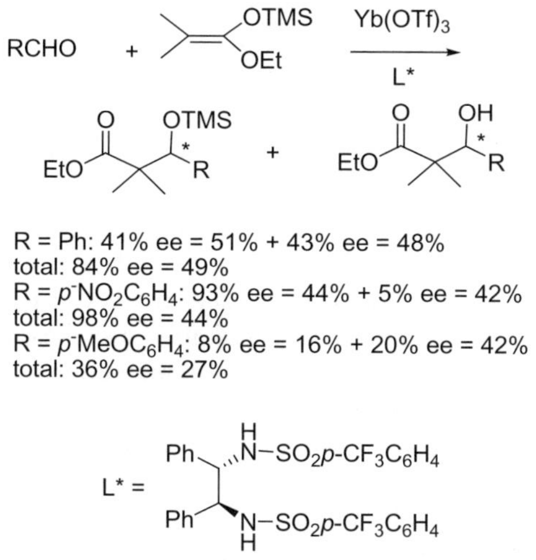

RCHO + (structure: OTMS, OEt) → Yb(OTf)$_3$, L*

EtO (O, OTMS, R) + EtO (O, OH, R)

R = Ph: 41% ee = 51% + 43% ee = 48%
total: 84% ee = 49%
R = p-NO$_2$C$_6$H$_4$: 93% ee = 44% + 5% ee = 42%
total: 98% ee = 44%
R = p-MeOC$_6$H$_4$: 8% ee = 16% + 20% ee = 42%
total: 36% ee = 27%

L* = (structure: Ph, H N-SO$_2$$p$-CF$_3C_6H_4$; Ph, N-SO$_2$$p$-CF$_3C_6H_4$ H)

Scheme 10.27 Yb-catalysed Mukaiyama aldol reactions with sulfonamide ligand.

been successfully employed for catalytic asymmetric nitroaldol reactions. Binaphthols or amino alcohols have typically been used in combination with Li-lanthanides or dialkylzinc. Only very recently have combinations of nitrogen-based ligands and transition metals emerged as catalyst systems for asymmetric nitroaldol reactions. More recently, several chiral sulfur-containing ligands have been introduced as potential ligands for catalytic asymmetric

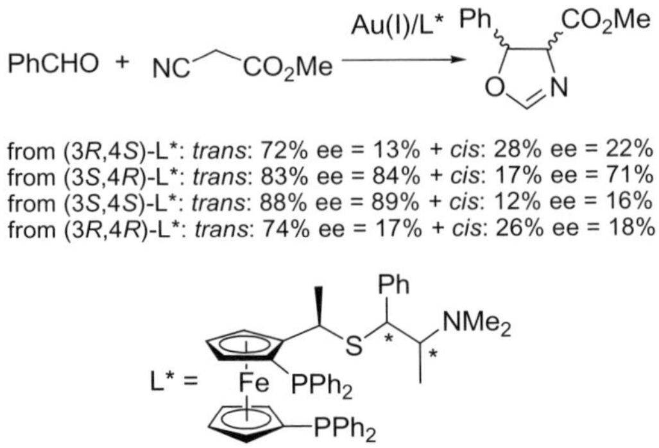

from (3*R*,4*S*)-L*: *trans*: 72% ee = 13% + *cis*: 28% ee = 22%
from (3*S*,4*R*)-L*: *trans*: 83% ee = 84% + *cis*: 17% ee = 71%
from (3*S*,4*S*)-L*: *trans*: 88% ee = 89% + *cis*: 12% ee = 16%
from (3*R*,4*R*)-L*: *trans*: 74% ee = 17% + *cis*: 26% ee = 18%

Scheme 10.28   Au-catalysed aldol reaction with sulfur-containing ferrocenylphosphine ligand.

nitroaldol reactions. As an example, an asymmetric nitroaldol reaction has been developed by Gao and Martell on the basis of using Zn(II) complexes of new chiral hydrophobic macrocyclic ligands, which contained chiral diamino and thiophene moieties synthesised by the Schiff-base condensation approach.[65] This novel type of ligand have proved to be efficient since the reaction between benzaldehyde and nitromethane, performed in the presence of the trimeric catalyst that was preformed from three equivalents of ZnEt$_2$ and one equivalent of the chiral ligand, afforded the nitroaldol product in a good yield and an enantioselectivity reaching 75% ee (Scheme 10.29).

In 2004, Xu *et al.* developed another approach involving chiral bis(thiazolines) possessing a diphenylamine backbone as linkage between both thiazoline rings as tridentate ligands in the Henry reaction.[66] The application of these new C_2-symmetric ligands to the Cu(II)-catalysed nitroaldol reaction of an α-keto ester proved their real efficiency compared to the corresponding well-known analogous bis(oxazolines) ligands (Scheme 10.30). Indeed, comparative studies demonstrated that bis(oxazolines)–copper complexes furnished moderate enantioselectivities ($\leq$60% ee) whereas the corresponding bis(thiazolines)–copper complexes gave slightly better enantioselectivities of up to 70% ee for the *tert*-leucinol derivative for example.

In 2005, these authors studied the influence of the metal nature on the enantioselectivity of this reaction.[67] It was found that the absolute configuration of the nitroaldol product depended on the nature of the Lewis acid used. Thus, when the Lewis acid used to prepare the catalytic complex was Cu(OTf)$_2$, the major product was the (*S*) enantiomer, whereas the (*R*) enantiomer was obtained as the major product when using ZnEt$_2$ as the Lewis acid (Scheme 10.31). This reversal of the absolute configuration of the product suggested that the potential ability of a tridentate coordination and a hydrogen donation of the NH group in the C_2-symmetric tridentate ligands and also in the bis(oxazolines) analogues played a crucial role in the enantioselective catalytic Henry reaction, since the replacement of the NH group by a CH$_2$ group resulted in the

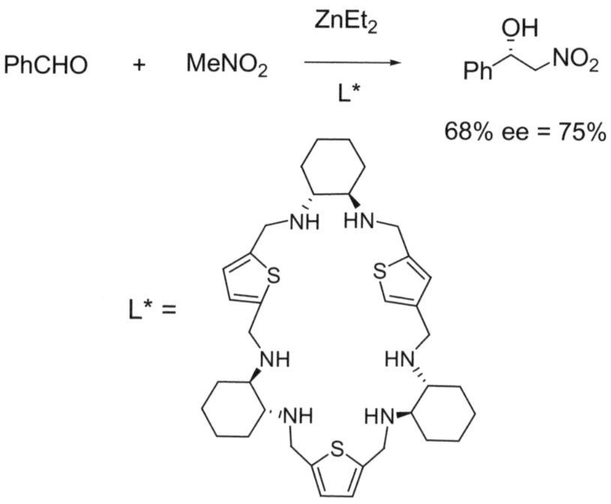

Scheme 10.29 Zn-catalysed Henry reaction with S/N macrocyclic ligand.

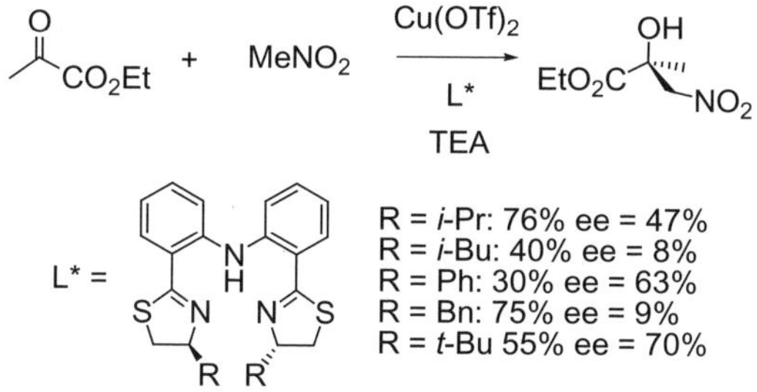

Scheme 10.30 Cu-catalysed Henry reaction with bis(thiazolines) ligands.

formation of the (*R*) enantiomer with Cu(OTf)$_2$, while with ZnEt$_2$ the reaction afforded a poor enantioselectivity.

Novel chiral thiolated amino alcohols have been recently synthesised and then evaluated by Vilaivan *et al.* as a potential new class of ligands for Cu-catalysed nitroaldol reactions.[68] Amino alcohol ligands bearing *N*-(2-alkyl-thio)benzyl substituents provided only modest enantioselectivities (22–46% ee) while those carrying *N*-2-thienylmethyl substituents provided better enantios-electivities of up to 75% ee for the nitroaldol reaction between *p*-nitro-benzaldehyde and nitromethane. A range of aromatic aldehydes were acceptable substrates giving moderate to high enantioselectivities of up to 88% ee, as shown in Scheme 10.32.

In addition, highly enantioselective nitroaldol reactions were performed by Bandini *et al.* by using a new class of C_2-symmetric oligothiophene ligands, in 2007.[69] Thus, associated to copper, these C_2-symmetric bis(amino) ligands allowed the synthesis of a wide range of enantiomerically enriched nitroalcohols

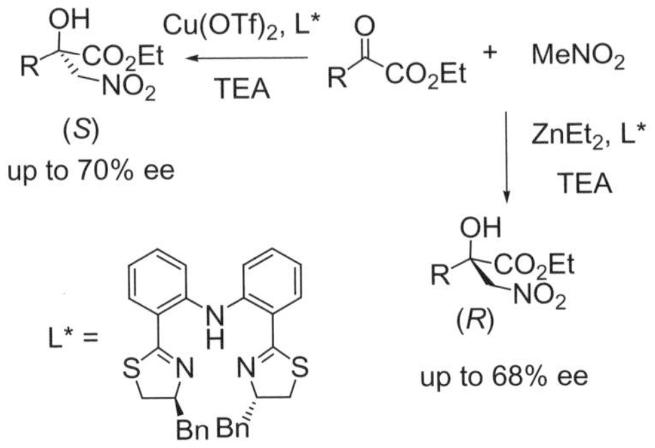

Scheme 10.31 Cu- and Zn-catalysed Henry reactions with bis(thiazolines) ligand.

to be achieved in high yields and enantioselectivities of up to 99% ee (Scheme 10.33). The generality in the scope (aliphatic, aromatic, heteroaromatic and α,β-unsaturated aldehydes) combined with the mildness of the reaction conditions (up to 0.1% of catalyst loading) have located this novel catalytic system among the most efficient procedures for the synthesis of chiral nitroalcohols and functionalised polycyclic aromatic compounds, such as the highly functionalised natural tetrahydro-isoquinoline depicted in Scheme 10.33. Indeed, the key step of the synthesis of this natural product was the Henry reaction of 3,5-dimethoxybenzaldehyde, giving the corresponding nitroaldol product in good yield and enantioselectivity, which was further converted into the targeted chiral tetrahydro-isoquinoline.

10.10 Mannich-Type Reaction

The catalytic enantioselective Mannich-type addition of enolate anion equivalents to imines represents an extremely powerful strategy for the preparation of chiral β-amino carbonyls, which are key structural units in biologically relevant compounds such as β-lactams and β-amino acids.[70] Despite the impressive progress achieved in this reaction in recent years there is still room for improvement, especially towards developing novel procedures displaying a wide structural scope with regard to the substitution at the two reaction partners, imine and enolate reagent. Thus, although there are some very efficient asymmetric metal-catalysed protocols for the addition of enolate reagents to *N*-aryl imines, *N*-acyl imines, *N*-acyl hydrazones and *N*-phosphinoyl imines, the use of sulfonyl imines as electrophiles is limited mainly to the case of the highly electronically activated α-tosyliminoesters. In this context, Carretero *et al.* have recently applied copper(I) complexes of Fesulphos ligands to promote the enantioselective Mannich-type addition of silyl enol ethers of ketones, esters and thioesters to *N*-sulfonyl aldimines.[71] The corresponding chiral β-amino

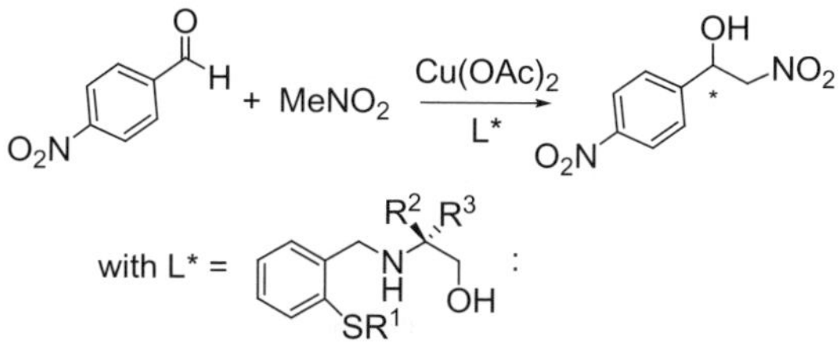

R^1 = *p*-ClC_6H_4, R^2 = Bn, R^3 = H: 90% ee = 30% (*R*)
R^1 = R^3 = Ph, R^2 = H: 89% ee = 46% (*S*)
R^1 = Ph, R^2 = *t*-Bu, R^3 = H: 90% ee = 0%
R^1 = Et, R^2 = H, R^3 = Ph: 88% ee = 29% (*S*)
R^1 = Me, R^2 = H, R^3 = Ph: 87% ee = 24% (*S*)

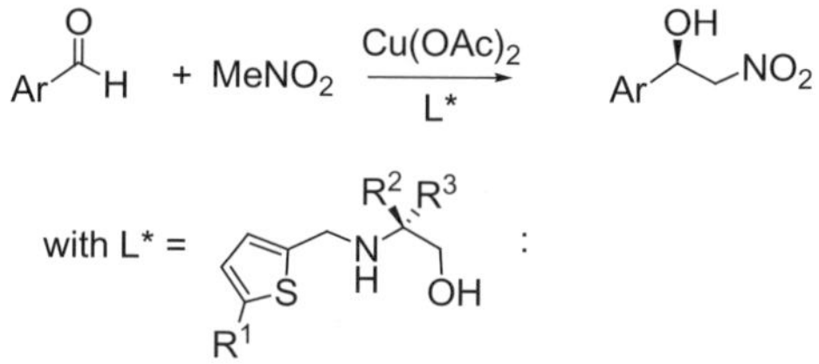

R^1 = R^2 = H, R^3 = Bn, Ar = *p*-$NO_2C_6H_4$: 87% ee = 53%
R^1 = R^2 = H, R^3 = Ph, Ar = *p*-$NO_2C_6H_4$: 97% ee = 75%
R^1 = Me, R^2 = H, R^3 = Ph, Ar = *p*-$NO_2C_6H_4$: 92% ee = 63%
R^1 = Br, R^2 = H, R^3 = Ph, Ar = *p*-$NO_2C_6H_4$: 97% ee = 66%
R^1 = R^2 = H, R^3 = Ar = Ph: 77% ee = 69%
R^1 = R^2 = H, R^3 = Ph, Ar = *p*-$CF_3C_6H_4$: 93% ee = 78%
R^1 = R^2 = H, R^3 = Ph, Ar = *o*-$MeOC_6H_4$: 85% ee = 75%
R^1 = R^2 = H, R^3 = Ph, Ar = *o*-$NO_2C_6H_4$: 98% ee = 85%
R^1 = R^2 = H, R^3 = Ph, Ar = *o*-$CF_3C_6H_4$: 91% ee = 88%
R^1 = R^2 = H, R^3 = Ph, Ar = *o*-FC_6H_4: 92% ee = 83%
R^1 = R^2 = H, R^3 = Ph, Ar = *p*-$NO_2C_6H_4$: 94% ee = 70%

Scheme 10.32 Cu-catalysed Henry reactions with thiolated amino alcohol ligands.

carbonyl derivatives were obtained in good yields and with moderate to good enantioselectivities of up to 93% ee, as shown in Scheme 10.34.

10.11 Claisen Rearrangement

The Claisen rearrangement has attracted much attention as an attractive tool for the construction of new carbon–carbon bonds.[72] Various catalytic systems have been developed to afford enantioselective versions of this process.[73] On the other hand, relatively few chiral sulfur-containing ligands have been investigated for this type of reaction. As an example, Taguchi *et al.* have

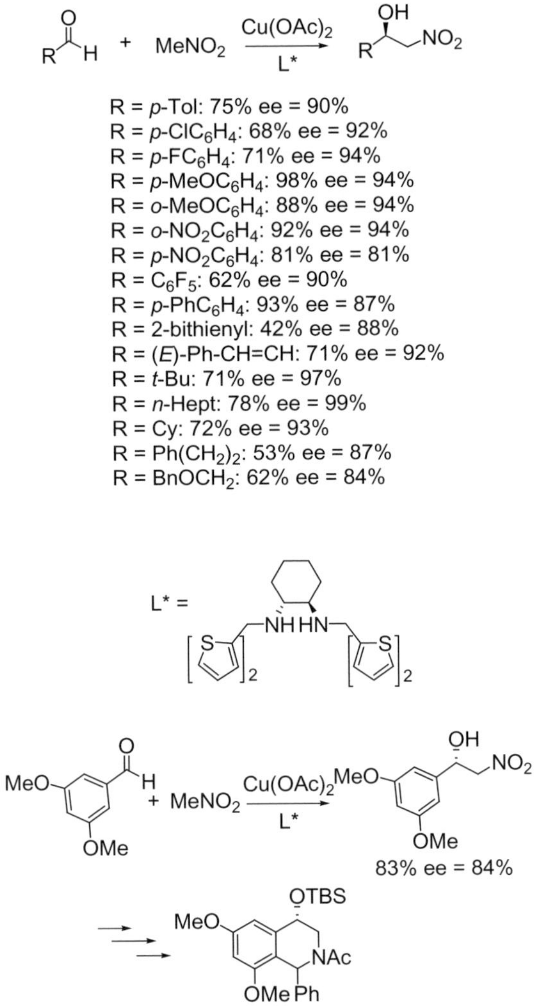
Cu(OAc)2
L*
MeNO2
R = p-Tol: 75% ee = 90%
R = p-ClC6H4: 68% ee = 92%
R = p-FC6H4: 71% ee = 94%
R = p-MeOC6H4: 98% ee = 94%
R = o-MeOC6H4: 88% ee = 94%
R = o-NO2C6H4: 92% ee = 94%
R = p-NO2C6H4: 81% ee = 81%
R = C6F5: 62% ee = 90%
R = p-PhC6H4: 93% ee = 87%
R = 2-bithienyl: 42% ee = 88%
R = (E)-Ph-CH=CH: 71% ee = 92%
R = t-Bu: 71% ee = 97%
R = n-Hept: 78% ee = 99%
R = Cy: 72% ee = 93%
R = Ph(CH2)2: 53% ee = 87%
R = BnOCH2: 62% ee = 84%
L* =
83% ee = 84%
OTBS
NAc

Scheme 10.33 Cu-catalysed Henry reactions with oligothiophene ligand.

reported the enantioselective aromatic Claisen rearrangement of catechol mono allylic ethers with a chiral sulfur-containing boron reagent as the Lewis acid.[74] The corresponding *ortho* substituted catechol derivatives were obtained in both high yields and enantioselectivities of up to 95% ee (Scheme 10.35). Moreover, this novel catalytic system was proved to avoid the problematic *para* rearrangement and abnormal Claisen rearrangement. The enantioselectivity of the process was explained by the formation of a rigid five-membered cyclic intermediate formed by reaction of the catechol mono allyl ether with the chiral boron reagent, followed by coordination of the allylic oxygen to the boron atom. The *Re* site of the benzene ring of the substrate could be shielded

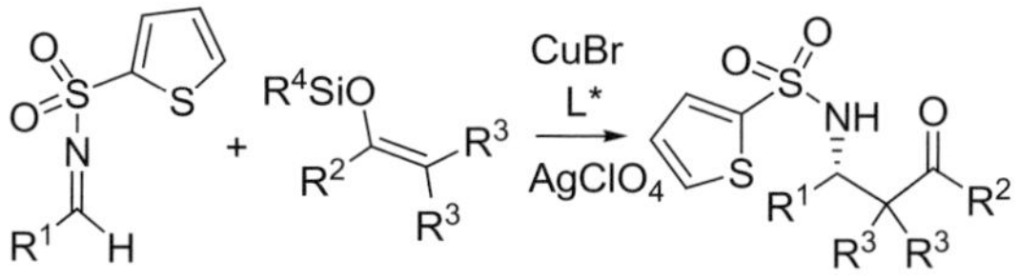

R¹ = Ph, R² = S-*t*-Bu, R³ = H, R⁴ = TBS: 80% ee = 91%
R¹ = *p*-FC₆H₄, R² = S-*t*-Bu, R³ = H, R⁴ = TBS: 60% ee = 82%
R¹ = *p*-MeOC₆H₄, R² = S-*t*-Bu, R³ = H, R⁴ = TBS: 58% ee = 81%
R¹ = *o*-Tol, R² = S-*t*-Bu, R³ = H, R⁴ = TBS: 77% ee = 93%
R¹ = *m*-MeOC₆H₄, R² = S-*t*-Bu, R³ = H, R⁴ = TBS: 91% ee = 83%
R¹ = *m*-ClC₆H₄, R² = S-*t*-Bu, R³ = H, R⁴ = TBS: 80% ee = 88%
R¹ = 1-Naph, R² = S-*t*-Bu, R³ = H, R⁴ = TBS: 86% ee = 88%
R¹ = 2-Naph, R² = S-*t*-Bu, R³ = H, R⁴ = TBS: 71% ee = 91%
R¹ = 2-Fu, R² = S-*t*-Bu, R³ = H, R⁴ = TBS: 87% ee = 49%
R¹ = (E)-Ph-CH=CH, R² = S-*t*-Bu, R³ = H, R⁴ = TBS: 40% ee = 71%
R¹ = (2-Fu)-CH=CH, R² = S-*t*-Bu, R³ = H, R⁴ = TBS: 83% ee = 60%
R¹ = Cy, R² = S-*t*-Bu, R³ = H, R⁴ = TBS: 70% ee = 76%
R¹ = Ph, R² = OMe, R³ = H, R⁴ = TBS: 75% ee = 80%
R¹ = Ph, R² = OMe, R³ = Me, R⁴ = TBS: 77% ee = 86%
R¹ = R² = Ph, R³ = H, R⁴ = TMS: 71% ee = 93%
R¹ = Ph, R² = *p*-MeOC₆H₄, R³ = H, R⁴ = TMS: 80% ee = 85%
R¹ = Ph, R² = 2-Naph, R³ = H, R⁴ = TMS: 90% ee = 86%

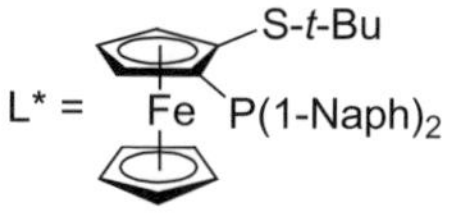

Scheme 10.34 Cu-catalysed Mannich-type reactions of *N*-sulfonyl imines with Fesulphos ligand.

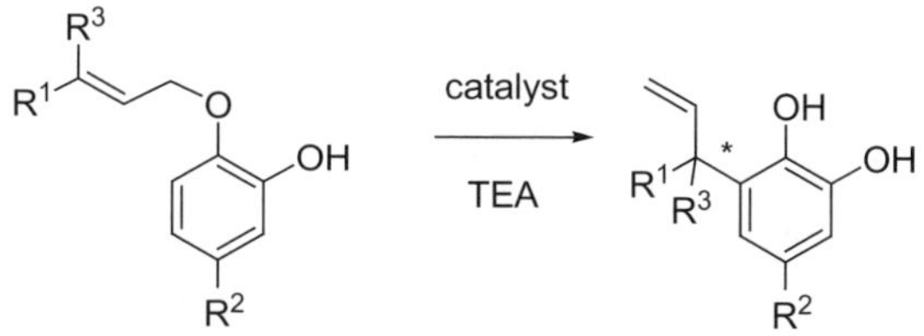

R¹ = *n*-Pr, R² = R³ = H: 89% ee = 94% (S)
R¹ = *n*-Pr, R² = Me, R³ = H: 97% ee = 93% (S)
R¹ = R² = H, R³ = *n*-Pr: 92% ee = 95% (R)

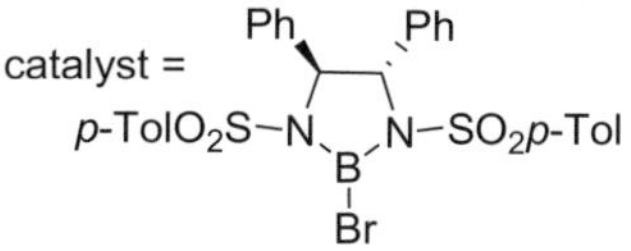

Scheme 10.35 Aromatic Claisen rearrangements of catechol mono allylic ethers with sulfur-containing boron catalyst.

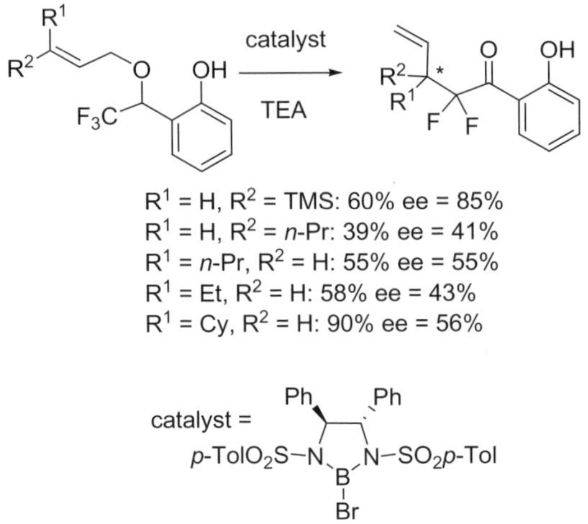

R^1 = H, R^2 = TMS: 60% ee = 85%
R^1 = H, R^2 = *n*-Pr: 39% ee = 41%
R^1 = *n*-Pr, R^2 = H: 55% ee = 55%
R^1 = Et, R^2 = H: 58% ee = 43%
R^1 = Cy, R^2 = H: 90% ee = 56%

Scheme 10.36   Claisen rearrangements of difluorovinyl allyl ethers with sulfurcontaining boron catalyst.

by one tolyl group of the sulfonamide ligand. Therefore, the approach of the allylic moiety should occur on the *Si* face giving rise to the (*S*)-product.

The scope of this methodology was extended to the enantioselective rearrangement of difluorovinyl allyl ethers by these authors, furnishing a novel powerful tool for the synthesis of chiral β-substituted α,α′-difluorocarbonyl compounds.[75] As shown in Scheme 10.36, moderate to good enantioselectivities of up to 85% ee were observed.

10.12   Epoxidation

The asymmetric synthesis of epoxides from carbonyl compounds using chiral sulfur ylide intermediates has emerged as a powerful method for not only creating C–C bonds with control of asymmetry, but also for generating a functionality suitable for further manipulation.[76] Two catalytic methods have been developed involving the reaction of a chiral sulfide with an alkyl halide in the presence of a base and an aldehyde or the reaction of a chiral sulfide with a diazo compound or a diazo precursor in the presence of a metal catalyst and an aldehyde. In both cases, the catalytic sulfide was regenerated during the catalytic cycle. The first method has been performed by a number of research groups in the presence of various sulfide structures, as depicted in Scheme 10.37. The highest enantioselectivities were obtained by Metzner[77] and Goodman,[78] using chiral *C*$_2$-symmetric sulfides **1** and **2**, respectively. Other sulfides such as **3** and **4** were employed by Shimizu[79] and Saito,[80] respectively, for a similar reaction performed between benzaldehyde and benzyl bromide, giving better diastereoselectivities but lower

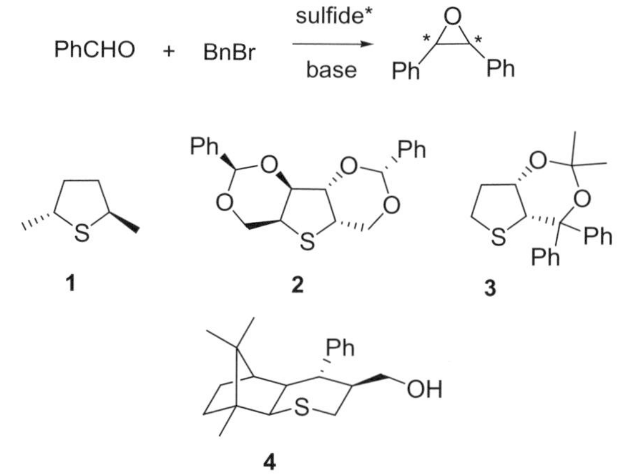

PhCHO + BnBr $\xrightarrow[\text{base}]{\text{sulfide*}}$

1   **2**   **3**

4

sulfide = **1**: 82% de = 80% ee = 85% (*S,S*)
sulfide = **2**: 41% de = 82% ee = 97% (*R,R*)
sulfide = **3**: 52% de = 100% ee = 78% (*R,R*)
sulfide = **4**: 63% de = 92% ee = 56% (*S,S*)

Scheme 10.37   Sulfide-catalysed epoxidation of benzaldehyde with BnBr.

enantioselectivities (Scheme 10.37). Although modest to high levels of enantio- and diastereoselectivities have been reported, the scope of this method remains somewhat limited. Indeed, only benzyl bromide and substituted allyl halides for the alkyl halide component, and aromatic and heteroaromatic aldehydes for the carbonyl-coupling partner, have mostly been employed.

The second method of asymmetric epoxidation has been extensively studied by Aggarwal *et al.*,[81,82] and is based on the reaction, under neutral conditions, of a chiral sulfide with an *in situ* generated diazo compound in the presence of a metal catalyst and an aldehyde. Scheme 10.38 summarises the best results obtained using a chiral bicyclic sulfide, prepared from camphorsulfonyl chloride, with a range of aromatic, heteroaromatic and α,β-unsaturated aldehydes, as well as aliphatic substrates in the presence of various tosylhy- drazone salts. In addition, this methodology was successfully applied to the synthesis of biologically important β-hydroxy-δ-lactones, such as (+)-pre- lactone B.[83]

Although, this process has quite a broad scope, it also has, like most catalytic processes, its limitations. As an example, aldehydes with basic groups (*e.g.* pyridylcarboxaldehydes) were poor electrophiles and, moreover, α,β-unsatu- rated hydrazones were poor carbene precursors. As an alternative strategy, these authors have considered the use of a stoichiometric process, ideally involving the efficient recovery of the chiral transfer sulfur reagent. Hence, high levels of selectivity were obtained through the initial formation of a sulfonium salt, as for the reaction between 2-pyridinecarboxaldehyde and the benzylsul- fonium salt of the same sulfide, as depicted in Scheme 10.38 (88% yield,

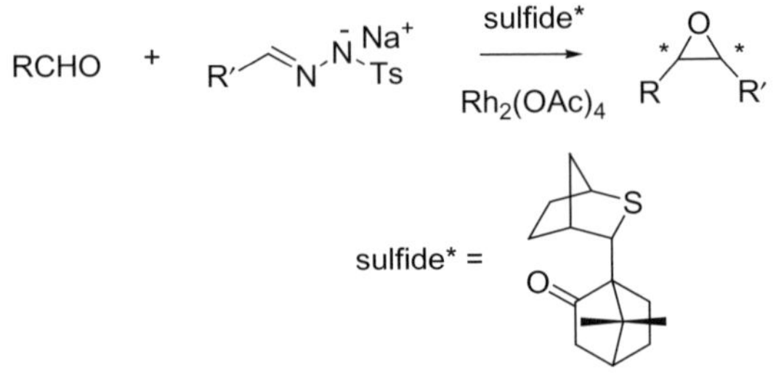

R = R' = Ph: 82% *trans:cis* > 98:2 ee = 94%
R = p-NO$_2$C$_6$H$_4$, R' = Ph: 75% *trans:cis* > 98:2 ee = 92%
R = p-MeOC$_6$H$_4$, R' = Ph: 68% *trans:cis* > 98:2 ee = 92%
R = Fu, R' = Ph: 60% *trans:cis* > 98:2 ee = 91%
R = (E)-Ph-CH=CH, R' = Ph: 70% *trans:cis* > 98:2 ee = 87%
R = Ph, R' = p-MeOC$_6$H$_4$: 95% *trans:cis* = 80:20 ee = 93%
R = Ph, R' = p-CNC$_6$H$_4$: 70% *trans:cis* = 80:20 ee = 64%
R = Ph, R' = o-MeOC$_6$H$_4$: 70% *trans:cis* > 98:2 ee = 93%
R = Ph, R' = Fu: 53% *trans:cis* = 90:10 ee = 61%

Scheme 10.38 Rh-catalysed epoxidations of aldehydes by tosylhydrazones with sulfide.

trans:cis = 98:2, ee = 99%). In addition, this methodology was applied to the synthesis of the anti-inflammatory agent, CDP-840.[84]

On the other hand, Seki *et al.* have developed a catalytic asymmetric synthesis of glycidic amides by the reaction of diazoacetamides with aromatic aldehydes in the presence of 20% molar equivalent of a chiral binaphthyl sulfide and 10% molar equivalent of copper(II)-acetylacetonate, providing enantioselectivities of up to 64% ee.[85] More recently, Metzner *et al.* have reported that ferrocenyl derivatives,[86] bearing an adjacent sulfur atom included in a fused ring and exhibiting planar and central chiralities, could be used as a catalytic source of asymmetric sulfonium ylides.[87] A one-pot reaction was achieved, involving the addition of an aldehyde, benzyl bromide, 20% molar equivalent of the ferrocenyl sulfide and sodium iodide in a mixture of *t*-butanol and water. The best results (up to 94% ee) were observed with the chiral sulfide depicted in Scheme 10.39, bearing a *t*-butyl group on the carbon adjacent to the nitrogen atom and that was entirely recovered. On the other hand, an enantioselectivity of only 67% ee was obtained by using the planar *tert*-butylsulfenylferrocene depicted in Scheme 10.39 as the catalyst for the epoxidation of benzaldehyde.

Finally, with the aim of discovering novel chiral oxomolybdenum catalysts able to perform enantioselective alkene epoxidations, Kühn *et al.* have reported the exploration of the catalytic behaviour of a series of dioxomolybdenum(VI) complexes with chiral *cis*-8-phenylthiomenthol ligands derived from (+)-pulegone.[88] Therefore, the epoxidation of *cis*-β-methylstyrene using *t*-butyl-hydroperoxide as the oxidant and performed in the presence of (+)-(2R,5R)-2-[1-methyl-1-(phenylthio)ethyl]-5-methylcyclohexanone oxime as the ligand, did not produce, however, a significant optical induction in these conditions.

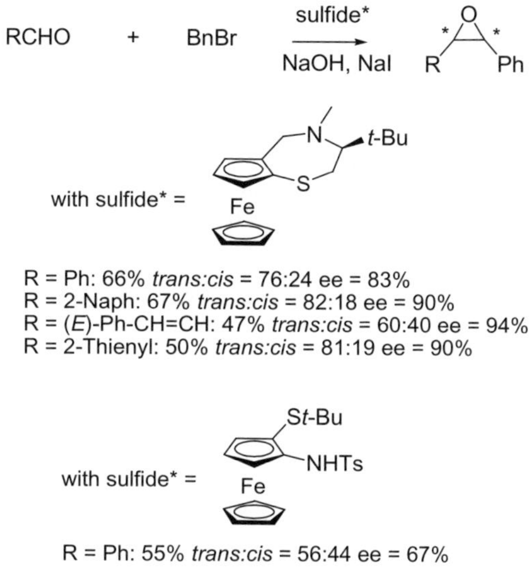

RCHO + BnBr →[sulfide*][NaOH, NaI] (epoxide: R—CH—CH—Ph with epoxide oxygen)

with sulfide* = (ferrocenyl sulfide structure with N–t-Bu)

R = Ph: 66% *trans:cis* = 76:24 ee = 83%
R = 2-Naph: 67% *trans:cis* = 82:18 ee = 90%
R = (E)-Ph-CH=CH: 47% *trans:cis* = 60:40 ee = 94%
R = 2-Thienyl: 50% *trans:cis* = 81:19 ee = 90%

with sulfide* = (ferrocenyl sulfide structure with St-Bu and NHTs)

R = Ph: 55% *trans:cis* = 56:44 ee = 67%

Scheme 10.39   Ferrocenyl sulfide-catalysed epoxidations of aldehydes.

10.13   Silylcyanation

Chiral cyanohydrins are important chiral building blocks for a wide variety of chiral products such as α-hydroxy acids, α-hydroxy aldehydes, α-hydroxy ketones, β-hydroxy amines, α-amino alcohols and α-amino acid derivatives.[89] In recent years, a number of synthetic methods have been reported employing enzymes, synthetic peptides and chiral metal complexes.[90] In particular, the hydrosilylation is often preferred as a method of reduction over the hydrogenation owing to the milder conditions needed to carry out the reaction. Of the chiral metal complexes reported so far to be used in the asymmetric hydrosilylation reaction, titanium-based Lewis acids have attracted the most interest, involving various chiral ligands including TADDOLs, BINOLs, peptides, Schiff bases and others. Usually, trimethylsilylcyanide is used as the cyanide source. Its asymmetric addition to the aldehyde followed by desilylation of the intermediate cyanosilyl ether gives the corresponding cyanohydrin in a good yield. In 1995, Bolm and Müller developed the enantioselective trimethylsilylcyanation of aldehydes in the presence of a titanium catalyst derived from a chiral sulfoximine.[91] As summarised in Scheme 10.40, a variety of aldehydes including substituted aromatic, aliphatic and α,β-unsaturated aldehydes were silylcyanated in high yields and with a good enantiocontrol (up to 91% ee). In an attempt to improve the enantioselectivity, the ligand structure was varied.[92] Increasing the size of the alkyl substituent at the sulfoximine sulfur had only a small effect on the extent of the asymmetric induction in the standard trimethylsilylcyanation of benzaldehyde. Moreover, lower yields and longer reaction times were observed.

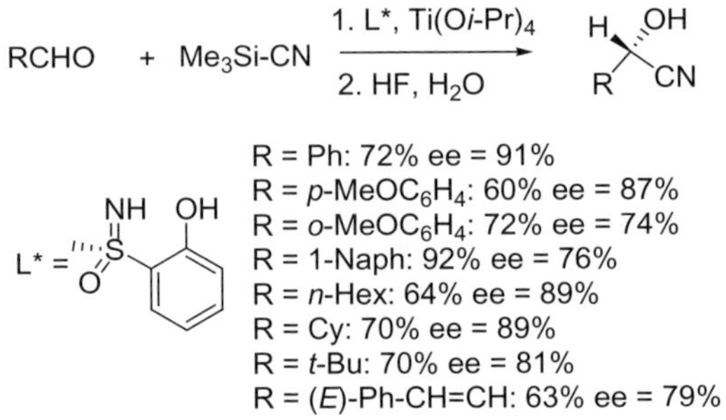

RCHO + Me₃Si-CN $\xrightarrow[\text{2. HF, H}_2\text{O}]{\text{1. L*, Ti(O}i\text{-Pr)}_4}$ cyanohydrin product

R = Ph: 72% ee = 91%
R = p-MeOC₆H₄: 60% ee = 87%
R = o-MeOC₆H₄: 72% ee = 74%
R = 1-Naph: 92% ee = 76%
R = n-Hex: 64% ee = 89%
R = Cy: 70% ee = 89%
R = t-Bu: 70% ee = 81%
R = (E)-Ph-CH=CH: 63% ee = 79%

Scheme 10.40 Silylcyanations with sulfoximine ligand.

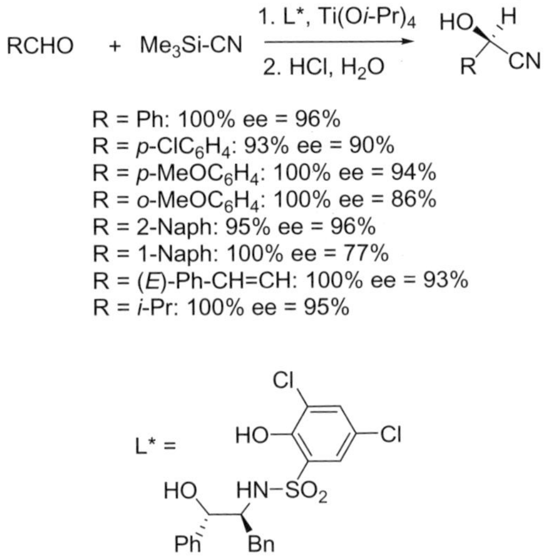

RCHO + Me₃Si-CN $\xrightarrow[\text{2. HCl, H}_2\text{O}]{\text{1. L*, Ti(O}i\text{-Pr)}_4}$ cyanohydrin product

R = Ph: 100% ee = 96%
R = p-ClC₆H₄: 93% ee = 90%
R = p-MeOC₆H₄: 100% ee = 94%
R = o-MeOC₆H₄: 100% ee = 86%
R = 2-Naph: 95% ee = 96%
R = 1-Naph: 100% ee = 77%
R = (E)-Ph-CH=CH: 100% ee = 93%
R = i-Pr: 100% ee = 95%

Scheme 10.41 Silylcyanations with sulfonylated β-amino alcohol ligand.

Similar reactions were undertaken by Choi *et al.* in the presence of a new family of *N*-sulfonylated β-amino alcohols possessing two stereocentres as the chiral ligands.[93] In using the chiral sulfonylated β-amino alcohol ligand depicted in Scheme 10.41, the asymmetric addition of Me₃SiCN to a wide range of aldehydes afforded the corresponding cyanohydrins in both excellent yields and enantioselectivities of up to 96% ee.

In addition, Rowlands has involved chiral sulfoxide-containing ligands for the catalytic addition of Me₃SiCN to aldehydes.[94] The ligand structure was based on a phenolic oxazoline scaffold with introduction of the sulfur substituent via cysteine derivatives. The best enantioselectivities of up to 61% ee were obtained with the bulkiest *tert*-butyl substituted ligand (Scheme 10.42). The effect of the sulfoxide configuration was studied, showing that the use of

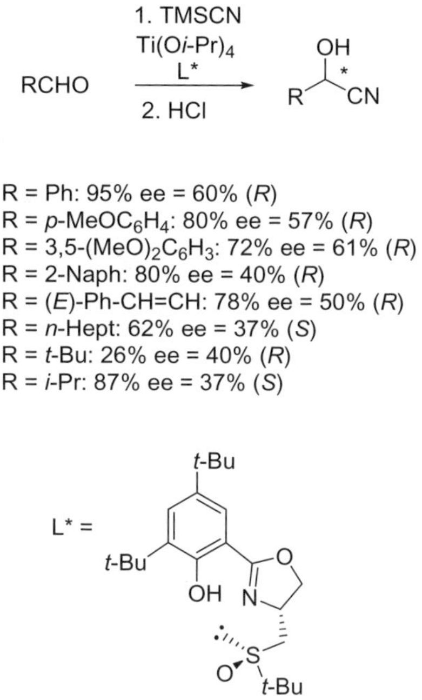

RCHO $\xrightarrow[\text{2. HCl}]{\begin{array}{c}\text{1. TMSCN}\\\text{Ti(O}i\text{-Pr)}_4\\\text{L*}\end{array}}$

R = Ph: 95% ee = 60% (*R*)
R = *p*-MeOC$_6$H$_4$: 80% ee = 57% (*R*)
R = 3,5-(MeO)$_2$C$_6$H$_3$: 72% ee = 61% (*R*)
R = 2-Naph: 80% ee = 40% (*R*)
R = (*E*)-Ph-CH=CH: 78% ee = 50% (*R*)
R = *n*-Hept: 62% ee = 37% (*S*)
R = *t*-Bu: 26% ee = 40% (*R*)
R = *i*-Pr: 87% ee = 37% (*S*)

Scheme 10.42 Silylcyanations with sulfoxide ligand.

the opposite configuration on the sulfoxide moiety resulted in a reversed enantioselectivity. Moreover, the investigation of the analogous ligand without the sulfoxide moiety derived from leucinol and phenylalaninol showed an important reduction in both the activity and enantioselectivity. These results clearly indicated the crucial role played by the sulfoxide moiety in the silylcyanation reaction. The variation of the aldehydes as substrates showed that the catalyst was sensitive to both steric and electronic effects, and the electron-donating substituents gave the best results.

10.14 Hydrosilylation

The asymmetric catalytic hydrosilylation of ketones or alkenes with organosilanes is a versatile method, providing optically active compounds such as alcohols and alkanes.[95] Asymmetric hydrosilylation has been an active field of research in the last twenty years.[96] The most studied reaction has probably been the hydrosilylation of acetophenone to yield the corresponding silyl ether that, when hydrolysed, gives the corresponding enantiomerically enriched 1-phenylethanol. Most of the chiral ligands developed for the asymmetric hydrosilylation of ketones are N- and P-containing compounds, possessing either C_1- or C_2-symmetry. Mixed P/N and N/N ligands have also played a dominant role amongst the heterodonor ligands. On the other hand, chiral S/P

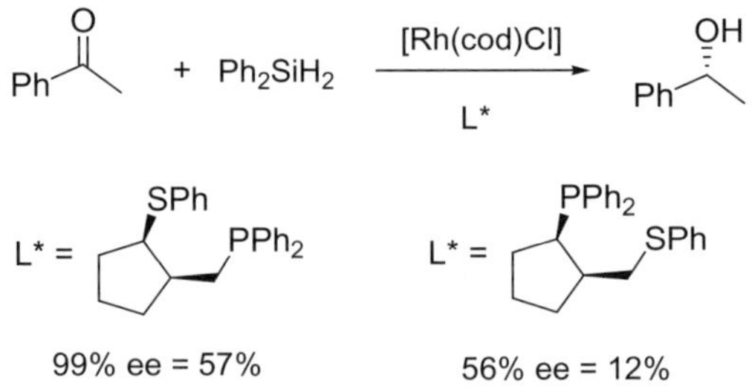

Scheme 10.43 Hydrosilylation of acetophenone with thioether phosphine ligands.

ligands have scarcely been used to activate rhodium complexes in order to induce chirality in the enantioselective hydrosilylation of ketones. As an example, Achiwa *et al.* described, in 1998, new ligands based on a chiral cyclopentane sulfide–phosphine backbone and depicted in Scheme 10.43.[97] Their corresponding rhodium complexes proved to be efficient, but only moderate enantioselectivities of up to 57% ee were observed for the hydrosilylation of acetophenone.

Another approach in the use of chiral S/P ligands for the hydrosilylation reaction of ketones was proposed more recently by Evans *et al.*[98] Thus, in 2003, these workers studied the application of new chiral thioether-phosphinite ligands to enantioselective rhodium-catalysed ketone hydrosilylation processes. For a wide variety of ketones, such as acyclic aryl alkyl and dialkyl ketones as well as cyclic aryl alkyl ketones and also cyclic keto esters, the reaction gave high levels of enantioselectivity of up to 99% ee (Scheme 10.44).

In 2005, Diéguez *et al.* introduced D-xylose-derived thioether-phosphinites as ligands for the Rh-catalysed hydrosilylation of ketones.[99] Several ligands of this type bearing substituents with different steric and electronic demands on the sulfur atom have been investigated in the hydrosilylation of ketones, giving high activities and good enantioselectivities of up to 90% ee, as shown in Scheme 10.45. The results showed that the enantioselectivity depended strongly on the steric properties of the substituent on the thioether moiety and the steric properties of the substrate. A bulky group on the thioether moiety of the ligand had a positive effect on the enantioselectivity. Moreover, the enantioselectivities were also higher when an *ortho* substituent in the aryl moiety of the ketone was present.

The number of asymmetric hydrosilylations of ketones involving chiral S/N ligands is higher than that of similar reactions using chiral S/P ligands. In 1983, Brunner *et al.* reported the synthesis of a novel type of thiazoline ligands obtained by condensation of L-cysteine ester and 2-pyridinecarbaldehyde.[100] When these ligands were combined with rhodium and applied to catalyse the hydrosilylation of acetophenone, they led to the formation of the expected alcohol in a quantitative yield and an enantioselectivity of up to 97% ee (Scheme 10.46). It was surprising that, despite the fact that the ligand was used

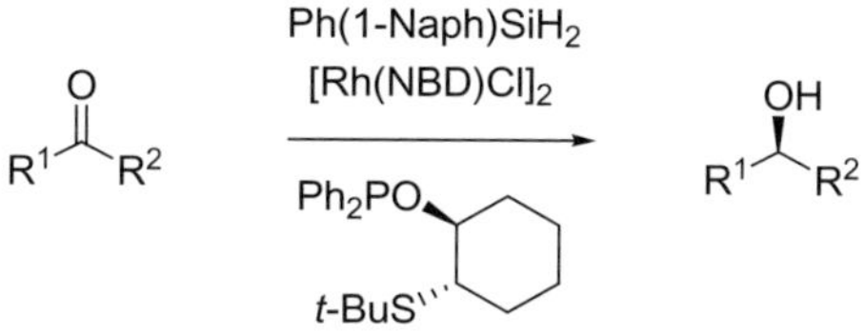

R^1 = Ph, R^2 = Me: 80% ee = 95%
R^1 = o-Tol, R^2 = Me: 98% ee = 95%
R^1 = Ts, R^2 = Me: 90% ee = 92%
R^1 = o-MeOC$_6$H$_4$, R^2 = Me: 90% ee = 95%
R^1 = p-MeOC$_6$H$_4$, R^2 = Me: 56% ee = 88%
R^1 = o-ClC$_6$H$_4$, R^2 = Me: 90% ee = 98%
R^1 = p-ClC$_6$H$_4$, R^2 = Me: 95% ee = 85%
R^1 = 1-Naph, R^2 = Me: 99% ee = 98%
R^1 = 2-Naph, R^2 = Me: 99% ee = 95%
R^1 = Ph, R^2 = Et: 95% ee = 94%
R^1 = Ph, R^2 = i-Bu: 95% ee = 94%
R^1 = Ph, R^2 = Bn: 75% ee = 94%
R^1 = c-Hex, R^2 = Me: 90% ee = 92%
R^1 = t-Bu, R^2 = Me: 85% ee = 91%
R^1 = Me, R^2 = (Me)$_2$CCO$_2$Me: 94% ee = 99%

Scheme 10.44 Rh-catalysed hydrosilylations of ketones with thioether-phosphinite ligand.

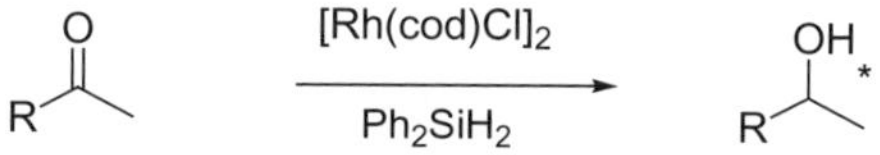

R = Ph: 90% ee = 86%
R = p-FC$_6$H$_4$: 72% ee = 84%
R = p-MeOC$_6$H$_4$: 98% ee = 78%
R = p-CF$_3$C$_6$H$_4$: 98% ee = 72%
R = m-MeOC$_6$H$_4$: 95% ee = 87%
R = o-MeOC$_6$H$_4$: 82% ee = 90%
R = 2-Naph: 85% ee = 88%

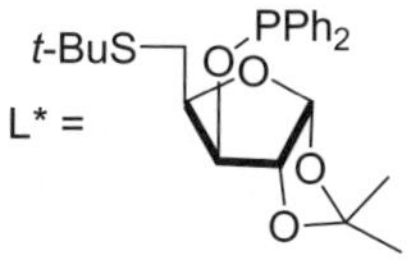

Scheme 10.45 Rh-catalysed hydrosilylations of ketones with D-xylose-derived thioether-phosphinite ligand.

as a 58:42 mixture of two diastereomers, the synthesis of 1-phenylethanol was almost stereospecific.

In 1991, Helmchen *et al.* studied this reaction in the presence of C_2-symmetric bis(thiazolines) ligands prepared in a few simple steps from amino acids and

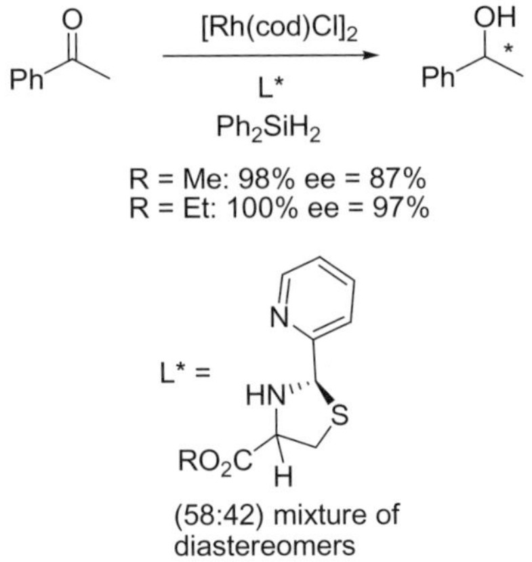

Scheme 10.46 Rh-catalysed hydrosilylation of acetophenone with thiazoline ligands.

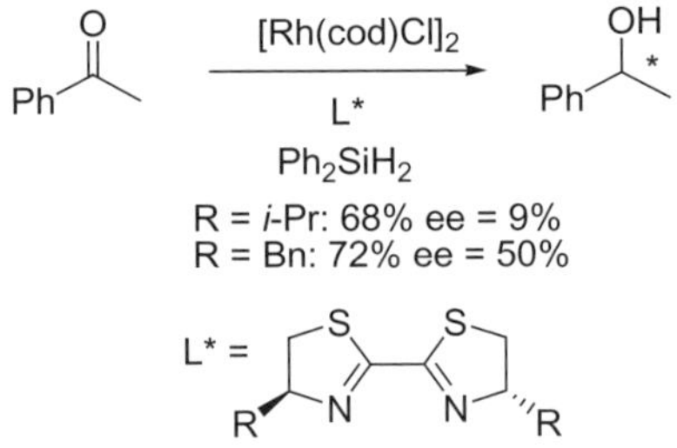

Scheme 10.47 Rh-catalysed hydrosilylation of acetophenone with bis(thiazolines) ligands.

depicted in Scheme 10.47.[101] A good yield but a moderate enantioselectivity were obtained, whereas the corresponding bis(oxazolines) ligands were found to induce up to 84% ee in the same reaction.

Other S/N ligands such as sulfur-containing diferrocenyl dichalcogenides have proved to be efficient ligands of rhodium to promote the hydrosilylation of acetophenone.[102] The most efficient ligand depicted in Scheme 10.48 has allowed, however, only moderate yield and enantioselectivity ($\leq$31% ee). Better results were obtained by using the corresponding diselenide ligands with an enantioselectivity of up to 85% ee for the same reaction.

Other metals such as iridium have also been combined to chiral sulfur-containing ligands in order to induce the chirality in the hydrosilylation of ketones. Therefore, Lemaire *et al.* have described the use of several chiral thiourea ligands for the iridium-catalysed hydrosilylation of acetophenone.[103] The best but moderate enantioselectivity (52% ee) was observed with the use of a C_2-symmetric monothiourea ligand (Scheme 10.49) while the employment of

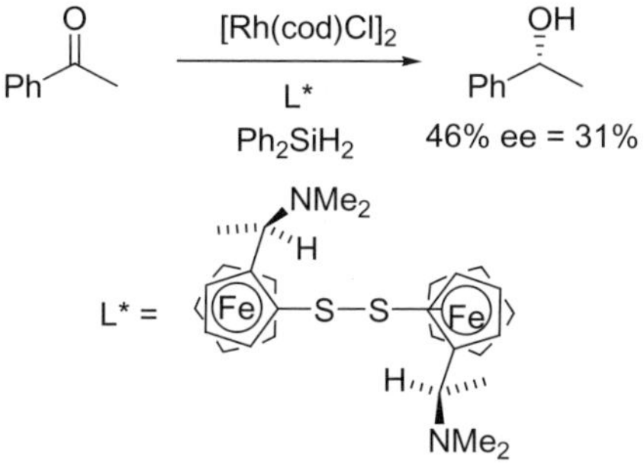

Scheme 10.48 Rh-catalysed hydrosilylation of acetophenone with diferrocenyl dichalcogenide ligand.

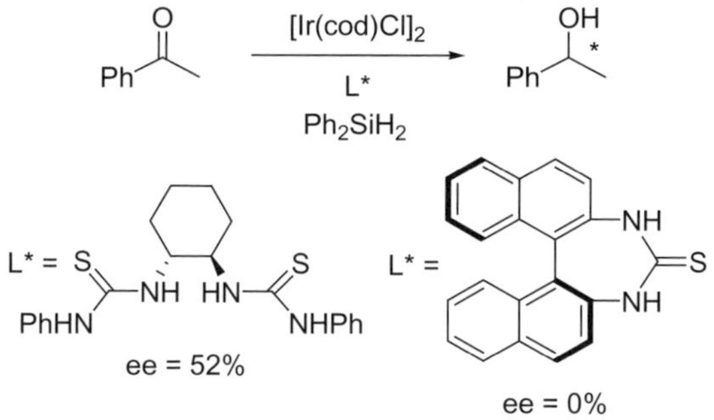

Scheme 10.49 Ir-catalysed hydrosilylation of acetophenone with thiourea ligands.

dithiourea ligands such as that depicted in Scheme 10.49 led to the expected alcohol with no chiral induction despite a good conversion. These authors have checked that the use of $[Rh(cod)Cl]_2$ as catalyst instead of iridium catalyst also led to disappointing results under similar conditions.

In 2005, Riant *et al.* reported the synthesis of a new air-stable S/N-chelating zinc catalyst, depicted in Scheme 10.50, which was fully characterised by all spectroscopic methods. This complex, prepared from the corresponding ferrocene oxazoline, was applied to the enantioselective hydrosilylation of ketones in the presence of polymethylhydrosiloxane, PMHS, providing modest enantioselectivities ($\leq 55\%$ ee).[104]

In addition, Faller and Chase have reported the use of chiral tridentate S/N/P ligands for the rhodium-catalysed hydrosilylation of ketones.[105] The best ligand, which provided an enantioselectivity of up to 64% ee, was that bearing the shortest reach to the metal to give a tridentate ligand, as shown in Scheme 10.51.

Examples of asymmetric hydrosilylation of alkenes performed in the presence of chiral sulfur-containing ligands are much rarer in literature than

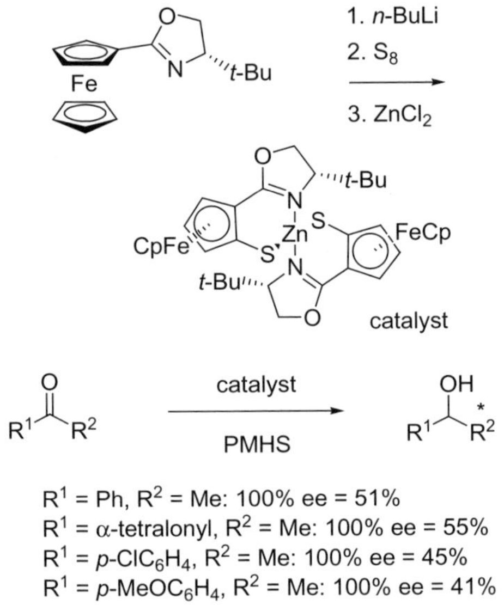
1. n-BuLi
2. S8
3. ZnCl2
t-Bu
Fe
CpFe
FeCp
Zn
catalyst
PMHS
R1 = Ph, R2 = Me: 100% ee = 51%
R1 = α-tetralonyl, R2 = Me: 100% ee = 55%
R1 = p-ClC6H4, R2 = Me: 100% ee = 45%
R1 = p-MeOC6H4, R2 = Me: 100% ee = 41%

Scheme 10.50 Zn-catalysed hydrosilylations of ketones with S/N-chelating zinc catalyst.

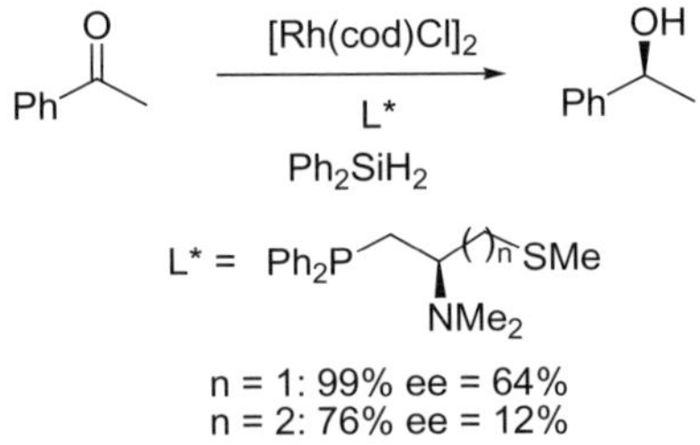
[Rh(cod)Cl]2
L*
Ph2SiH2
L* = Ph2P
SMe
NMe2
n = 1: 99% ee = 64%
n = 2: 76% ee = 12%

Scheme 10.51 Rh-catalysed hydrosilylation of acetophenone with tridentate S/N/P ligands.

those involving the hydrosilylation of ketones. In 2001, a reaction of this type was described by Gladiali *et al.*[106] Thus, the hydrosilylation of alkenes such as styrene was studied in the presence of palladium complexes and chiral S/P-heterodonor ligands having a binaphthalene backbone (Scheme 10.52). These phosphanyl sulfide ligands gave moderate yields and enantioselectivities of up to 51% ee. Moreover, it was noted that both the conversion and the enantioselectivity in this reaction were strongly dependent on the nature of the substituent group of the sulfur atom.

In addition, various chiral (β-*N*-sulfonylaminoalkyl)phosphine ligands were earlier employed by Achiwa *et al.* for the asymmetric palladium-catalysed hydrosilylations of cyclopentadiene and styrene, affording the corresponding

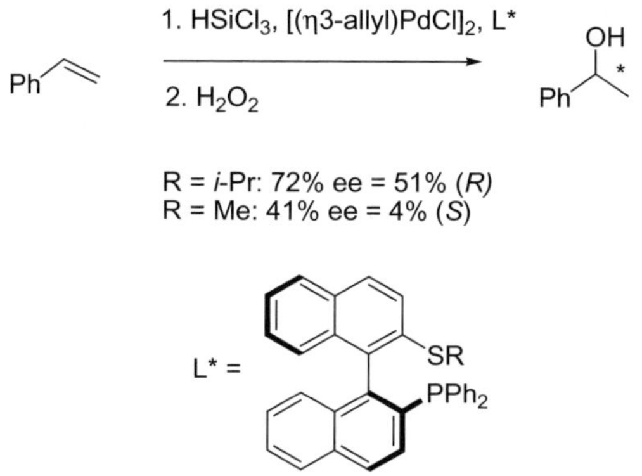

Scheme 10.52 Pd-catalysed hydrosilylation of styrene with S/P-heterodonor ligands.

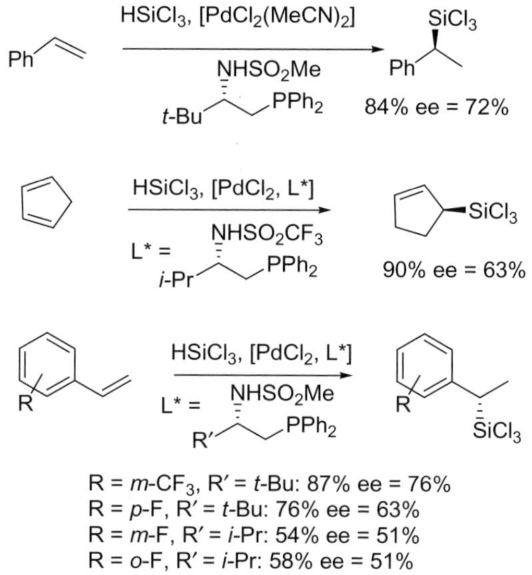

R = *m*-CF$_3$, R′ = *t*-Bu: 87% ee = 76%
R = *p*-F, R′ = *t*-Bu: 76% ee = 63%
R = *m*-F, R′ = *i*-Pr: 54% ee = 51%
R = *o*-F, R′ = *i*-Pr: 58% ee = 51%

Scheme 10.53 Pd-catalysed hydrosilylations of alkenes with (β-*N*-sulfonylami-noalkyl)phosphine ligands.

trichlorosilylated products in high yields and with enantioselectivities of up to 72% ee (Scheme 10.53).[107] This methodology was extended to the hydro-silylation of trifluoromethyl- and fluorostyrenes, providing in similar condi-tions the corresponding fluorinated products with enantioselectivities of up to 76% ee. The electronic effects of the sulfonamide groups were considered to play important roles in the enantioselections.

10.15 Borane Reduction

The asymmetric borane reduction has attracted much attention owing to its usefulness in preparing optically active secondary alcohols.[108] The development of the borane-mediated enantioselective reduction of ketones was initiated by Itsuno *et al.* in the 1980s.[109] In particular, the first application of a chiral sulfur-containing ligand to this type of reactions was reported by Johnson and Stark as early as in 1979, who involved a chiral sulfoximine as a ligand.[110] Thus, the use of this ligand in the borane reduction of acetophenone allowed the corresponding alcohol to be isolated in a good yield and with a high enantioselectivity of up to 82% ee (Scheme 10.54).

Fourteen years later, Bolm and Felder introduced β-hydroxy sulfoximines as ligands in the asymmetric borane reduction of a variety of ketones, affording the corresponding alcohols in high yields and with good enantioselectivities of up to 93% ee (Scheme 10.55).[111] In this case, $BH_3.SMe_2$ was used as the hydride source. Highly reactive ketones, such as trichloroacetophenone, gave much

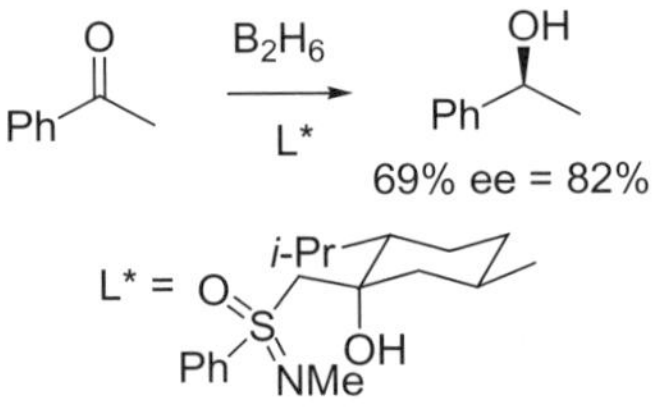
B2H6
L*
OH
Ph
69% ee = 82%
L* =
i-Pr
O
S
Ph
NMe
OH

Scheme 10.54 Borane reduction of acetophenone with sulfoximine ligand.

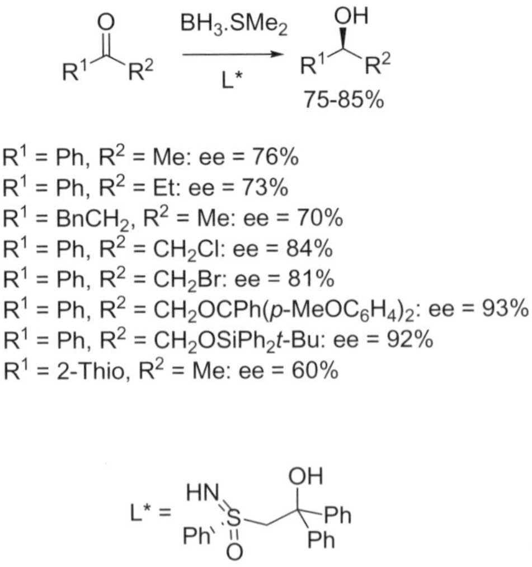
O
R1
R2
BH3.SMe2
L*
OH
75-85%
R1 = Ph, R2 = Me: ee = 76%
R1 = Ph, R2 = Et: ee = 73%
R1 = BnCH2, R2 = Me: ee = 70%
R1 = Ph, R2 = CH2Cl: ee = 84%
R1 = Ph, R2 = CH2Br: ee = 81%
R1 = Ph, R2 = CH2OCPh(p-MeOC6H4)2: ee = 93%
R1 = Ph, R2 = CH2OSiPh2t-Bu: ee = 92%
R1 = 2-Thio, R2 = Me: ee = 60%
L* =
HN
S
Ph
O
OH
Ph
Ph

Scheme 10.55 Borane reductions of ketones with β-hydroxy sulfoximine ligand.

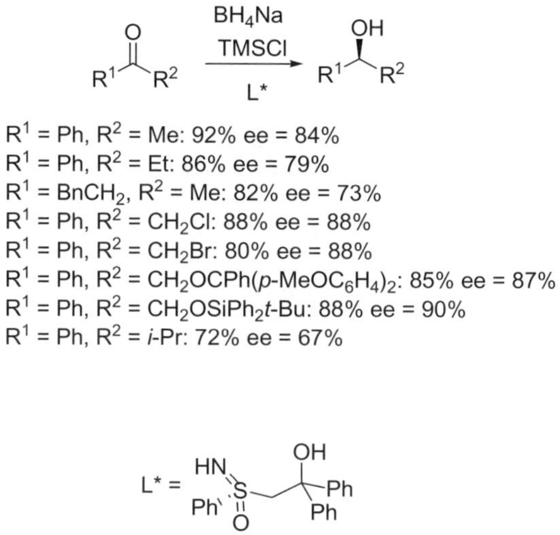

R^1 = Ph, R^2 = Me: 92% ee = 84%
R^1 = Ph, R^2 = Et: 86% ee = 79%
R^1 = BnCH$_2$, R^2 = Me: 82% ee = 73%
R^1 = Ph, R^2 = CH$_2$Cl: 88% ee = 88%
R^1 = Ph, R^2 = CH$_2$Br: 80% ee = 88%
R^1 = Ph, R^2 = CH$_2$OCPh(p-MeOC$_6$H$_4$)$_2$: 85% ee = 87%
R^1 = Ph, R^2 = CH$_2$OSiPh$_2$$t$-Bu: 88% ee = 90%
R^1 = Ph, R^2 = i-Pr: 72% ee = 67%

Scheme 10.56   Borane reductions of ketones with β-hydroxy sulfoximine ligand.

lower enantioselectivities (8% ee) presumably due to competition from the uncatalysed reduction.

In order to avoid the use of a rather expensive and potentially dangerous borane complex, Bolm *et al.* have developed an improved procedure for the borane reduction of ketones, which involved two inexpensive reagents namely NaBH$_4$ and TMSCl.[112] The reduction of a series of ketones was examined applying these novel reaction conditions and the same β-hydroxy sulfoximine ligand to that described above (Scheme 10.56). For most ketones, both the level of asymmetric induction and the yield compared favorably to the precedent results.

In 1994, the scope of this β-hydroxy sulfoximine ligand was extended to the borane reduction of ketimine derivatives by these workers.[113] The corresponding chiral amines were formed with enantioselectivities of up to 72% ee, as shown in Scheme 10.57. It was found that the *N*-substituent of the ketimine had a major influence on the asymmetric induction, with a ketoxime thioether (SPh) being the most successful substrate.

In 1995, Kang *et al.* demonstrated that a thiazazincolidine complex (Scheme 10.58) was an excellent catalyst for enantioselective reductive differentiation of the carbonyl groups of *meso N*-phenylimides.[114] This thiazazincolidine complex was prepared *in situ* from (1*R*,2*S*)-1-phenyl-2-(1-piperidino)-1-propanethiol and ZnEt$_2$. When treated with this complex in combination with bis(2,6-dimethylphenoxy)borane (BDMPB) as the reducing agent, *meso N*-phenylimides afforded the corresponding hydroxy lactams with high yields and enantioselectivities of up to 99% ee. These lactams were further converted into the corresponding chiral lactones by treatment with NaBH$_4$ followed by acid-catalysed lactonisation.

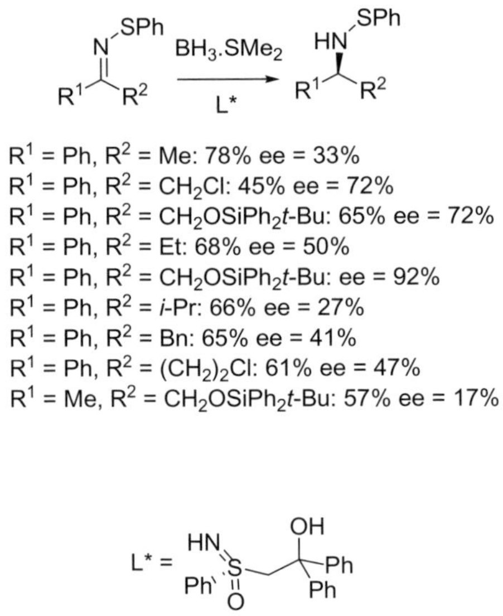

R^1 = Ph, R^2 = Me: 78% ee = 33%
R^1 = Ph, R^2 = CH$_2$Cl: 45% ee = 72%
R^1 = Ph, R^2 = CH$_2$OSiPh$_2$t-Bu: 65% ee = 72%
R^1 = Ph, R^2 = Et: 68% ee = 50%
R^1 = Ph, R^2 = CH$_2$OSiPh$_2$t-Bu: ee = 92%
R^1 = Ph, R^2 = i-Pr: 66% ee = 27%
R^1 = Ph, R^2 = Bn: 65% ee = 41%
R^1 = Ph, R^2 = (CH$_2$)$_2$Cl: 61% ee = 47%
R^1 = Me, R^2 = CH$_2$OSiPh$_2$t-Bu: 57% ee = 17%

Scheme 10.57   Borane reductions of ketimine derivatives with β-hydroxy sulfoximine
ligand.

In 1999, this methodology was applied to the synthesis of unnatural biologically active (+)-5-*epi*-nojirimycin-δ-lactam, a potent and selective glycosidase
inhibitor.[115] The key step of this synthesis was the asymmetric reduction of a
cyclic triacetyloxy *meso* imide under the same conditions to those described
above, which resulted in the formation of the corresponding hydroxy δ-lactam
in good yield and enantioselectivity of 85% ee (Scheme 10.59).

Other S/N ligands have been investigated in the enantioselective catalytic
reduction of ketones with borane. Thus, Mehler and Martens have reported the
synthesis of sulfur-containing ligands based on the L-methionine skeleton and
their subsequent application as new chiral catalysts for the borane reduction of
ketones.[116] The *in situ* formed chiral oxazaborolidine catalyst has been used in
the reduction of aryl ketones, providing the corresponding alcohols in nearly
quantitative yields and high enantioselectivities of up to 99% ee, as shown in
Scheme 10.60.

Finally, the enantioselective borane reduction of acetophenone was also
investigated in the presence of novel C_2-symmetric bis-β-*primary*- and *sec*-
amino alcohols derived from D-cysteine.[117] The application of the bis-β-
primary-amino alcohol ligands to the reduction of acetophenone generally gave
both good activities and enantioselectivities of up to 82% ee, but an important
decrease in the enantioselectivity was observed when the analogous β-*sec*-
amino alcohols were used (Scheme 10.61). In this approach, the secondary
amino group with a low sterical substitution (R= Et) seemed to be the
best combination, affording the expected alcohol with an enantioselectivity of
69% ee.

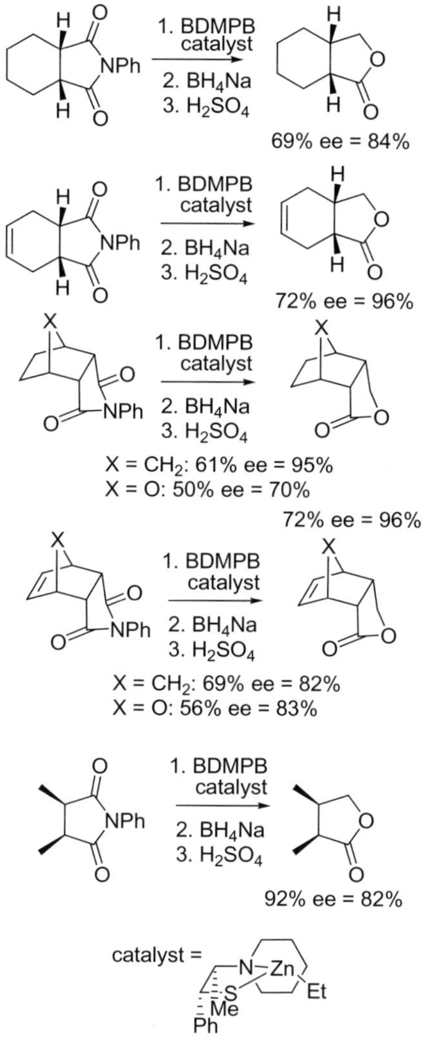

Scheme 10.58 Borane reductions of *meso* N-phenylimides with thiazazincolidine complex.

Chiral S/O ligands constitute the second type of chiral sulfur-containing ligands that have been used to promote enantioselective borane reductions. As an example, Kagan *et al.* have developed several chiral 1,2- and 1,3-hydro-xythiols derived from (R)-camphor, (1S)-(+)-10-camphorsulfonyl chloride, cysteine and cystine derivatives.[118] These novel compounds were evaluated as ligands in the borane reduction of ketones. It was shown that the nature of the reducing agent had an influence on the result of the reaction since the use of BH$_3$. THF instead of catecholborane allowed both the activities and enantio-selectivities to be higher for the reduction of acetophenone. These authors also

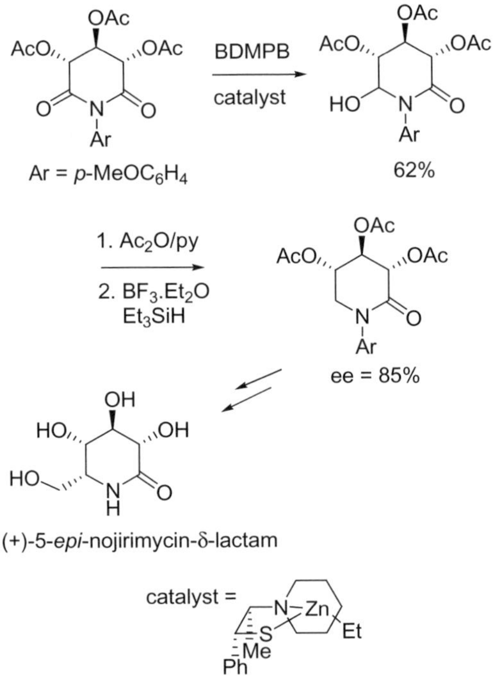

Scheme 10.59 Synthesis of (+)-5-*epi*-nojirimycin-δ-lactam.

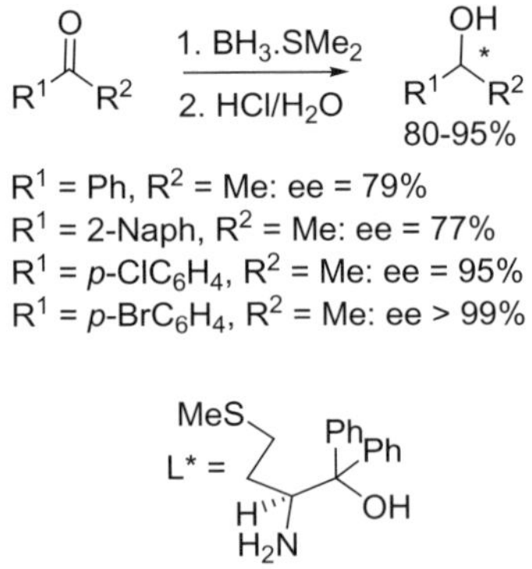

Scheme 10.60 Borane reductions of ketones with L-methionine-derived S/N ligand.

evaluated the influence of the nitrogen substitution of the ligand on the enantioselectivity. Thus, the results collected in Scheme 10.62 demonstrated that this substitution played a crucial role, since no enantioselectivity was observed by using the *N,N*-dimethylcysteinol ligand. The best enantioselectivities were obtained with (*R*)-camphor-derived 1,3-hydroxythiol ligands.

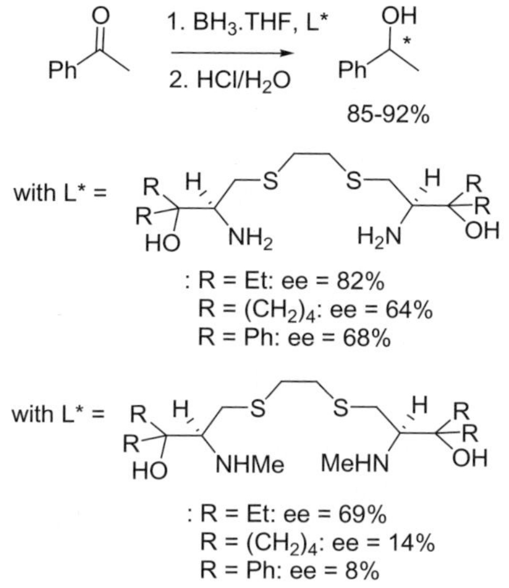

Scheme 10.61 Borane reduction of acetophenone with D-cysteine-derived bis-β-amino alcohol ligands.

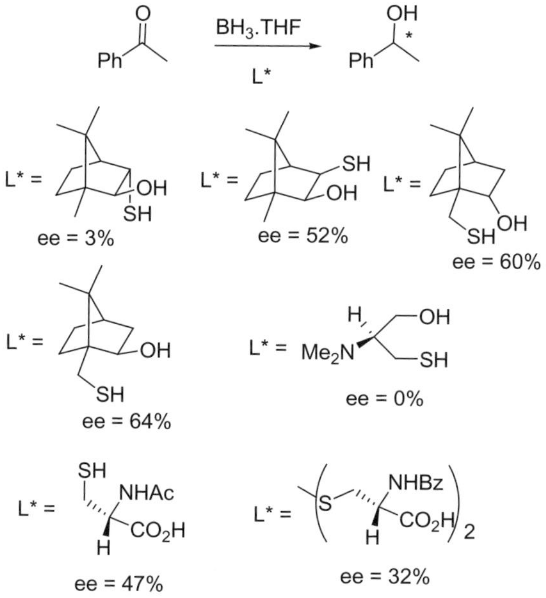

Scheme 10.62 Borane reduction of acetophenone with 1,2- and 1,3-hydroxythiol ligands.

Several other chiral S/O ligands have been involved by Yang and Lee in this type of reactions, such as [(1*R*,2*S*,3*R*)-3-mercaptocamphan-2-ol)], MerCO, which produced, when applied to aryl methyl ketones, the corresponding 1-aryl ethyl alcohols in enantioselectivities of up to 92% ee (Scheme 10.63).[119]

Another approach has involved a heterogeneous version for the borane reduction of acetophenone by elemental boron.[120] According to preliminary studies dealing with Ni-supported oxazoborolidines prepared by reaction of an amino alcohol with elemental boron in amorphous nickel boride (NiB$_2$), Molvinger and Court prepared a nickel-supported oxathiaborolidine by reaction of (1*R*,2*S*,3*R*)-3-mercaptocamphan-2-ol) with amorphous NiB$_2$, applying the same procedure (Scheme 10.64). Using the resulting heterogeneous catalyst, only poor enantioselectivities were observed in the borane reduction of acetophenone. However, when the nickel surface was at first passivated by poisoning with 3-methylthiophene, the reaction of (1*R*,2*S*,3*R*)-3-mercaptocamphan-2-ol with nickel boride afforded the oxathiaborolidine

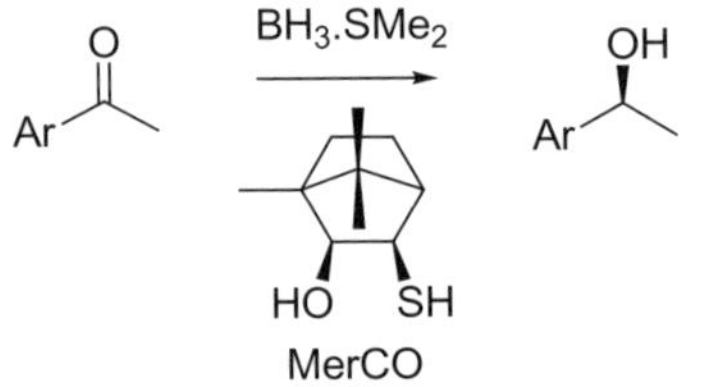

Ar = Ph: 96% ee = 87%
Ar = 2-Naph: 92% ee = 75%
Ar = *p*-MeOC$_6$H$_4$: 88% ee = 70%
Ar = *p*-ClC$_6$H$_4$: 90% ee = 80%
Ar = *p*-NO$_2$C$_6$H$_4$: 94% ee = 92%
Ar = 1-Naph: 98% ee = 64%

Scheme 10.63   Borane reductions of aryl methyl ketones with MerCO ligand.

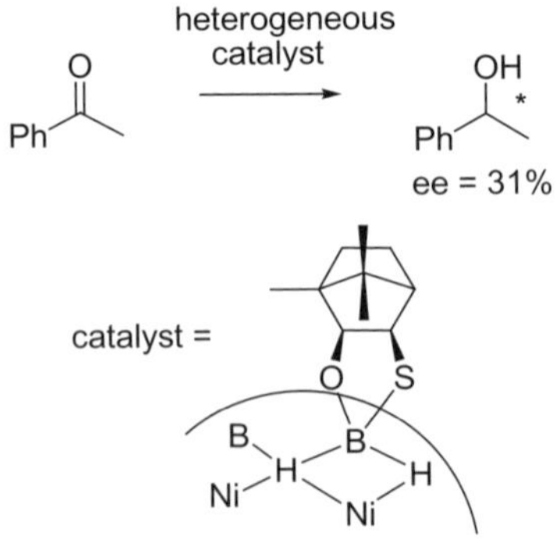

Scheme 10.64   Heterogeneous borane reduction of acetophenone with MerCO ligand.

anchored to the nickel nanoparticules, and moderate but better enantioselectivities of up to 31% ee were observed. These results clearly indicated a competitive interaction between the sulfur atoms on both the chiral ligand and methylthiophene with a nearly complete recovery of the surface by methylthiophene. The S/O ligand was consequently preferentially coordinated to the anchored boronic reducing species, leading to a more selective catalyst.

In 2000, Woodward *et al.* reported that $LiGaH_4$, in combination with the S/O-chelate, 2-hydroxy-2'-mercapto-1,1'-binaphthyl ($MTBH_2$), formed an active catalyst for the asymmetric reduction of prochiral ketones with catecholborane as the hydride source (Scheme 10.65).[121] The enantioface differentiation was on the basis of the steric requirements of the ketone substituents. Aryl *n*-alkyl ketones were reduced in enantioselectivities of 90–93% ee, whereas alkyl methyl ketones (*e.g.* *i*-Pr, Cy, *t*-Bu) gave lower enantioselectivities of 60–72% ee.

A plethora of chiral ligands have been employed for the development of asymmetric catalytic olefin hydroboration processes.[122] Much of the early groundwork relied on the application of homobidentate P/P ligands. However, it has been the advent of heterobidentate P/N ligands in particular that has advanced the scope of this process in terms of asymmetric induction. In contrast to borane reductions of ketones performed in the presence of chiral sulfur-containing ligands, the hydroboration of alkenes performed in the presence of this type of ligands has been scarcely developed.[123] An example has been

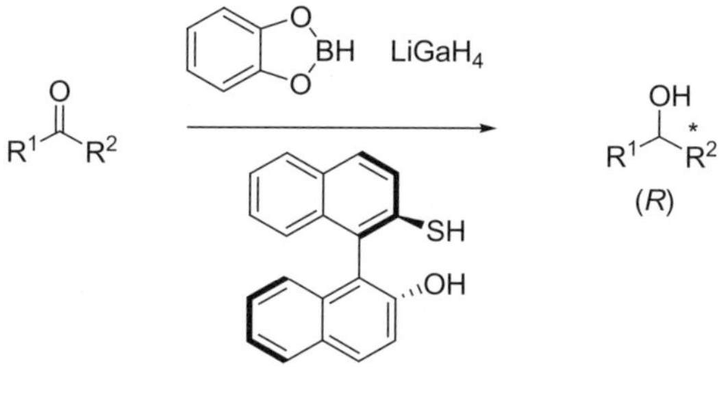

R^1 = Ph, R^2 = Me: 90% ee = 90%
R^1 = Ph, R^2 = Et: 96% ee = 93%
R^1 = Ph, R^2 = *n*-Bu: 80% ee = 92%
R^1 = Ph, R^2 = *i*-Pr: 65% ee = 24%
R^1 = Ph, R^2 = *i*-Bu: 65% ee = 92%
R^1 = *p*-BrC_6H_4, R^2 = Me: 80% ee = 87%
R^1 = Ts, R^2 = Me: 95% ee = 87%
R^1 = Ph, R^2 = CH_2Br: 60% ee = 70%
R^1 = 2-Fu, R^2 = *n*-Hex: 76% ee = 81%
R^1 = 2-$C_{10}H_7$, R^2 = Me: 83% ee = 73%
R^1 = *i*-Pr, R^2 = Me: 81% ee = 69%
R^1 = *i*-Bu, R^2 = Me: 93% ee = 46%
R^1 = Cy, R^2 = Me: 72% ee = 72%
R^1 = *t*-Bu, R^2 = Me: 76% ee = 79%

Scheme 10.65   Ga-catalysed borane reductions of ketones with $MTBH_2$ ligand.

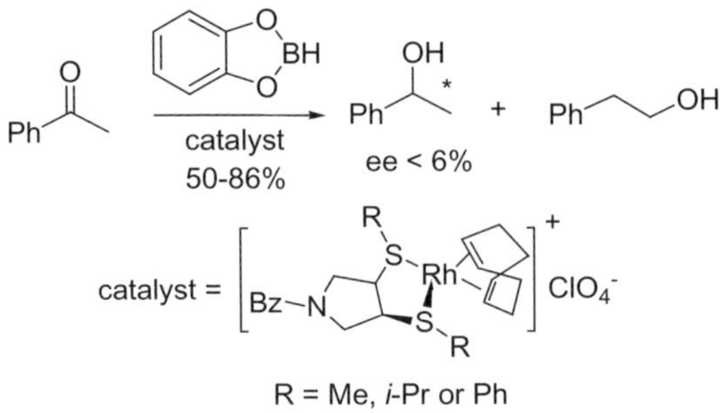

Scheme 10.66 Hydroboration of styrene with S/S ligands.

developed by Claver *et al.* by using S/S ligands based on degus R. These ligands were applied to the hydroboration of styrene with catecholborane, providing a mixture of the expected branched alcohol along with the corresponding linear alcohol with a very poor regioselectivity (ca. 50%) and a moderate yield (50–86%). Moreover, the desired branched alcohol was isolated with a very poor enantioselectivity (Scheme 10.66).

10.16 Pauson–Khand Reaction

In the past two decades, the Pauson–Khand reaction has attracted much attention from the synthetic chemistry community.[124] This reaction constitutes an interesting way to synthesise cyclopentane-based compounds from alkenes or alkynes by using stoichiometric amounts of the prochiral dicobalt hexacarbonyl complex ($Co_2(CO)_8$). Although some progress has been achieved lately on the way to convert the intramolecular Pauson–Khand reaction into a catalytic enantioselective process, the intermolecular version of this process has been omitted from most of these advances. Consequently, the remaining challenge probably lies in the development of a general catalytic asymmetric version.[125] In this context, an efficient asymmetric version of the intermolecular Pauson–Khand reaction of alkynes with norbornadiene was recently reported by Verdaguer *et al.* based on the use of pulegone-derived S/P ligands, such as PuPHOS and CyPHOS (Scheme 10.67).[126] These chiral bidentate ligands were designed with the aim of driving the coordination of the sulfur atom in the cobalt–alkyne cluster to the cobalt atom not bound to the phosphorus atom. In this context, the distance between the phosphorus and the sulfur atoms in the ligand design was crucial. The corresponding chiral cobalt complexes were prepared by a ligand exchange reaction with hexacarbonyldicobalt–alkyne in toluene. However, this approach yielded a diastereomeric mixture, and the major isomers were isolated by isomerisation–crystallisation sequences. These chiral catalysts were further investigated in the intermolecular Pauson–Khand reaction of alkynes with norbornadiene, providing the corresponding enones with enantioselectivities of up to 97% ee and with good yields of up to 93%.

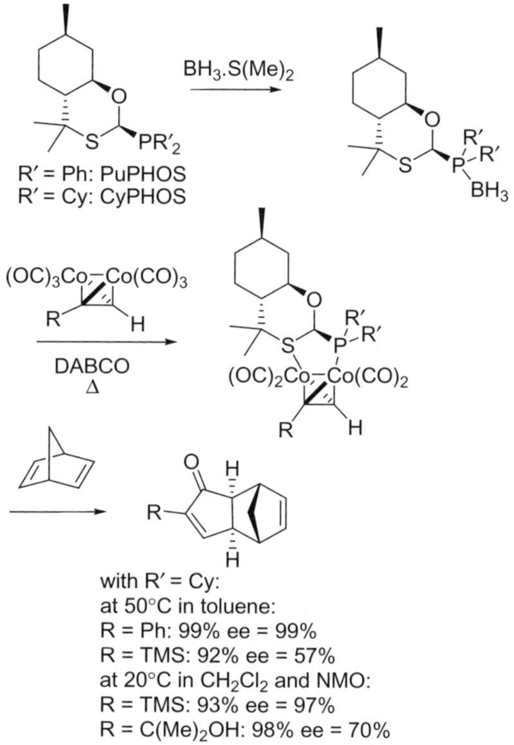
BH3.S(Me)2
O
S
PR'2
R' = Ph: PuPHOS
R' = Cy: CyPHOS
R'
R'
P
BH3
(OC)3Co–Co(CO)3
R
H
DABCO
Δ
(OC)2Co–Co(CO)2
with R' = Cy:
at 50°C in toluene:
R = Ph: 99% ee = 99%
R = TMS: 92% ee = 57%
at 20°C in CH2Cl2 and NMO:
R = TMS: 93% ee = 97%
R = C(Me)2OH: 98% ee = 70%

Scheme 10.67 Intermolecular Pauson–Khand reactions with PuPHOS and CyPHOS.

The intermolecular Pauson–Khand reaction of the resulting S/P-cobalt complexes with norbornadiene was studied under thermal and *N*-oxide activation conditions. Thus, heating the diastereomerically pure complex (R = Ph, R′ = Cy) with ten equivalents of norbornadiene at 50 °C in toluene afforded the corresponding *exo*-cyclopentenone in a quantitative yield and with an enantioselectivity of 99% ee. Under similar conditions, the analogous trimethylsilyl complex (R = TMS, R′ = Cy) afforded the expected product in a high yield but with a lower enantioselectivity of 57% ee. In order to increase this enantioselectivity, these authors performed this reaction at room temperature in dichloromethane as the solvent and in the presence of NMO, which allowed an enantioselectivity of 97% ee to be reached. These authors assumed that the thermal activation promoted the isomerisation of the S/P ligand leading to a nonstereoselective process.

Two S/P ligands derived from camphor, CamPHOS and MeCamPHOS were also developed by these authors for the diastereoselective coordination to alkyne–hexacarbonyldicobalt complexes (Scheme 10.68).[127] These two ligands were converted in good yields into their borane-protected forms. The influence of the alkyne group (R) on their coordination to dicobalt–hexacarbonyl–alkyne complexes was evaluated. It was shown that MeCamPHOS ligand provided a

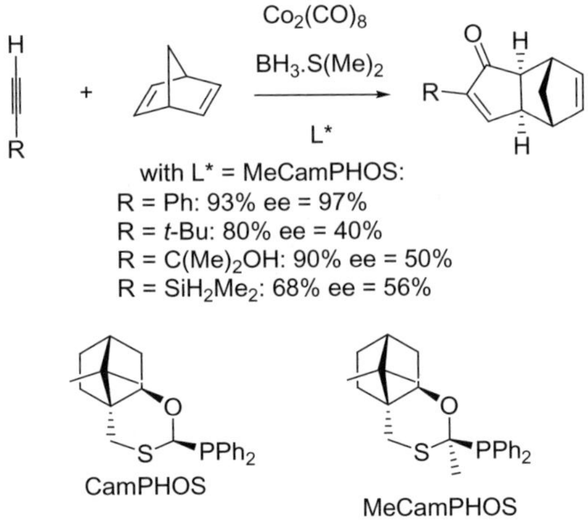

Scheme 10.68   Intermolecular Pauson–Khand reactions with MeCamPHOS.

high diastereoselectivity of up to 90% de. This result could be explained by a thermodynamic equilibration yielding the more stable isomer. The resulting chiral complexes were further investigated in the intermolecular Pauson–Khand reaction with norbornadiene, yielding the corresponding products in good yields but with opposite absolute configurations. This behavior was explained on the basis that the two ligands led to pseudoenantiomeric tetra-carbonyl complexes, which was confirmed by circular dichroism analysis. The best results are collected in Scheme 10.68.

In the same context, Verdaguer *et al.* studied, in 2005, the intermolecular Pauson–Khand reaction of amido-alkynes with norbornadiene in the presence of PuPHOS and CamPHOS-derived ligands.[128] These workers demonstrated that a unique methine moiety, attached to the three heteroatoms (O,P,S) of these ligands, served as a strong hydrogen-bond donor. A nonclassical hydrogen bonding of this methine with an amido-carbonyl acceptor provided a completely diastereoselective ligand exchange process between the alkyne dicobalthexacarbonyl complex and the ligand. This weak contact has been studied by means of X-ray analysis, ^{1}H NMR and quantum-mechanical calculations, revealing that the present interaction felt in the range of strong C–H$\cdots$O=C bonds. The hydrogen-bond bias obtained in the ligand exchange process has been further exploited in the asymmetric Pauson–Khand reaction with norbornadiene to produce the corresponding cyclization adducts in enantioselectivities of up to 94% ee (Scheme 10.69).

A novel family of P/N/S/O ligands such as *N*-diphenylphosphino *tert*-butyl-sulfinamides was reported by Verdaguer *et al.*, in 2007.[129] These authors have begun to study the coordination of these ligands towards dicobaltcarbonyl-alkyne complexes. Heating the ligands with the dicobalthexacarbonyl complex of 2-methyl-3-butyn-2-ol at 65 °C in toluene provided a diastereomeric mixture of the corresponding bridged complexes. Most interestingly, it was found that the

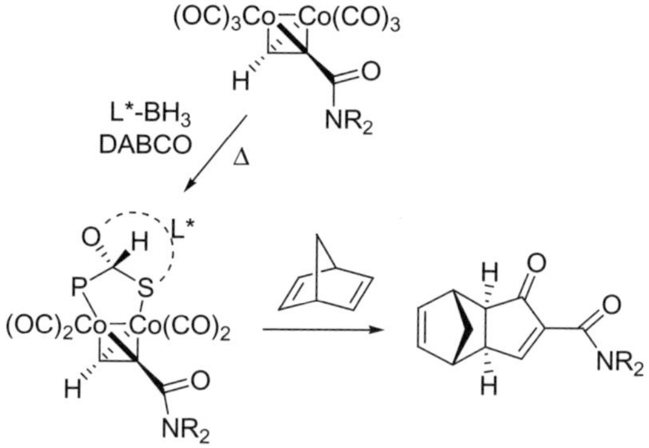

R	L*	conditions	yield (%)	ee (%)
Et	PuPHOS	thermal	33	90
Et	PuPHOS	NMO	81	73
Et	CamPHOS	thermal	47	94
Et	CamPHOS	NMO	81	80
i-Pr	CamPHOS	thermal	91	91
i-Pr	CamPHOS	thermal	72	79
Et	TolCamPHOS	thermal	47	93
Et	CyCamPHOS	thermal	87	87

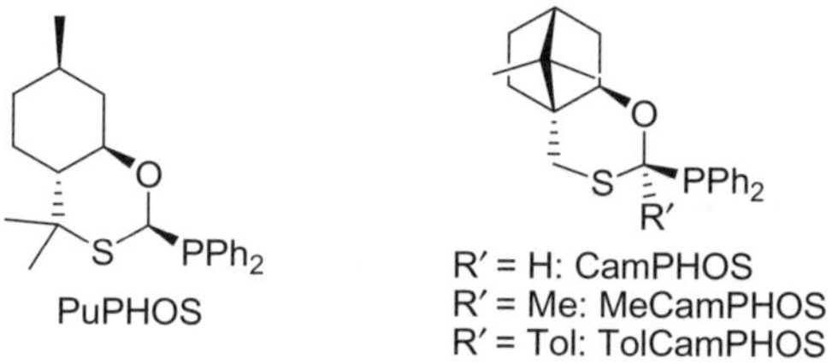

Scheme 10.69 Intermolecular Pauson–Khand reactions of amido-alkynes with PuPHOS and CamPHOS-derived ligands.

N-benzyl ligand gave enhanced reaction yields and diastereoselectivities of up to 85% de. In order to determine whether the P/N/S/O ligands acted as P/S or P/O ligands, an X-ray analysis of the major diastereomer was carried out, showing that the sulfinamide moiety was unprecedently bound to the metal centre through the sulfur atom. After purification by crystallisation, the diastereomerically pure complexes were reacted with norbornadiene to lead to the corresponding Pauson–Khand products with quantitative yields and unprecedented excellent enantioselectivities (Scheme 10.70). A complete stereocontrol in the cyclisation was observed by using the *N*-benzyl ligand, while the involvement of the ligand bearing an electron-withdrawing *p*-F-benzyl group showed a decrease in the enantioselectivity (78% ee). This behaviour could be explained on the basis that electron-deficient phosphines are more prone to isomerisation within the bimetallic complex, thus jeopardizing the stereochemical integrity of the initial dicobalt–P/N/S/O complex.

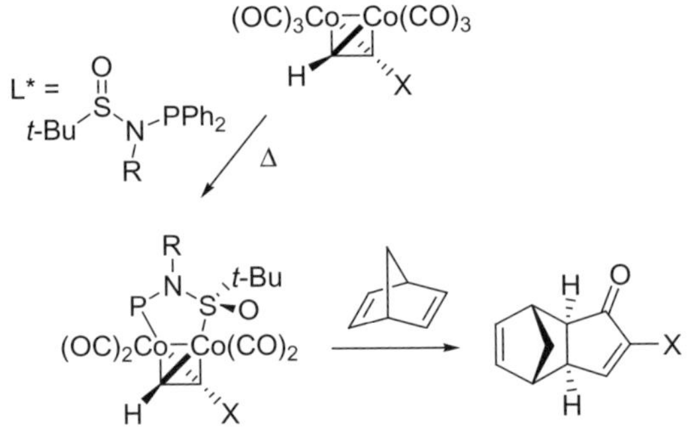

Scheme 10.70   Intermolecular Pauson–Khand reactions with P/N/S/O ligands.

R	X	conditions	yield (%)	ee (%)
Bn	C(Me)$_2$O	thermal	95	92
Bn	C(Me)$_2$O	NMO	87	93
Bn	TMS	NMO	99	97
Bn	Ph	20°C	99	99
Bn	CH$_2$OTBDPS	20°C	99	92
Bn	CH$_2$OH	NMO	65	73
p-MeOC$_6$H$_4$CH$_2$	TMS	NMO	99	99
p-FC$_6$H$_4$CH$_2$	TMS	NMO	99	78

10.17   Radical C–C Bond-Forming Reaction

Free-radical reactions have received renewed interest because of relatively recent discoveries demonstrating that the stereochemistry of these transformations could be controlled.[130] Relatively few reports have been published to date concerning radical C–C bond-forming reactions performed in the presence of chiral ligands. In 2000, Hiroi and Ishii reported the first example of asymmetric synthesis via radicals using chiral sulfoxides as chiral ligands.[131] Hence, asymmetric induction reaching 83% ee was observed in the intermolecular radical C–C bond-forming reaction of N-arylsulfonyl-α-bromocarboxamides using chiral sulfoxides in the presence of a Lewis acid such as Mg(OTf)$_2$ (Scheme 10.71). The intramolecular version of these radical reactions was investigated but gave, however, much lower enantioselectivities (≤5% ee).

10.18   Ring Opening of Mesoheterobicyclic Alkenes

Among recently described new Pd-catalysed enantioselective reactions, the ring opening of *meso* oxabicyclic alkenes with dialkyl zinc reagents in the presence of chiral P/P and P/N ligands reported by Lautens *et al.* constitutes a synthetically outstanding C–C bond-forming desymmetrization reaction.[132]

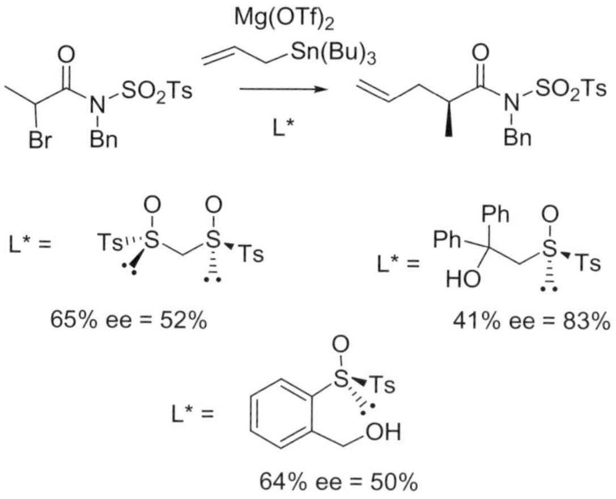

Scheme 10.71 Radical C–C bond-forming reaction of sulfonamide with sulfoxide ligands.

According to mechanistic evidence,[133] this alkylative ring-opening reaction proceeds by coordination and subsequent enantioselective carbopalladation of the alkene, with a cationic alkyl palladium intermediate, $[L_2PdR]^+$, formed *in situ* under the reaction conditions employed. In 2004, the Fesulphos ligands, successfully developed by Carretero *et al.* as palladium chelates for the Tsuji–Trost reaction,[134] or as copper chelates for Diels–Alder reactions,[135] were also proven to be efficient for the palladium-catalysed enantioselective ring opening of *meso* heterobicyclic alkenes.[136] A wide variety of this type of ligands was prepared bearing either bulky alkyl or aryl substituents on the sulfur atom. In combination with a palladium precatalyst, these S/N ligands were investigated as catalysts for the ring-opening addition of $ZnMe_2$ to 7-oxabenzonorborna-diene, providing the corresponding alcohol with a good yield and an enan-tioselectivity of up to 94% ee (Scheme 10.72). This methodology was extended to an analogous substituted substrate leading to the corresponding product in even higher enantioselectivity of up to 99% ee. A significant increase in the reaction rate occurred when a catalytic amount of $NaB(ArF)_4$ was added as a chloride scavenger in the reaction mixture, allowing a complete conversion within 10–30 minutes to be observed. The alkylative ring opening of less reactive substrates, such as azabenzonorbornadienes and nonaromatic [2.2.1]oxabicyclic alkenes, was also investigated, providing within a few hours and at room temperature, the corresponding ring-opened products in an enantioselectivity of up to 99% ee, as shown in Scheme 10.72. According to computational studies and X-ray analyses, these authors proposed that the high enantioselectivity displayed by these Fesulphos ligands could be explained by assuming a combination of electronic and steric factors, such as a strong *trans* effect of the phosphine moiety, together with a major steric hindrance imposed by the stereogenic sulfur atom.

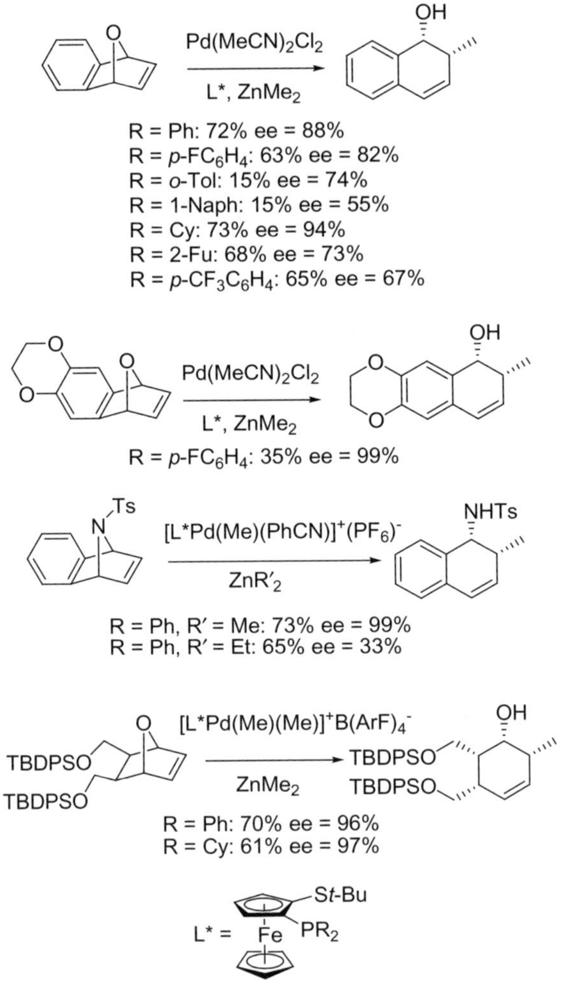
Pd(MeCN)2Cl2
L*, ZnMe2
OH
R = Ph: 72% ee = 88%
R = p-FC6H4: 63% ee = 82%
R = o-Tol: 15% ee = 74%
R = 1-Naph: 15% ee = 55%
R = Cy: 73% ee = 94%
R = 2-Fu: 68% ee = 73%
R = p-CF3C6H4: 65% ee = 67%
Pd(MeCN)2Cl2
L*, ZnMe2
OH
R = p-FC6H4: 35% ee = 99%
Ts
N
[L*Pd(Me)(PhCN)]+(PF6)-
ZnR'2
NHTs
R = Ph, R' = Me: 73% ee = 99%
R = Ph, R' = Et: 65% ee = 33%
TBDPSO
TBDPSO
[L*Pd(Me)(Me)]+B(ArF)4-
ZnMe2
OH
TBDPSO
TBDPSO
R = Ph: 70% ee = 96%
R = Cy: 61% ee = 97%
St-Bu
L* = Fe PR2

Scheme 10.72 Ring openings of *meso* heterobicyclic alkenes with Fesulphos ligands.

10.19 Bisalkoxycarbonylation of Olefins

At the beginning of the 1970s a convenient procedure was described for converting olefins into substituted butanedioates, namely through a Pd(II)-catalysed bisalkoxycarbonylation reaction.[137] So far various catalytic systems have been applied to this process, but it took twenty years before the first examples of an enantioselective bisalkoxycarbonylation of olefins were reported.[138] Ever since, the asymmetric bisalkoxycarbonylation of alkenes catalysed by palladium complexes bearing chiral ligands has attracted much attention.[139] The products of these reactions are important intermediates in the syntheses of pharmaceuticals such as 2-arylpropionic acids, the most important class of

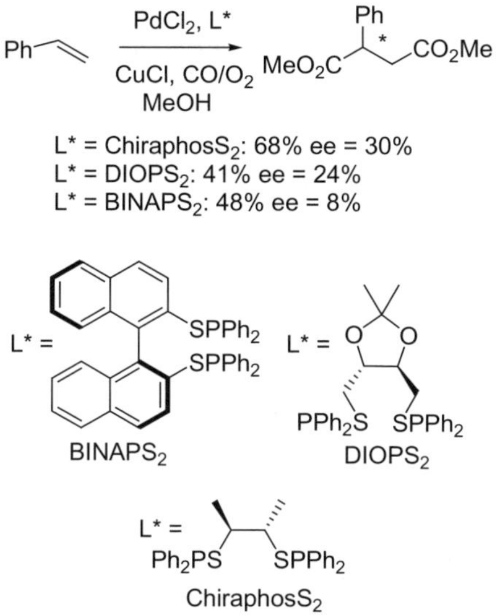

L* = ChiraphosS$_2$: 68% ee = 30%
L* = DIOPS$_2$: 41% ee = 24%
L* = BINAPS$_2$: 48% ee = 8%

Scheme 10.73 Pd-catalysed bisalkoxycarbonylation of styrene with biphosphine sulfide ligands.

nonsteroidal anti-inflammatory drugs. In 1998, Saigo *et al.* demonstrated that chiral biphosphine sulfides could be applied as ligands to the palladium-catalysed bisalkoxycarbonylation of olefins.[140] Several chiral sulfur-containing ligands such as DIOSP$_2$, BINAPS$_2$ and ChiraphosS$_2$ were prepared from commercially available enantiopure corresponding biphosphines, DIOP, BINAP and Chiraphos, respectively. As shown in Scheme 10.73, the Pd-catalysed bisalkoxycarbonylation of styrene afforded, in the presence of these ligands, the expected phenyl succinate, even though the enantioselectivity was low ($\leq$30% ee).

10.20 C–H Insertion

O-Heterocyclic ring systems are prominent features of numerous naturally occurring compounds and the development of simple but effective methods for their stereocontrolled synthesis is a goal of considerable practical importance. The past thirty years have witnessed a significant increase in the use of diazocarbonyl compounds as precursors in carbon–carbon bond-forming reactions. Selective activation of unfunctionalised C–H bonds represents a major challenge in organic synthesis.[141] One notable method is the intramolecular carbenoid insertion into C–H bonds that has assumed strategic importance in organic synthesis.[142] Transition-metal compounds are effective catalysts for carbenoid formation from diazocarbonyl compounds.[143] In this context,

McKervey and Ye have developed chiral sulfur-containing dirhodium carboxylates that have been subsequently employed as catalysts for asymmetric
intramolecular C–H insertion reactions of γ-alkoxy-α-diazo-β-keto esters.[144]
These reactions produced the corresponding *cis*-2,5-disubstituted-3(2H)-
furanones with diastereoselectivities of up to 47% de. Moreover, when a chiral
γ-alkoxy-α-diazo-β-keto ester containing the menthyl group as a chiral auxiliary was combined with rhodium(II) benzenesulfoneprolinate catalyst, a
considerable diastereoselectivity enhancement was achieved with the de value
being more than 60% (Scheme 10.74).

In the same area, good levels of enantioselectivity have been achieved in
intramolecular C–H insertion reactions of α-diazocarbonyl compounds

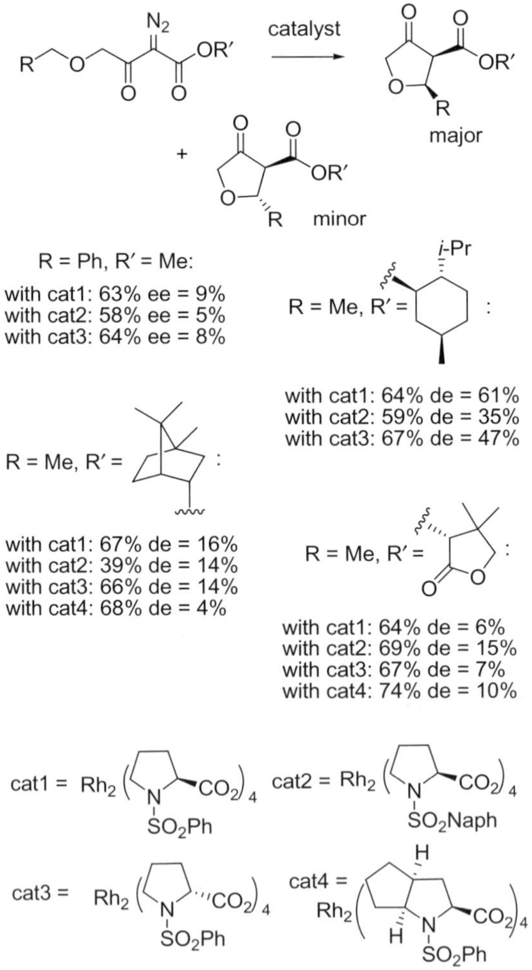

R = Ph, R′ = Me:

with cat1: 63% ee = 9%
with cat2: 58% ee = 5%
with cat3: 64% ee = 8%

R = Me, R′ = (*i*-Pr menthyl group):

with cat1: 64% de = 61%
with cat2: 59% de = 35%
with cat3: 67% de = 47%

R = Me, R′ = (bornyl group):

with cat1: 67% de = 16%
with cat2: 39% de = 14%
with cat3: 66% de = 14%
with cat4: 68% de = 4%

R = Me, R′ = (lactone group):

with cat1: 64% de = 6%
with cat2: 69% de = 15%
with cat3: 67% de = 7%
with cat4: 74% de = 10%

cat1 = $Rh_2(\text{N-SO}_2Ph\text{-CO}_2)_4$ cat2 = $Rh_2(\text{N-SO}_2Naph\text{-CO}_2)_4$

cat3 = $Rh_2(\text{N-SO}_2Ph\text{-CO}_2)_4$ cat4 = $Rh_2(\text{N-SO}_2Ph\text{-CO}_2)_4$

Scheme 10.74 Rh-catalysed intramolecular C–H insertions of γ-alkoxy-α-diazo-β-
keto esters with sulfonamide ligands.

catalysed by similar catalysts to those described above.[145] These reactions furnished predominantly the corresponding *cis*-disubstituted chiral chromanones through the use of rhodium(II) carboxylates including chiral sulfonamide ligands as the catalysts. A competition between the C–H insertion and the sigmatropic rearrangement, the latter producing the corresponding benzofuranones, was observed with precursors containing a proximal *O*-allyl side chain. While the chromanone was uniformly produced with the *cis*-geometry, the *cis:trans* ratio did vary somewhat with the nature of the catalyst employed, that containing the *N*-phenylsulfonylproline (cat1 in Scheme 10.75) giving the highest stereoselectivity (93:7). The enantioselectivities were higher in the *cis*-isomer than in the *trans*-isomer and were markedly catalyst dependent, with this catalyst producing the highest value of 60% ee for the *cis*-chromanone.

In contrast to the intramolecular carbenoid C–H insertion, the intermolecular version has not been greatly developed and has been for a long time regarded as a rather inefficient and unselective process. In this context, Davies and Hansen have developed asymmetric intermolecular carbenoid C–H insertions catalysed by rhodium(II) (*S*)-*N*-(*p*-dodecylphenyl)sulfonylprolinate.[146] Therefore, these catalysts were found to induce asymmetric induction in the decomposition of aryldiazoacetates performed in the presence of cycloalkanes,

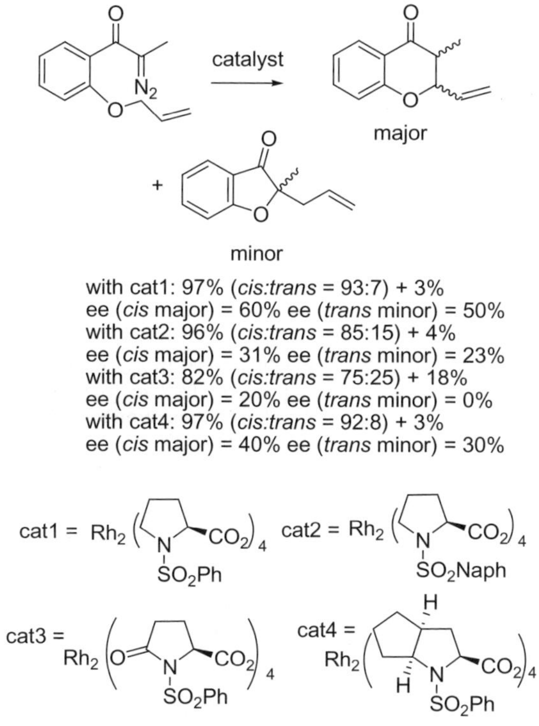

with cat1: 97% (*cis:trans* = 93:7) + 3%
ee (*cis* major) = 60% ee (*trans* minor) = 50%
with cat2: 96% (*cis:trans* = 85:15) + 4%
ee (*cis* major) = 31% ee (*trans* minor) = 23%
with cat3: 82% (*cis:trans* = 75:25) + 18%
ee (*cis* major) = 20% ee (*trans* minor) = 0%
with cat4: 97% (*cis:trans* = 92:8) + 3%
ee (*cis* major) = 40% ee (*trans* minor) = 30%

Scheme 10.75 Rh-catalysed intramolecular C–H insertions of α-diazo-ketone with sulfonamide ligands.

resulting in the formation of the corresponding C–H insertion products in yields ranging from 55 to 96% and moderate to high enantioselectivities of up to 95% ee (Scheme 10.76). However, the efficiency of this aryldiazoacetate system for asymmetric C–H insertion into cycloalkanes was in sharp contrast to the results that were obtained with more traditional carbenoid precursors, such as ethyl diazoacetate that did not generate the expected product, or methyl diazoacetate, which led to the corresponding C–H insertion product but with

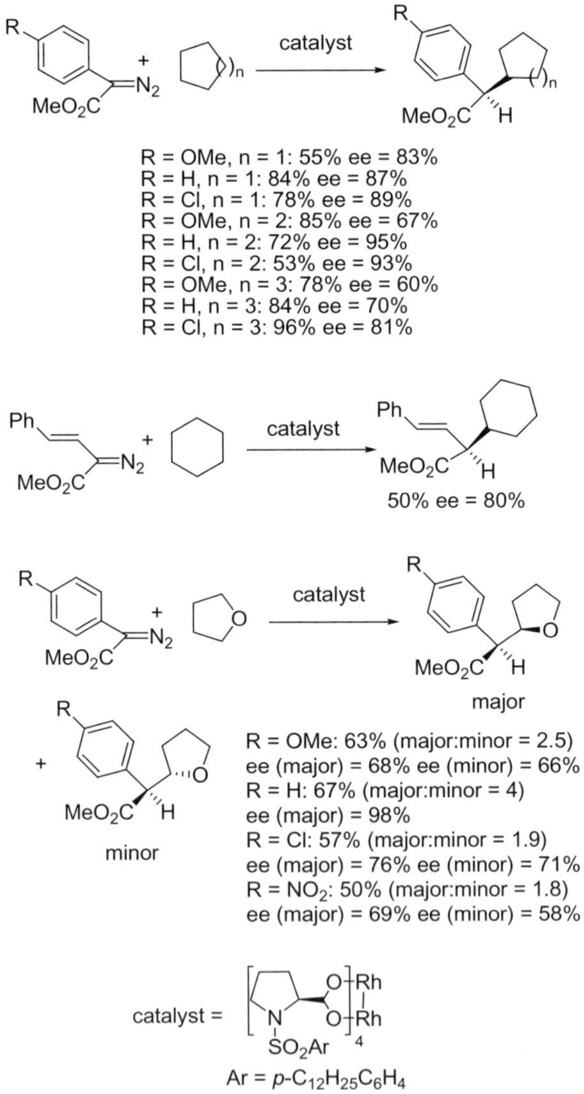

R = OMe, n = 1: 55% ee = 83%
R = H, n = 1: 84% ee = 87%
R = Cl, n = 1: 78% ee = 89%
R = OMe, n = 2: 85% ee = 67%
R = H, n = 2: 72% ee = 95%
R = Cl, n = 2: 53% ee = 93%
R = OMe, n = 3: 78% ee = 60%
R = H, n = 3: 84% ee = 70%
R = Cl, n = 3: 96% ee = 81%

50% ee = 80%

major

R = OMe: 63% (major:minor = 2.5)
ee (major) = 68% ee (minor) = 66%
R = H: 67% (major:minor = 4)
ee (major) = 98%
R = Cl: 57% (major:minor = 1.9)
ee (major) = 76% ee (minor) = 71%
R = NO$_2$: 50% (major:minor = 1.8)
ee (major) = 69% ee (minor) = 58%

minor

catalyst =

Ar = p-C$_{12}$H$_{25}$C$_6$H$_4$

Scheme 10.76 Rh-catalysed intermolecular C–H insertions of α-diazoacetates into cycloalkanes and tetrahydrofuran with sulfonamide ligand.

only 3% ee of enantioselectivity. On the other hand, the reaction with a vinylcarbenoid precursor was very effective, resulting in the formation of the corresponding C–H insertion product in an enantioselectivity of 83% ee (Scheme 10.76). Moreover, the extension of the methodology to tetrahydrofuran as the carbenoid trap illustrated that the C–H insertion of aryldiazoacetates also held promise for a good regioselectivity. The reaction led to the formation of a diastereomeric mixture of products formed by C–H insertion into the methylene group adjacent to the ether (Scheme 10.76). The enantioselectivities were found to be generally lower than for the cycloalkanes with the highest ee value of 98%.

As another extension of this process, Davies *et al.* have developed highly regio-, diastereo- and enantioselective C–H insertions of methyl aryldiazoacetates into cyclic *N*-Boc-protected amines catalysed by rhodium(II) (*S*)-*N*-(*p*-dodecylphenyl)sulfonylprolinate.[147] The best results were obtained in the case of the C–H insertion of methyl aryldiazoacetates into *N*-Boc-pyrrolidine, which gave, in all cases, a diastereoselectivity and an enantioselectivity greater than 90% de and 90% ee respectively (Scheme 10.77). The synthetic utility of this method was demonstrated by means of a two-step asymmetric synthesis of a novel class of C_2-symmetric amines.

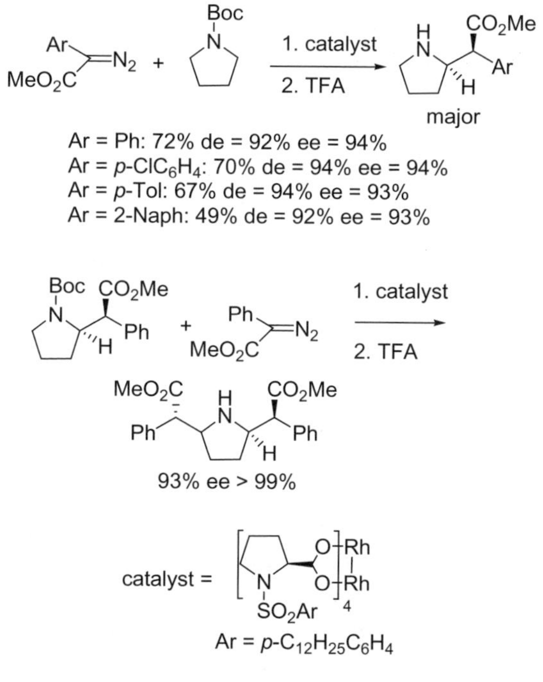
Boc
Ar
MeO2C
=N2 +
1. catalyst
2. TFA
H
CO2Me
Ar
H
major
Ar = Ph: 72% de = 92% ee = 94%
Ar = p-ClC6H4: 70% de = 94% ee = 94%
Ar = p-Tol: 67% de = 94% ee = 93%
Ar = 2-Naph: 49% de = 92% ee = 93%
Boc CO2Me
Ph
H
+
Ph
MeO2C
=N2
1. catalyst
2. TFA
MeO2C
H
CO2Me
Ph
Ph
H
93% ee > 99%
catalyst =
O
Rh
O
Rh
SO2Ar
4
Ar = p-C12H25C6H4

Scheme 10.77 Rh-catalysed intermolecular C–H insertions of methyl aryldiazoacetates into *N*-Boc-pyrrolidines with sulfonamide ligand.

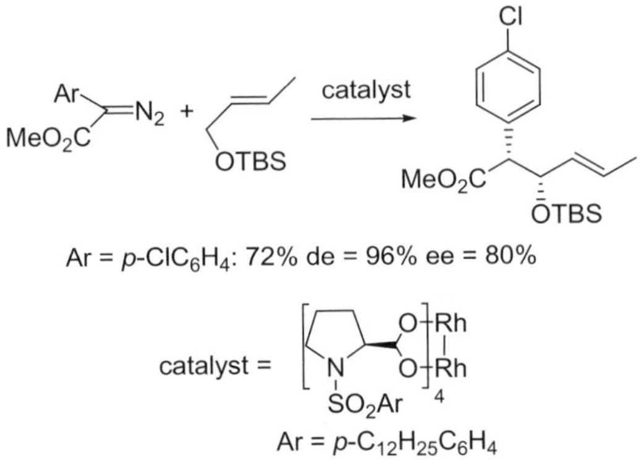

$$Ar = p\text{-}ClC_6H_4: 72\% \quad de = 96\% \quad ee = 80\%$$

$$\text{catalyst} = \quad Ar = p\text{-}C_{12}H_{25}C_6H_4$$

Scheme 10.78   Rh-catalysed intermolecular C–H insertion of methyl aryldiazoacetate into allylsilyl ether with sulfonamide ligand.

A further example of highly diastereoselective and enantioselective C–H insertions performed in similar conditions to those described above was the reaction between aryldiazoacetates and allylsilyl ethers, yielding β-hydroxy ester derivatives that are equivalents to aldol products.[148] An illustrative reaction between an aryldiazoacetate and *trans*-2-butenylsilyl ether is shown in Scheme 10.78. This reaction led to the diastereoselective formation of the equivalent of a *syn*-aldol product in both high yield and enantioselectivity.

Finally, Davies *et al.* have shown that the asymmetric Si–H insertion of aryldiazoacetates into *tert*-butyldiphenylsilane was also possible in conditions similar to those described above.[149] As shown in Scheme 10.79, the reaction of methyl phenyldiazoacetate performed at −78 °C generated the corresponding Si–H insertion product in a moderate yield and a high enantioselectivity of 85% ee. Similar reactions involving a series of vinyldiazoacetates have resulted in the successful formation of the corresponding allylsilanes in enantioselectivities ranging from 77 to 95% ee (Scheme 10.79).

10.21   Hydroamination and Amination

Due to its marked atom economy, the intramolecular hydroamination of alkenes represents an attractive process for the catalytic synthesis of nitrogen-containing organic compounds.[150] Moreover, the nitrogen heterocycles obtained by hydroamination/cyclisation processes are frequently found in numerous pharmacologically active products. The pioneering work in this area was reported by Marks *et al.* who have used lanthanocenes to perform hydroamination/cyclisation reactions in 1992.[151] These reactions can be performed in an intermolecular fashion and transition metals are by far the more efficient catalysts for promotion of these transformations *via* activation of the

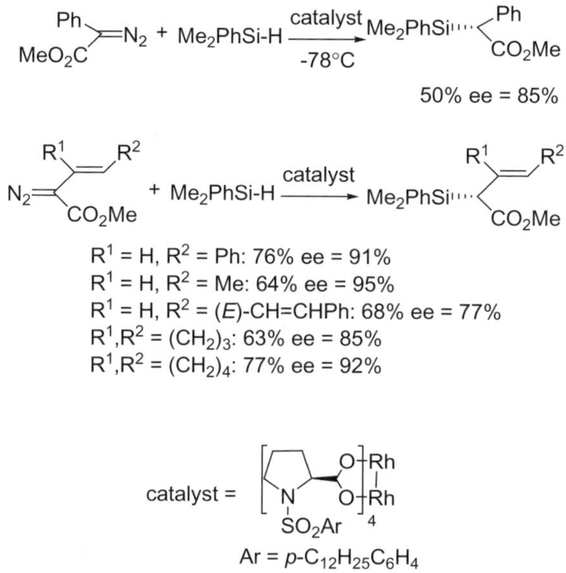
catalyst
-78°C
50% ee = 85%
catalyst
R1 = H, R2 = Ph: 76% ee = 91%
R1 = H, R2 = Me: 64% ee = 95%
R1 = H, R2 = (E)-CH=CHPh: 68% ee = 77%
R1,R2 = (CH2)3: 63% ee = 85%
R1,R2 = (CH2)4: 77% ee = 92%
catalyst =
Ar = p-C12H25C6H4

Scheme 10.79 Rh-catalysed intermolecular Si–H insertions of methyl phenyldiazoacetate and vinyldiazoacetates into allylsilyl ether with sulfonamide ligand.

carbon–carbon multiple bonds.[152] Noncyclopentadienyl rare-earth complexes and various transition-metal complexes have recently been reported to promote these reactions.[150b] So far, the asymmetric version of the intramolecular hydroamination reaction of alkenes has been rarely studied and only a few chiral catalysts have been reported to be able to induce such reactions.[153] In almost all cases, the efficient catalysts were based on lanthanides associated with various types of chiral ligands, such as binaphthol,[154] binaphthylamine,[155] or bis(oxazolines).[156] In addition, several chiral sulfur-containing ligands have recently been demonstrated to be efficient in inducing chirality in hydroamination/cyclisation of aminoalkenes. As an example, Kim and Livinghouse have obtained unprecedently high enantioselectivities of up to 87% ee by using chiral dithiol ligands bearing a binaphthylamine unit for the asymmetric yttrium-catalysed hydroamination of various aminoalkenes, providing the corresponding cyclic amines (Scheme 10.80).[157]

In 2003, Livinghouse *et al.* also reported that chelating bis(thiophosphonic amidates) complexes of lanthanide metals, such as yttrium or neodymium, were able to catalyse intramolecular alkene hydroaminations.[158] These complexes were prepared by attachment of the appropriate ligands to the metals by direct metalation with $Ln[N(TMS)_2]_3$. When applied to the cyclisation of 2-amino-5-hexene, these catalysts led to the formation of the corresponding pyrrolidine as a mixture of two diastereomers in almost quantitative yields and diastereoselectivities of up to 88% de (Scheme 10.81).

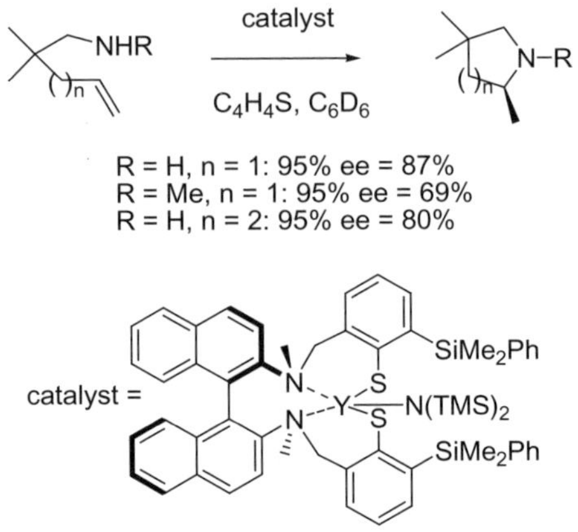

R = H, n = 1: 95% ee = 87%
R = Me, n = 1: 95% ee = 69%
R = H, n = 2: 95% ee = 80%

Scheme 10.80   Y-catalysed hydroaminations of aminoalkenes with binaphthylamine-derived dithiol ligand.

The synthesis of a series of chiral organophosphine oxide/sulfide-substituted binaphtholate ligands has recently been reported by Marks and Yu and their corresponding lanthanide complexes characterized.[159] These complexes, generated *in situ* from Ln[N(TMS)$_2$]$_3$, cleanly catalysed enantioselective intramolecular hydroamination/cyclisation of 1-amino-2,2-dimethyl-4-pentene albeit with a low enantioselectivity of 7% ee (Scheme 10.82).

In another area, chiral magnesium bis(sulfonamides) complexes have been prepared by Evans and Nelson by reaction of the corresponding bis(sulfonamides) with MgMe$_2$.[160] These complexes have been further used as catalysts for the enantioselective amination of *N*-acyloxazolidinones by di-*tert*-butyl azodicarboxylate, playing as the electrophilic nitrogen source. In all cases of substrates, the corresponding hydrazides were formed in high yields and good to high enantioselectivities, as shown in Scheme 10.83.

In 2005, Khiar *et al.* reported the preparation of novel phosphinite thioglycosides derived from D-galactose showing that bulky alkyl thioglycosides, such as *t*-butyl thioglycoside, allowed the synthesis of aminated products in enantioselectivities of up to 90% ee in palladium-catalysed allylic substitutions in the presence of benzylbromide.[161] The importance of the stereochemistry at the anomeric centre was shown, since the analogous α-thioglycoside led to the racemic product. Dynamic and NOE NMR studies revealed that the complex between the ligand and the palladium was obtained as a single diastereomer, showing an efficient control of the sulfur configuration. In addition, these authors demonstrated that it was possible to generate the corresponding opposite enantiomers in a good enantioselectivity (up to 78% ee) by using a phosphinite thioglycoside derived from D-arabinose as ligand of the same reaction.[162] Thus, even though derived from natural D-sugars, the two ligands behaved as enantiomers in Pd-catalysed asymmetric substitutions using benzylamine as the nucleophile (Scheme 10.84).

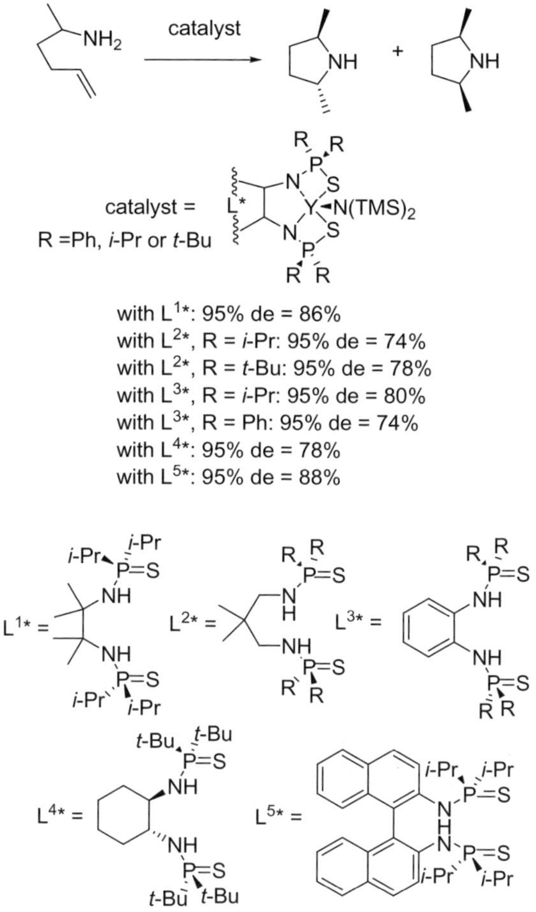

Scheme 10.81 Y-catalysed hydroaminations of 2-amino-5-hexene with bis(thiophosphonic amidates) ligands.

10.22 Reissert Reaction

The Reissert reaction[163] is the reaction of quinolines with acid chloride and potassium cyanide leading to the corresponding cyano-1,2-dihydroquinolines, which are further hydrolysed into the corresponding acids.[164] This reaction is also successful with isoquinolines and most pyridines. So far, enantioselective versions of this type of reactions are very rare in the literature. In 2004, Shibasaki *et al.* achieved the first catalytic enantioselective Reissert reaction of pyridine derivatives through the development of new Lewis acid (Et$_2$AlCl)-Lewis base chiral S/O- catalysts (Scheme 10.85).[165] Both sulfoxides and phosphine sulfides have played the role of efficient Lewis bases, providing high regio- and enantioselectivities of up to 96% ee.

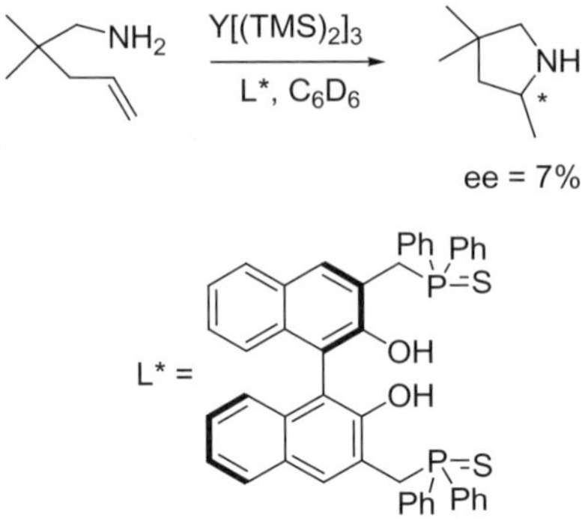

Scheme 10.82   Y-catalysed hydroamination of 1-amino-2,2-dimethyl-4-pentene with bis(thiophosphonic amidates) ligand.

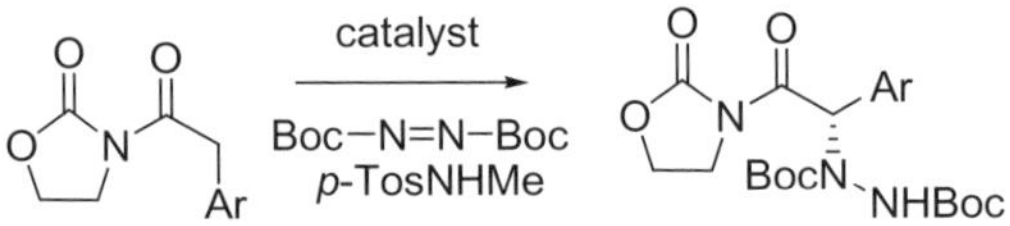

Ar = Ph: 92% ee = 86%
Ar = *p*-FC₆H₄: 97% ee = 90%
Ar = *p*-MeOC₆H₄: 93% ee = 86%
Ar = 2-Naph: 87% ee = 82%
Ar = 3-Cl-4-MeOC₆H₃: 84% ee
Ar = 3,4-(OCH₂O)C₆H₃: 85% ee = 82%

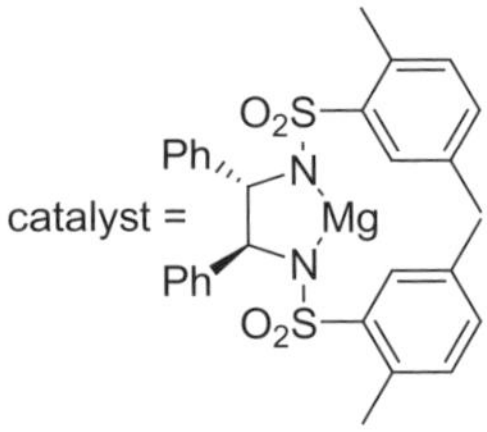

Scheme 10.83   Mg-catalysed aminations of *N*-acyloxazolidinones with bis(sulfonamides) ligand.

10.23   Aziridination

Aziridines are the nitrogen analogues of epoxides and exhibit similar reactivity patterns as electrophilic reagents.[166] Since they undergo highly regio- and stereoselective transformations, they are useful building blocks for organic synthesis. In addition, a number of aziridines exhibit various biological properties that make them attractive synthetic targets in their own right. Until ten years ago, the field of asymmetric catalytic aziridination was relatively underdeveloped in comparison to that of asymmetric epoxidation or cyclopropanation of olefins.[167] In 2001, Aggarwal *et al.* employed chiral [2.2.2] bicyclic

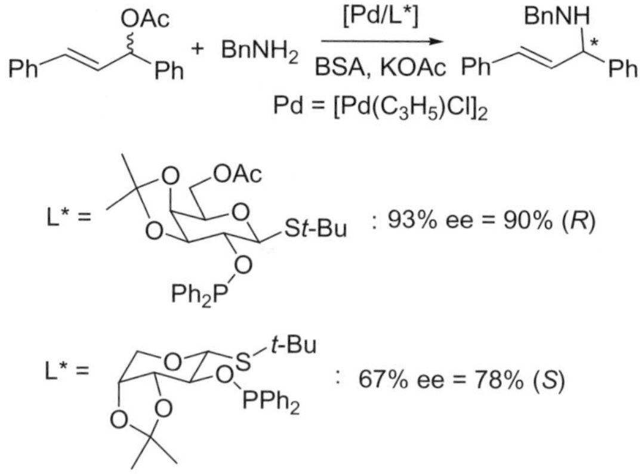
OAc
+ BnNH2
[Pd/L*]
BSA, KOAc
Pd = [Pd(C3H5)Cl]2
BnNH
Ph
Ph
L* =
OAc
St-Bu
Ph2P
: 93% ee = 90% (R)
t-Bu
PPh2
: 67% ee = 78% (S)

Scheme 10.84 Pd-catalysed allylic amination with phosphinite thioglycoside ligands.

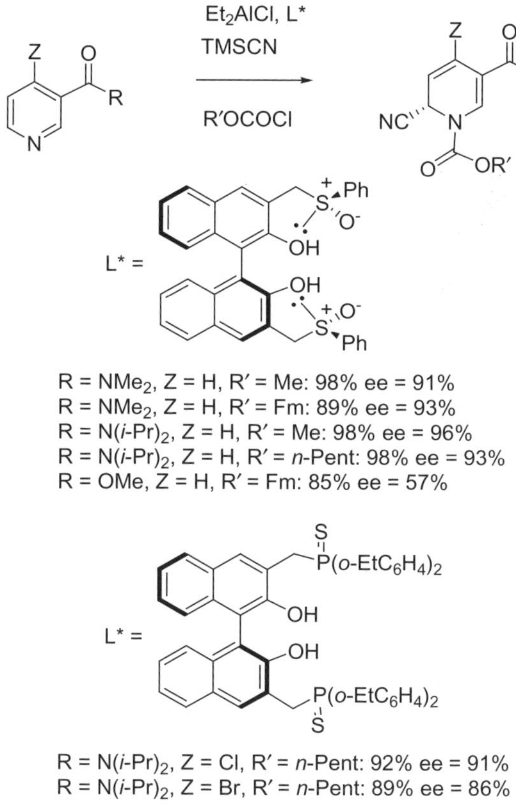
Et2AlCl, L*
TMSCN
R′OCOCl
Z
R
NC
OR′
L* =
Ph
OH
R = NMe2, Z = H, R′ = Me: 98% ee = 91%
R = NMe2, Z = H, R′ = Fm: 89% ee = 93%
R = N(i-Pr)2, Z = H, R′ = Me: 98% ee = 96%
R = N(i-Pr)2, Z = H, R′ = n-Pent: 98% ee = 93%
R = OMe, Z = H, R′ = Fm: 85% ee = 57%
P(o-EtC6H4)2
R = N(i-Pr)2, Z = Cl, R′ = n-Pent: 92% ee = 91%
R = N(i-Pr)2, Z = Br, R′ = n-Pent: 89% ee = 86%

Scheme 10.85 Al-catalysed Reissert reactions of pyridine derivatives with S/O-ligands.

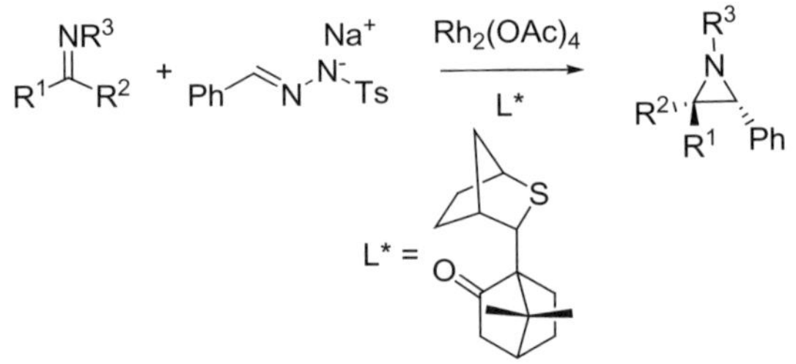

R[1] = Ph, R[2] = H, R[3] = SES: 75% *trans:cis* = 2.5:1 ee = 94%
R[1] = Ph, R[2] = H, R[3] = Ts: 68% *trans:cis* = 2.5:1 ee = 98%
R[1] = Ph, R[2] = H, R[3] = SO$_2$C$_{10}$H$_7$: 75% *trans:cis* = 3:1 ee = 97%
R[1] = *p*-ClC$_6$H$_4$, R[2] = H, R[3] = SES: 82% *trans:cis* = 2:1 ee = 98%:81%
R[1] = *p*-MeOC$_6$H$_4$, R[2] = H, R[3] = SES: 60% *trans:cis* = 2.5:1 ee = 92%:78%
R[1] = *n*-Hex, R[2] = H, R[3] = SES: 50% *trans:cis* = 2.5:1 ee = 98%:89%
R[1] = *t*-Bu, R[2] = H, R[3] = Ts: 53% *trans:cis* = 2:1 ee = 73%:95%
R[1] = 3-Fu, R[2] = H, R[3] = Ts: 72% *trans:cis* = 8:1 ee = 95%
R[1] = *(E)*-Ph-CH=CH, R[2] = H, R[3] = SES: 59% *trans:cis* = 8:1 ee = 94%

Scheme 10.86   Rh-catalysed aziridinations of imines with [2.2.2] bicyclic sulfide ligand.

sulfides as ligands in a highly efficient asymmetric rhodium-catalysed azir-idination of various imines derived from aromatic, heteroaromatic, unsaturated and even aliphatic aldehydes and ketones (Scheme 10.86).[168] As an application of this methodology, a synthesis of the side chain of taxol was developed.[169]

10.24   Grignard Crosscoupling

The transition-metal-catalysed crosscoupling between an organometallic species and an aryl or an alkenyl halide constitutes a powerful synthetic approach and one of the most straightforward methods for carbon–carbon bond formation.[170] The involvement of chiral sulfur-containing ligands in this type of reaction has opened the way to the preparation of a wide range of these ligands and their efficient application to many other carbon–carbon bond formations. The asymmetric version of the Grignard crosscoupling process has been widely developed by using optically active phosphine ligands.[171] With the aim of developing ligands that are more stable than phosphine ligands, Kellogg *et al.* have investigated the use of various chiral sulfides as efficient chiral ligands for catalytic enantioselective Grignard crosscouplings.[172] These novel chiral sulfur-containing ligands were evaluated for the Ni(II)-catalysed Grignard cross-coupling of 1-phenylethylmagnesium chloride with vinyl bromide, providing 3-phenyl-1-butene in a low enantioselectivity. In all cases of the flexible sulfide ligands tested and depicted in Scheme 10.87, the enantioselectivity never

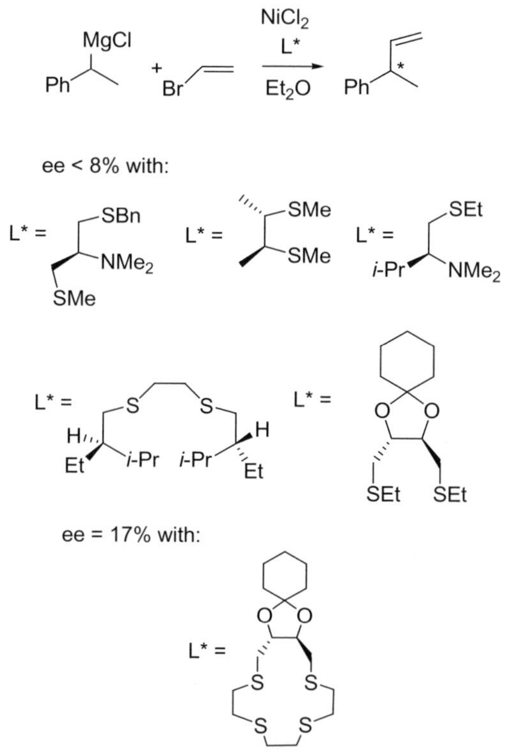

Scheme 10.87 Grignard crosscoupling of 1-phenylethylmagnesium chloride and vinyl bromide with sulfide ligands.

exceeded 8% ee. In order to improve the enantioselectivity of the process, these authors decided to increase the rigidity of the ligand by synthesising a ligand bearing a fourteen-membered ring (Scheme 10.87). In the presence of this ligand, the reaction provided the corresponding product in a quantitative yield and a slightly improved enantioselectivity of 17% ee. In spite of these moderate results, these first studies have clearly demonstrated the potential of chiral sulfur-containing ligands in the transition metal-catalysed carbon–carbon bond formation.

In 1986, these workers reported the synthesis of a series of C_2-symmetric macrocycles containing sulfide and amino linkages as coordination sites, which were further examined as ligands for the Ni-catalysed crosscoupling of 1-phenylethylmagnesium chloride with vinyl bromide.[173] In most cases, the catalysts were active, providing the expected product in good yields but the enantioselectivity remained low to moderate (≤46% ee) as shown in Scheme 10.88. The best result (46% ee) was obtained with (6R,15R)-6,15-bis-(dimethylamino)-1,4,8,13-tetrathiahexadecane derived from L-cysteine, which provided suitable square-planar coordination for the presumed nickel(0) and diorganonickel intermediates in the crosscoupling.

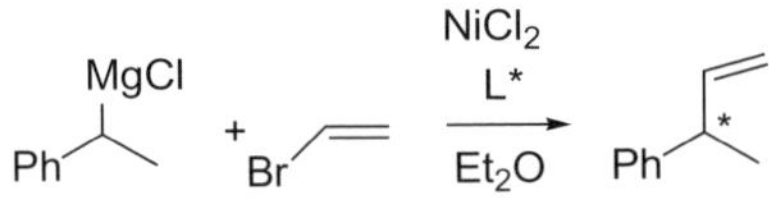

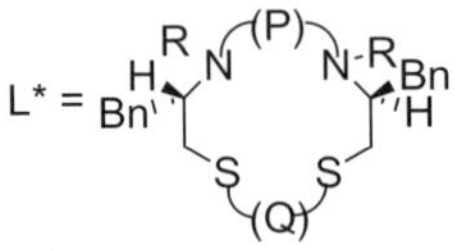

R = Et, P = Q = (CH$_2$)$_2$: 80% ee = 0%
R = Et, P = (CH$_2$)$_2$, Q = (CH$_2$)$_3$: 88% ee = 5% (*R*)
R = Et, P = (CH$_2$)$_2$, Q = (CH$_2$)$_4$: 83% ee = 0%
R = Et, P = (CH$_2$)$_4$, Q = (CH$_2$)$_3$: 88% ee = 5% (*R*)
R = Me, P = (CH$_2$)$_4$, Q = (CH$_2$)$_3$: 90% ee = 15% (*R*)
R = Me, P = (CH$_2$)O(CH$_2$)$_2$, Q = (CH$_2$)S(CH$_2$)$_2$S(CH$_2$)$_2$:
90% ee = 8% (*R*)

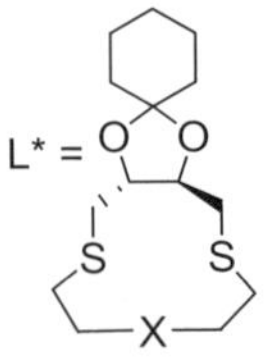

X = S: 95% ee = 3% (*S*)
X = S(CH$_2$)$_2$S: 87% ee = 10% (*S*)
X = S(CH$_2$)$_3$S: 95% ee = 2% (*S*)
X = S(CH$_2$)$_2$S(CH$_2$)$_2$S: 95% ee = 4% (*S*)

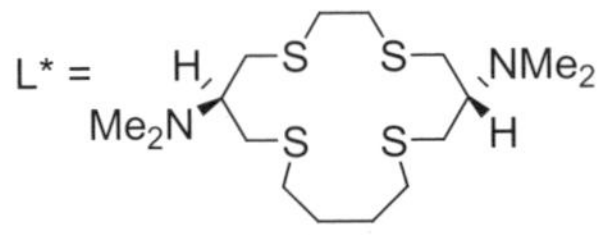

50% ee = 46% (*R*)

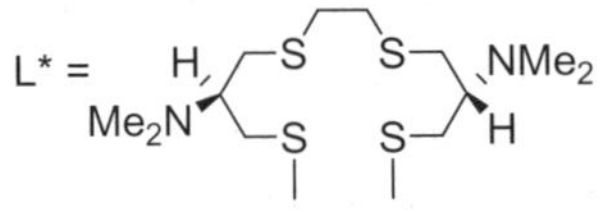

88% ee = 48% (*S*)

Scheme 10.88   Grignard crosscoupling of 1-phenylethylmagnesium chloride and vinyl
bromide with C$_2$-symmetric macrocyclic sulfide ligands.

More recently, Bäckvall *et al.* have reported the use of arenethiolatocopper(I)
as a catalyst for the analogous substitution reaction of Grignard reagents with
allylic substrates.[174] In this case, the crosscoupling reaction could occur in an
α(S$_N$2) or γ(S$_N$2′) manner, depending on the reaction conditions. In all cases,
the γ-product was isolated as the sole product with moderate to quantitative

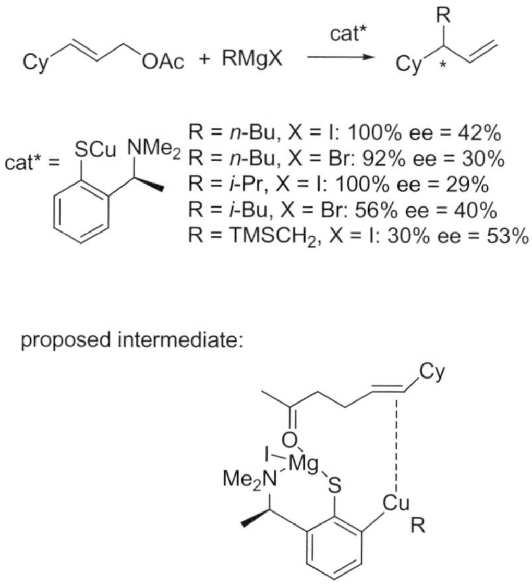

Scheme 10.89 Arenethiolatocopper(I)-catalysed Grignard crosscouplings of allylic substrate.

yields and enantioselectivities of up to 53% ee. In the course of optimising the reaction, these authors have noticed that the catalysis actually only started after the addition of at least one equivalent of the Grignard reagent (relative to the catalyst amount). This observation has led the authors to propose the key intermediate depicted in Scheme 10.89.

These reactions were also studied by these authors in the presence of a ferrocenylthiolate as ligand, providing selectively the expected γ-product in good yields and enantioselectivities of up to 64% ee in the presence of CuI (Scheme 10.90).[175] Actually, this ligand was employed as the lithium derivative, which was stable in the solid state under argon. Neutral thioether ligands were not suitable for these reactions, showing the importance of the anionic coordination to copper. Indeed, in these conditions, the transformation afforded racemic products. In addition, these authors have investigated a ferrocenyl oxazoline thiol as ligand, but this ligand combined with CuI proved less efficient since the observed enantioselectivity was of only 10% ee (Scheme 10.90).

In addition, Griffin and Kellogg have examined various sulfur-containing amino acids, S-alkylated amino phosphines derived from commercially available L-methionine and D-penicillamine.[176] The crosscoupling reaction was performed between vinyl bromide and 1-phenylethyl chloride magnesium or 2-octylchloride magnesium in the presence of these ligands and nickel chloride, providing excellent yields but variable enantiomeric excesses that reached only 14% ee when 2-octylchloride magnesium was used. On the other hand, the reaction involving 1-phenylethyl chloride magnesium gave better enantioselectivities of up to 70% ee (Scheme 10.91).[177] In these reactions, the authors

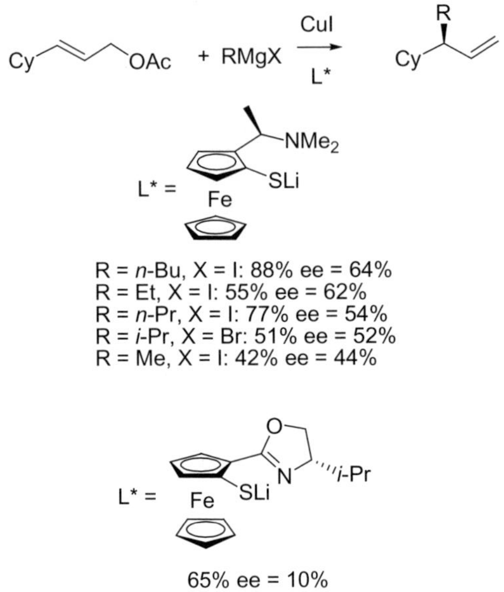

R = *n*-Bu, X = I: 88% ee = 64%
R = Et, X = I: 55% ee = 62%
R = *n*-Pr, X = I: 77% ee = 54%
R = *i*-Pr, X = Br: 51% ee = 52%
R = Me, X = I: 42% ee = 44%

65% ee = 10%

Scheme 10.90 Cu-catalysed Grignard crosscouplings of allylic substrates with fer-rocenylthiolate ligands.

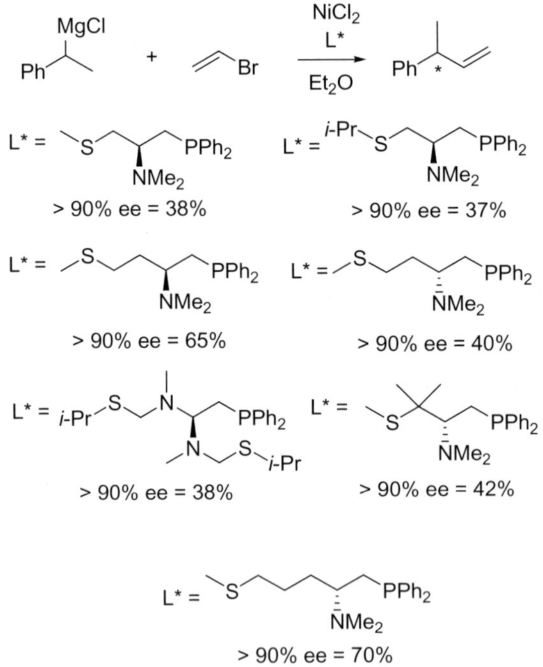

Scheme 10.91 Grignard crosscoupling of 1-phenylethylmagnesium chloride and vinyl bromide with *S*-alkylated amino phosphine ligands.

have demonstrated an important intramolecular participation of the sulfur atom. The different ligands were indeed derived from amino acids in which the sulfur atom was separated from the chiral carbon centre by a chain of one, two or three carbon atoms. The results collected in Scheme 10.91 show the influence of the carbon linker bridge between the sulfur and the phosphine atoms with the best results obtained in the case of a three-carbon-atom chain. It was proposed that the other sulfur-containing ligands had side chains too short to allow an effective participation of the sulfur atom in the nickel coordination.

Even if only moderate enantioselectivities have been obtained with chiral sulfur-containing ligands, the asymmetric Grignard crosscoupling reaction has allowed the potential of these novel ligands to provide chiral products to be highlighted. Indeed, these pioneering studies involving chiral sulfur-containing ligands have opened the way to the preparation of a wide range of this type of new ligands and their application to many carbon–carbon bond formations.

10.25 Conclusions

This chapter is well representative of the extensive efforts that have been made for the development of novel chiral sulfur-containing ligands applicable in a broad variety of reaction types. Numerous homodonor S/S ligands as well as heterodonor ligands such as S/P, S/N or S/O chelates have been implicated in a wide range of reactions with the best results generally obtained in recent years. Several S/P ligands have been especially highly efficient to induce chirality, such as thioether-phosphite ligands applied by Claver's group in 2000 to the asymmetric hydroformylation. Another particularly successful example was described in 2003 by Evans's group who involved thioether-phosphinite ligands in the rhodium-catalysed hydrosilylation of ketones. In addition, Fesulphos ligands have provided excellent enantioselectivities when used in asymmetric 1,3-dipolar cycloadditions, even when supported on polystyrene, as was demonstrated by Carretero's group in 2007. These authors also found, in 2006, that these ligands induced high enantioselectivities in Mannich-type reactions.

On the other hand, numerous S/N ligands have also been investigated to promote a wide range of reactions. A large variety of structures have been prepared, such as aminothioethers, aminothiols, aminothiophene derivatives, aminosulfoximines, bis(thiazolines), bis(sulfonamides), aminosulfoxides, bis(thioureas), sulfur-containing oxazolines and sulfur-containing amino-ferrocenyl derivatives. Among these ligands, aminosulfoximines, developed by Bolm's group in 2005, provided excellent enantioselectivities in Cu-catalysed Mukaiyama-type aldol reactions. More recently, Bandini's group has observed a high induction in Cu-catalysed Henry reactions performed in the presence of oligothiophene ligands.

On the other hand, S/O ligands have been developed to a much lesser extent, although the high efficiency of BINOL-derived hydroxysulfoxides was demonstrated by Shibasaki's group in 2004 for the Reissert reaction.

Some interesting results have also been observed by using homodonor S/S ligands, but these ligands remained generally less well performing than their corresponding nitrogen- or phosphorus-containing counterparts. Some ligands of this type, namely dithioether-pyrrolidine ligands, have given, however, excellent enantioselectivities in the asymmetric hydroformylation developed in 1999 by Claver's group. Much more work has been generally performed for the synthesis and subsequent application in asymmetric catalysis of heterodonor ligands.

General Conclusion

Asymmetric catalysis is a topic of increasing interest and is one of the most important focal areas in organic synthesis. In particular, the development of chiral ligands for asymmetric catalytic reactions is a subject of considerable interest in the field of asymmetric synthesis. Extensive efforts have been made for the development of new advantageous chiral ligands applicable in a broad variety of reaction types and their preparation continues to be an important area of synthetic organic research. Indeed, the ligand design is becoming an increasingly important part of the synthetic activity in chemistry.[178] This is, of course, because of the subtle control that ligands exert on the metal centre to which they are coordinated. The impressive number of reports dealing with the use of chiral sulfur-containing ligands for asymmetric catalysis is representative of the success of such ligands to promote numerous catalytic transformations. Nowadays, these ligands have become fairly renowned competitors to more usual phosphorus- or nitrogen-containing ligands. Indeed, chiral sulfur-containing ligands represent a highly efficient class of ligands applicable in a broad variety of reaction types, also in view of some peculiar properties, which differentiate them from more popular ligands such as phosphorus- or nitrogen-containing ligands. In particular, the stereoelectronic assistance of organosulfur functionalities and the possibility of stereocontrol through stereogenic sulfur atoms can provide interesting results in many applications. The key advantages of these new types of ligands are their easy synthesis, mostly starting from readily available commercial compounds, and their high stability, which allows easy storage and handling, especially compared to phosphine derivatives. Whereas the synthesis of chiral phosphorus ligands is often complex and difficult, the preparation of chiral sulfur compounds is much more convenient. Consequently, a wide diversity of chiral sulfur-containing ligands, more than forty different classes, is easily available either directly from the chiral pool or by facile modifications of other heteroatomic ligands. Homodonor S/S ligands as well as heterodonor S/N, S/P or S/O ligands have been involved in asymmetric catalysis. S/N ligands are by far the most developed and have been successfully used in almost all types of reactions cited in this book. For example, these ligands gave the best results for the Tsuji–Trost reaction, which is by far the most explored catalytic test with chiral sulfur ligands. In spite of advantages such as their stability and easy synthesis, these ligands remain, however, apart from relatively few exceptions, generally less active and enantioselective than their analogous N/P-counterparts for the Tsuji–Trost reaction. In the same context, S/N ligands have given the best results for the enantioselective addition of organozinc reagents to aldehydes, giving in most cases better enantioselectivities than their amino alcohol ligand analogues. Moreover, this type of ligand (sulfonamides) remains the only one able to efficiently catalyse the synthesis of chiral tertiary alcohols by condensation of organozinc reagents to a broad range of ketones. Various other S/N ligands, such as bis(sulfoximines), bis(sulfinyl)imidoamidines, sulfur-containing amino ferrocenyl derivatives, or bis(thiazolines) have given excellent results when applied to Diels–Alder reactions. Other S/N ligands such as bis(oxazolines)thiophenes have been successfully involved in the cyclopropanation of alkenes based on metal-decomposition of diazo compounds. In addition, the hydrogen transfer and a number of miscellaneous reactions depicted in the last chapter of the book were the most successful by using S/N ligands.

Due to the huge success of chiral diphosphines as ligands, numerous analogous S/P ligands have been recently prepared and investigated in asymmetric catalysis. These ligands proved more versatile than the S/S-ones and promoted efficiently various transformations, such as the Tsuji–Trost reaction, or the conjugate addition of various organometallic reagents to both cyclic and acyclic enones, providing results comparable to those of the well-known phosphorus ligands. The best enantioselectivities obtained for the hydrogenation of alkenes were also observed by using S/P ligands. In addition, even if there are relatively few examples of asymmetric Heck reactions in which chiral sulfur-containing ligands have been successfully employed; spectacular results have been very recently described by using other S/P ligands such as benzothiophene analogues of BINAP (BITIANP).

On the other hand, S/O ligands have been developed to a lesser extent, but their efficient use as chiral ligands was proven in the enantioselective addition of diethylzinc to aldehydes and also in the copper-catalysed asymmetric conjugate addition.

Homodonor S/S ligands were the first sulfur-containing ligands developed for asymmetric catalysis and, in particular, for Grignard crosscoupling reactions. Although the results remained moderate in terms of enantioselectivity, the works reported by Kellogg's group in the 1980s were an essential proof of concept for the start of this chemistry. These ligands have been also investigated for the Tsuji–Trost reaction, but they remained generally less well performing than their nitrogen- or phosphorus-containing counterparts, since they provided both moderate activity and enantioselectivity. Moreover, applied to other asymmetric reactions such as hydroformylation or hydrogen transfer, these ligands gave much less success.

This book clearly demonstrates that transition-metal complexes of chiral sulfur ligands are nowadays powerful catalysts in a considerable number of reactions, although, they have been generally less investigated than complexes with other donor atoms, since the interest in this chemistry is much more recent. In this book, the author has attempted to systematise the role that chiral sulfur-containing ligands play in asymmetric catalysis. Over the last ten years, the potential of chiral sulfur-containing ligands in asymmetric catalysis has been widely developed and explored. Many of the sulfur-containing ligands have only recently appeared in the literature, and it is clear that their application in asymmetric catalysis will undoubtedly increase in the near future for a number of additional asymmetric reactions, due in part to their easy synthesis. Moreover, due to their great stability, asymmetric heterogeneous catalysis seems to be a good way to develop the potential of these ligands in an economic and environmentally friendly manner. In particular, due to their great diversity, S/N ligands probably have the most promising future for performing catalytic enantioselective C–C bond formations, which remain a challenge in terms of activity, enantioselectivity and catalyst loading and recycling. It must be noted that in an increasing number of transformations, the sulfur ligands overtake other ligands in terms of both activity and enantioselectivity.

References

1. (a) M. Beller, B. Cornils, C. D. Frohning and C. W. Kohlpaintner, *J. Mol. Catal.*, 1995, **104**, 17–85; (b) C. D. Frohning and C. W. Kohlpaintner, in: *Applied Homogeneous Catalysis with Organometallic Compounds*, ed. B. Cornils, W. A. Herrmann, Springer Verlag, Weinheim, 1996, pp. 29–90; (c) F. Agbossou, J. F. Carpentier and A. Mortreux, *Chem. Rev.*, 1995, **95**, 2485–2506; (d) F. Ungvàry, *Coord. Chem. Rev.*, 1998, **170**, 245–281; (e) S. Gladiali, J. C. Bayon and C. Claver, *Tetrahedron: Asymmetry*, 1995, **6**, 1453–1474; (f) C. Claver, M. Dieguez, O. Pamies and S. Castillon, *Top. Organomet. Chem.*, 2006, **18**, 35–64.

2. M. Dieguez, O. Pamies and C. Claver, *Tetrahedron: Asymmetry*, 2004, **15**, 2113–2122.

3. C. Claver, S. Castillon, N. Ruiz, G. Delogu, D. Fabbri and S. Gladiali, *Chem. Commun.*, 1993, 1833–1834.

4. A. M. Masdeu-Bulto, A. Orejon, S. Castillon and C. Claver, *Tetrahedron: Asymmetry*, 1995, **6**, 1885–1888.

5. M. A. Casado, J. J. Pérez-Torrente, M. A. Ciriano and L. A. Oro, *Organometallics*, 1999, **18**, 3035–3044.

6. O. Pàmies, M. Diéguez, G. Net, A. Ruiz and C. Claver, *Organometallics*, 2000, **19**, 1488–1496.

7. M. Diéguez, A. Ruiz, C. Claver, M. M. Pereira, T. Flor, J. C. Bayon, M. E. S. Serra and A. M. d'A. Rocha Gonsalves, *Inorg. Chim. Acta*, 1999, **295**, 64–70.

8. J. A. J. Breuzard, M. L. Tommasino, F. Touchard, M. Lemaire and M. C. Bonnet, *J. Mol. Catal.*, 2000, **156**, 223–232.

9. R. Huisgen, *Angew. Chem., Int. Ed. Engl.*, 1963, **10**, 565–598.

10. (a) C. Najera and J. M. Sansano, *Angew. Chem., Int. Ed. Engl.*, 2005, **44**, 6272–6276; (b) M. Pichon and B. Figadere, *Tetrahedron: Asymmetry*, 1996, **7**, 927–964.

11. (a) G. Broggini, G. Molteni, A. Terraneo and G. Zecchi, *Heterocycles*, 2003, **59**, 823–858; (b) I. N. N. Namboothiri and A. Hassner, *Top. Curr. Chem.*, 2001, **216**, 1–49.

12. (a) S. Karlsson and H.-E. Högberg, *Org. Prep. Proc. Int.*, 2001, **33**, 103–172; (b) K. V. Gothelf and K. A. Jorgensen, *Chem. Rev.*, 1998, **98**, 863–909; (c) S. Kanemasa, *Synlett*, 2002, **9**, 1371–1387; (d) K. V. Gothelf, *Synthesis*, 2002, 211–247; (e) G. Broggini Molteni, A. Terraneo and G. Zecchi, *Heterocycles*, 2003, **59**, 823–858; (f) C. Najera and J. M. Sansano, *Curr. Org. Chem.*, 2003, **7**, 1105–1150; (g) H. Pellissier, *Tetrahedron*, 2007, **63**, 3235–3285.

13. J.-P. G. Seerden, M. M. M. Kuypers and H. W. Scheeren, *Tetrahedron: Asymmetry*, 1995, **6**, 1441–1450.

14. A. Heckel and D. Seebach, *Chem. Eur. J.*, 2002, **8**, 560–572.

15. H. M. Davies, B. Xiang, N. Kong and D. G. Stafford, *J. Am. Chem. Soc.*, 2001, **123**, 7461–7462.

16. S. Husinec and V. Savic, *Tetrahedron: Asymmetry*, 2005, **16**, 2047–2061.

17. L. M. Harwood and R. J. Vickers, in *Synthetic Applications of Products*, ed. A. Padwa, W. Pearson, Wiley and Sons, New York, 2002, Chapter 3.

18. L. M. Stanley and M. P. Sibi, *Chem. Rev.*, 2008, **108**, 2887–2902.

19. (a) J. Priego, O. Garcia Mancheno, S. Cabrera, R. Gomez Arrayas, T. Llamas and J. C. Carretero, *Chem. Commun.*, 2002, 2512–2513; (b) S. Cabrera, R. Gomez Arrayas and J. C. Carretero, *Angew. Chem., Int. Ed. Engl.*, 2004, **43**, 3944–3947.

20. S. Cabrera, R. Gomez Arrayas and J. C. Carretero, *J. Am. Chem. Soc.*, 2005, **127**, 16394–16395.

21. Y. Oderaotoshi, W. Cheng, S. Fujitomi, Y. Kasano, S. Minakata and M. Komatsu, *Org. Lett.*, 2003, **5**, 5043–5046.

22. (a) M. Heitbaum, F. Glorius and I. Escher, *Angew. Chem., Int. Ed. Engl.*, 2006, **45**, 4732–4762; (b) P. McMorn and G. J. Hutchings, *Chem. Soc. Rev.*, 2004, **33**, 108–122; (c) N. E. Leadbeater and M. Marco, *Chem. Rev.*, 2002, **102**, 3217–3274; (d) C. A. McNamara, M. J. Dixon and M. Bradley, *Chem. Rev.*, 2002, **102**, 3275–3300; (e) Q. Fan, Y.-M. Li and A. S. C. Chan, *Chem. Rev.*, 2002, **102**, 3385–3466; (f) *Chiral Catalyst Immobilization and Recycling*, ed., D. E. de Vos, I. F. Vankelecom, P. A. Jacobs, Wiley-VCH, Weinheim, 2000.

23. B. Martin-Matute, S. I. Pereira, E. Pena-Cabrera, J. Adrio, A. M. S. Silva and J. C. Carretero, *Adv. Synth. Catal.*, 2007, **349**, 1714–1724.

24. W. Zeng and Y.-G. Zhou, *Tetrahedron Lett.*, 2007, **48**, 4619–4622.

25. M. Shi and J.-W. Shi, *Tetrahedron: Asymmetry*, 2007, **18**, 645–650.

26. H. Staudinger, *Ber. Dtsch. Chem. Ges.*, 1905, **38**, 1735–1739.

27. (a) H. Staudinger, *Justus Liebigs Ann. Chem.*, 1907, **356**, 51–123; (b) T. T. Tidwell, *Eur. J. Org. Chem.*, 2006, **1**, 563–576; (c) T. T. Tidwell in *Ketenes*, John Wiley and Sons, Hoboken, New Jersey, 1995 and 2nd edn in 2006; (d) J. A. Hyatt and P. W. Raynolds, *Org. React.*, 1994, **45**, 159–646; (e) C. M. Temperley in *Ketenes, Their Cumulene Analogues, and Their S, Se, and Te Analogues* in *Comprehensive Organic Functional Group*

Transformations II, Vol. 3, ed. A. R. Katritzky, R. J. K. Taylor, Elsevier, Amsterdam, 2005, pp 573–603.

28. Y. Tamai, H. Yoshiwara, M. Someya, J. Fukumoto and S. Miyano, *J. Chem. Soc., Chem. Commun.*, 1994, 2281–2282.

29. V. Nevalainen, *Tetrahedron: Asymmetry*, 1993, **4**, 2517–2530.

30. (a) K. Mikami and M. Shimizu, *Chem. Rev.*, 1992, **92**, 1021–1050; (b) M. L. Clarke and M. B. France, *Tetrahedron*, 2008, **64**, 9003–9031.

31. K. Mikami, E. Sawa and M. Terada, *Tetrahedron: Asymmetry*, 1991, **2**, 1403–1412.

32. M. Langner, P. Rémy and C. Bolm, *Synlett*, 2005, **5**, 781–784.

33. G. M. Coppola and H. F. Schuster, in α-*Hydroxy Acids in Enantioselective Syntheses*, VCH, Weinheim, 1997.

34. D. A. Evans, S. W. Tregay, C. S. Burgey, N. A. Paras and T. Vojkovsky, *J. Am. Chem. Soc.*, 2000, **122**, 7936–7943.

35. H.-J. Federsel, A. N. Collins, G. N. Sheldrake and J. Crosby, in *Chirality in Industry II*, John Wiley and Sons, Chichester, 1997, pp. 225–244.

36. (a) S. Kobayashi and H. Ishitani, *Chem. Rev.*, 1999, **99**, 1069–1094; (b) T. Vilaivan, W. Bhanthumnavin and Y. Sritana-Anant, *Curr. Org. Chem.*, 2005, **9**, 1315–1392; (c) G. K. Friestad and A. K. Mathies, *Tetrahedron*, 2007, **63**, 2541–2569; (d) K.-i. Yamada and K. Tomioka, *Chem. Rev.*, 2008, **108**, 2874–2886.

37. (a) M. Shi and W. S. Sui, *Chirality*, 2000, **12**, 574–580; (b) M. Shi and W. S. Sui, *Tetrahedron: Asymmetry*, 2000, **11**, 773–779; (c) M. Shi and W. S. Sui, *Tetrahedron: Asymmetry*, 2000, **12**, 835–841.

38. J.-M. Brunel, *Chem. Rev.*, 2005, **105**, 857–897.

39. K. Ding, X. Li, B. Ji, H. Guo and M. Kitamura, *Curr. Org. Synth.*, 2005, **2**, 499–545.

40. (a) I. Ojima, (ed.), in *Catalytic Asymmetric Synthesis*, VCH Publications, New York, 1993; (b) E. N. Jacobsen, A. Pfaltz, H. Yamamoto, ed., in *Comprehensive Asymmetric Catalysis*, Springer-Verlag, Heidelberg, 2001.

41. C.-J. Wang and M. Shi, *J. Org. Chem.*, 2003, **68**, 6229–6237.

42. M. Shi and C.-J. Wang, *Adv. Synth. Catal.*, 2003, **345**, 971–973.

43. M. Shi and W. Zhang, *Tetrahedron: Asymmetry*, 2003, **14**, 3407–3414.

44. B. Liu, J. Liu, X. Jia, L. Huang, X. Li and A. S. C. Chan, *Tetrahedron: Asymmetry*, 2007, **18**, 1124–1128.

45. S. E. Denmark and J. Fu, *Chem. Rev.*, 2003, **103**, 2763–2793.

46. (a) A. Cunningham and S. Woodward, *Synlett*, 2002, **1**, 43–44; (b) A. Cunningham, V. Mokal-Parekh, C. Wilson and S. Woodward, *Org. Biomol. Chem.*, 2004, **2**, 741–748.

47. (a) G. J. Rowlands and W. K. Barnes, *Chem. Commun.*, 2003, 2712–2713; (b) A. Massa, A. V. Malkov, P. Kocovsky and A. Scettri, *Tetrahedron Lett.*, 2003, **44**, 7179–7181.

48. S. Kobayashi, C. Ogawa, H. Konishi and M. Sugiura, *J. Am. Chem. Soc.*, 2003, **125**, 6610–6611.

49. E. M. Carreira, A. Fettes and C. Marti, *Org. React.*, 2006, **67**, 1–216.

50. (a) S. G. Nelson, *Tetrahedron: Asymmetry*, 1998, **9**, 357–389; (b) E. M. Carreira, in *Comprehensive Asymmetric Catalysis*, Vol. 1, ed. E. N. Jacobsen, A. Pfaltz, H. Yamamoto, Springer, Berlin, 1999, pp 997–1065.

51. M. Shibasaki and M. Kanai, *Chem. Rev.*, 2008, **108**, 2853–2873.

52. (a) A. Schwartz and H. E. Van Wart, *Prog. Med. Chem.*, 1992, **29**, 271–334; (b) G. M. Coppola and H. F. Schuster, in *α-Hydroxy Acids in Enantioselective Synthesis*, VCH, Weinheim, 1997.

53. (a) I. Denissova and L. Barriault, *Tetrahedron*, 2003, **59**, 10105–10146; (b) K. Fuji, *Chem. Rev.*, 1993, **93**, 2037–2066; (c) S. F. Martin, *Tetrahedron*, 1980, **36**, 419–460.

54. M. Langner and C. Bolm, *Angew. Chem., Int. Ed. Engl.*, 2004, **43**, 5984–5987.

55. M. Langner, P. Rémy and C. Bolm, *Chem. Eur. J.*, 2005, **11**, 6254–6265.

56. (a) D. A. Evans, D. W. C. MacMillan and K. R. Campos, *J. Am. Chem. Soc.*, 1997, **119**, 10859–10860; (b) D. A. Evans, M. C. Kozlowski, D. W. C. Burgey and D. W. C. MacMillan, *J. Am. Chem. Soc.*, 1997, **119**, 7843–7894; (c) D. A. Evans, D. W. Burgey, M. C. Kozlowski and S. W. Tregay, *J. Am. Chem. Soc.*, 1999, **121**, 686–699.

57. P. Rémy, M. Langner and C. Bolm, *Org. Lett.*, 2006, **8**, 1209–1211.

58. T. Mukaiyama, H. Asanuma, I. Hachiya, T. Harada and S. Kobayashi, *Chem. Lett.*, 1991, 1209–1212.

59. K. Uotsu, H. Sasai and M. Shibasaki, *Tetrahedron: Asymmetry*, 1995, **6**, 71–74.

60. A. Togni and R. Häusel, *Synlett*, 1990, 633–635.

61. L. Henry, *C. R. Hebd. Seances Acad. Sci.*, 1895, **120**, 1265–1268.

62. (a) G. Rosini, in *Comprehensive Organic Synthesis*, ed. B. M. Trost, Pergamon, Oxford, 1996; (b) F. A. Luzzio, *Tetrahedron*, 2001, **57**, 915–945; (c) N. Ono, in *The Nitro Group in Organic Synthesis*, Wiley-VCH, New York, 2001.

63. C. Palomo, M. Oiarbide and A. Mielgo, *Angew. Chem., Int. Ed. Engl.*, 2004, **43**, 5442–5444.

64. H. Sasai, T. Suzuki, S. Arai and M. Shibasaki, *J. Am. Chem. Soc.*, 1992, **114**, 4418–4420.

65. J. Gao and A. E. Martell, *Org. Biomol. Chem.*, 2003, **1**, 2801–2806.

66. S.-F. Lu, D.-M. Du, S.-W. Zhang and J. Xu, *Tetrahedron: Asymmetry*, 2004, **15**, 3433–3441.

67. D.-M. Du, S.-F. Lu, T. Fang and J. Xu, *J. Org. Chem.*, 2005, **70**, 3712–3715.

68. W. Mansawat, I. Saengswang, P. U-prasitwong, W. Bhanthumnavin and T. Vilaivan, *Tetrahedron Lett.*, 2007, **48**, 4235–4238.

69. M. Bandini, F. Piccinelli, S. Tommasi, A. Umani-Ronchi and C. Ventrici, *Chem. Commun.*, 2007, 616–618.

70. (a) M. Arend, B. Westermann and N. Risch, *Angew. Chem., Int. Ed. Engl.*, 1998, **37**, 1044–1070; (b) S. Kobayashi and H. Ishitani, *Chem. Rev.*, 1999, **99**, 1069–1094; (c) A. Cordova, *Acc. Chem. Res.*, 2004, **37**, 102–112; (d) M. M. B. Marques, *Angew. Chem., Int. Ed. Engl.*, 2006, **45**, 348–352;

(e) J. M. M. Verkade, L. J. C. vanHemert, P. J. L. M. Quaedflieg and F. P. J. T. Rutjes, *Chem. Soc. Rev.*, 2008, **37**, 29–41.

71. A. S. Gonzales, R. G. Arrayas and J. C. Carretero, *Org. Lett.*, 2006, **8**, 2977–2980.

72. (a) S. J. Rhoads, N. R. Raulins, *Org. React.*, Wiley, New York, 1974, **22**, 1–252; (b) P. Wipf, in *Comprehensive Organic Synthesis*, ed. B. M. Trost, I. Fleming, Pergamon Press, Oxford, 1991, **Vol. 5**, Chapter 7.2, 827–873; (c) R. P. Luts, *Chem. Rev.*, 1984, **84**, 205–247; (d) G. B. Bennett, *Synthesis*, 1977, 589–606; (e) F. E. Ziegler, *Chem. Rev.*, 1988, **88**, 1426–1452; (f) S. Blechert, *Synthesis*, 1989, 71–82; (g) A. N. M. Castro, *Chem. Rev.*, 2004, **104**, 2939–3002; (h) D. Chatterjee, *Coord. Chem. Rev.*, 2008, **252**, 176–198.

73. H. Ito and T. Taguchi, *Chem. Soc. Rev.*, 1999, **28**, 43–50.

74. H. Ito, A. Sato and T. Taguchi, *Tetrahedron Lett.*, 1997, **38**, 4815–4818.

75. H. Ito, A. Sato, T. Kobayashi and T. Taguchi, *Chem. Commun.*, 1998, 2441–2442.

76. V. K. Aggarwal and C. L. Winn, *Acc. Chem. Res.*, 2004, **37**, 611–620.

77. (a) K. Julienne, P. Metzner and V. Henyron, *J. Chem. Soc., Perkin Trans. I*, 1999, 731–735; (b) J. Zanardi, C. Leriverend, D. Aubert, K. Julienne and P. Metzner, *J. Org. Chem.*, 2001, **66**, 5620–5623.

78. C. L. Winn, B. Bellanie and J. M. Goodman, *Tetrahedron Lett.*, 2002, **43**, 5427–5430.

79. R. Hayakawa and M. Shimizu, *Synlett*, 1999, 1328–1330.

80. T. Saito, D. Akiba, M. Sakairi and S. Kanazawa, *Tetrahedron Lett.*, 2001, **42**, 57–59.

81. V. K. Aggarwal and C. L. Winn, *Acc. Chem. Res.*, 2004, **37**, 611–620.

82. (a) V. K. Aggarwal, E. Alonso, I. Bae, G. Hynd, K. M. Lydon, M. J. Palmer, M. Patel, M. Porcelloni, J. Richardson, R. A. Stenson, J. R. Studley, J.-L. Vasse and C. L. Winn, C. L. *J. Am. Chem. Soc.*, 2003, **125**, 10926–10940; (b) V. K. Aggarwal, R. Angelaud, D. Bihan, P. Blackburn, R. Fieldhouse, S. Fonquerna, J. G. Ford, G. Hynd, E. Jones, R. V. H. Jones, P. Jubault, M. J. Palmer, P. D. Ratcliffe and H. Adams, *J. Chem. Soc., Perkin Trans. I*, 2001, 2604–2622; (c) V. K. Aggarwal, E. Alonso, G. Hynd, K. M. Lydon, M. J. Palmer, M. Porcelloni and J. R. Studley, *Angew. Chem., Int. Ed. Engl.*, 2001, **40**, 1430–1433.

83. V. K. Aggarwal, I. Bae and H.-Y. Lee, *Tetrahedron*, 2004, **60**, 9725–9733.

84. V. K. Aggarwal, I. Bae, H. Y. Lee, J. Richardson and D. T. Williams, *Angew. Chem., Int. Ed. Engl.*, 2003, **42**, 3274–3278.

85. R. Imashiro, T. Yamanaka and M. Seki, *Tetrahedron: Asymmetry*, 1999, **10**, 2845–2851.

86. R. G. Arrayas, J. Adrio and J. C. Carretero, *Angew. Chem., Int. Ed. Engl.*, 2006, **45**, 7674–7715.

87. (a) S. Minière, V. Reboul, R. G. Arrayas, P. Metzner and J. C. Carretero, *Synthesis*, 2003, 2249–2254; (b) S. Minière, V. Reboul and P. Metzner, *Arkivoc*, 2005, **6**, 161–177; (c) S. Minière, V. Reboul, P. Metzner, M. Fochi and B. F. Bonini, *Tetrahedron: Asymmetry*, 2004, **15**, 3275–3280.

88. I. S. Gonçalves, A. M. Santos, C. C. Romao, A. D. Lopes, J. E. Rodri-guez-Borges, M. Pillinger, P. Ferreira, J. Rocha and F. E. Kühn, *J. Organomet. Chem.*, 2001, **626**, 1–10.

89. (a) R. J. H. Gregory, *Chem. Rev.*, 1999, **99**, 3649–3682; (b) F. Effenber-ger, *Angew. Chem., Int. Ed. Engl.*, 1994, **33**, 1555–1560; (c) C. G. Kruse, in *Chirality in Industry*, ed. A. N. Collins, G. N. Schedrake, J. Crosby, Wiley, Chichester, 1992, Chap. 14; (d) N.-u. H. Khan, R. I. Kureshy, S. H. R. Abdi, S. Agrawal and R. V. Jasra, *Coord. Chem. Rev.*, 2008, **252**, 593–623.

90. M. North, *Synlett*, 1993, 807–820.

91. C. Bolm and P. Müller, *Tetrahedron Lett.*, 1995, **36**, 1625–1628.

92. C. Bolm, P. Müller and K. Harms, *Acta Chem. Scand.*, 1996, **50**, 305–315.

93. J.-S. Yiu, H.-M. Gau and C. K. Choi, *Chem. Commun.*, 2000, 1963–1964.

94. G. J. Rowlands, *Synlett*, 2003, 236–240.

95. (a) H. Nishiyama, in *Comprehensive Asymmetric Catalysis*, ed. E. N. Jacobsen, A. Pfaltz, H. Yamamoto, **Vol. 1**, Springer, Berlin, 1999, pp. 267–289; (b) R. Noyori, in *Asymmetric Catalysis in Organic Synthesis*, Wiley, New York, 1994; (c) *Catalytic Asymmetric Synthesis*, I. Ojima, ed., Wiley-VCH, New York, 2000; (d) *Comprehensive Asymmetric Catalysis*, ed., E. N. Jacobsen, A. Pfaltz, H. Yamamoto, Springer, Berlin, 1999, Vol. 1.

96. H. Brunner, H. Nishiyama and K. Itoh, in *Catalytic Asymmetric Synthe-sis*, ed. I. Ojima, VCH, New York, 1993, pp. 302–322.

97. M. Hiraoka, A. Nishikawa, T. Morimoto and K. Achiwa, *Chem. Pharm. Bull.*, 1998, **46**, 704–706.

98. D. A. Evans, F. E. Michael, J. S. Tedrow and K. R. Campos, *J. Am. Chem. Soc.*, 2003, **125**, 3534–3543.

99. M. Diéguez, O. Pamies and C. Claver, *Tetrahedron: Asymmetry*, 2005, **16**, 3877–3880.

100. (a) H. Brunner, G. Riepl and H. Weitzer, *Angew. Chem., Int. Ed. Engl.*, 1983, **22**, 331–332; (b) H. Brunner, R. Becker and G. Riepl, *Organome-tallics*, 1984, **3**, 1354–1359.

101. G. Helmchen, A. Krotz, K.-T. Ganz and D. Hansen, *Synlett*, 1991, 257–259.

102. (a) Y. Nishibayashi, J. D. Singh, K. Segawa, S.-i. Fukuzawa and S. Uemura, *J. Chem. Soc., Chem. Commun.*, 1994, 1375–1376; (b) Y. Nishibayashi, J. D. Singh, K. Segawa, S.-i. Fukuzawa, K. Ohe and S. Uemura, *Organometallics*, 1996, **15**, 370–379.

103. I. Karamé, M. L. Tommasino and M. Lemaire, *J. Mol. Catal. A*, 2003, **196**, 137–143.

104. S. Gérard, Y. Pressel and O. Riant, *Tetrahedron: Asymmetry*, 2005, **16**, 1889–1891.

105. J. W. Faller and K. J. Chase, *Organometallics*, 1994, **13**, 989–992.

106. S. Gladiali, S. Medici, G. Pirri, S. Pulacchini and D. Fabbri, *Can. J. Chem.*, 2001, **79**, 670–678.

107. S. Sakuraba, T. Okada, T. Morimoto and K. Achiwa, *Chem. Pharm. Bull.*, 1995, **43**, 927–934.

376   *Chapter 10*

108. (a) L. Deloux and M. Srebnik, *Chem. Rev.*, 1993, **93**, 763–784; (b) M. M. Midland, *Chem. Rev.*, 1989, **89**, 1553–1561.
109. (a) A. Hirao, S. Itsuno, S. Nakahama and N. Yamazaki, *J. Chem. Soc., Chem. Commun.*, 1981, 315–317; (b) S. Itsuno, A. Hirao, S. Nakahama and N. Yamazaki, *J. Chem. Soc., Perkin Trans I*, 1983, 1673–1676; (c) S. Itsuno, K. Ito, A. Hirao and S. Nakahama, *J. Chem. Soc., Chem. Commun.*, 1983, 3469–470.
110. C. R. Johnson and C. J. Stark, *Tetrahedron Lett.*, 1979, **49**, 4713–4716.
111. C. Bolm and M. Felder, *Tetrahedron Lett.*, 1993, **34**, 6041–6044.
112. C. Bolm, A. Seger and M. Felder, *Tetrahedron Lett.*, 1993, **34**, 8079–8080.
113. C. Bolm and M. Felder, *Synlett*, 1994, 655–656.
114. J. Kang, J. W. Lee, J. I. Kim and C. Pyun, *Tetrahedron Lett.*, 1995, **36**, 4265–4268.
115. J. Kang, C. W. Lee, G. J. Lim and B. T. Cho, *Tetrahedron: Asymmetry*, 1999, **10**, 657–660.
116. T. Mehler and J. Martens, *Tetrahedron: Asymmetry*, 1993, **4**, 1983–1986.
117. M. Kossenjans and J. Martens, *Tetrahedron: Asymmetry*, 1998, **9**, 1409–1417.
118. J.-C. Fiaud, F. Mazé and H. B. Kagan, *Tetrahedron: Asymmetry*, 1998, **9**, 3647–3655.
119. T.-K. Yang and D.-S. Lee, *Tetrahedron: Asymmetry*, 1999, **10**, 405–409.
120. K. Molvinger and J. Court, *Tetrahedron: Asymmetry*, 2001, **12**, 1971–1973.
121. A. J. Blake, A. Cunningham, A. Ford, S. J. Teat and S. Woodward, *Chem. Eur. J.*, 2000, **6**, 3586–3594.
122. (a) C. M. Crudden and D. Edwards, *Eur. J. Org. Chem.*, 2003, 4695–4712; (b) A.-M. Carroll, T. P. O'sullivan and P. J. Guiry, *Adv. Synth. Catal.*, 2005, **347**, 609–631.
123. M. Diéguez, A. Ruiz, C. Claver, M. M. Pereira, M. T. Flor, J. C. Bayon, M. E. S. Serra and A. M. d'A. Rocha Gonzalves, *Inorg. Chim. Acta*, 1999, **295**, 64–70.
124. (a) Y. K. Chung, *Coord. Chem. Rev.*, 1999, **188**, 297–341; (b) K. M. Brummond and J. L. Kent, *Tetrahedron*, 2000, **56**, 3263–3283.
125. S. T. Ingate and J. Marco-Contelles, *Org. Prep. Proc. Int.*, 1998, **30**, 121–143.
126. (a) X. Verdaguer, A. Lledo, C. Lopez-Mosquera, M. A. Maestro, M. A. Pericas and A. Riera, *J. Org. Chem.*, 2004, **69**, 8053–8061; (b) X. Verdaguer, A. Moyano, M. A. Pericas, A. Riera, M. A. Maestro and J. Mahia, *J. Am. Chem. Soc.*, 2000, **122**, 10242–10243.
127. X. Verdaguer, M. A. Pericas, A. Riera, M. A. Maestro and J. Mahia, *Organometallics*, 2003, **22**, 1868–1877.
128. J. Solà, A. Riera, X. Verdaguer and M. A. Maestro, *J. Am. Chem. Soc.*, 2005, **127**, 13629–13633.
129. J. Solà, M. Revés, A. Riera and X. Verdaguer, *Angew. Chem., Int. Ed. Engl.*, 2007, **46**, 5020–5023.
130. (a) D. P. Curran, N. A. Porter and B. Giese, in *Stereochemistry of Radical Reactions*, VCH, New York, 1995; (b) G. Bar and P. F. Parsons, *Chem.*

Soc. Rev., 2003, **32**, 251–263; (c) H. Miyabe, M. Veda and T. Naito, *Synlett*, 2004, **7**, 1140–1157.

131. K. Hiroi and M. Ishii, *Tetrahedron Lett.*, 2000, **41**, 7071–7074.

132. M. Lautens, K. Fagnou and S. Hiebert, *Acc. Chem. Res.*, 2003, **36**, 48–58.

133. M. Lautens, S. Hiebert and J.-L. Renaud, *J. Am. Chem. Soc.*, 2001, **123**, 6834–6839.

134. (a) J. Priego, O. G. Mancheno, S. Cabrera, R. G. Arrayas, T. Llamas and J. C. Carretero, *Chem. Commun.*, 2002, 2512–2513; (b) O. G. Mancheno, J. Priego, S. Cabrera, R. G. Arrayas, T. Llamas and J. C. Carretero, *J. Org. Chem.*, 2003, **68**, 3679–3686.

135. (a) O. G. Mancheno, R. G. Arrayas and J. C. Carretero, *J. Am. Chem. Soc.*, 2004, **126**, 456–457; (b) O. G. Mancheno, R. G. Arrayas and J. C. Carretero, *Organometallics*, 2005, **24**, 557–561.

136. (a) S. Cabrera, R. G. Arrayas and J. C. Carretero, *Angew. Chem., Int. Ed. Engl.*, 2004, **43**, 3944–3947; (b) S. Cabrera, R. G. Arrayas, I. Alonso and J. C. Carretero, *J. Am. Chem. Soc.*, 2005, **127**, 17938–17947.

137. R. F. Heck, *J. Am. Chem. Soc.*, 1972, **94**, 2712–2716.

138. S. C. A. Nefkens, M. Sperrle and G. Consiglio, *Angew. Chem.*, 1993, **105**, 1837–1838.

139. C. Godard, B. K. Munoz, A. Ruiz and C. Claver, *Dalton Trans.*, 2008, 853–860.

140. M. Hayashi, H. Takezaki, Y. Hashimoto, K. Takaoki and K. Saigo, *Tetrahedron Lett.*, 1998, **39**, 7529–7532.

141. B. A. Arndtsen, R. G. Bergman, T. A. Mobley and T. H. Peterson, *Acc. Chem. Res.*, 1995, **28**, 154–162.

142. (a) S. D. Burke, P. A. Grieco, *Org. React.*, New York, 1979, **26**, 361–475; (b) M. P. Doyle, *Chem. Rev.*, 1986, **86**, 919–940; (c) J. Adams and D. M. Spero, *Tetrahedron*, 1991, **47**, 1765–1808; (d) A. Padwa and K. E. Krumpe, *Tetrahedron*, 1992, **48**, 5385–5453; (e) T. Ye and M. A. McKervey, *Chem. Rev.*, 1994, **94**, 1091–1160.

143. (a) G. Maas, *Top. Curr. Chem.*, 1987, **137**, 75–253; (b) H. M. L. Davies and R. E. J. Beckwith, *Chem. Rev.*, 2003, **103**, 2861–2903.

144. T. Ye and M. A. McKervey, *Tetrahedron Lett.*, 1994, **35**, 7269–7272.

145. T. Ye, C. F. Garcia and M. A. McKervey, *J. Chem. Soc., Perkin Trans. 1*, 1995, **1**, 1373–1379.

146. (a) H. M. L. Davies and T. Hansen, *J. Am. Chem. Soc.*, 1997, **119**, 9075–9076; (b) H. M. L. Davies and E. G. Antoulinakis, *J. Organomet. Chem.*, 2001, **617–618**, 47–55; (c) J. Adams, M.-A. Poupart, L. Greainer, C. Schaller, R. Quimet and R. Frenette, *Tetrahedron Lett.*, 1989, **30**, 1749–1752.

147. H. M. L. Davies, T. Hansen, D. W. Hopper and S. A. Panaro, *J. Am. Chem. Soc.*, 1999, **121**, 6509–6510.

148. H. M. L. Davies, *Eur. J. Org. Chem.*, 1999, 2459–2469.

149. H. M. L. Davies, T. Hansen, J. Rutberg and P. R. Bruzinski, *Tetrahedron Lett.*, 1997, **38**, 1741–1744.

150. (a) T. E. Mueller, K. C. Hultzsch, M. Yus, F. Foubelo and M. Tada, *Chem. Rev.*, 2008, **108**, 3795–3892; (b) A. L. Odom, *Dalton Trans.*, 2005, 225–233; (c) S. Hong and T. J. Marks, *Acc. Chem. Res.*, 2004, **37**, 673–686; (d) S. Doye, *Synlett*, 2004, 1653–1672; (e) F. Pohlki and S. Doye, *Chem. Soc. Rev.*, 2003, 104–114; (f) T. E. Mueller, *Rec. Res. Dev. Org. Chem.*, 2002, **6**, 625–647; (g) T. E. Mueller and M. Beller, *Chem. Rev.*, 1998, **98**, 675–704.

151. M. R. Gagné, C. L. Stern and T. J. Marks, *J. Am. Chem. Soc.*, 1992, **114**, 275–294.

152. (a) M. Nobis and B. Driessen-Hölscher, *Angew. Chem., Int. Ed. Engl.*, 2001, **40**, 3983–3985; (b) J. J Brunet and D. Neibecker, in *Catalytic Heterofunctionalization from Hydroamination to Hydrozirconation*, ed. A. Togni, H. Grutzmacher, Wiley-VCH, Weinheim, Germany, 2001, pp. 91–141.

153. (a) K. C. Hultzsch, *Adv. Synth. Catal.*, 2005, **347**, 367–391; (b) I. Aillaud, J. Collin, J. Hannedouche and E. Schulz, *Dalton Trans*, 2007, 5105–5118.

154. D. V. Gribkov, K. C. Hultzsch and F. Hampel, *J. Am. Chem. Soc.*, 2006, **128**, 3748–3759.

155. D. Riegert, J. Collin, A. Meddour, E. Schulz and A. Trifonov, *J. Org. Chem.*, 2006, **71**, 2514–2517.

156. S. Hong, S. Tian, M. V. Metz and T. J. Marks, *J. Am. Chem. Soc.*, 2003, **125**, 14768–14783.

157. J. Y. Kim and T. Livinghouse, *Org. Lett.*, 2005, **7**, 1737–1739.

158. Y. K. Kim, T. Livinghouse and Y. Horino, *J. Am. Chem. Soc.*, 2003, **125**, 9560–9561.

159. X. Yu and T. J. Marks, *Organometallics*, 2007, **26**, 365–376.

160. D. A. Evans and S. G. Nelson, *J. Am. Chem. Soc.*, 1997, **119**, 6452–6453.

161. N. Khiar, B. Suàrez, M. Stiller, V. Valdivia and I. Fernàndez, *Phosphorus, Sulfur, and Silicon*, 2005, **180**, 1253–1258.

162. N. Khiar, B. Suarez, V. Valdivia and I. Fernandez, *Synlett*, 2005, **19**, 2963–2967.

163. A. Reissert, *Ber. Dtsch. Chem. Ges.*, 1905, **38**, 1603–1614.

164. (a) E. Mosettig, *Org. React.*, 1954, **8**, 218–257; (b) W. E. McEwen and R. L. Cobb, *Chem. Rev.*, 1955, **55**, 511–549.

165. E. Ichikawa, M. Suzuki, K. Yabu, M. Albert, M. Kanai and M. Shibasaki, *J. Am. Chem. Soc.*, 2004, **126**, 11808–11809.

166. A. Padwa and S. S. Murphree, in *Progress in Heterocyclic Chemistry*, ed. G. W. Gribble, T. L. Gilchrist, Elsevier Science, Oxford, 2000, **Vol. 12**, Chapter 4.1, p 57.

167. (a) D. Tanner, *Angew. Chem., Int. Ed. Engl.*, 1994, **33**, 599–619; (b) H. M. I. Osborn and J. Sweeney, *Tetrahedron: Asymmetry*, 1997, **8**, 1693–1715; (c) A.-H. Li, L.-X. Dai and V. K. Aggarwal, *Chem. Rev.*, 1997, **97**, 2341–2372; (d) P. Müller and C. Fruit, *Chem. Rev.*, 2003, **103**, 2905–2919.

168. V. K. Aggarwal, E. Alonso, G. Fang, M. Ferrara, G. Hynd and M. Porcelloni, *Angew. Chem., Int. Ed. Engl.*, 2001, **40**, 1433–1436.

169. V. K. Aggarwal and J.-L. Vasse, *Org. Lett.*, 2003, **5**, 3987–3990.

170. (a) M. Ogasawara and T. Hayashi, in *Catalytic Asymmetric Synthesis*, ed. I. Ojima, Wiley-VCH, New York, 2000, pp. 651–674; (b) T. Hayashi, in *Catalytic Asymmetric Synthesis*, ed. I. Ojima, VCH Pub, New York, 1993, pp. 325–365.

171. (a) T. Hayashi and M. Kumada, in *Asymmetric Synthesis, Chiral Catalysis*, ed. J. D. Morrison, Academic Press, Inc., London, 1985, **Vol. 5**, Chapter 5, p 148; (b) T. J. Colacot, *Chem. Rev.*, 2003, **103**, 3101–3118.

172. M. Lemaire, J. Buter, B. K. Vriesema and R. M. Kellogg, *J. Chem. Soc., Chem. Commun.*, 1984, 309–310.

173. B. K. Vriesema, M. Lemaire, J. Buter and R. M. Kellogg, *J. Org. Chem.*, 1986, **51**, 5169–5177.

174. G. J. Meuzelaar, A. S. E. Karlström, M. van Klaveren, E. S. M. Persson, A. del Villar, G. van Koten and J.-E. Bäckvall, *Tetrahedron*, 2000, **56**, 2895–2903.

175. A. S. E. Karlström, F. F. Huerta, G. J. Meuzelaar and J.-E. Bäckvall, *Synlett*, 2001, 923–926.

176. J. H. Griffin and R. M. Kellogg, *J. Org. Chem.*, 1985, **50**, 3261–3266.

177. B. K. Vriesema and R. M. Kellogg, *Tetrahedron Lett.*, 1986, **27**, 2049–2052.

178. P. Braunstein and F. Naud, *Angew. Chem., Int. Ed. Engl.*, 2001, **40**, 680–699.

Subject Index